Developments in Geochemistry 11

Geological Sequestration of Carbon Dioxide

Thermodynamics, Kinetics, and Reaction Path Modeling

Developments in Geochemistry

In this series:

Vol. 1: Fluids in the Earth's Crust
W.S. Fyfe, N.J. Price and A.B. Thompson

Vol. 2: Rare Earth Element Geochemistry
P. Henderson (Editor)

Vol. 3: Helium Isotopes in Nature
B.A. Mamyrin and I.N. Tolstikhin

Vol. 4: Structure and Properties of Silicate Melts
B.O. Mysen

Vol. 5: Radioanalysis in Geochemistry
H.A. Das, A. Faanhof and H.A. van der Sloot

Vol. 6: Diversity of Environmental Biogeochemistry
J. Berthelin

Vol. 7: Geochemistry and Fluid Flow
L.W. Lake, S.L. Bryant and A.N. Araque-Martinez

Vol. 8: Geochemical and Tectonic Evolution of Arc-Backarc Hydrothermal Systems
N. Shikazono

Vol. 9: High-Pressure Geochemistry and Mineral Physics: Basics for Planetology and Geo-Material Science
S. Mitra

Vol. 10: Silicate Glasses and Melts: Properties and Structure
B. Mysen and P. Richet

Developments in Geochemistry 11

Geological Sequestration of Carbon Dioxide

Thermodynamics, Kinetics, and Reaction Path Modeling

Luigi Marini

Laboratory of Geochemistry, University of Genova, Genova, Italy

ELSEVIER

Amsterdam – Boston – Heidelberg – London – New York – Oxford – Paris
San Diego – San Francisco – Singapore – Sydney – Tokyo

Elsevier
Radarweg 29, PO Box 211, 1000 AE Amsterdam, The Netherlands
The Boulevard, Langford Lane, Kidlington, Oxford OX5 1GB, UK

First edition 2007

Library of Congress Cataloging-in-Publication Data
A catalog record for this book is available from the Library of Congress

British Library Cataloguing in Publication Data
A catalogue record for this book is available from the British Library

ISBN-13: 978-0-444-52950-3
ISBN-10: 0-444-52950-0

Printed and bound in The Netherlands

07 08 09 10 11 10 9 8 7 6 5 4 3 2 1

Table of Contents

Preface

Dear human reader, let me briefly explain how this book was conceived. I live in Viareggio, which is called the Pearl of the Tyrrhenian sea, since most people knows Viareggio for its sandy beaches and the carnival parade of papier-mâché carts. During the carnival there is no access to the downtown beach, otherwise people go to the parade without paying the ticket, which is a very popular practice in Italy. On the contrary, in summertime there is too much access to the beaches, which become places of very high density, in terms of inhabitants per square meter, of course.

According to my wife Claudia, who presumably knows me rather well, I have a character more similar to bears than to human beings and, indeed, I must admit that I do not like crowded places. But I like very much to walk along the strand, at least when it is not a high-density place.

Two years ago, in a cold winter morning, I and my young daughter Costanza were walking along the strand looking for shells, mostly pelecypoda. Suddenly, we noticed that the flat-sea surface was ploughed by the dorsal fin of a huge fish and . . . it was coming at remarkably high speed towards us. In a few seconds the fish reached the strand and stopped. It was a marvellous tuna! We were surprised and paralysed but our surprise grew considerably when the tuna said:

– Hi, Prof. Marini, I was looking for you. I am the very same tuna who saved Pinocchio and his father Geppetto from the shark. It happened not far from here, one hundred and fifty years ago or so. Tunas are very long-lived. But, who is the nice girl accompanying you ?

– She is my daughter Costanza, she is . . .

I fell silent as soon as I realised that I was talking to a tuna. It was unbelievable, but the tuna was there. The silence was interrupted by Costanza who asked the tuna:

– Mr. Tuna, may I pat your head ?

– Yes, you can.

– You are cold, but your skin is very smooth and . . . slippery.

I intervened in their conversation:

– I am sorry, Mr. Tuna why are you looking for me ?

– Dear Prof. Marini, I am here on behalf of all the tunas, to discuss with you a very important matter. First of all, I must say that we are very worried. We read that humans intend to dispose carbon dioxide into the deep ocean waters. We know that humans are very selfish, but we are indignant by their absence of consideration for other God's creatures and we are seriously worried for our own lives. Can you imagine the risks associated with carbon dioxide disposal into deep ocean waters ?

– Did you say . . . you read ?

– Of course, we know almost everything of the human world and now, thanks to the web, it is much easier than before to learn what humans think and intend to do.

– Well, but why are you looking for me ? I mean, why me . . . instead of someone else? Today, the scientific world is mostly based in the USA and even though President Bush does not want to sign the Kyoto Protocol for the reduction of CO_2 emissions into the atmosphere, the USA scientists receive a lot of money to carry out researches on CO_2 disposal. The European funds on this topic are less, and the money spent by Italy is practically insignificant.

– For a very simple reason: I was swimming in the Northern Tyrrhenian sea. So, it was very simple to reach your hometown and contact you. I would like to learn more on geological carbon dioxide sequestration, mineral carbonation, and related subjects because these are much more appealing options than disposal in ocean waters, at least for tunas. Would you like to write a book on this subject?

I realised that I had many potential readers, though tunas, and that was a great satisfaction. I was also very curious . . . I have never discussed with a fish . . . it seemed to be a very stimulating experience. I said:

– Well Mr. Tuna, I am sure that a geochemist is the right person for you, I am certainly not the best geochemist neither of the world nor of Italy, but I am probably the best one available on the Viareggio strand in this cold morning. Costanza, would you mind if I discuss a little bit with Mr. Tuna?

– Dad, . . . you said a little bit. I know you, I will go on looking for shells without you, don't worry.

– Well Mr. Tuna, what do you know about carbon dioxide ?

– I documented myself . . . carbon dioxide was first identified by Joseph Black, when I was very young. He was the first one who performed experiments of CO_2 production and CO_2 sequestration. It is amazing to realise that it happened about two hundred and fifty years ago! I also read that one of the possible uses of carbon dioxide was early discovered by Joseph Priestly, a nonconformist priest who lived, by chance, near a brewery. Thanks to his privileged location, he could get a big supply of carbon dioxide and this fact induced him to study this gas. The practical result was the production of soda water or seltzer.

I stopped Mr. Tuna saying:

– That is really funny ! A priest found the way to produce a soft drink using a by-product of beer production. That was a giant step for the redemption of mankind! I am sorry, Mr. Tuna . . . please go ahead.

– I also read "*La Géochimie*" by Vladimir Ivanovič Vernadsky. Here is the book. At that time it was already clear that atmospheric carbon dioxide enters the atmosphere through different sources, including the *exhalations of anthropogenic gases produced through combustion reactions, calcination of limestones, fermentation reactions, etc. The amounts of CO_2 thus produced by humans becomes larger and larger and it must be taken into account in the geochemical history of carbon. . . . Thus civilised man displaces the state of equilibrium and represents a new geological force, whose importance becomes larger and larger in the geochemical history of all the chemical elements.*

– I also know "*La Géochimie*". It is a collection of lectures given by Vernadsky at the Sorbonne in Paris in 1922–1923 and, therefore, it gives a clear picture of the understanding of geochemical cycles some eighty years ago, based on a thorough review of the existing literature by a great geochemist. Mr. Tuna, I assure you that I will do my best, but you should not expect something similar to the Vernadsky book.

–Just write your book, I will read it with interest.

He disappeared as suddenly as he came. I started to write and . . . here is the book. I am sure Mr. Tuna and his friends will find the way to read it.

I would like to dedicate this book to those who taught me geology and geochemistry, namely the late Prof. Riccardo Assereto, Prof. Franco Tonani, and Prof. Harold C. Helgeson.

I am indebted with Prof. Giulio Ottonello for the numerous interesting conversations, which have helped to clarify my understanding of several geochemical aspects.

Last but not the least, I must acknowledge the valuable help I received from Marina Accornero, who is presently working on her PhD Thesis. This book has greatly benefitted from her competent and careful review. Marina read critically the manuscript and found mistakes and unclear parts. Of course, I am the sole responsible of those that may still be present.

Viareggio,
November 11, 2005

Chapter 1

Why We Should Care: The Impact of Anthropogenic Carbon Dioxide on the Carbon Cycle

1.1. Carbon dioxide: from its discovery to the understanding of its role

Carbon dioxide was first identified around the middle of the 18th century by Joseph Black (1728–1799), a Scottish, in the framework of his studies to get the degree in medicine at the University of Edinburgh. Results of Black's chemical investigations were published in 1756 under the title *Experiments upon Magnesia Alba, Quick-lime, and Some Other Alcaline Substances* (Leicester, 1956).

Black demonstrated that *magnesia alba* (an hydroxycarbonate of magnesium, possibly hydromagnesite, as considered in the following reactions) developed a gas upon heating and consequently transformed in *magnesia calcinata* (magnesium oxide):

$$4MgCO_3 \cdot Mg(OH)_2 \cdot 4H_2O \rightarrow 5MgO + 5H_2O + 4CO_{2(g)}. \tag{1-1}$$

Upon acid attack, driven for example by HCl, both *magnesia calcinata* and *magnesia alba* produced the same salts:

$$MgO + 2HCl \rightarrow MgCl_2 + H_2O \tag{1-2}$$

$$4MgCO_3 \cdot Mg(OH)_2 \cdot 4H_2O + 10HCl \rightarrow 5MgCl_2 + 10H_2O + 4CO_{2(g)}, \tag{1-3}$$

but effervescence was observed only in the reaction involving *magnesia alba*, due to CO_2 development.

It was possible to obtain again *magnesia alba* by reacting *magnesia calcinata* with the so-called *alkalies*, i.e., sodium or potassium carbonate:

$$5MgO + 4Na_2CO_3 + 9H_2O \rightarrow 4MgCO_3 \cdot Mg(OH)_2 \cdot 4H_2O + 8NaOH \tag{1-4}$$

Black also observed that similar experiments carried out with limestone produced quicklime and the same gas. Again, the original limestone could be obtained through reaction of quicklime with *alkalies*. The gas was called *fixed air* by Black because it was fixed in solid form by magnesia and quicklime. Black was thus the first one who realized experiments of CO_2 production and CO_2 sequestration.

These results were really something new in that it was possible to combine chemically a gas with a solid to produce a new solid compound with different properties. The effect of this discovery is well evident in the words of John Robinson, Black's colleague, who wrote the introduction of the conferences of chemistry held by Black and published posthumously in 1803.

He discovered that a cubic inch of marble was made up by pure quicklime for about half of its weight and by as much air to fill a six-gallons wine container . . . What could be more singular than to prove that a substance as thin as air can exist in the form of hard stone and that its presence determines such a change in the properties of stone?

Joseph Priestly (1733–1804) not only triggered soda water production in 1772, but also isolated and studied a series of gases, including NO, N_2O_5, CO, SO_2, HCl, NH_3 and O_2, and observed that the latter gas was emitted by green plants exposed to light. This observation, confirmed and broadened by Jan Ingenhousz (1730–1799) and Jean Senebier (1742–1809), represented the base of all subsequent studies of photosynthesis, which demonstrated the important role of atmospheric carbon dioxide as a source of carbon for plants.

The first successful results in this direction were obtained by T. de Saussure, a Genevese, who carried out his studies in 1797–1804, but he did not fully realize the huge impact of his discovery. Such importance was partly recognized some 20 to 30 years later by the Russian physicist G. Parrot and the French botanist A. Brongniart, but their theories were wrong. Only the researches carried out by J. Boussingault in 1830–1840 led him to finally establish the role of atmospheric carbon dioxide in the formation of living matter. Afterwards, some further improvements concerned the existence of microbes, which can bypass CO_2 and use the carbon of CH_4 and CO, and the complex reactions of aquatic organisms, which are able to use bicarbonate ion and carbonic acid, as shown by Raspail in 1833.

Today we know that carbon gain by plants is strictly linked to the operation of Rubisco (ribulose–1,5 bisphosphate carboxylase–oxygenase), which catalyses the competitive processes of carboxylation (leading to photosynthesis) and oxygenation (leading to photorespiration). Both CO_2 and O_2 compete for the first acceptor molecule ribulose bisphosphate at their binding sites on Rubisco and are, consequently, mutually competitive inhibitors (Lawlor, 1993). A rise in O_2 tends to inhibit the carboxylation reaction of Rubisco, suppressing photosynthesis, whereas a rise in CO_2 inhibits the oxygenase reaction, suppressing photorespiration.

1.2. The short-term carbon cycle

The main reservoirs of the short-term carbon cycle are (Fig. 1.1) as follows: (a) the oceans that store ~38,000 $\times$ 10^{15} g C, (b) the terrestrial biosphere, whose total C mass, ~2,000 $\times$ 10^{15} g C, is distributed between the terrestrial plants (1/4) and soil (3/4), and (c) the atmosphere, which is the smallest reservoir with ~730 $\times$ 10^{15} g C (Prentice et al., 2001).

1.2.1. The terrestrial biosphere

In higher plants, a total amount of ~270 $\times$ 10^{15} g C a^{-1} of CO_2 diffuses from the atmosphere into the leaves through the stomata, their small pores, and reaches the sites of

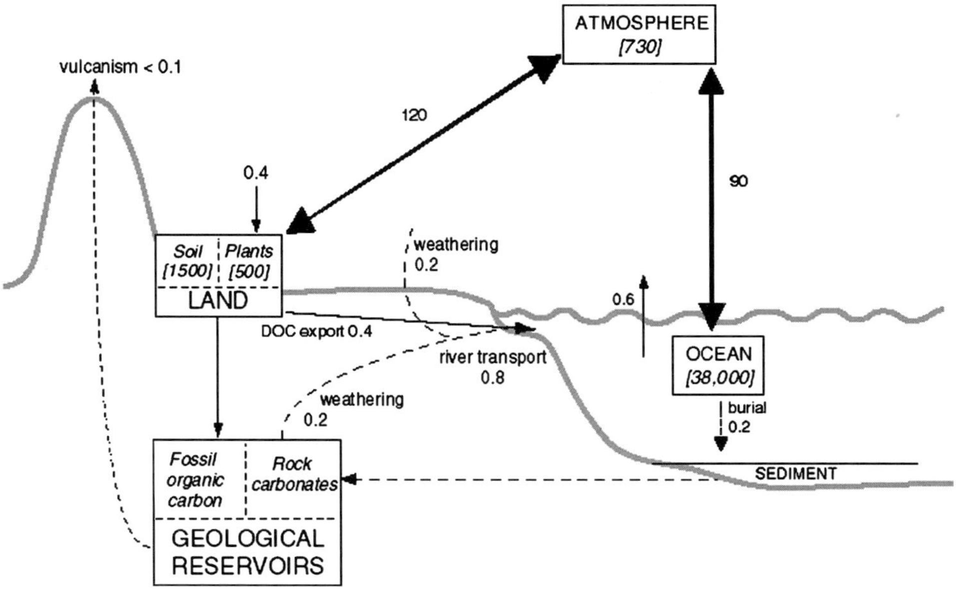

Figure 1.1. Schematic representation of the short-term carbon cycle. Reservoirs in Gt C = 10^{15} g C. Fluxes in Gt C $\times$ a^{-1}. (Reproduced with permission from I. C. Prentice et al., 2001. Copyright © 2001 by the Intergovernmental Panel on Climate Change 2001.)

photosynthesis. The mass of CO_2 converted into carbohydrates, which is called gross primary production (GPP), is only ~120 $\times$ 10^{15} g C a^{-1} (Ciais et al., 1997), as most CO_2 diffuses back to the atmosphere without being involved in photosynthesis.

The difference between GPP and autotrophic respiration (i.e., the back conversion of carbohydrates to CO_2) represents the annual plant growth or net primary production (NPP). This amounts globally to ~60 $\times$ 10^{15} g C a^{-1} (Saugier and Roy, 2001).

The carbon fixed as NPP is returned to the atmosphere by both heterotrophic respiration (HR by bacteria, fungi and herbivores) and combustion processes during fires. The difference between NPP and HR is called net ecosystem production (NEP) and represents the mass of carbon either gained or lost by the ecosystem, without considering additional losses, such as fires, harvesting, erosion and transport processes by rivers. The NEP global amount is ~10 $\times$ 10^{15} g C a^{-1} (Bolin et al., 2000). Subtraction of these additional losses from NEP gives the carbon accumulated by the terrestrial biosphere or net biome production (NBP), which represents the net land uptake from the atmosphere (Schulze and Heimann, 1998). Based on atmospheric O_2 and CO_2 concentrations, Prentice et al. (2001) have evaluated NBP of -0.2×10^{15} g C a^{-1} during the 1980s and of -1.4×10^{15} g C a^{-1} during the 1990s, with an uncertainty of $\pm$ 0.7 $\times$ 10^{15} g C a^{-1}.

At steady state, NBP is zero. These non-zero NBP, which are explained by transient changes mainly related to human activities, natural noise and climate oscillations, indicate that the terrestrial biosphere globally represents a carbon sink at present.

1.2.2. The processes in the oceans

The oceans store a mass of carbon, which is ~50 times larger than the carbon mass in the atmosphere, and the two-way CO_2 exchange between the atmosphere and the surface ocean is ~90 × 10^{15} g C a^{-1} (Fig. 1.2). Carbon dioxide is exchanged between these two reservoirs on a time scale of several hundred years to more than a thousand years, as this is the time scale of the vertical exchanges between surface- and deep-water masses in the oceanic reservoir.

In spite of this kinetic constraint, the oceans represent a large sink for anthropogenic CO_2, due to both its relatively high solubility and reactions with seawater chemical constituents. A primary role is played by the reaction of dissolved CO_2 with CO_3^{2-} ions to produce HCO_3^- ions:

$$H_2CO_3^* + CO_3^{2-} \rightarrow 2\,HCO_3^- \qquad (1\text{-}5)$$

where $H_2CO_3^*$ indicates the so-called apparent carbonic acid, comprising both true carbonic acid $H_2CO_3^*$ and true dissolved molecular $CO_{2(aq)}$. Because of this homogeneous

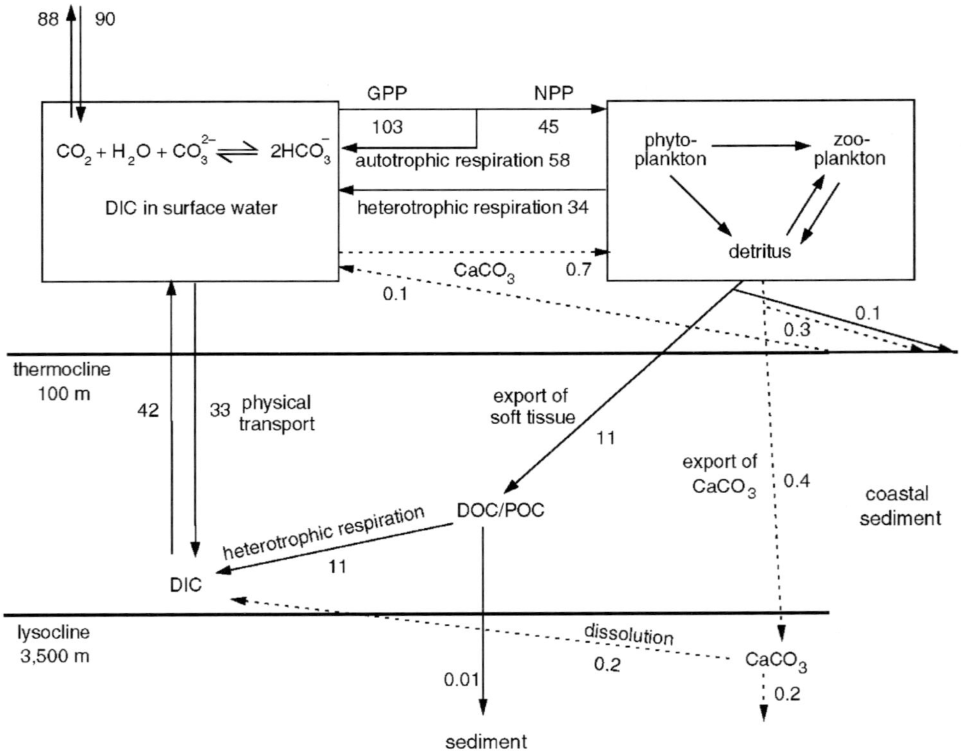

Figure 1.2. The cycling of carbon in the ocean. (Reproduced with permission from I. C. Prentice et al., 2001. Copyright © 2001 by the Intergovernmental Panel on Climate Change 2001.)

reaction, however, less CO_3^{2-} is available to react with the added CO_2 and an increasing fraction of the added CO_2 remains as H_2CO_3*, preventing further conversion to HCO_3^- ion.

Natural transfers of CO_2 between the atmosphere and the ocean are controlled by the temperature changes of surface waters and biological production and respiration. Both circulation of oceanic waters and sinking of C-rich particles transfer carbon within the oceans, from spatially separated source and sink regions.

Cooling of surface waters favours CO_2 uptake due to increased CO_2 solubility at low temperatures and vice versa, determining seasonal and regional patterns in the atmosphere–seawater transfer of CO_2 (Watson et al., 1995).

Seasonal and regional patterns of CO_2 fluxes are also caused by biological processes. The GPP of oceanic phytoplankton, ~103×10^{15} g C a^{-1} (Bender et al., 1994), is partly reconverted into dissolved inorganic carbon (DIC) by autotrophic respiration, leaving an NPP of ~45×10^{15} g C a^{-1} (Longhurst et al., 1995). The organic carbon thus produced is either consumed by zooplankton or becomes detritus. A fraction of organic carbon dissolves in ocean water (DOC) and part of it is oxidized by bacterial activity. Dead organisms and detritus constitute the particulate organic carbon (POC). Both sinking of POC and transfer of DOC make up a downward flux of organic carbon called *export production*, which ranges from 10×10^{15} to 20×10^{15} g C a^{-1} (Falkowski et al., 1998), and is close to 11×10^{15} g C a^{-1} according to Schlitzer (2000). Most export production is converted back into DIC *via* heterotrophic respiration in deep waters, whereas only a small fraction of the export production, ~0.1×10^{15} g C a^{-1}, is incorporated into the sediments (Gattuso et al., 1998). Upwelling of deep DIC-rich waters transports DIC into the shallow oceanic layer, causing its re-equilibration with the atmospheric CO_2. This overall biological mechanism, known as *biological pump*, has a buffering effect on both DIC concentrations in deep oceanic waters and atmospheric CO_2 levels, which are lowered by ~200 μmol/mol (Sarmiento and Toggweiler, 1984).

Upwelling deep waters also supply potentially limiting nutrients, such as nitrate, phosphate, silicate and iron, which control the overall productivity of the ocean.

In shallow waters, marine organisms precipitate solid $CaCO_3$ that accumulates in sediments, in part upon sinking at depth, in part in shallow waters (e.g., coral reefs). The production of $CaCO_3$ at a global scale is ~0.7×10^{15} g C a^{-1}. Precipitation of $CaCO_3$ causes a decrease in CO_3^{2-}, driving reaction (eq. (1-5)) to the left and increasing the concentration of dissolved CO_2, which is partly released into the atmosphere. The overall effect of biological activity on the P_{CO_2} of surface oceanic waters is conveniently expressed through the ratio between the export production and the export of $CaCO_3$ (i.e. the net downward flux of solid $CaCO_3$, ~0.4×10^{15} g C a^{-1} globally), which is known as *rain ratio* (e.g., Broecker and Peng, 1982).

Continental areas discharge both organic and inorganic carbon into the ocean through river (and subordinately groundwater) transport. This carbon is partly derived from natural sources, corresponding to ~0.8×10^{15} g C a^{-1} at a global scale, and in part from anthropogenic activities, ~0.1×10^{15} g C a^{-1} (Fig. 1.1). Half of the natural carbon is organic and half inorganic, whereas the anthropogenic carbon is mostly organic (Meybeck, 1993). The CO_2 released from the ocean in response to the natural river input amounts to ~0.6×10^{15} g C a^{-1}.

At present, the biologically driven carbon cycle in the oceans is close to steady state. Experiments have shown that the overall productivity of the ocean is not controlled by CO_2 concentrations, apart from few exceptions (Riebesell et al., 1993). In addition, the ratios among potentially limiting nutrients and DIC in deep oceanic waters are close to the so-called Redfield ratios, i.e. the nutritional requirements of marine organisms (Redfield et al., 1963). Therefore, in general, it is virtually impossible to increase the atmosphere–ocean transfer of carbon acting on the productivity rate of the ocean.

1.3. Atmospheric CO_2 concentration

Time changes in atmospheric CO_2 concentration have been monitored during the last 45–50 a. Determinations on trapped air bubbles in Antarctic ice cores allow to reconstruct the variations in atmospheric CO_2 concentration during the last 420,000 a. Estimates of atmospheric CO_2 during the Earth's history have been obtained through modelling of the long-term geochemical carbon cycle. Although the uncertainties of these data grow considerably with increasing time before present, the available picture leaves no doubt on the anomalous values of atmospheric CO_2 concentration attained soon after the onset of the industrial revolution as well as on its continuous increase.

1.3.1. Direct determinations

Continuous direct measurements of atmospheric CO_2 concentration have been carried out at the Mauna Loa and South Pole stations since 1957 (Keeling et al., 1995; Keeling and Whorf, 2003). Further determinations have been performed by the global network developed in the 1970s (Conway et al., 1994; Keeling et al., 1995). Values of the $^{13}C/^{12}C$ isotopic ratio of atmospheric CO_2 are available since 1977 (Francey et al., 1995; Keeling et al., 1995; Trolier et al., 1996). Changes in atmospheric O_2 concentrations of few μmol/mol against a background value of 209,000 μmol/mol have also been measured on a regular basis since the early 1990s (Keeling and Shertz, 1992; Keeling et al., 1993, 1996; Bender et al., 1996; Battle et al., 2000).

Available data of atmospheric CO_2 concentration (Fig. 1.3) show an annual cyclical oscillation of ~6 μmol/mol, which results from excess photosynthesis during spring and summer and excess respiration during autumn and winter. The yearly average CO_2 concentration experienced a progressive increase from 315 μmol/mol in 1958 to 367 μmol/mol in 1999.

These data allow one to compute the rate of rise in atmospheric CO_2, which was 3.3×10^{15} g C a^{-1} from 1980 to 1989 and 3.2×10^{15} g C a^{-1} from 1990 to 1999. The uncertainties associated with these estimates are $\pm 0.1 \times 10^{15}$ g C a^{-1}.

The increase in atmospheric CO_2 is unanimously attributed to the burning of fossil fuels, coal, natural gas and oil, and subordinately to the production of cement. The annual global emission of CO_2 from burning of fossil fuels and cement production has been evaluated for the period of time from 1751 through 1999 (Marland et al., 2000). Average values were $5.4 \pm 0.3 \times 10^{15}$ g C a^{-1} during the 1980s and $6.3 \pm 0.4 \times 10^{15}$ g C a^{-1} during the 1990s.

Removal of carbon-bearing vegetation through forest clearing for agricultural lands and harvesting of wood has also determined an increase in atmospheric CO_2. This source

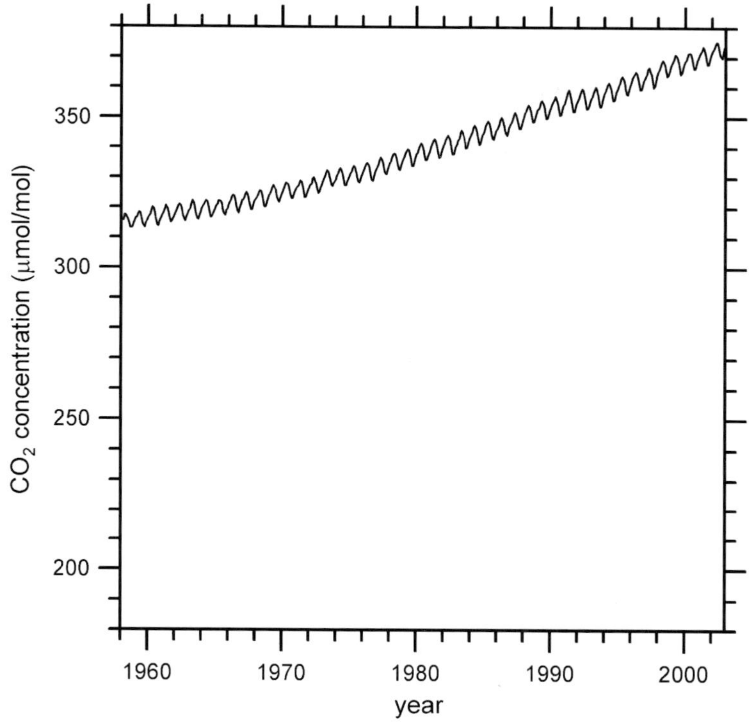

Figure 1.3. Atmospheric CO_2 concentration in recent years based on direct determinations at Mauna Loa Observatory. Data from Keeling and Whorf (2003).

is equivalent to more than half of the fossil fuel source according to Sarmiento and Gruber (2002), but there is no general consensus on this evaluation.

Since anthropogenic emissions are responsible for the rise in atmospheric CO_2, it is instructive to compare the rate of these two phenomena (Fig. 1.4). This comparison shows that the rates of increase in atmospheric CO_2 are less than the emission, suggesting that the emitted CO_2 is partly taken up by the two carbon reservoirs acting as sinks, i.e. the oceanic waters and the terrestrial ecosystems (see Section 1.2). The relative importance of these two reservoirs can be established based on O_2 measurements and the $^{13}C/^{12}C$ isotopic ratio of atmospheric CO_2. The existence of a large inter-annual variability in the rate of atmospheric CO_2 increase is also evident in Fig. 1.4.

The effect of fossil fuel burning on the O_2/CO_2 atmospheric ratio can be established based on the stoichiometry of the combustion reactions (Keeling, 1988). It corresponds to the line labelled "fossil fuel burning" in Fig. 1.5. Carbon dioxide dissolution in oceanic waters has no effects on atmospheric O_2. In contrast, biospheric uptake implies a release of O_2 due to excess photosynthesis over respiration and other oxidative processes. Assuming that biological O_2 uptake in the oceans has not changed in the considered lapse of time because of nutrient limitations, the biospheric uptake is attributed to the terrestrial ecosystems. The effect of degassing of O_2 upon heating of the oceans (Levitus et al., 2000) is also shown in Fig. 1.5.

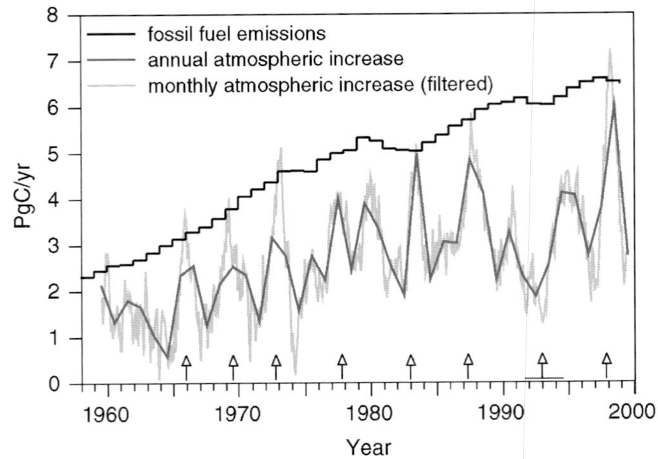

Figure 1.4. Comparison of fossil fuel emissions and the rate of rise in atmospheric CO_2 concentration. Vertical arrows indicate El Niño events. (Reproduced with permission from I. C. Prentice et al., 2001. Copyright © 2001 by the Intergovernmental Panel on Climate Change 2001.)

Use of the $^{13}C/^{12}C$ isotopic ratio of atmospheric CO_2 to discriminate land and ocean uptakes is based on the significant carbon isotope fractionation during the photosynthesis by C_3 plants. This process depletes biospheric carbon in ^{13}C by ~18% with respect to the atmosphere. On the contrary, small fractionations occur during ocean–atmosphere CO_2 exchanges. Variations in the $^{13}C/^{12}C$ isotopic ratio of atmospheric CO_2, however, are not simply relatable with changes in the biospheric uptake, but depend also on the turnover time of carbon in the ocean and in the terrestrial ecosystem, due to the continuous input of light carbon into the atmosphere through fossil fuel burning (Keeling et al., 1980). A further problem is due to the fractionation by C_4 plants, which is lower than that by C_3 plants. Therefore, the relative importance of C_3- and C_4-photosynthesis has to be assessed. Use of disequilibrium fractionations leads to results consistent with those based on O_2 measurements (Langenfelds et al., 1999; Battle et al., 2000).

Based on the use of atmospheric O_2 and CO_2 concentrations and the isotopic signature of atmospheric CO_2, the ocean–atmosphere flux and the land–atmosphere flux were estimated to be $-1.9 \pm 0.6 \times 10^{15}$ g C a^{-1} and $-0.2 \pm 0.7 \times 10^{15}$ g C a^{-1}, respectively, during the 1980s and $-1.7 \pm 0.5 \times 10^{15}$ g C a^{-1} and $-1.4 \pm 0.7 \times 10^{15}$ g C a^{-1}, respectively, during the 1990s (Prentice et al., 2001). The minus sign indicates that land and ocean act as sinks. Considering the uncertainties in these fluxes, the effect of the ocean sink in the 1990s was the same as that during the 1980s, whereas the land sink has taken up more carbon during the 1990s than during the 1980s, perhaps due to a lower rate of tropical deforestation, at least in part. In fact, the land–atmosphere flux obtained from observational data is the sum of a positive land-use flux (a source), mainly governed by tropical deforestation, and a residual negative term (a sink), called "missing sink" or residual-terrestrial sink. During the 1980s, the role of this residual-terrestrial sink, with a flux of -1.9×10^{15} g C a^{-1}, was similar to that of the oceanic sink, although the reported range

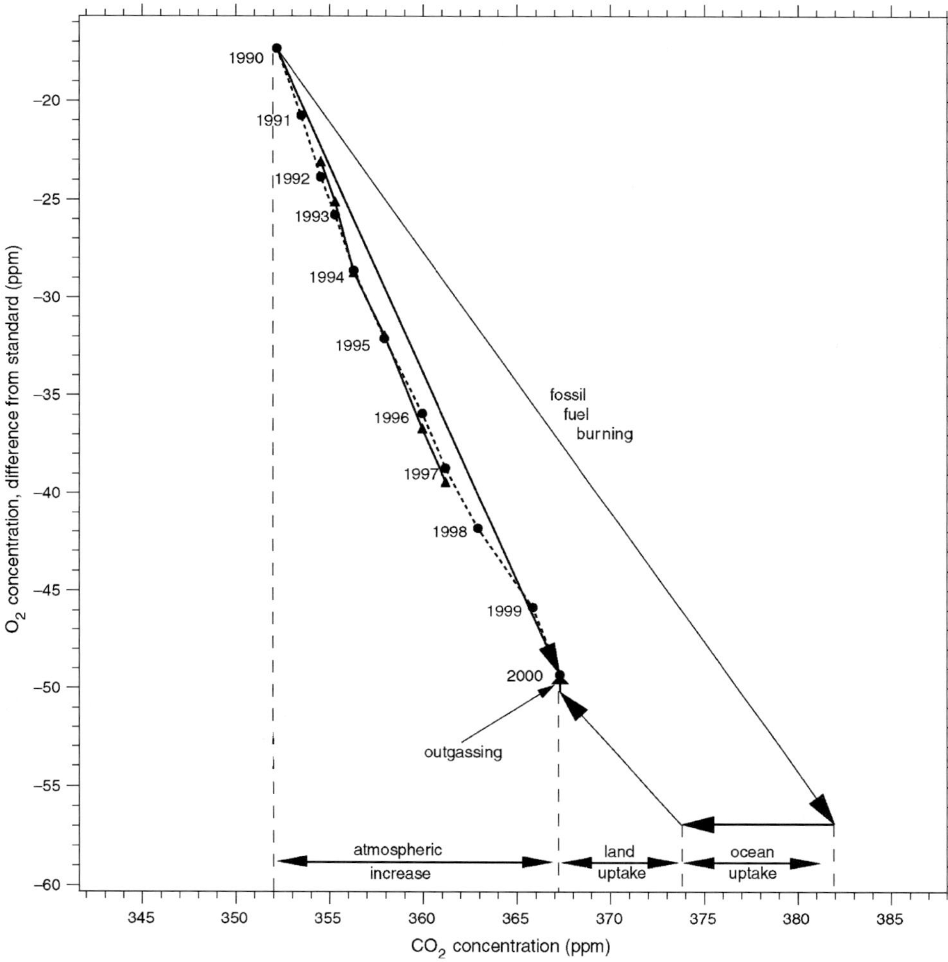

Figure 1.5. Diagram of annual average CO_2 and O_2 concentrations used to partition fossil fuel uptake between land and ocean sinks. (Reproduced with permission from I. C. Prentice et al., 2001. Copyright © 2001 by the Intergovernmental Panel on Climate Change 2001.)

is very large, from -3.8×10^{15} to $+ 0.3 \times 10^{15}$ g C a^{-1}. The nature of the residual terrestrial sink is still a matter of debate, with the land-use changes (e.g., forest re-growth) and the fertilization by atmospheric CO_2 and atmospheric nitrogen deposition as the most likely explanations. This unresolved riddle has important consequences, in that the global CO_2 uptake capacity expected for land-use changes is much lower than that for fertilization effects (Sarmiento and Gruber, 2002).

The inter-annual variability in the rate of increase in atmospheric CO_2 (Fig. 1.4) cannot be explained by the fossil fuel emissions, which do not experience this kind of variability. Consequently, the explanation must be related with short-term changes in either the ocean–atmosphere flux or the land–atmosphere flux or both. These changes could be caused by climatic changes, as suggested by the correlation with the El Niño events (e.g., Bacastow, 1976).

1.3.2. CO_2 concentration in the air bubbles of Antarctic ice

Atmospheric CO_2 concentration was close to 280 ± 5 μmol/mol from 1000 to 1800 AD (Fig. 1.6), as shown by high-detail Antarctic ice cores (Siegenthaler et al., 1988; Neftel et al., 1994; Barnola et al., 1995; Etheridge et al., 1996, 1998).

The Taylor Dome Antarctic ice core indicates that atmospheric CO_2 concentration varied between 260 and 280 μmol/mol during the past 11,000 a (Smith et al., 1999; Indermühle et al., 1999), and similar values are suggested by ice core BH7 near Vostok (Peybernès et al., 2000). The causes of these CO_2 changes are unknown.

Study of the Vostok Antarctic ice core indicates that during the last four glacial–inter-glacial cycles, i.e. during the past 420,000 a, atmospheric CO_2 concentration oscillated

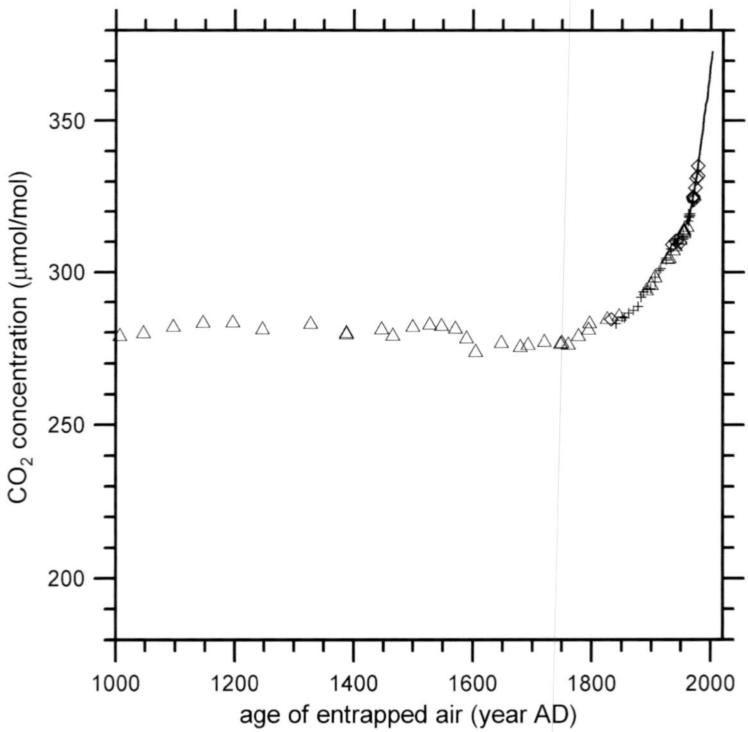

Figure 1.6. Atmospheric carbon dioxide concentration during the past 1000 a, based on the Antarctic ice cores Law Dome DE08 (crosses), DE08-2 (diamonds), and DSS (triangles) (data from Etheridge et al., 1998 and references therein). Recent atmospheric determinations at Mauna Loa Observatory (data from Keeling and Whorf, 2003) are also shown (solid line).

between 180 and 300 μmol/mol (Fig. 1.7; data from Barnola et al., 2003), and it was high during the inter-glacial periods and low during the glacials (Petit et al., 1999; Fischer et al., 1999). In spite of these large changes, atmospheric CO_2 concentration was always lower than at present.

At the onset of glacials, the lags of 2,000–4,000 a in the CO_2 decrease suggest that low atmospheric CO_2 concentrations were not the cause of glaciations, although they might have strengthened the global cooling (e.g., Fischer et al., 1999).

The $^{13}C/^{12}C$ isotopic ratio of deep-sea sediments suggests that the terrestrial biosphere stores an excess of 300×10^{15} to 700×10^{15} g C during the inter-glacials with respect to the glacials (Shackleton, 1977). Since these figures do not explain the difference in atmospheric CO_2 between glacials and inter-glacials, the oceans probably played a pivotal role, although it has not been understood yet.

1.3.3. Atmospheric CO_2 during the Earth's history

The long-term geochemical carbon cycle has been modelled by Berner and coworkers (Berner, 1997 and references therein) with the aim to predict the atmospheric CO_2 concentration in the geological past.

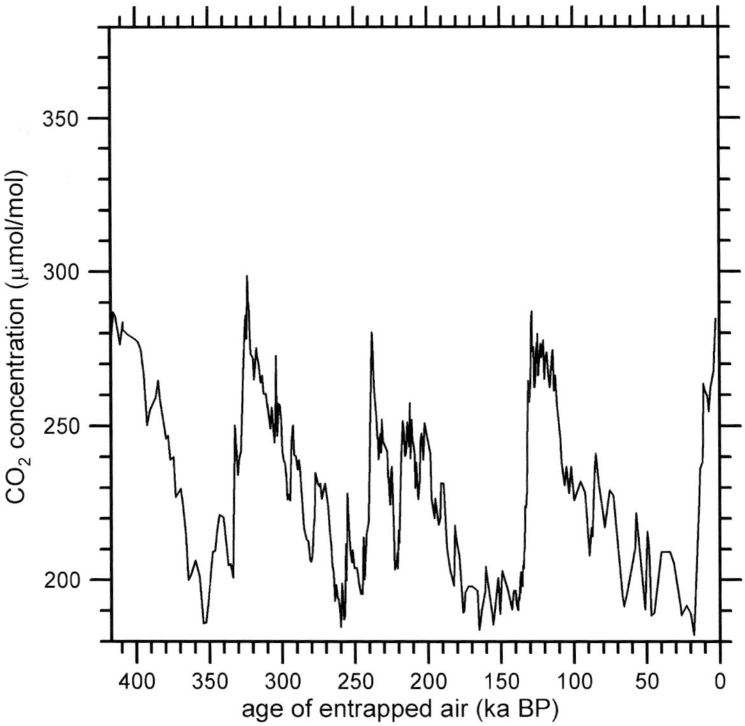

Figure 1.7. Carbon dioxide concentration in the air bubbles trapped in the Vostok Antarctic ice core (data from Barnola et al., 2003 and references therein).

On long time scales, atmospheric CO_2 concentration is assumed to be controlled by the balance between the CO_2-consuming geochemical processes, such as weathering of silicates and burial of organic matter in sedimentary rocks, and the release of CO_2 into the atmosphere through Earth degassing, including diagenesis, metamorphism, magmatism and mantle degassing.

Modelling of the long-term geochemical carbon cycle indicates that very high atmospheric CO_2 concentrations, greater than 3,000 μmol/mol, occurred from Cambrian to Silurian (600–400 Ma BP) and during the Jurassic (195–140 Ma BP). More recently, atmospheric CO_2 concentration experienced a marked decrease after ~60 Ma BP and was probably less than 300 μmol/mol during the latter 20 Ma (Pagani et al., 1999a; Pearson and Palmer 1999). These low CO_2 levels might have been the primary cause of the expansion of C_4 plants from 7 to 5 Ma BP (Pagani et al., 1999b).

These results agree with the findings of Arrhenius (1896), on the occurrence of *global warming* in the past in response to increases in atmospheric CO_2 concentration, which were shared by several other authors including Högbom (1894), Chamberlin (1898), Callendar (1938), Plass (1956), Budyko and Ronov (1979) and Fischer (1983).

1.4. Carbon cycle modelling and prediction of future atmospheric CO_2 concentrations

The short-term carbon cycle (Section 1.2) has been modelled by several authors (see Prentice et al., 2001 and references therein) taking into account the complex interactions among relevant processes. The aim of this exercise is to evaluate the possible responses of the carbon cycle to climate changes and to predict possible future variations in fluxes between different reservoirs. Modelling has been carried out separately for the land and the ocean.

Process-based terrestrial models include: (a) terrestrial biogeochemical models for the simulation of C, N and H_2O fluxes among terrestrial ecosystems and (b) dynamic global vegetation models that relate the processes of interest for the carbon cycle with variations in the structure and composition of the terrestrial ecosystems.

Process-based ocean models take into account the carbon chemistry, the CO_2 exchange with the atmosphere, and the physical and biogeochemical transport processes in the ocean.

These carbon cycle models have been developed and tested against available data and then used for future projections, until the end of the century, assuming reasonable emissions scenarios. Oceanic models indicate that ocean uptake increases with time but the uptake/emissions ratio decreases with time. Terrestrial model suggests that the rate of uptake by the land increases until 2050 and stabilizes afterwards. Both the ocean uptake and the terrestrial uptake decrease if climate-change feedbacks are taken into account in modelling. To quantify these effects coupled models are needed.

To this purpose, Prentice et al. (2001) used two simplified, fast models (ISAM and Bern-CC) for prediction of both future atmospheric CO_2 concentrations under given emissions scenarios and future emissions under different CO_2 stabilisation scenarios. Ocean and terrestrial climate-change feedbacks are taken into account in both models. In particular, the predicted atmospheric CO_2 concentration in 2100 ranges from 541 to 963 μmol/mol based on the Bern-CC model and between 549 and 970 μmol/mol according to the ISAM model, depending on the considered emissions scenario.

According to Prentice et al. (2001), "*anthropogenic CO_2 emissions are virtually certain to be the dominant factor determining CO_2 concentrations throughout the 21st century*. The importance of anthropogenic emissions is underlined by the expectation that the proportion of emissions taken up by both ocean and land will decline at high atmospheric CO_2 concentrations (even if absolute uptake by the ocean continues to rise). There is considerable uncertainty in projections of future CO_2 concentration, because of uncertainty about the effects of climate change on the processes determining ocean and land uptake of CO_2. These uncertainties do not negate the main finding that anthropogenic emissions will be the main control."

In addition these authors recognize that "the possibilities for enhancing natural sinks have to be placed in perspective: a rough upper bound for the reduction in CO_2 concentration that could be achieved by enhancing terrestrial carbon uptake through land-use change over the coming century is 40 to 70 ppm [μmol/mol], to be considered against a two to four times larger potential for increasing CO_2 concentration by deforestation, and a >400 ppm [μmol/mol] range among the emissions scenarios."

1.5. Conclusive remarks

As shown above, today atmospheric CO_2 concentrations are more than 30% higher than before the industrial revolution and present levels were never attained during the past 20 million years. This growth in atmospheric CO_2 has brought about an increase in the trapping of infrared radiations emitted from the Earth surface, with respect to the pre-industrial situation. Indeed, more than half of this increase in infrared absorption is attributed to CO_2, while the remaining part is probably controlled by other atmospheric gases, chiefly CH_4, N_2O and chlorofluorocarbons.

The effects of this increased infrared absorption on the Earth climate involve several feedback mechanisms. However, there is little doubt that increased trapping enhances the *atmospheric greenhouse effect*, thus leading to *global warming*. It probably explains most of the 0.6 ± 0.2°C temperature increases that took place during the last century (Prentice et al., 2001). Nevertheless, humans cannot wait for a definite answer on this topic. Anthropic CO_2 inputs into the atmosphere have to be drastically reduced, and several strategies have to be undertaken to this purpose, as we do not have a single magic option.

Reduction in fossil-energy use is probably the wiser option, but its acceptance is unlikely, since civilized man acts as *Homo faber* rather than *Homo sapiens*, as already recognized by Vernadsky (1924).

Strengthening of natural carbon sinks, for instance through reforestation and stimulation of the oceanic *biological pump* (by addition of missing nutrients, such as Fe), has been proposed as a possible tool. But these measures alone will not be able to limit the increase in atmospheric CO_2 assuming business-as-usual emissions of ~$1{,}450 \times 10^{15}$ g C a^{-1} over this century.

Injection of CO_2 in deep oceanic waters might temporarily reduce atmospheric CO_2 concentrations, but it is not a final solution. In addition, it represents a technological challenge and involves high environmental risks linked to sudden release of CO_2.

Sequestration of CO_2 through injection into deep geological reservoirs and mineral carbonation represent two other options and the subject of the rest of this book. In principle, geological storage of CO_2 involves three different processes, namely (Hitchon, 1996): (1) trapping as gas or supercritical fluid below a low-permeability caprock, a process termed *hydrodynamic trapping*; (2) dissolution into deep waters or *solubility trapping*; and (3) precipitation of secondary carbonates brought about by dissolution of primary silicates and Al-silicates upon injection of CO_2 into a deep aquifer. The last process, or *mineral sequestration*, is especially attractive as the virtually permanent CO_2 fixation in form of carbonates into relatively deep geological formations prevents its return to the atmosphere.

Silicate minerals can be reacted with CO_2, also in a chemical plant, producing carbonates under controlled temperature and pressure conditions. The process is known as mineral carbonation and is also treated in this book, although its primary focus is the geological sequestration of CO_2.

The following chapters are dedicated to the reactants, i.e., CO_2 and primary silicates and Al–silicates, and products, i.e., the carbonate minerals. Their thermodynamic properties and the kinetics of relevant dissolution/precipitation reactions will be reviewed. A brief digression on basic thermodynamic concepts is required first.

Chapter 2

The Thermodynamic Background

Injection of high-pressure CO_2 in a relatively deep geological reservoir and mineral carbonation are typical phase-equilibrium problems. These can be solved by application of thermodynamics as long as consideration is restricted to reactions between CO_2 and H_2O, although not all these reactions are instantaneous. For instance, at low temperatures and pressures, the hydration of dissolved CO_2 to true carbonic acid, $H_2CO_3^\circ$, has a half-time of ~6 min (Jones et al., 1964). This is the time it takes for converting 50% of dissolved CO_2 to $H_2CO_3^\circ$ (see Section 6.1.2).

Geological CO_2 disposal is a relatively simple problem if we consider that rocks are unreactive: under this assumption, CO_2 distributes itself between an aqueous liquid phase and a gaseous phase, with gas solubility depending on temperature, pressure and the composition of the two phases.

However, the problem is much more complicated as dissolution/precipitation reactions of rock-forming minerals do occur and have to be considered. Apart from the complexity determined by the larger number of chemical components involved in the system of interest, with respect to the previous assumption, kinetics is of fundamental importance and brings about a lot of complications. In fact, if it is reasonable to assume that gas dissolution and homogeneous chemical reactions in the aqueous solution attain the equilibrium condition instantaneously or nearly so, the same assumption is certainly false for heterogeneous dissolution/precipitation reactions of solid phases. Hence, the rate constants of these reactions and the surface areas of minerals have to be known and properly inserted into thermodynamic models. Even in this case, however, thermodynamics plays a pivotal role in solving the phase-equilibrium problem of interest.

This chapter is devoted to recall the pertinent, fundamental thermodynamic concepts. The first part of the following presentation (Sections 2.1–2.3) is mostly taken from Prausnitz et al. (1999) to whom the reader is addressed for further details. Discussion on kinetic concepts is presented later in Chapter 6.

2.1. The chemical potential

For a *homogeneous* system, i.e., a system consisting of a single phase made up of different chemical components, the chemical potential of chemical component i, μ_i, is equal to the partial derivative of the Gibbs free energy of the system, G, with respect to the moles n_i

of the *i*th component, at constant pressure P, temperature T, and composition of all constituents except i:

$$\mu_i = \left(\frac{\partial G}{\partial n_i}\right)_{P,T,n_j}. \tag{2-1}$$

The chemical potential of component i is equal to its partial molar Gibbs free energy, $\bar{G}_i$, because the independent variables temperature and pressure, which are chosen for defining the partial molal properties, are also the fundamental independent variables for G, as indicated by the relation:

$$dG = -SdT + VdP, \tag{2-2}$$

where S is the entropy and V the volume.

It is important to recall the difference between molar properties and partial molar properties. The molar properties are defined for pure substances and are indicated here by means of lower case letters: for instance, g stands for the molar Gibbs free energy, i.e., the Gibbs free energy of one mole of the considered pure substance.

Partial molar properties, such as the partial molal Gibbs free energy, are defined for a specified component i of a given phase and are indicated here by the symbol of the property with a superscripted bar, e.g., $\bar{G}_i$. Partial molar properties can be defined only for extensive properties (e.g., G, S, V,. . .) through differentiation at constant temperature and pressure.

For a *heterogeneous* system, i.e., consisting of two or more phases, the equilibrium distribution of a given chemical component i between two phases α and β is described by

$$\mu_{i,\alpha} = \mu_{i,\beta}, \tag{2-3}$$

where $\mu_{i,\alpha}$ and $\mu_{i,\beta}$ are the chemical potentials of the *i*th chemical component in the two phases α and β, respectively. This equation was proven by J. Willard Gibbs in 1875.

For applying equation (2-3) to a given system of interest, it is crucial to know how $\mu_{i,\alpha}$ and $\mu_{i,\beta}$ are related not only to temperature and pressure but also to the chemical composition of the system, which can be expressed through the molar fractions of different chemical components in each of the two phases α and β. To obtain these relationships, it is convenient to introduce two additional functions: the activity and the fugacity.

For instance, if phase α is a gas (e.g., a CO_2-rich gas phase) and phase β is a liquid (e.g., an aqueous solution containing some dissolved CO_2), then equation (2-3) can be rewritten as follows:

$$\frac{\Gamma_{i,G} \cdot y_{i,G} \cdot P}{f_i^o} = \gamma_{i,L} \cdot x_{i,L}, \tag{2-4}$$

where:

P is total gas pressure; $\Gamma_{i,G}$ and $y_{i,G}$ are the fugacity coefficient and the molar fraction of the *i*th component in the gas phase G, respectively;

$\gamma_{i,L}$ and $x_{i,L}$ are the activity coefficient and the molar fraction of the *i*th component in the liquid phase *L*, respectively; and
f_i° is the fugacity of component *i* in a condition of reference which is called standard state.

Both the discussion on how equation (2-4) is obtained from equation (2-3) and further details on the variables it contains are presented later in Sections 2.3 and 3.7. Here, it is important to underscore that although equations (2-3) and (2-4) have the same rank, equation (2-4) is less abstract than equation (2-3), as it contains the pressure *P* and two compositional variables of interest, $y_{i,G}$ and $x_{i,L}$.

In addition, equation (2-4) contains two other variables, $\Gamma_{i,G}$ and $\gamma_{i,L}$, both describing how real mixtures deviate from ideal mixtures. Ideal mixtures have peculiar, limiting properties, and both $\Gamma_{i,G}$ and $\gamma_{i,L}$ approach 1 when real mixtures approach the ideal behaviour. In general, however, we have to know the relations describing how $\Gamma_{i,G}$ and $\gamma_{i,L}$ depend on temperature, pressure and composition, or

$$\Gamma_{i,G} = f(T, P, y_1, y_2, y_3, \ldots) \tag{2-5}$$

and

$$\gamma_{i,L} = f(T, P, x_1, x_2, x_3, \ldots). \tag{2-6}$$

This subject will be dealt with in Chapters 3 and 4.

2.2. The standard state

As discussed above, one of the main tasks of phase-equilibrium thermodynamics is to establish the dependence of the chemical potential of a given substance upon the measurable variables: temperature, pressure and composition. However, there is a difficulty (or an inconvenience) in performing this exercise, in that absolute values of the chemical potential cannot be measured or computed. For instance, a reference value of zero cannot be given to the chemical potential in the complete absence of the considered substance, since it would be necessary to carry out an experiment of creation of matter and measure the chemical potential related to such process, which is impossible for humans. Looking at this problem from a different point of view, we are able to calculate variations in the chemical potential ensuing from variations in temperature, pressure and composition, but not absolute values of the chemical potential. This fact stems from the nature of the relations linking the chemical potential and the independent measurable variables: since these relations are differential equations, their integration leads to differences. For instance, let us consider the differential equation linking the chemical potential of a given substance *k* to temperature and pressure:

$$\mathrm{d}\mu_k = -s_k \mathrm{d}T + v_k \mathrm{d}P, \tag{2-7}$$

where s_k and v_k are the molar entropy and the molar volume of substance k, respectively. Integration of equation (2-7) leads to

$$\mu_{k,T,P} = \mu_{k,Tr,Pr} - \int_{Tr}^{T} s_k \mathrm{d}T + \int_{Pr}^{P} v_k \mathrm{d}P, \tag{2-8}$$

where T and P are the temperature and pressure of interest and Tr and Pr are the temperature and pressure of a given arbitrarily but suitably chosen reference state, respectively. If the thermal and volumetric properties of substance k are known over the temperature interval Tr to T and the pressure interval Pr to P, respectively, then it is possible to compute the two integrals $\int_{Tr}^{T} s_k \, \mathrm{d}T$ and $\int_{Pr}^{P} v_k \, \mathrm{d}P$, but the term $\mu_{k,Tr,Pr}$, i.e., the chemical potential at the reference state is still unknown. Consequently, the chemical potential at the temperature and pressure of interest can only be defined with respect to the value it assumes at the reference state, which is usually called *standard state*.

This represents a constant that cancels out in the calculation of the change in chemical potential resulting from any variation in the independent measurable variables: not only temperature and pressure (as discussed in previous lines) but also composition.

The chemical potential at standard state is usually indicated by the superscript ° and it is related to the activity, a_k , by the following equation:

$$\mu_k = \mu_k^o + RT \cdot \ln a_k . \tag{2-9}$$

In geochemistry, the standard-state convention usually adopted for solids and liquids is that of pure phases (i.e., stoichiometric minerals or pure liquids, including water) at all pressures and temperatures. For gases, the standard-state convention is generally the hypothetical perfect gas at 1 bar and any specified temperature. For aqueous species other than water, the standard-state convention is usually the *hypothetical* one molal solution referenced to infinite dilution at any pressure and temperature. The term *hypothetical* indicates that the adopted standard-state convention has nothing to do with the real one molal solution.

Independent of the adopted convention, from equation (2-9), it follows that activity is 1 under the standard-state condition. Therefore, pure solids and liquids have unit activity at all pressures and temperatures. Pure gases have unit activity at 1 bar and any temperature. Aqueous solutes have unit activity in the *hypothetical* one molal solution at any pressure and temperature.

Other needed conventions to compute the chemical potentials of elements and substances are: (1) the chemical potential of any chemical element in its most stable phase is equal to zero under standard-state conditions; (2) the chemical potential of the proton and of the electron in aqueous solution is equal to zero under standard-state conditions.

Since any compound may be considered as the product of a chemical reaction involving chemical elements as reactants, and since the chemical potential of any chemical element at standard state is equal to zero (Convention 1), the standard-state chemical potential for a compound is equal to the standard-state Gibbs free energy of the reaction of formation of the compound from its constituent chemical elements. The standard-state chemical potentials of ionic solutes are defined based on Convention 2, referring to suitable reactions involving the proton and/or the electron.

Further explanations are given in many textbooks, such as Nordstrom and Muñoz (1985), Anderson and Crerar (1993), and Ottonello (1997), as well as by Johnson et al. (1992) and references therein.

2.3. Fugacity and activity

Fugacity was first defined by Gilbert Newton Lewis in 1901, who tried to derive an auxiliary function closer to the real world than the abstract chemical potential. He first considered the chemical potential for a pure, ideal gas k. Based on equation (2-7), it turns out that:

$$\left(\frac{\partial \mu_k}{\partial P}\right)_T = v_k. \qquad \text{(2-10)}$$

From the ideal gas law:

$$v_k = \frac{RT}{P}. \qquad \text{(2-11)}$$

Insertion of equation (2-11) in equation (2-10) and integration at constant temperature from pressure P° to pressure P leads to

$$\mu_k - \mu_k^o = RT \cdot \ln \frac{P}{P^\circ}. \qquad \text{(2-12)}$$

This equation tells us that the isothermal transition from pressure P° to pressure P corresponds to a variation in the chemical potential equal to RT times the natural logarithm of the pressure ratio P/P°. In other words, the change in the abstract chemical potential is related to the change in a physically measurable variable, the pressure. Since equation (2-12) is true only for a pure, ideal gas, Lewis defined a new function, called fugacity, by analogy with equation (2-12). He stated that for any chemical component i in any phase, either a *gas*, a *liquid* or a *solid*, *pure* or *mixed*, *ideal* or *not ideal*, the change in chemical potential during an isothermal transition from fugacity f_i^o to fugacity f_i is

$$\mu_i - \mu_i^o = RT \cdot \ln \frac{f_i}{f_i^o}. \qquad \text{(2-13)}$$

where either μ_i^o or f_i^o is arbitrarily chosen, but not both, since one is fixed by the other.

The fugacity is equal to the pressure for a pure, ideal gas and it is equal to the partial pressure $P \times y_i$ for the ith component of a mixture of ideal gases. Since both pure gases and gas mixtures approach the behaviour of ideal gases at very low pressures (as we will verify for CO_2 in the next chapter), the following limits are adopted to complete the definition of fugacity:

$$\lim_{P\to 0}\frac{f}{P}=1 \qquad \text{(for pure gases)} \tag{2-14}$$

and

$$\lim_{P\to 0}\frac{f_i}{P\cdot y_i}=1 \quad \text{(for gas mixtures).} \tag{2-15}$$

Hence, the fugacity is a sort of either corrected pressure or corrected partial pressure, with corrections related in some way to the deviations from the ideal behaviour. The ratio f_i/f_i^o was termed *activity* by Lewis. This explains equation (2-9); see above. Since equation (2-13) was derived at constant temperature, the temperature of the standard state must be equal to that of the state of interest.

Based on the definition of fugacity it is straightforward to derive a new formulation for the equilibrium distribution of a given chemical component i between two phases α and β,

$$f_{i,\alpha}=f_{i,\beta}, \tag{2-16}$$

which is fully equivalent to equation (2-3), whereas the condition $a_{i,\alpha}=a_{i,\beta}$ is valid only if the standard state is the same for all the phases (Prausnitz et al., 1999).

Comparison of equations (2-8) and (2-13) suggests that fugacities can be obtained (at constant temperature) from the integral $\int v\mathrm{d}P$. This requires measuring the molar volumes down to sufficiently low pressures. However, under these conditions the molar volume becomes very large and difficult to be measured. In addition, from the point of view of calculus, the integral $\int v\mathrm{d}P$ is difficult to evaluate for $v\to\infty$. To overcome these difficulties, it is convenient to define the function:

$$v_{\text{ideal}}-v_{\text{real}}=\frac{RT}{P}-v, \tag{2-17}$$

which leads to the following relation:

$$\ln f=\ln P-\frac{1}{RT}\int_{P\to 0}^{P}\left(\frac{RT}{P}-v\right)\mathrm{d}P. \tag{2-18}$$

The integration in equation (2-18) is easier to perform than $\int v\mathrm{d}P$, as the difference $RT/P - v$ tends to a finite value for $P\to 0$. Equation (2-18) was also suggested as an alternative way to define fugacity (Tunell, 1931).

The fugacity of a pure gas is related to pressure through the following equation:

$$f=\Gamma\cdot P, \tag{2-19}$$

where Γ is the fugacity coefficient of the gas. For a gas mixture, the fugacity of the *i*th gaseous component is related to its partial pressure by

$$f_i = \Gamma_i \cdot P \cdot y_i. \tag{2-20}$$

2.4. The study of chemical equilibrium

Thermodynamics is typically used to investigate chemical equilibria of multi-component systems. However, many natural systems are far from chemical equilibrium, either initially or during their evolution towards a final state of stable equilibrium, which can be eventually attained if enough time is given. Let us think, for instance, of a system made up of an olivine crystal that crystallized at magmatic temperatures and was brought to the Earth surface by a volcanic eruption. This system is obviously far from equilibrium, under the pressure and temperature conditions prevailing at the Earth's surface, and the conditions of stable equilibrium will be approached (and possibly attained) through a quite complex series of irreversible chemical reactions. At first look it might seem that equilibrium thermodynamics is not the best tool for the investigations of chemical reactions occurring in such natural systems. However, this is not true. It has been shown (Helgeson, 1967, 1968, 1979; Helgeson et al., 1969, 1970) that the redistribution of chemical components in multi-phase, multi-component systems taking place during irreversible chemical reactions can be successfully modelled by breaking the overall reaction path into a series of partial equilibrium states, which can be conveniently described by chemical equilibrium thermodynamics.

This discussion will be resumed in Chapter 7. For the moment, we will see a way to describe the condition of chemical equilibrium for a given reaction. The equation we need is

$$\Delta_r G = \Delta_r G^\circ + RT \cdot \ln \prod_i a_i^{v_i}, \tag{2-21}$$

where $\Delta_r G$ and $\Delta_r G^\circ$ are the Gibbs free energy and the standard-state Gibbs free energy of the reaction, respectively. The $\prod_i a_i^{v_i}$ term involves the activities of reactants and products of the reaction, each elevated to the corresponding stoichiometric coefficient, v_i. This is a negative number for reactants and a positive number for products.

But, where does equation (2-21) come from? The Gibbs free energy of the considered reaction, $\Delta_r G$, is given by the sum of the chemical potentials of all the reactants and products, each multiplied by their stoichiometric coefficient. For instance, for the carbonation reaction of the Mg_2SiO_4 component (forsterite) of olivine:

$$CO_{2(g)} + 0.5Mg_2SiO_{4\ \text{olivine}} = MgCO_{3\ \text{magnesite}} + 0.5SiO_{2\ \text{quartz}}, \tag{2-22}$$

the chemical potentials of reactants and products are (from equation (2-9)) as follows:

$$\mu_{CO_{2(g)}} = \mu^o_{CO_{2(g)}} + RT \cdot \ln a_{CO_{2(g)}} \tag{2-23}$$

$$\mu_{\text{olivine}} = \mu^{o}_{\text{olivine}} + RT \cdot \ln a_{\text{Mg}_2\text{SiO}_4,\text{olivine}} \tag{2-24}$$

$$\mu_{\text{magnesite}} = \mu^{o}_{\text{magnesite}} + RT \cdot \ln a_{\text{MgCO}_3,\text{magnesite}} \tag{2-25}$$

$$\mu_{\text{quartz}} = \mu^{o}_{\text{quartz}} + RT \cdot \ln a_{\text{SiO}_2,\text{quartz}}. \tag{2-26}$$

For olivine, the notation $a_{\text{Mg}_2\text{SiO}_4,\text{olivine}}$ indicates the activity of the Mg_2SiO_4 component (forsterite) in the olivine solid mixture. The same holds for the other mineral phases. Again, for gaseous CO_2, the notation $a_{\text{CO}_{2(g)}}$ stands for the activity of CO_2 in the gas mixture. The Gibbs free energy of the reaction is:

$$\Delta_r G = \mu_{\text{magnesite}} + 0.5 \cdot \mu_{\text{quartz}} - \mu_{\text{CO}_{2(g)}} - 0.5 \cdot \mu_{\text{olivine}}. \tag{2-27}$$

Substitution of equations (2-23) to (2-26) into equation (2-27) and rearrangement lead to

$$\begin{aligned}\Delta_r G = &\left(\mu^{o}_{\text{magnesite}} + 0.5 \cdot \mu^{o}_{\text{quartz}} - \mu^{o}_{\text{CO}_{2(g)}} - 0.5 \cdot \mu^{o}_{\text{olivine}}\right) \\ &+ RT \cdot \left(\ln a_{\text{MgCO}_3,\text{magnesite}} + 0.5 \cdot \ln a_{\text{SiO}_2,\text{quartz}} - \ln a_{\text{CO}_{2(g)}} - 0.5 \cdot \ln a_{\text{Mg}_2\text{SiO}_4,\text{olivine}}\right).\end{aligned} \tag{2-28}$$

Equation (2-28) corresponds to

$$\Delta_r G = \Delta_r G^\circ + RT \cdot \ln\left(\frac{a_{\text{MgCO}_3,\text{magnesite}} \cdot a^{0.5}_{\text{SiO}_2,\text{quartz}}}{a_{\text{CO}_{2(g)}} \cdot a^{0.5}_{\text{Mg}_2\text{SiO}_4,\text{olivine}}}\right). \tag{2-29}$$

It must be underscored that this is equation (2-21) specifically written for reaction (2-22).

If the reaction of interest in the considered system has attained the state of chemical equilibrium, then $\Delta_r G = 0$ and

$$\Delta_r G^\circ = -RT \cdot \ln\left(\frac{a_{\text{MgCO}_3,\text{magnesite}} \cdot a^{0.5}_{\text{SiO}_2,\text{quartz}}}{a_{\text{CO}_{2(g)}} \cdot a^{0.5}_{\text{Mg}_2\text{SiO}_4,\text{olivine}}}\right)_{\text{equilibrium}}. \tag{2-30}$$

In general terms, equation (2-30) is written as follows:

$$\Delta_r G^\circ = -RT \cdot \ln\left(\prod_i a_i^{v_i}\right)_{\text{equilibrium}} = -RT \cdot \ln K. \tag{2-31}$$

The activity product ratio at equilibrium, $\left(\prod_i a_i^{v_i}\right)_{\text{equilibrium}}$, is usually indicated with K and is called thermodynamic equilibrium constant, or simply equilibrium constant, of the reaction of interest. Note that since *activities* are *dimensionless* quantities measuring the deviations of the chemical potentials from their standard-state values, the *thermodynamic equilibrium constant* is also *dimensionless*.

Equation (2-31) is one of the most useful equations in chemical thermodynamics (e.g., Nordstrom and Muñoz, 1985; Anderson and Crerar, 1993) and certainly deserves some remarks. Equation (2-31) tells us that the equilibrium constant depends on a difference between standard-state Gibbs free energies only, and consequently on temperature and pressure only, and does not depend on the composition of the system. Again, although the left-hand-side term of equation (2-31) is simply a difference between Gibbs free energies, its right-hand-side term is something totally distinct: the activity product ratio under equilibrium conditions.

To deal with the left-hand-side term of equation (2-31), $\Delta_r G^\circ$, we have still to clarify how to compute the changes in Gibbs free energy with temperature and pressure. This is the subject of the next section.

2.5. Changes in Gibbs free energy with temperature and pressure

Let us suppose that we want to compute $\Delta_r G^o_{T,P}$, i.e., the standard Gibbs free energy of a given reaction, for instance the carbonation reaction of forsterite (2-22), at the *T*, *P* conditions of interest. In general, these *T*, *P* conditions will be different from those chosen for reference, which are usually Tr = 298.15 K and Pr = 1 bar. Then we start to write

$$\begin{aligned}\Delta_r G^o_{T,P} &= \Delta_r G^o_{Tr,Pr} + \int_{Tr}^{T} \frac{\partial \Delta_r G^o}{\partial T} \mathrm{d}T + \int_{Pr}^{P} \frac{\partial \Delta_r G^o}{\partial P} \mathrm{d}P \\ &= \Delta_r G^o_{Tr,Pr} - \int_{Tr}^{T} \Delta_r S^o \mathrm{d}T + \int_{Pr}^{P} \Delta_r V^o \mathrm{d}P,\end{aligned} \quad (2\text{-}32)$$

where $\Delta_r G^o_{Tr,Pr}$ is the standard Gibbs free energy of the considered reaction at the reference temperature Tr and reference pressure Pr, whereas $\Delta_r S^o$ and $\Delta_r V^o$ indicate the standard entropy of reaction and the standard volume of reaction, respectively (cf. equations (2-32) and (2-2)).

2.5.1. Pressure effect

We start to deal with the integral $\int_{Pr}^{P} \Delta_r V^o \mathrm{d}P$, which can be split into two parts, one for the solids and one for the fluid phase (CO_2 in the case of reaction 2-22), as follows:

$$\int_{Pr}^{P} \Delta_r V^o \mathrm{d}P = \int_{Pr}^{P} \Delta_s V^o \mathrm{d}P - \int_{Pr}^{P} \Delta_r v^o_{CO_2} \mathrm{d}P. \quad (2\text{-}33)$$

For reaction (2-22), $\Delta_s V^o = v^o_{MgCO_3} + 0.5 v^o_{SiO_2} - 0.5 v^o_{Mg_2SiO_4}$. The reason for this split is that the volume of solids is virtually pressure independent (at least for our purposes) whereas the volume of gases changes dramatically with pressure (see Chapter 3). Assuming that the reaction volume for solids is constant, and referring to equation (2-13), we can write:

$$\int_{Pr}^{P} \Delta_r V^o \mathrm{d}P = \Delta_s V^o (P-1) - RT \cdot \ln\left(\frac{f_{CO_2,P}}{f_{CO_2,Pr}}\right). \tag{2-34}$$

where f is the fugacity of the specified gas (e.g., CO_2) at the specified pressure, either P or the reference pressure Pr, usually 1 bar. We will see how to calculate fugacities in the next chapter.

2.5.2. Temperature effect

To compute the standard Gibbs free energy of a given reaction (e.g., the carbonation reaction of forsterite, equation (2-22)) at any given temperature $T \neq Tr$, we have to know the standard entropy of reaction $\Delta_r S°$ as indicated in equation (2-32). The temperature dependence of the standard entropy of reaction is given by the following expression:

$$\Delta_r S_T^o - \Delta_r S_{Tr}^o = \int_{Tr}^{T} \frac{\Delta_r C_P^o}{T} \mathrm{d}T, \tag{2-35}$$

where $\Delta_r C_P^o$ is the standard heat capacity at constant pressure of the considered reaction (i.e., the sum of the heat capacities of reactants and products, each multiplied by the corresponding stoichiometric coefficient). Heat capacities are usually expressed as simple temperature functions, which can be easily integrated. Among such temperature functions, one of the most popular is the Maier–Kelley equation (Maier and Kelley, 1932):

$$\Delta_r C_P^o = \Delta a + \Delta b \cdot T - \Delta c \cdot T^{-2}, \tag{2-36}$$

Use of the Maier–Kelley equation for expressing the standard heat capacity at constant pressure of the considered reaction leads to (Helgeson et al., 1978):

$$\Delta_r S_T^o = \Delta_r S_{Tr}^o + \Delta a \cdot \ln\frac{T}{Tr} + \Delta b(T - Tr) + \frac{\Delta c}{2}\left(\frac{1}{T^2} - \frac{1}{Tr^2}\right). \tag{2-37}$$

and

$$\begin{aligned}\Delta_r G_T^o = {} & \Delta_r G_{Tr}^o - \Delta S_{Tr}^o (T - Tr) + \Delta a \cdot \left(T - Tr - T \cdot \ln\frac{T}{Tr}\right) \\ & + \frac{\left(\Delta c - \Delta b \cdot T \cdot Tr^2\right) \cdot \left(T - Tr\right)^2}{2 \cdot T \cdot Tr^2}\end{aligned} \tag{2-38}$$

This is the needed equation to compute the standard Gibbs free energy of a given reaction at any given temperature $T \neq Tr$.

It is useful to recall that the standard heat capacity at constant pressure is also involved in the following relation:

$$\Delta_r H_T^o - \Delta_r H_{Tr}^o = \int_{Tr}^{T} \Delta_r C_P^o \, \mathrm{d}T, \tag{2-39}$$

expressing the temperature dependence of the standard enthalpy of a given reaction. Referring to the Maier–Kelley equation, for expressing the standard heat capacity at constant pressure of the considered reaction, one obtains:

$$\Delta_r H_T^o = \Delta_r H_{Tr}^o + \Delta a(T - Tr) + \frac{\Delta b}{2}(T^2 - Tr^2) + \Delta c\left(\frac{1}{T} - \frac{1}{Tr}\right). \tag{2-40}$$

The enthalpy of reaction is needed to compute the thermodynamic equilibrium constant at any given temperature $T \neq Tr$. Again, use of the Maier–Kelley equation leads to (Anderson and Gerar, 1993):

$$\begin{aligned}\ln K_T = \ln K_{Tr} &- \frac{\Delta_r H_{Tr}^o}{R}\left(\frac{1}{T} - \frac{1}{Tr}\right) + \frac{\Delta a}{R}\left(\ln\frac{T}{Tr} + \frac{Tr}{T} - 1\right) \\ &+ \frac{\Delta b}{2R}\left(T + \frac{Tr^2}{T} - 2\cdot Tr\right) + \frac{\Delta c}{2R}\left(\frac{-T^2 - Tr^2 + 2\cdot T\cdot Tr}{T\cdot Tr^2}\right)\end{aligned} \tag{2-41}$$

2.5.3. Calculation of the thermodynamic properties of reactions at high temperatures and pressures

In the above-mentioned text we have recalled the equations needed to carry out thermodynamic equilibrium calculations between solids and gases whereas aqueous species have been purposely ignored, as they require a separate treatment, which will be discussed later in Chapter 4.

From previous discussion it is evident that the computation of the Gibbs free energy and the thermodynamic equilibrium constant of a single reaction of interest is rather simple, but also rather tedious, provided that a set of internally consistent thermodynamic data is available. The task becomes very cumbersome and very tedious if we are interested in computing the properties of many reactions.

Today geochemists are lucky, as accurate values of the enthalpy, entropy, Gibbs free energy, and heat capacity for many substances of geo-chemical interest (gases, minerals, and aqueous species) have been critically examined, checked for internal consistency, and tabulated thanks to the efforts of Helgeson and collaborators (e.g., Helgeson et al., 1978; Sverjensky, 1987; Shock and Helgeson, 1988, 1990; Tanger and Helgeson, 1988; Shock et al., 1989; Shock and Koretsky, 1993, 1995; Shock, 1995; Sverjensky et al., 1997).

Moreover, a software code, SUPCRT92, has been developed by Johnson et al. (1992) for computing the thermodynamic properties of reactions from 1 to 5,000 bar and 0 to 1,000°C, which is enough for our purposes.

As we will see in Chapter 7, the software packages used for geochemical modelling, such as EQ3/6, require a thermodynamic data-file in which the equilibrium constants of all the reactions of interest involving aqueous species, minerals, and gases are stored. In the default thermodynamic data-file of version 7.2b of EQ3/6, the equilibrium constants are reported for 1.013 bar total pressure from 0 to 100°C and the total pressures of steam/liquid water equilibrium co-existence from 100 to 300°C. This pressure–temperature grid, however, does not match the P, T values expected in the geological sequestration of CO_2 (see Section 3.1). Consequently, the thermodynamic data-file must be recalculated and the most obvious choice is that of variable temperatures and constant total pressure. For instance, version 8.0 of EQ3/6 includes thermodynamic data-files for constant total pressures of 0.5, 1, 2 and 5 kbar. However, unless we are interested in a total pressure of 500 bar (1, 2 and 5 kbar are very high), it is necessary to calculate the thermodynamic properties of reactions at the total pressure of interest, as recently underscored by Allen et al. (2005). SUPCRT92 is what we need to perform these calculations.

Chapter 3

Carbon Dioxide and CO_2-H_2O Mixtures

3.1. The geological sequestration of CO_2: What happens?

Let us suppose to inject high-pressure CO_2 into a system made up of a relatively deep aquifer hosting an aqueous solution, most likely of comparatively high salinity. Let us assume, for the moment, that aquifer rocks are totally unreactive, which is a reasonable assumption if we are interested in the short-term behaviour of our system. What happens?

To answer this question we have first to evaluate T, P conditions that will depend on the depth of the aquifer as well as on the local geothermal gradient and pressure gradient. For economical reasons the depth of the aquifer may be in the order of some kilometres. Taking a surface temperature of 15°C and an average geothermal gradient of 33°C km^{-1}, temperature will be 48°C at 1 km and 114°C at 3 km. Assuming an hydrostatic gradient of 100 bar km^{-1}, pressure will be 101 bar at 1 km and 301 bar at 3 km. Therefore, to investigate the sequestration potential of our deep aquifer we have to compute, initially, the thermodynamic properties of CO_2-H_2O mixtures up to temperatures slightly in excess of 100°C and pressures of some hundred bars. We will show that, under most of these conditions, a CO_2-rich gas or liquid phase and an H_2O-rich liquid phase typically coexist. The influence of dissolved salts on the mutual solubilities of CO_2 and H_2O has also to be taken into account in a second step.

Furthermore, if we want to predict the long-term behaviour of our system we have to consider the reactions between fluids and aquifer rocks, but this is the subject of subsequent chapters. This chapter is focused on CO_2-H_2O fluids. Let us start by taking into account pure CO_2.

3.2. The *P–T* phase diagram of CO_2

The prominent characteristics of the *P–T* phase diagram of CO_2 (Fig. 3.1) are the triple point, at -56.57 ± 0.03°C, 5.185 ± 0.005 bar (Angus et al., 1976), and the critical point close to 31°C, 74 bar. Recent evaluations of the critical temperature are 31.03 ± 0.04°C (Suehiro et al., 1996), 30.978 ± 0.015°C (Span and Wagner, 1996), 30.95 ± 0.1°C (Weber, 1989), 31.2 ± 0.4°C (Li and Kiran, 1988), 31.05 ± 0.02°C (Morrison, 1981) and 31.08°C (Efremova and Shvarts, 1972). Recent critical pressure estimates are 73.80 ± 0.15 bar (Suehiro et al., 1996), 73.773 ± 0.003 bar (Span and Wagner, 1996), 73.40 ± 0.50 bar (Li and Kiran, 1988) and 73.825 bar (Angus et al., 1976). Uncertainties are from Lemmon et al. (2003).

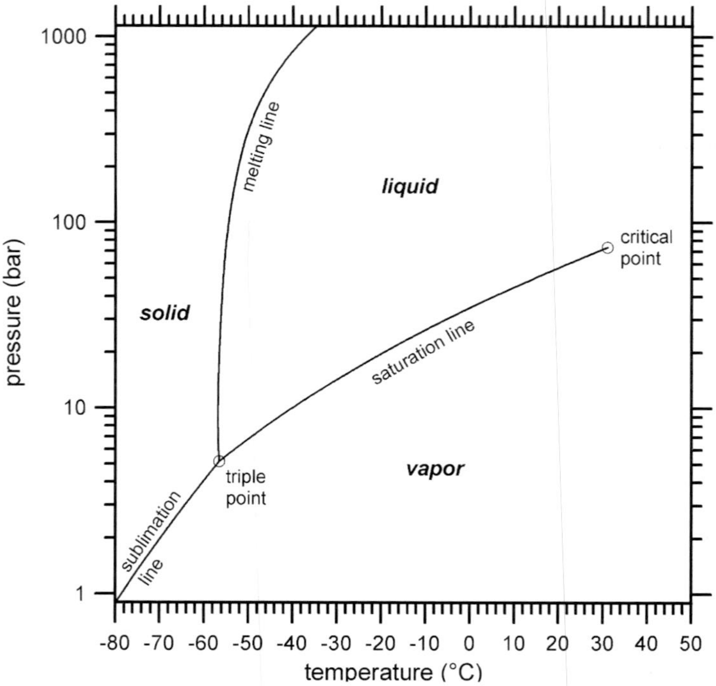

Figure 3.1. *P–T* phase diagram of CO_2 (data from Span and Wagner, 1996).

The univariant vapour–liquid curve (or saturation curve) extending from the triple point to the critical point is reproduced within sufficient accuracy (typically <0.4%) by the following equation:

$$\log P_{sat} = \frac{-863.6}{T_{sat}} + 4.705, \tag{3-1}$$

where P_{sat} is the vapour pressure in bar and T_{sat} is in Kelvin, based on the data by Span and Wagner (1996). From this dataset, the following equation is obtained for the univariant melting curve (or solid–liquid curve):

$$P_{melt} = 523.18 - 51.547\,T_{melt} + 0.22695\,T_{melt}^{\;2}, \tag{3-2}$$

where P_{melt} is in bar and T_{melt} is in Kelvin. This curve corresponds to pressures much higher and temperatures much lower than those of interest for the geological sequestration of CO_2 (Fig. 3.1). Although liquid CO_2 can be produced below the critical temperature, at pressures of some ten bars or higher, it is evident that CO_2 is a supercritical fluid (i.e., a gas that cannot be liquefied regardless of the pressure exerted on it) under most *P*, *T* conditions of relevance for the geological sequestration of CO_2.

3.3. The equation of state for a pure gas

Since CO_2 is a gas under most *P*, *T* conditions of interest, a good starting point for our discussion is the ideal gas law. This is the simplest equation of state (EOS), i.e., the simplest relation linking pressure, temperature and volume of the fluid, but it holds for the ideal (perfect) gas only. Referring to 1 mole of such a gas, the ideal gas law is written as

$$P = \frac{RT}{v}, \tag{3-3}$$

where *v* represents the molar volume and *R* is the gas constant. *R* is related to the Boltzmann's constant k_B (1.38066×10^{-23} J K^{-1}) through the equation $R = k_B \times N_A$, where $N_A = 6.02214 \times 10^{23}$ mol^{-1} is the Avogadro's number, i.e., the number of molecules in a mole of substance. Therefore, *R* is equal 8.31451 J mol^{-1} K^{-1} or 1.98721 cal mol^{-1} K^{-1} or 83.1451 cm^3 bar mol^{-1} K^{-1}, depending on the units of measurement.

The ideal gas law was first introduced by Benoit Paule Emile Clapeyron in 1834 in the work *Mémoire sur la puissance motrice de la chaleur*. The ideal gas is a theoretical gaseous substance made up of infinitesimally small molecules that do not interact with one another. Although this limiting condition is approached by real gases at relatively low pressures and high temperatures, the ideal gas law is not satisfactory to describe the *P–T–v* relationship of real gases.

Therefore, scientists tried to derive alternative EOS. In 1873, J.D. van der Waals suggested, in his Ph.D. dissertation (Rowlinson, 1988), to use the following relationship:

$$P = \frac{RT}{v-b} - \frac{a}{v^2}, \tag{3-4}$$

which incorporates the two adjustable parameters (or correcting terms) *a* and *b*. The parameter *b* is the effective volume of the molecules contained in a mole of gas, and the difference *v–b* represents the volume available for the movements of gas molecules, i.e., the unoccupied or free volume per mole. The second term on the right-hand side of equation (3-4) is a correction for intermolecular attractions, which are proportional to $1/v^2$, with the *a* coefficient representing a sort of proportionality constant.

Many EOS deriving from the van der Waals equation were proposed since 1873 and, similar to the van der Waals equation, most of these EOS take into account repulsive and attractive effects. One of these EOS was suggested by Redlich and Kwong (1949):

$$P = \frac{RT}{v-b} - \frac{a}{T^{0.5}v(v+b)}. \tag{3-5}$$

The first term on the right-hand side of equation (3-5) is the same of equation (3-4), whereas the second term was modified, although "there is no really good theoretical justification for the two changes in the denominator of the *a*-term but the agreement with observed data is considerably improved" (Redlich, 1976). In spite of these changes, the Redlich–Kwong EOS has only two adjustable parameters as the van der Waals equation.

Since 1949, the Redlich–Kwong EOS was further modified to explain the properties of CO_2 and other gases at elevated pressures and temperatures. For this purpose, *a* and *b* parameters were considered as a function of either temperature or pressure or both. For instance, in the modified Redlich–Kwong EOS of Kerrick and Jacobs (1981), *b* is constant and equal to 29.0 $cm^3\ mol^{-1}$, whereas *a* is a function of both *T* and *V*, as follows:

$$P = \frac{RT(1+y+y^2-y^3)}{v(1-y)^3} - \frac{a}{T^{0.5}v(v+b)} \tag{3-6}$$

where:

$$y = \frac{b}{4v}$$

$$a(T,v) = c + \frac{d}{v} + \frac{e}{v^2}$$

$$c = \left(290.78 - 0.30276 \cdot T + 0.0014774 \cdot T^2\right) \times 10^6 \text{ bar cm}^2\ \text{K}^{0.5}\ \text{mol}^{-2}$$

$$d = \left(-8374 + 19.437 \cdot T - 0.008148 \cdot T^2\right) \times 10^6 \text{ bar cm}^3\ \text{K}^{0.5}\ \text{mol}^{-3}$$

$$e = \left(76600 - 133.9 \cdot T + 0.1071 \cdot T^2\right) \times 10^6 \text{ bar cm}^4\ \text{K}^{0.5}\ \text{mol}^{-4}$$

Experimental data of CO_2 and H_2O at elevated pressures and temperatures are generally reproduced to better than $\pm 1\%$ by the EOS of Kerrick and Jacobs. This improvement in fitting the experimental data, with respect to the original Redlich–Kwong equation, is due to the presence of 10 parameters in the Kerrick and Jacobs EOS, against the 2 adjustable parameters of the Redlich–Kwong equation. Note that the EOS of Kerrick and Jacobs was proposed for the elevated pressures and temperatures typical of metamorphic environments and it is not applicable to the relatively low temperatures and high-pressure conditions expected for the geological sequestration of CO_2.

The *a* and *b* parameters of the van der Waals and Redlich–Kwong EOS can be accurately estimated by fitting the available *P–T–v* data for the considered gas. Alternatively, they can be obtained from the *P–T– v* data at the critical point, where both the first and second partial derivatives of *P* with respect to *V*, at constant *T*, are equal to zero. To understand this fact, it is enough to look at the critical isotherm in the *P–v* diagram for the fluid of interest (Fig. 3.2). Indeed, the critical isotherm is horizontal at the critical point, which also represents an inflection point of the critical isotherm.

At the critical point, the van der Waals equation (3-4) can be written as follows:

$$P_c = \frac{RT_c}{v_c - b} - \frac{a}{v_c^2}, \tag{3-7}$$

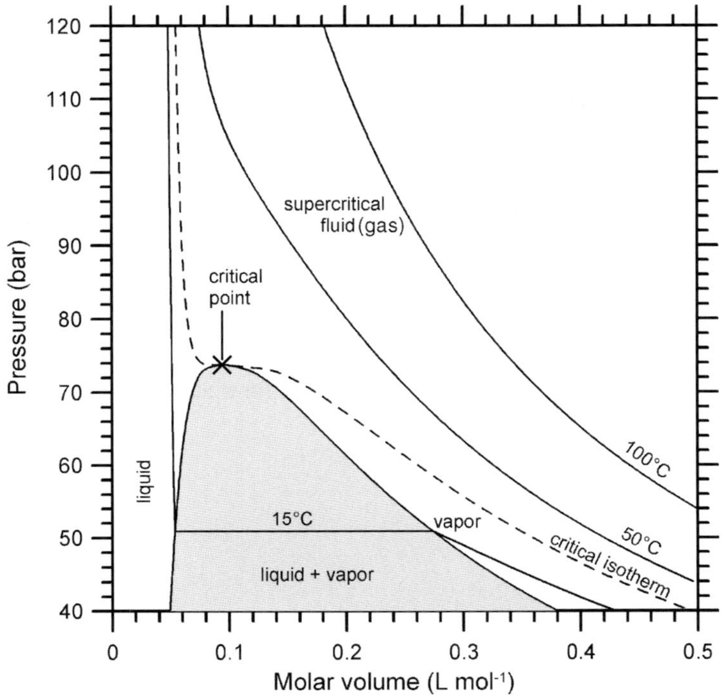

Figure 3.2. Simplified *P–v* diagram for CO_2 (data from Lemmon et al., 2003).

where P_c, T_c and v_c are the critical data. Taking its first and second partial derivatives at constant *T* we write

$$\left(\frac{\partial P}{\partial v}\right)_{T_c} = \frac{-RT_c}{(v_c - b)^2} + \frac{2a}{v_c^3} = 0, \tag{3-8}$$

and

$$\left(\frac{\partial^2 P}{\partial v^2}\right)_{T_c} = \frac{2RT_c}{(v_c - b)^3} - \frac{6a}{v_c^4} = 0. \tag{3-9}$$

Equations (3-7), (3-8) and (3-9) are solved for *a* and *b* parameters, obtaining

$$a = \frac{27 \cdot R^2 T_c^2}{64 \cdot P_c}, \tag{3-10}$$

and

$$b = \frac{RT_c}{8 \cdot P_c}. \tag{3-11}$$

Application of the same procedure to the Redlich–Kwong equation (3-5) gives

$$P_c = \frac{RT_c}{v_c - b} - \frac{a}{T_c^{0.5} v_c (v_c + b)} \tag{3-12}$$

$$\left(\frac{\partial P}{\partial v}\right)_{T_c} = \frac{-RT_c}{(v_c - b)^2} + \frac{a}{T_c^{0.5}}\left[\frac{1}{v_c^2(v_c + b)} + \frac{1}{v_c(v_c + b)^2}\right] = 0 \tag{3-13}$$

$$\left(\frac{\partial^2 P}{\partial v^2}\right)_{T_c} = \frac{2RT_c}{(v_c - b)^3} - \frac{2a}{T^{0.5}}\left[\frac{1}{v_c^3(v_c + b)} + \frac{1}{v_c^2(v_c + b)^2} + \frac{1}{v_c(v_c + b)^3}\right] = 0 \tag{3-14}$$

Then equations (3-12), (3-13) and (3-14) are solved for a and b:

$$a = 0.4275 \frac{R^2 T_c^{2.5}}{P_c} \tag{3-15}$$

and

$$b = 0.0866 \frac{RT_c}{P_c} \tag{3-16}$$

The a and b parameters thus obtained (either equations (3-10) and (3-11) or equations (3-15) and (3-16)) are then inserted in the corresponding EOS, which can be used to compute the molar volume as a function of P, T. Below, we will do these calculations for CO_2 and will also compute two parameters, the compressibility factor (defined in the next section) and the fugacity coefficient (defined in Section 2.3), that describe the deviations from the ideal gas behaviour. Before doing these exercises, however, it is worth mentioning that the van der Waals equation, the Redlich–Kwong equation and their modifications are not the only type of EOS available to describe the P–T–v relations of gases. A distinct approach led to the formulation of the so-called virial EOS:

$$Z = \frac{Pv}{RT} = 1 + \frac{B}{v} + \frac{C}{v^2} + \frac{D}{v^3} + \cdots, \tag{3-17}$$

which was proposed in 1901 by Heike Kamerlingh Onnes (who incidentally won the Nobel Prize in Physics in 1913 "for his investigations on the properties of matter at low temperatures, which led, inter alias, to the production of liquid helium"). Equation (3-17) might seem nothing else than a simple polynomial expansion of the ideal gas law, in which the so-called first virial coefficient is the gas constant R. Indeed, equation (3-17) reduces to the ideal gas law for $P \rightarrow 0$. The temperature-dependent coefficients $B(T)$, $C(T)$ and $D(T)$ are the so-called second, third and fourth virial coefficients. Several years after it was first introduced, the virial EOS was derived through statistical mechanics (Mayer and Mayer, 1940) and it was shown that the virial coefficients are not empirical parameters, but are linked to non-ideal molecular interactions.

3.4. The molar volume of pure CO_2 and related thermodynamic properties

We begin to compute the molar volume of CO_2 at 50°C by means of the different EOS presented in the previous section. Computation by means of the ideal gas law (equation (3-3)) is straightforward, whereas use of the van der Waals and Redlich–Kwong EOS requires first to recast equations (3-4) and (3-5) as cubic equations in terms of volume, thus obtaining:

$$v^3 - \left(b + \frac{RT}{P}\right)v^2 + \left(\frac{a}{P}\right)v - \frac{ab}{P} = 0 \tag{3-18}$$

and

$$v^3 - \frac{RT}{P}v^2 - \left(\frac{RTb}{P} - \frac{a}{T^{0.5}P} + b^2\right)v - \frac{ab}{T^{0.5}P} = 0 \tag{3-19}$$

respectively. These cubics can be solved by trial and error, i.e. by inserting a tentative solution (a reasonable value of v) and computing the value of the left-hand-side term (let us call it the function), which most likely will be different from zero, since it is virtually impossible to find the exact solution by chance at the first attempt. Then the tentative solution is changed in a series of equal (relatively large) steps until the function changes in sign. At this point it is necessary to go back at the previous tentative solution, decrease the step (e.g., 10 times), and proceed again towards the exact solution. The process is repeated until the molar volume is computed with the desired approximation. Of course, this exercise cannot be done without the help of a software code. The described approach is rather crude but it works fine.

Other ways to solve the cubic do exist, such as the so-called Cardan's solution, which was proposed by the Italian physician and mathematician Girolamo Cardano (Jerome Cardan) in 1545 in his *Ars Magna*, a treatise on algebra. Recently, the Cardan's solution was improved by Nickalls (1993).

Whatever the way chosen to find the roots, the cubic equations (3-18) and (3-19) yield one root above the critical point, but two or three roots below it. In the latter case, the minimum root gives the molar volume of the liquid phase whereas the molar volume of the gas phase (or better of the vapour phase) is given by the maximum root (Spycher et al., 2003). If we use the trial-and-error approach to solve the cubic and are interested in computing the molar volume of the gas phase, it is advisable to start from an approximated value greater than the expected maximum root.

The molar volumes of CO_2 computed by means of the different equations of state discussed in the previous section are compared in the log v vs. log P plot (Fig. 3.3), also showing the molar volumes recommended by NIST (Lemmon et al., 2003). Two Redlich–Kwong equations, differing for the a and b parameters, are displayed in this diagram. In one case a and b parameters were derived from the critical data, and they are both constant, with $a = 6.44 \times 10^7$ bar cm^6 K$^{0.5}$ mol^{-2} and $b = 29.6$ cm^3 mol^{-1}. The a and b parameters obtained from fitting of available data (Spycher et al., 2003) are used in the

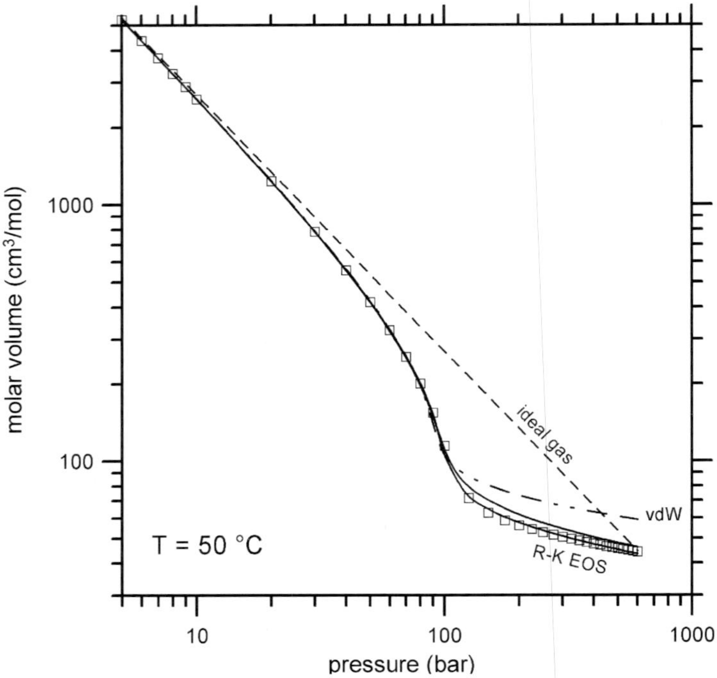

Figure 3.3. Log–log plot showing the molar volume of CO_2 computed at different pressures by means of the ideal gas law, the van der Waals equation and two Redlich–Kwong equations (the upper one in which parameters *a* and *b* are both constant and the lower one in which parameter *b* only is constant whereas parameter *a* is a function of temperature). The molar volumes recommended by NIST (data from Lemmon et al., 2003) are also shown (squares).

other Redlich–Kwong equation. In this case, *b* is still constant, 27.80 ± 0.01 cm^3 mol^{-1}, whereas *a* varies with temperature, as expressed by the following equation:

$$a = 7.54 \times 10^7 - 4.13 \times 10^4 \times T\,(\mathrm{K}) \tag{3-20}$$

which holds in the temperature range 10−107°C and gives $a = 6.31 \times 10^7$ bar cm^6 K$^{0.5}$ mol^{-2} at 25°C and $a = 6.00 \times 10^7$ bar cm^6 K$^{0.5}$ mol^{-2} at 100°C. The molar volumes shown in Fig. 3.3 refer to the temperature of 50°C and variable pressures, from 1 to 600 bar.

In this graphical space, the ideal gas law identifies a straight line with slope −1. In particular, at *P* of 1 bar, the molar volume of the ideal gas is numerically equal to the $R \times T$ product, 26,863 cm^3 mol^{-1}. As expected, the volumes computed by the van der Waals equation and the two Redlich–Kwong equations are very close to those given by the ideal gas law at low pressures (less than 10−20 bar), but these equations of state become steeper than the ideal gas law and deviate progressively from it from 20 to 100 bar. At even higher pressure values, the van der Waals equation and the two Redlich–Kwong equations become flatter than the ideal gas law and cross it at 430−600 bar approximately. The molar volumes recommended by NIST are satisfactorily fitted by the two Redlich–Kwong equations, especially by that with *a* as function of temperature. However, in Fig. 3.3, the differences between the distinct equations are somewhat obscured by the use of the logarithmic scale on the two axes.

In order to magnify the deviations from ideality, we compute the compressibility factor *Z*, which is defined by the following equation:

$$Z = \frac{Pv}{RT}, \qquad (3\text{-}21)$$

and is obviously equal to 1 for a perfect gas. Again, the compressibility factor is calculated at 50°C and variable pressures and results are reported against pressure in Fig. 3.4, together with the molar volumes recommended by NIST (Lemmon et al., 2003). As indicated by this plot, CO_2 shows a marked deviation from ideality at the considered temperature of 50°C, especially from some tens to some hundreds bars. The van der Waals equation and the two Redlich–Kwong equations are in mutual agreement and consistently reproduce the NIST data below 100 bar, whereas the Redlich–Kwong equations are better than the van der Waals equation above this pressure value. As expected, the equation that fits the NIST data at best is the Redlich–Kwong equation in which parameter *b* only is constant whereas

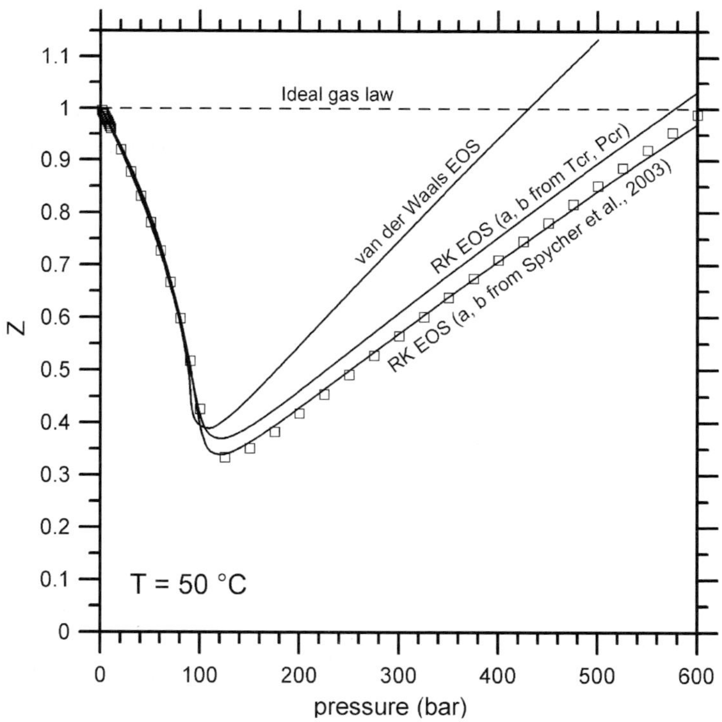

Figure 3.4. Compressibility factor of CO_2 computed at 50°C and variable pressures by means of the van der Waals equation and two Redlich–Kwong equations (the upper one in which parameters *a* and *b* are both constant and the lower one in which parameter *b* only is constant whereas parameter *a* is a function of temperature). The molar volumes recommended by NIST (data from Lemmon et al., 2003) are also shown (squares).

parameter a is a function of temperature. This EOS was used to compute the compressibility factor at different temperatures in the range of interest (from 25 to 125°C each 25°C) and, again, the NIST data are satisfactorily reproduced in all these cases (Fig. 3.5a).

Knowing the compressibility factors of CO_2 computed by means of the Redlich–Kwong equation in which parameter b only is constant whereas parameter a is a function of temperature, we calculate now the fugacity coefficient of CO_2, Γ, by means of the following equation (Prausnitz et al., 1999):

$$\ln \Gamma = \int_0^P \frac{Z-1}{P}\,\mathrm{d}P, \tag{3-22}$$

which is equivalent to equation (2-18). Equation (2-18) can be, in fact, rearranged as follows:

$$\begin{aligned} RT\ln\left(\frac{f}{P}\right) &= \int_0^P \left(v - \frac{RT}{P}\right)\mathrm{d}P = \int_0^P \left(\frac{ZRT}{P} - \frac{RT}{P}\right)\mathrm{d}P \\ &= RT\int_0^P \frac{Z-1}{P}\,\mathrm{d}P \end{aligned} \tag{3-23}$$

To compute the fugacity coefficient first we compute the ratio $(Z-1)/P$ at different pressures. Then we fit these data to a polynomial equation which can be easily integrated. At 100°C and 125°C this exercise is easily carried out, whereas at lower temperatures the $(Z-1)/P$ function has a singular point and has to be broken down into two parts. Results are reported in Fig. 3.5b, which shows that fugacity coefficients decrease with increasing pressure and decreasing temperature, as expected, attaining values close to 0.2 at 25°C and 600 bar.

The computation of fugacity coefficients and fugacities is a very critical point, since this information is used to compute the Gibbs free energy of reactions at the pressure of interest, as we have seen in Section 2.5, as well as to compute phase equilibria, as we will see in Section 3.7. Fugacities are also involved in geochemical modelling (see Chapter 7).

Since in most natural systems we will have to deal with mixtures rather than pure phases, it is convenient to examine the CO_2-H_2O system at this point of the discussion.

3.5. The CO_2-H_2O system

The CO_2-H_2O phase diagram of Fig. 3.6 shows the different phases stable at low temperatures and relatively low pressures. These comprise: (1) a solid, non-stoichiometric CO_2–clathrate–hydrate with formula close to $CO_2 \cdot 7.5\ H_2O$ (H in Fig. 3.6); (2) a CO_2-bearing water-rich liquid (aqueous phase), labelled L_{aq}; (3) a CO_2-rich liquid phase (L_{CO2}); and (4) a CO_2-rich vapour phase (V). In the T–P space of Fig. 3.6 there is one quadrupole point (Q_1), at 9.77°C and 44.60 bar, where these four phases (H, L_{aq}, L_{CO_2} and V) coexist. Q_1 is also the point of intersection of the three stable curves H+L_{aq}+V, L_{aq}+L_{CO_2}+V and H+L_{aq}+L_{CO_2} and the metastable curve H+L_{CO_2}+V. A second quadrupole point (Q_2),

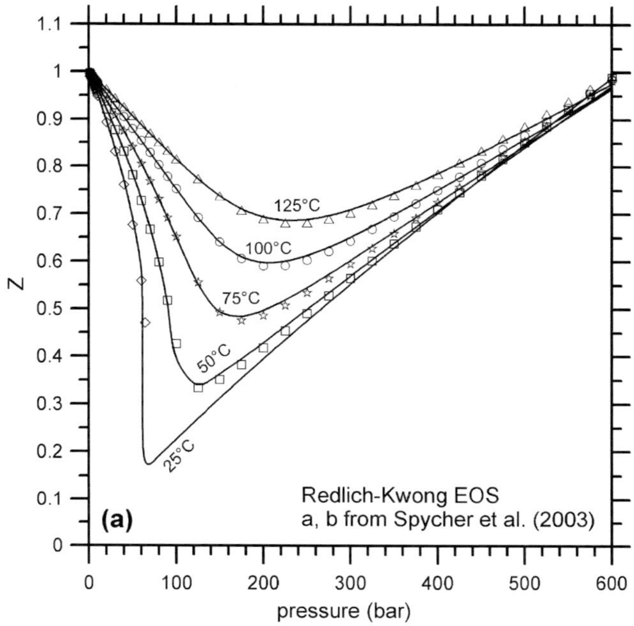

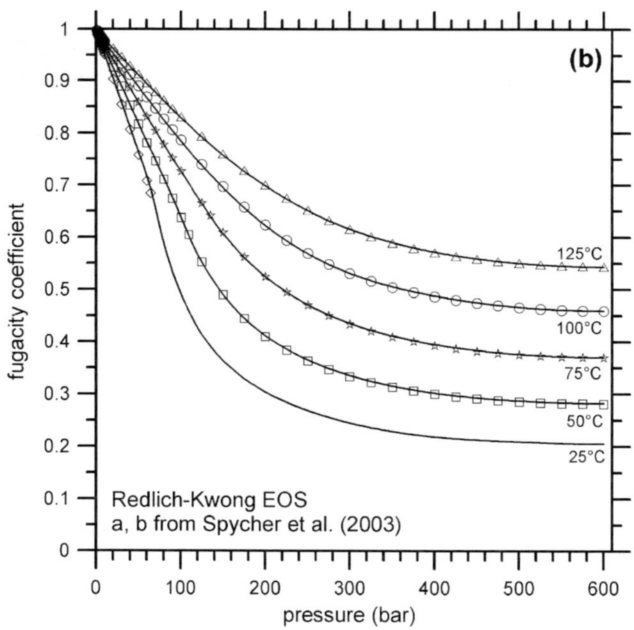

Figure 3.5 (a) Compressibility factor of CO_2 and (b) fugacity coefficient of CO_2 computed at 25, 50, 75, 100 and 125°C and variable pressures, by means of the Redlich–Kwong equation in which parameter b only is constant whereas parameter a is a function of temperature. The molar volumes recommended by NIST (data from Lemmon et al., 2003) are also shown with different symbols depending on temperature, as follows: 25°C = diamonds; 50°C = squares; 75°C = stars; 100°C = circles; 125°C = triangles. Reprinted from Spycher et al. (2003), Copyright (2003), adapted with permission from Elsevier.

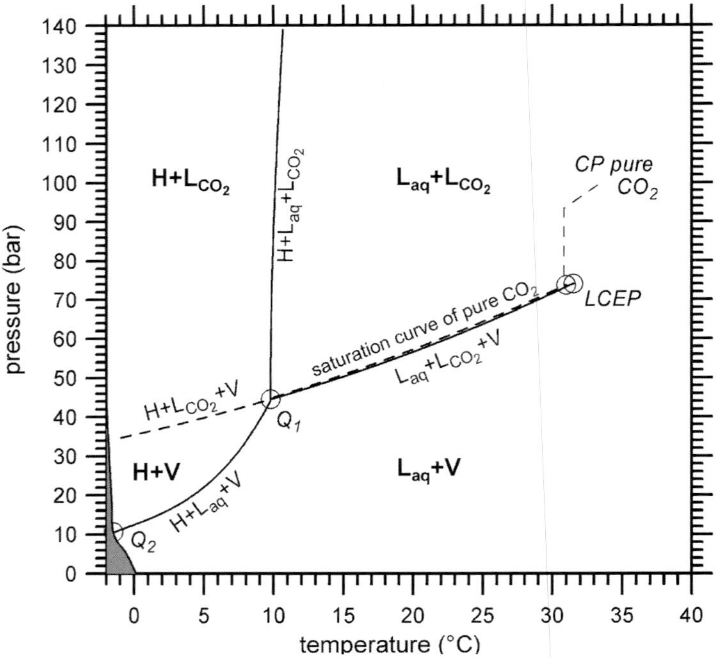

Figure 3.6. Phase diagram for the CO_2-H_2O binary at low temperatures and relatively low pressures (data from Ng and Robinson, 1985; Fan and Guo, 1999; Wendland et al., 1999). The grey area represents the field of stability of ice.

characterized by the coexistence of H, L_{aq}, V and ice limits the extension of the $H+L_{aq}+V$ curve, but it is of negligible interest for our purposes.

In Fig. 3.6, the three-phase equilibrium curves $L_{aq}+L_{CO_2}+V$ and $H+L_{aq}+V$ have been constrained following Wendland et al. (1999), whereas the $H+L_{aq}+L_{CO_2}$ curve is conveniently described by the following equation, derived by Spycher et al. (2003) on the basis of the data of Ng and Robinson (1985) and Fan and Guo (1999):

$$\frac{P}{P_{Q_1}} = 1 + 32.33 \cdot \left(\frac{T}{T_{Q_1}} - 1 \right)^{1/2} + 91.169 \cdot \left(\frac{T}{T_{Q_1}} - 1 \right), \tag{3-24}$$

where T_{Q_1} and P_{Q_1} are the temperature (in Kelvin) and pressure of the quadrupole point Q_1 (see above). Since this function is based on data extending up to 140 bar, it should not be extrapolated above this value, which represents the upper limit of the pressure axis in Fig. 3.6.

The three-phase equilibrium curve $L_{aq}+L_{CO_2}+V$ is situated very close to the saturation curve of pure carbon dioxide and ends at the lower critical end point (LCEP) of the CO_2-H_2O binary system. The LCEP was positioned at 31.05°C and 73.9 bar by Song and Kobayashi (1987), but this result was recently revised by Wendland et al. (1999), who suggested 31.48°C and 74.11 bar. Also the LCEP, therefore, almost coincides with the critical point of pure CO_2.

From Fig. 3.6, it is evident that at the *T*, *P* conditions of interest for the geological sequestration of CO_2, a H_2O-rich liquid phase coexists with a CO_2-rich phase, either a vapour or a liquid phase depending on pressure, below the LCEP and above 10−12°C. This distinction between the vapour and liquid CO_2-rich phases vanishes above the LCEP, where a CO_2-rich gas is present instead.

3.6. The equations of state for CO_2-H_2O gas mixtures

The simplest way to treat gas mixtures is to assume ideal behaviour. Under this hypothesis, the fugacity of the *i*th component in a multi-component gas mixture is equal to its partial pressure P_i, which is given by the Dalton's law:

$$f_i = P_i = y_i\ P_{\mathrm{total}}, \tag{3-25}$$

where y_i denotes the mole fraction of the *i*th component in the gas mixture. Unfortunately, the relationship $f_i = P_i$ is reasonable only at very low total pressures, for real gases.

A first step forward is done through the use of the Lewis fugacity rule:

$$f_i = y_i f_{i,\mathrm{pure}}, \tag{3-26}$$

where $f_{i,\mathrm{pure}}$ represents the fugacity of pure gas *i*. The Lewis fugacity rule evidently assumes that fugacities do not depend on the composition of gas mixtures. In other words, the fugacity of CO_2 is postulated to be the same in CO_2-H_2O, CO_2-H_2, CO_2-CH_4 and any other gas mixture, although it is well known that molecular interactions are different from system to system. In particular, the Lewis fugacity rule is expected to give a poor description of fugacities for mixtures containing a highly polarized component, such as water.

A giant step forward consists in using one of the equations of state discussed in Section 3.3, such as the van der Waals, the Redlich–Kwong, the modified Redlich–Kwong and the virial EOS. Apart from the case of the virial equation, in the other three cases, application of these EOS to gas mixtures rather than to pure gases requires the estimation of *a* and *b* parameters by means of the so-called mixing rules (Prausnitz et al., 1999). In particular, for the two-component CO_2-H_2O gas mixture (which is the subject of our interest), the repulsive *b* parameter, which is a measure of the volume of gas molecules, can be obtained by averaging the contributions of the two components:

$$b_{\mathrm{mix}} = y_{\mathrm{CO_2}} \cdot b_{\mathrm{CO_2}} + y_{\mathrm{H_2O}} \cdot b_{\mathrm{H_2O}}. \tag{3-27}$$

The *a* parameter, which describes the effects of molecular attraction, is computed by averaging these effects between all the possible types of molecular pairs, which reduces to

$$a_{\mathrm{mix}} = y^2_{\mathrm{CO_2}} a_{\mathrm{CO_2}} + 2 y_{\mathrm{H_2O}} y_{\mathrm{CO_2}} a_{\mathrm{H_2O-CO_2}} + y^2_{\mathrm{H_2O}} a_{\mathrm{H_2O}} \tag{3-28}$$

for the CO_2-H_2O gas mixture. Since we are dealing with a CO_2-rich gas (or liquid) phase coexisting with a H_2O-rich liquid, it is reasonable to assume infinite H_2O dilution in the CO_2-rich phase (King et al., 1992; Spycher et al., 2003), which means $y_{\mathrm{H_2O}} = 0$ and $y_{\mathrm{CO_2}} = 1$. Under this hypothesis, a_{mix} and b_{mix} parameters are equal to *a* and *b* parameters of pure CO_2. In other words, it is assumed that, in the gas phase, the relatively scarce H_2O molecules do not affect the much more abundant CO_2 molecules whereas the important effect of CO_2 molecules on H_2O molecules is considered.

However, the $a_{H_2O-CO_2}$ parameter is still needed to compute the fugacity coefficient of H_2O in the CO_2-H_2O gas mixture (see below). It can be classically derived by computing the following geometric mean:

$$a_{H_2O-CO_2} = \left(a_{H_2O} \cdot a_{CO_2}\right)^{1/2}. \tag{3-29}$$

This equation, or geometric-mean assumption, was suggested by Berthelot on empirical grounds and utilized by many researchers, including van der Waals. In 1930, it was shown by London that it is justified on theoretical grounds, at least under some conditions (Prausnitz et al., 1999). However, it is advisable to obtain the $a_{H_2O-CO_2}$ parameter by regression of available data on the gas mixture, rather than through equation (3-29). Numbers clarify this suggestion. For instance, the a parameter is 6.44×10^7 bar cm^6 K$^{0.5}$ mol^{-2} for CO_2 and 1.42×10^8 bar cm^6 K$^{0.5}$ mol^{-2} for H_2O, based on critical constants (equations (3-15) and (3-16)). Insertion of these values in equation (3-29) gives an $a_{H_2O-CO_2}$ parameter of 9.56×10^7 bar cm^6 K$^{0.5}$ mol^{-2}, which is significantly different from that estimated by Spycher et al. (2003) through regression of available data, 7.89×10^7 ($\pm 0.08 \times 10^7$) bar cm^6 K$^{0.5}$ mol^{-2}. Although use of equations (3-15) and (3-16) is not the best way to compute the a parameter for pure gases, as shown above, use of equation (3-29) makes things even worse.

We are now ready to use an EOS, for instance the Redlich–Kwong EOS, to compute fugacity coefficients in the CO_2-rich phase. For a given component k in a gas mixture of n components, the needed equation is (Spycher et al., 2003):

$$\begin{aligned}\ln \Gamma_k = {} & \ln\left(\frac{v}{v-b_{mix}}\right) + \frac{b_k}{v-b_{mix}} - \frac{2\sum_{i=1}^{n} y_i \cdot a_{ik}}{b_{mix} RT^{3/2}} \cdot \ln\left(\frac{v+b_{mix}}{v}\right) \\ & + \frac{a_{mix} b_k}{b_{mix}^2 RT^{3/2}}\left[\ln\left(\frac{v+b_{mix}}{v}\right) - \frac{b_{mix}}{v+b_{mix}}\right] - \ln Z,\end{aligned} \tag{3-30}$$

where v and Z are the molar volume and the compressibility factor of the gas mixture, respectively. In general, the molar volume v is obtained by means of equation (3-19), using a_{mix} instead of a and b_{mix} instead of b. However, under the hypothesis of infinite H_2O dilution in the CO_2-rich phase, since the a_{mix} and b_{mix} parameters are equal to a and b parameters of pure CO_2 (see above), v and Z are the volume and the compressibility factor of pure CO_2. Therefore, under the hypothesis of infinite H_2O dilution in the CO_2-rich phase, equation (3-30) reduces to the following relation for H_2O:

$$\begin{aligned}\ln \Gamma_{H_2O} = {} & \ln\left(\frac{v_{CO_2}}{v_{CO_2}-b_{CO_2}}\right) + \frac{b_{H_2O}}{v_{CO_2}-b_{CO_2}} - \frac{2a_{H_2O-CO_2}}{b_{CO_2} RT^{3/2}} \cdot \ln\left(\frac{v_{CO_2}+b_{CO_2}}{v_{CO_2}}\right) \\ & + \frac{a_{CO_2} b_{H_2O}}{b_{CO_2}^2 RT^{3/2}}\left[\ln\left(\frac{v_{CO_2}+b_{CO_2}}{v_{CO_2}}\right) - \frac{b_{CO_2}}{v_{CO_2}+b_{CO_2}}\right] - \ln Z_{CO_2}.\end{aligned} \tag{3-31}$$

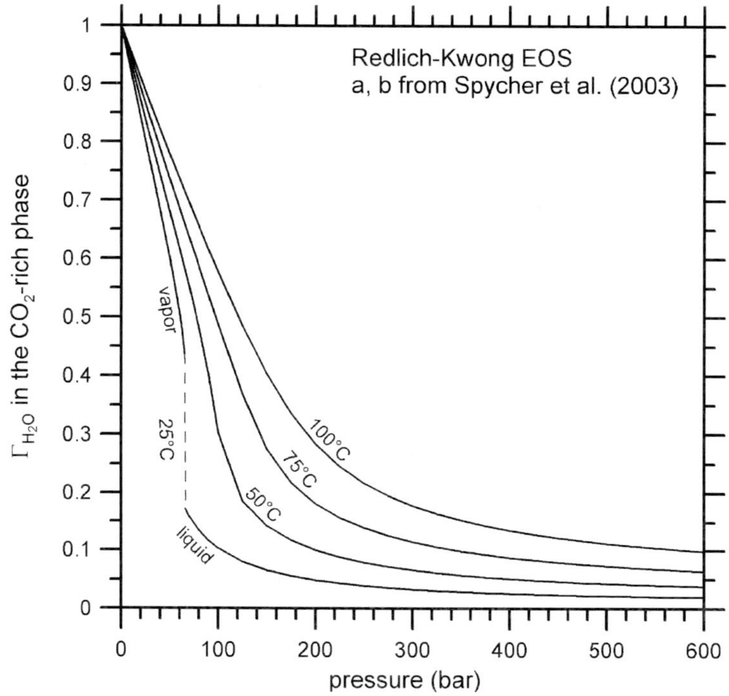

Figure 3.7. Fugacity coefficients of H_2O in the CO_2-rich phase, computed at 25, 50, 75 and 100°C and variable pressures, by means of the Redlich–Kwong equation with constant *b* and *a* as function of temperature and assuming infinite H_2O dilution, as suggested by Spycher et al. (2003).

In spite of the simplification linked to this hypothesis, the fugacity coefficients of H_2O in the CO_2-rich phase depends not only on pressure and temperature but also on the composition of the mixture, through the molecular interaction parameters $a_{H_2O-CO_2}$ and b_{H_2O}. Note that the parameter a_{H_2O} is not used. Results of calculations are shown in Fig. 3.7.

In contrast, the fugacity coefficient of CO_2 in the CO_2-rich phase is equal to that of pure CO_2, under the hypothesis of infinite H_2O dilution. This can be easily verified by solving equation (3-30) for CO_2.

3.7. Mutual solubilities of CO_2 and H_2O in CO_2-H_2O mixtures

The fugacity coefficients of CO_2 and H_2O in CO_2-H_2O mixtures can be used to compute the mutual solubilities of CO_2 and H_2O in such mixtures, following the approach of Spycher et al. (2003). This is based on the following equilibrium reactions:

$$H_2O_{(l)} \Leftrightarrow H_2O_{(g)} \tag{3-32}$$

and

$$CO_{2(aq)} \Leftrightarrow CO_{2(g)}, \tag{3-33}$$

whose thermodynamic equilibrium constants at any given *T*, *P* are

$$K_{H_2O,T,P} = \frac{a_{H_2O_{(g)}}}{a_{H_2O_{(l)}}} = \frac{f_{H_2O_{(g)}} / f^0_{H_2O_{(g)}}}{a_{H_2O_{(l)}}} \tag{3-34}$$

and

$$K_{CO_2,T,P} = \frac{a_{CO_{2(g)}}}{a_{CO_{2(aq)}}} = \frac{f_{CO_{2(g)}} / f^0_{CO_{2(g)}}}{a_{CO_{2(aq)}}}. \tag{3-35}$$

Since the standard state adopted for gases is unit fugacity of the hypothetical gas at 1 bar and any temperature, both $f^0_{H_2O_{(g)}}$ and $f^0_{CO_{2(g)}}$ are equal to 1.

The standard state for liquid water calls for unit activity of pure water at all pressures and temperatures. Since the solubility of CO_2 in the aqueous phase is relatively small at the *P*, *T* conditions of interest, the activity of water can be set equal to its mole fraction, x_{H_2O} (we use *x* to indicate mole fractions in the aqueous solution). The mole fraction of water, in turn, is equal to $1 - x_{CO_2}$, since H_2O and CO_2 are the only two components of the considered system.

The standard state for dissolved CO_2 calls for unit activity in a hypothetical one molal solution referenced to infinite dilution at any pressure and temperature. The mole fraction of dissolved CO_2 is related to its molal concentration through the following relation:

$$x_{CO_2} = \frac{m_{CO_2}}{m_{CO_2} + m_{H_2O}} = \frac{m_{CO_2}}{m_{CO_2} + 55.508}. \tag{3-36}$$

which is solved for m_{CO_2} obtaining:

$$m_{CO_2} = \frac{x_{CO_2} \cdot 55.508}{(1 - x_{CO_2})} \cong x_{CO_2} \cdot 55.508. \tag{3-37}$$

Taking the activity coefficient of dissolved CO_2, γ_{CO_2} equal to 1, which is reasonable for neutral solutes in dilute aqueous solutions, we have

$$a_{CO_2} = m_{CO_2} \cdot \gamma_{CO_2} \cong x_{CO_2} \cdot 55.508 \tag{3-38}$$

Therefore, equations (3-34) and (3-35) can be rewritten as follows:

$$K_{H_2O,T,P} \cong \frac{f_{H_2O_{(g)}}}{1 - x_{CO_2}} \tag{3-39}$$

and

$$K_{CO_2,T,P} \cong \frac{f_{CO_2(g)}}{x_{CO_2} \cdot 55.508}. \tag{3-40}$$

The dependence of these equilibrium constants on temperature is suitably defined by the following polynomial equations (T in °C):

$$\log K_{H_2O,T,P^\circ} = -2.209 + 0.03097 \cdot T - 0.0001098 \cdot T^2 + 2.048 \times 10^{-7} \cdot T^3 \tag{3-41}$$

$$\log K_{CO_2(g),T,P^\circ} = 1.189 + 0.01304 \cdot T - 5.446 \times 10^{-5} \cdot T^2 \tag{3-42}$$

$$\log K_{CO_{2(l)},T,P^\circ} = 1.169 + 0.01368 \cdot T - 5.380 \times 10^{-5} \cdot T^2 \tag{3-43}$$

whereas their dependence on pressure is taken into account through the following expression:

$$K_{T,P} = K_{T,P^\circ} \cdot \exp\left(\frac{(P - P^\circ) \cdot \overline{V}_i}{RT}\right) \tag{3-44}$$

where $\overline{V}_i$ is the average partial molar volume of the ith component in the pressure range P° to P and P° is taken equal to 1 bar for $T < 100$°C and to water saturation pressure for $T > 100$°C. Since $\overline{V}_i$ also varies with temperature, it has been averaged over the temperature interval of interest, obtaining 18.1 cm^3 mol^{-1} for H_2O and 32.6 cm^3 mol^{-1} for CO_2, irrespective of their physical state.

Let us recall equation (2-20)

$$f_i = \Gamma_i \cdot y_i \cdot P \tag{3-45}$$

where, as noted above, y_i denotes the mole fraction of the ith component in the gas mixture. Substitution of equation (3-45) into equations (3-39) and (3-40) leads to

$$f_{H_2O_{(g)}} = \Gamma_{H_2O} \times y_{H_2O} \times P = K_{H_2O,T,P} \times (1 - x_{CO_2}) \tag{3-46}$$

and

$$f_{CO_{2(g)}} = \Gamma_{CO_2} \times y_{CO_2} \times P = K_{CO_2,T,P} \times x_{CO_2} \times 55.508. \tag{3-47}$$

Solving equation (3-46) for y_{H_2O} and taking into account equation (3-44) we write

$$y_{H_2O} = \frac{K_{H_2O,T,P^\circ} \cdot (1 - x_{CO_2})}{\Gamma_{H_2O} \cdot P} \cdot \exp\left(\frac{(P - P^\circ) \cdot \overline{V}_{H_2O}}{RT}\right), \tag{3-48}$$

whereas from equations (3-47) and (3-44) we obtain

$$x_{CO_2} = \frac{\Gamma_{CO_2} \cdot (1 - y_{H_2O}) \cdot P}{K_{CO_2,T,P^\circ} \cdot 55.508} \cdot \exp\left(-\frac{(P - P^\circ) \cdot \overline{V}_{CO_2}}{RT}\right). \tag{3-49}$$

Solution of equations (3-48) and (3-49) is carried out by setting

$$A = \frac{K_{H_2O,T,P^\circ}}{\Gamma_{H_2O} \cdot P} \cdot \exp\left(\frac{(P-P^\circ)\cdot \bar{V}_{H_2O}}{RT}\right) \tag{3-50}$$

and

$$B = \frac{\Gamma_{CO_2} \cdot P}{K_{CO_2,T,P^\circ} \cdot 55.508} \exp\left(-\frac{(P-P^\circ)\cdot \bar{V}_{CO_2}}{RT}\right). \tag{3-51}$$

Therefore,

$$y_{H_2O} = \frac{1-B}{(1/A)-B} \tag{3-52}$$

and

$$x_{CO_2} = B\cdot(1-y_{H_2O}). \tag{3-53}$$

Again, since fugacity coefficients are computed under the hypothesis of infinite H_2O dilution in the CO_2-rich phase, equations (3-52) and (3-53) are also solved in a direct way.

In performing these calculations K_{CO_2,T,P° is computed by equation (3-42) when the temperature and the volume of the gas phase are above the critical parameters of CO_2, 31°C and 94 cm^3 mol^{-1}, respectively, whereas K_{CO_2,T,P° is computed by equation (3-43), when the temperature and the volume of the gas phase are below the critical parameters of CO_2. In this way, it is implicitly assumed that the phase boundary for the CO_2-rich phase does not differ from that of pure CO_2 (which is a reasonable assumption, see Fig. 3.6 and related discussion). It also neglects the very small *P–T* space where CO_2 gas, CO_2 liquid and H_2O liquid coexist (see Figure 3 in Spycher et al., 2003).

This methodology was specifically implemented for computing mutual solubilities of CO_2 and H_2O in CO_2-H_2O mixtures from 12 to 100°C and for pressures up to 600 bar, and it should not be applied outside these ranges (Spycher et al., 2003).

At this point, you might be curious to see how the predicted mutual solubilities of CO_2 and H_2O in CO_2-H_2O mixtures compare with experimental data. Although early experimental studies were chiefly focused on pressures and temperatures much higher than those of interest for the geological sequestration of CO_2, data have been produced also in the *P*, *T* intervals of interest to us, in recent years. Useful data are reported by Wiebe and Gaddy (1939, 1940, 1941), Tödheide and Franck (1963), Coan and King (1971), Briones et al. (1987), Song and Kobayashi (1987), D'Souza et al. (1988), Müller et al. (1988), Sako et al. (1991), King et al. (1992), Dohrn et al. (1993) and Bamberger et al. (2000).

Further data of interest on the solubility of CO_2 in liquid water are given by Bartholomé and Friz (1956), Matous et al. (1969), Zawisza and Malesinska (1981), Jackson et al. (1995), Teng et al. (1997), Rosenbauer et al. (2001) and Anderson (2002). Many other studies were performed on this subject, mainly at pressures less than 50 bar and neglecting the composition of the coexisting gas phase. Crovetto (1991), Carrol and

Mather (1992) and Diamond and Akinfiev (2003) assembled and reviewed the existing data to evaluate the solubility of CO_2 in water and other thermodynamic properties.

Selected experimental data and computed values of y_{H_2O} and x_{CO_2} are reported against pressure in the plots of Figs. 3.8 and 3.9, respectively. Each plot refers to a given temperature (25, 50, 75 and 100°C) and pressures up to 600 bar. Computed values of y_{H_2O} at 25°C, as well as at any sub-critical temperature, show a marked discontinuity coinciding with the small pressure interval of gas–liquid phase transition of the CO_2-rich phase. Water solubility in this CO_2-rich phase decreases markedly with pressure, below the transition pressure, whereas an opposite behaviour is evident above the transition pressure. Although this sharp discontinuity disappears at supercritical temperatures, its memory is

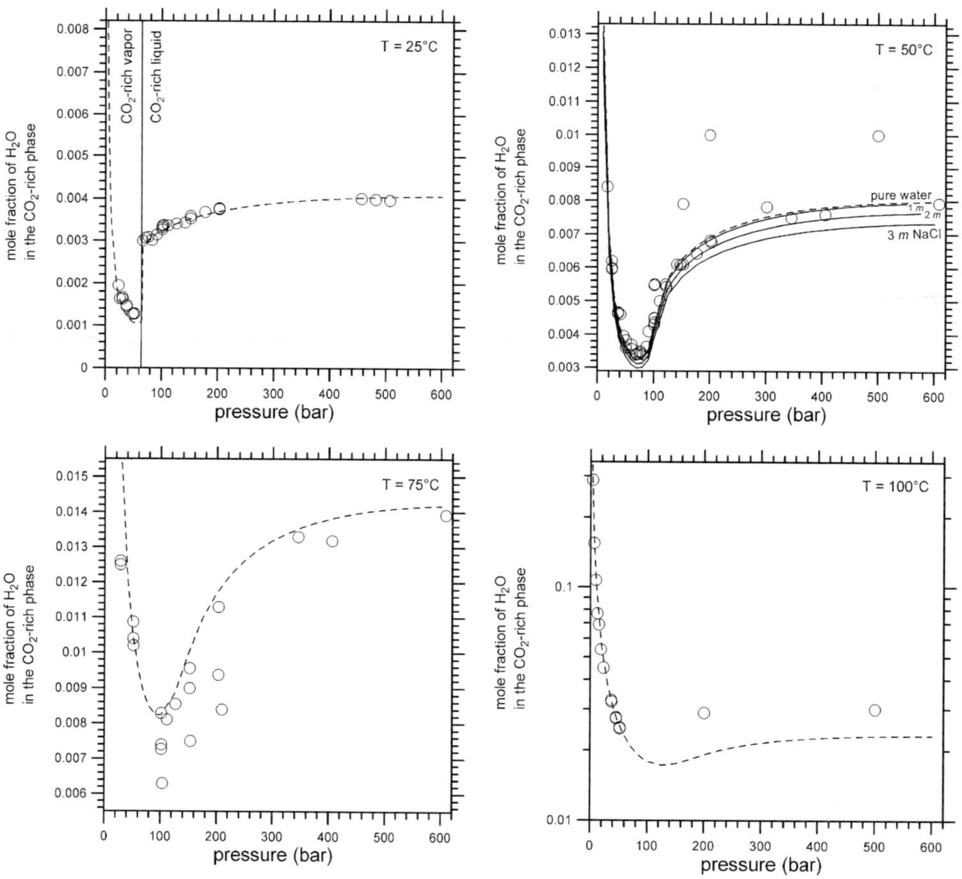

Figure 3.8. Mole fraction of H_2O (y_{H_2O}) in the CO_2-rich phase at 25, 50, 75 and 100°C and pressures up to 600 bar. Results of selected experiments (circles, see text for references) are compared with the theoretical data (dashed line) obtained by Spycher et al. (2003). The solid lines in the 50°C-plot refer to approximated computations for the CO_2-rich phase in equilibrium with 1*m*, 2*m* and 3*m* NaCl aqueous solutions. Reprinted from Spycher et al. (2003), Copyright (2003), adapted with permission from Elsevier.

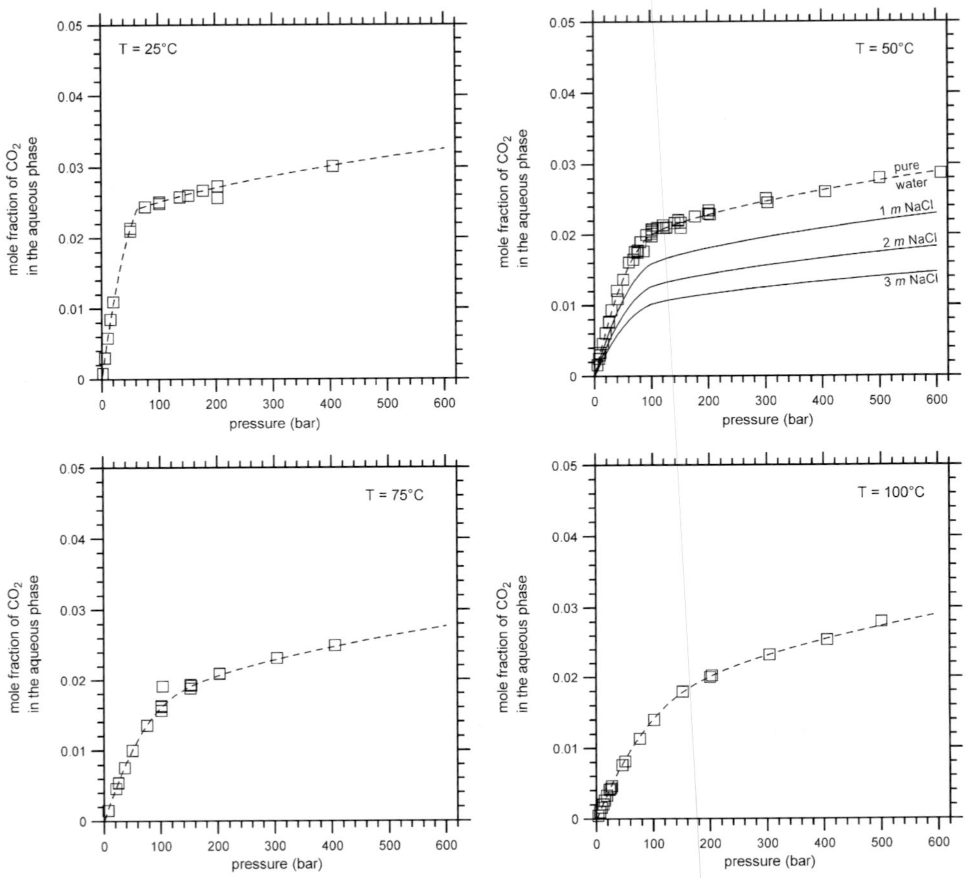

Figure 3.9. Mole fraction of CO_2 (x_{CO_2}) in the aqueous phase at 25, 50, 75 and 100°C and pressures up to 600 bar. Results of selected experiments (squares, see text for references) are compared with the theoretical data (dashed line) obtained by Spycher et al. (2003). The solid lines in the 50°C-plot refer to approximated computations for 1*m*, 2*m* and 3*m* NaCl aqueous solutions. Reprinted from Spycher et al. (2003), Copyright (2003), adapted with permission from Elsevier.

still present at 100°C. The computed y_{H_2O} values are in good agreement with the experimental data at 25°C and 50°C, whereas the correspondence between the two series of data is not as good at 75°C and 100°C. To understand the reasons of these discrepancies, a detailed review of experimental data would be necessary, but this is beyond our present objectives.

At sub-critical temperatures (e.g., 25°C), the solubility of CO_2 in the aqueous phase increases steeply with pressure below the gas–liquid phase transition of the CO_2-rich phase, whereas the x_{CO_2}–pressure curve is less steep above the transition pressure. In other words, the overall x_{CO_2}–pressure curve results from the superposition of two curves, one for liquid CO_2 above the transition pressure and another for gaseous CO_2 below this pressure value. At supercritical temperatures, the x_{CO_2}–pressure curve refers to the dissolution of gaseous CO_2, but a bend (corresponding to a sort of memory of the sharp break in the 25°C-curve) is still evident at all the considered temperatures. The computed x_{CO_2} values are in good agreement with the experimental data at all temperatures.

In addition to the solubility model of Spycher et al. (2003), models for the CO_2-H_2O binary system have been developed by other workers, including Spycher and Reed (1988), Duan et al. (1992), King et al. (1992), Shyu et al. (1997) and Bamberger et al. (2000), but they either do not cover or cover only in part the P, T range of interest for the geological sequestration of CO_2.

In particular, the virial EOS of Spycher and Reed (1988) is in terms of pressure and temperature, which is a great advantage for geochemists, but it cannot be used below 100°C.

The EOS proposed by Duan et al. (1992) is based on a fifth-order virial expansion in volume and can be used from 50 to 1,000°C and from 0 to 1,000 bar. This EOS is accessible on the website http://geotherm.ucsd.edu/geofluids/.

King et al. (1992) utilized a Redlich–Kwong EOS to reproduce with enough accuracy their solubility data as well as those by Wiebe and Gaddy (1940, 1941), from 15 to 40°C and up to a pressure of 500 bar.

Shyu et al. (1997) utilized a Peng–Robinson EOS and rather complex mixing rules to describe CO_2-H_2O mixtures from 25 to 350°C and up to 1,000 bar. The Peng–Robinson EOS is more complex than the Redlich–Kwong equation, and it reproduces the liquid–vapor boundary with good accuracy (see Prausnitz et al., 1999 for further details). A Peng–Robinson EOS was also used by Bamberger et al. (2000) to fit their data at 50, 60 and 80°C and up to 141 bar.

3.8. Impact of dissolved salts on the mutual solubilities of CO_2 and H_2O

To assess the impact of dissolved salts on the mutual solubilities of CO_2 and H_2O, the model of Spycher et al. (2003) can be extended to the $CO_2-H_2O-NaCl$ system, by changing the activity of liquid water and the activity coefficient of dissolved CO_2, which were assumed to be unity for the CO_2-H_2O binary. To a first approximation, we will use the simplified approach of Spycher et al. (2003) and try to check the consistency of obtained results afterwards.

Again, we start from equations (3-34), (3-44) and (3-45), to derive the following expression for y_{H_2O}:

$$y_{H_2O} = \frac{K_{H_2O,T,P^\circ} \cdot a_{H_2O}}{\Gamma_{H_2O} \cdot P} \cdot \exp\left(\frac{(P-P^\circ)\cdot \overline{V}_{H_2O}}{RT}\right). \qquad (3\text{-}54)$$

This equation is then solved by inserting the values of a_{H_2O} suggested by Helgeson (1969) for 1, 2 and 3 mol kg^{-1} NaCl aqueous solutions (see Table 3.1). This means to assume that the effect of dissolved carbon dioxide on the activity of water is negligible, an assumption that has to be properly verified afterwards.

Instead of equation (3-36), which applies to the CO_2-H_2O binary, the mole fraction of CO_2 in the aqueous phase is

$$x_{CO_2} = \frac{m_{CO_2}}{m_{CO_2} + m_{H_2O} + 2 \cdot m_{NaCl}} = \frac{m_{CO_2}}{m_{CO_2} + 55.508 + 2 \cdot m_{NaCl}} \tag{3-55}$$

for the CO_2-H_2O–NaCl ternary, where NaCl molality is multiplied by two, since we assume complete dissociation of dissolved NaCl into Na^+ and Cl^- ions. Recalling that $a_{CO_2} = m_{CO_2} \times \gamma_{CO_2}$, equation (3-55) is rearranged as follows:

$$a_{CO_2} = \frac{x_{CO_2} \gamma_{CO_2} (55.508 + 2 \cdot m_{NaCl})}{1 - x_{CO_2}}. \tag{3-56}$$

We now go back to equations (3-35), (3-44) and (3-45), to write

$$a_{CO_2} = \frac{\Gamma_{CO_2} \cdot (1 - y_{H_2O}) \cdot P}{K_{CO_2,T,P^\circ}} \cdot \exp\left(- \frac{(P - P^\circ) \cdot \overline{V}_{CO_2}}{RT} \right). \tag{3-57}$$

It is implicitly assumed that the mole fraction of NaCl in the CO_2-rich phase is negligible, which is a reasonable hypothesis at least when this phase is a gas, i.e., either above 31°C at any pressure or below 31°C at low pressures (see Fig. 3.6 and related discussion). Then we solve equations (3-56) and (3-57) for x_{CO_2}, inserting the y_{H_2O} value computed through equation (3-54) and the activity coefficients of dissolved CO_2 derived by Helgeson (1969) for 1, 2 and 3 mol kg^{-1} NaCl aqueous solutions, based on the experimental data of Ellis and Golding (1963, see Table 3.1).

Values of y_{H_2O} and x_{CO_2} computed at 50°C referring to 1, 2 and 3 mol kg^{-1} NaCl aqueous solutions are reported in Figs. 3.8 and 3.9. Inspection of these plots reveals that dissolved NaCl has a relatively minor effect on the composition of the CO_2-rich phase, at least when this phase is a gas. On the other hand, the effect of dissolved NaCl on the composition of the aqueous phase is important and increases with NaCl concentration. This is

TABLE 3.1

Activity of water and activity coefficient of dissolved CO_2 for 1, 2 and 3 mol kg^{-1} NaCl aqueous solutions (from Helgeson, 1969) at the temperatures of interest (values at 75°C have been obtained by interpolation)

m_{NaCl}	Temperature			
	25°C	50°C	75°C	100°C
Activity of water ($a_{H_2O_{(l)}}$)				
1	0.9669	0.9667	0.9667	0.9669
2	0.9316	0.9308	0.9308	0.9315
3	0.8932	0.8919	0.8918	0.893
Activity coefficient of dissolved CO_2 (γ_{CO_2})				
1	1.27	1.24	1.22	1.20
2	1.57	1.50	1.46	1.44
3	1.93	1.80	1.75	1.74

the well-known *salting-out effect*, which is directly related to the activity coefficient of dissolved CO_2. Since CO_2 activity is fixed by chemical equilibrium (equation (3-33)) and γ_{CO_2} increases with the salinity of the aqueous phase (see Table 3.1), m_{CO_2} and x_{CO_2} must necessarily decline, as shown in Fig. 3.9.

To test the reliability of this approach, we now compute the activity of water for the 3 mol kg^{-1} NaCl aqueous solution at 600 bar and 50°C by means of the speciation code EQ3NR (Wolery, 1992), inserting the CO_2 fugacity as input datum. Under these conditions, CO_2 fugacity is (see Figs 3.5b and 3.9 for the values of Γ_{CO_2} and y_{CO_2}, respectively):

$$f_{CO_2(g)} = \Gamma_{CO_2} \cdot y_{CO_2} \cdot P = 0.282 \times 0.993 \times 600\,\text{bar} = 168\,\text{bar}, \tag{3-58}$$

which is much less than CO_2 partial pressure, 596 bar. Results of the EQ3NR code indicates that the aqueous solution is dominated by dissociated Na^+ and Cl^- ions, which represent the 82.8% of total dissolved Na and Cl, respectively, with equal molalities of 2.48. The weight of the $NaCl^\circ$ ion pair is comparatively minor, with a molal concentration of 0.516. At the pH of this aqueous solution, 2.87, CO_2 is by far the main carbonate species, with a molality of 1.995. Indeed, the molal concentration of HCO_3^- ion is 0.00178 only. Interestingly, EQ3NR computes a water activity of 0.8727, which is only 2% lower than the value of Helgeson (1969), 0.8919. The similarity between these two water activity values is due to the fact that dissolved neutral species, such as aqueous CO_2, are not involved in the EQ3NR computation of water activity. Therefore, the approximated approach used to assess the impact of dissolved NaCl on the mutual solubilities of CO_2 and H_2O cannot be strengthened (nor weakened) by speciation calculations.

Recently, the model of Spycher et al. (2003) was extended by Spycher and Pruess (2005) to compute the mutual solubilities of CO_2 and H_2O, again in a non-iterative way, for aqueous solutions up to 4 mol kg^{-1} $CaCl_2$ and 6 mol kg^{-1} NaCl. The accuracy of computed values compares with that of experimental data.

3.9. The plot of pressure versus enthalpy for carbon dioxide

Two important thermodynamic properties of carbon dioxide, namely its enthalpy and its density (i.e., the inverse of the molar volume), are displayed at different T, P conditions in the plot of Fig. 3.10.

As already mentioned (see Chapter 2), since the absolute enthalpy of substances cannot be determined, geochemists generally work with the enthalpy of formation from the elements. However, when dealing with a single pure substance, there is no need for a reaction of reference and it is sufficient to refer its thermodynamic properties to a suitable reference state. For water, this is either liquid water at 0°C (Denbigh, 1971) or liquid water at the triple point, i.e., 273.16 K, 0.006113 MPa. The latter reference state, with $S_{H_2O(l),triple} = 0, G_{H_2O(l),triple} = 0$, was adopted by the 5th International Conference on the Properties of Steam (Helgeson and Kirkham, 1974a).

In the NIST Standard Reference Database (Lemmon et al., 2003), one of the four reference state conventions can be selected by the user: (a) the "normal boiling point"

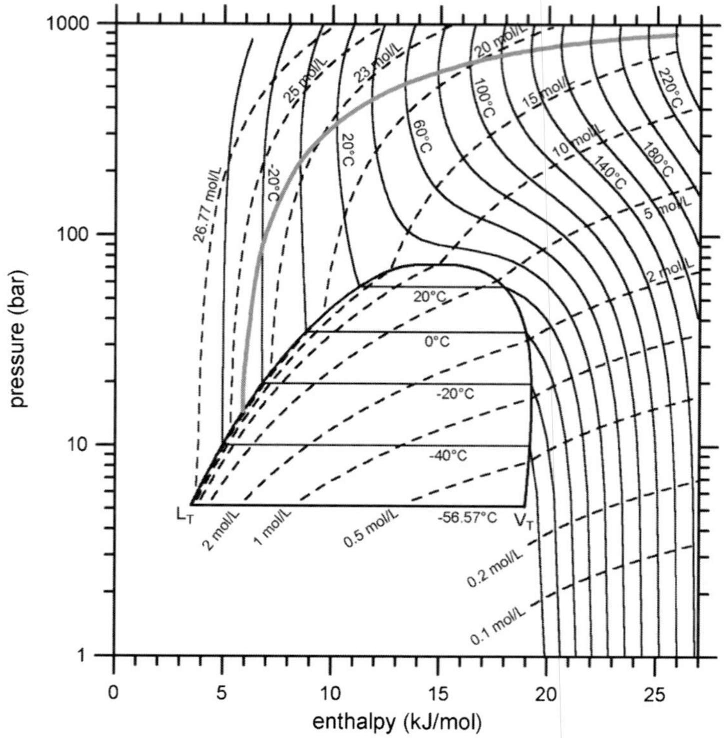

Figure 3.10. Plot of pressure vs. enthalpy for carbon dioxide, showing the saturation curve (heavy solid line), relevant isotherms (thin solid lines) and isochores (dashed lines) and the Joule–Thompson inversion curve (thick gray curve). Points V_T and L_T refer to the vapour and the liquid at the triple point (−56.57°C, 5.185 bar), respectively. Data from Lemmon et al. (2003).

convention, which fixes the enthalpy and entropy to zero for the saturated liquid at the normal boiling point temperature; (b) the ASHRAE (American Society of Heating, Refrigerating and Air-Conditioning Engineers) convention, which sets to zero the enthalpy and entropy for the saturated liquid at −40°C (233.15 K); (c) the IIR (International Institute of Refrigeration) convention, which adopts the reference state values of 200 kJ kg^{-1} and 1 kJ (kg K)$^{-1}$ for enthalpy and entropy, respectively, for the saturated liquid at 0°C; (d) the reference state of zero internal energy, zero entropy for the saturated liquid at 273.16 K.

The latter choice was adopted for constructing the pressure vs. enthalpy plot of Fig. 3.10. This plot is a sort of simplified Mollier diagram, which is normally utilized by chemical engineers to quantify physico-chemical phenomena. We will use this plot to understand what occurs during CO_2 compression, a process which is relevant for the geological sequestration of CO_2 through injection of pressurized CO_2 into a deep aquifer.

Everybody has had empirical, direct perception of what occurs during gas compression. A typical situation is experienced by the cyclist who feels her/his hands warm when pumping the tires of the bike by means of a hand-held pump. This rather common experiment suggests that gaseous substances heat during compression and cool during expansion.

But this is not always true, as ascertained by Joule and Thompson in 1853. Indeed, each gaseous substance has an interval of *P*, *T* conditions in which it heats on compression and another where it cools on compression. For isenthalpic processes, these two intervals are divided by the Joule–Thompson inversion curve, which is shown in Fig. 3.10 for CO_2.

Let us suppose to compress carbon dioxide isenthalpically, initially at 1 bar, 20°C. Based on Fig. 3.10, it is evident that the enthalpy of CO_2, under these initial conditions, is somewhat higher than 22 kJ mol^{-1}, being equal to 22.076 kJ mol^{-1} for the sake of precision. If enthalpy is kept constant at this value throughout the compression, temperature increases attaining ~40°C at 20 bar, ~60°C at 41 bar, ~80°C at 71 bar, ~100°C at 102 bar, ~120°C at 147 bar and so on. It is evident that CO_2 heats up considerably during this process, since we remain well below the Joule–Thompson inversion curve, unless pressures close to 800 bar are attained (which are probably too high for our purposes). Therefore, it is very likely that pressurized CO_2 enters the aquifer at a relatively high temperature, possibly higher than that of the aquifer itself. This injection of hot CO_2 might have important influences on the heat balance of the aquifer, at least in the immediate surroundings of the injection point, and consequently on the dissolution of CO_2 in the aqueous phase and on the dissolution/precipitation reactions of minerals.

Chapter 4

The Aqueous Electrolyte Solution

4.1. The important role of aqueous electrolyte solutions

Werner F. Giggenbach, one of the most brilliant geochemists who ever studied magmatic and hydrothermal systems, in one of his last works (Giggenbach, 1997) wrote that "according to the early geochemist Agricola (1556), medieval alchemists had a well-known saying: *non reagent nisi soluti*. A geochemical translation is: nothing much happens if there isn't a fluid, a reaction medium, allowing minerals components to be transported and to interact with one another".

Among the fluids, aqueous electrolyte solutions are widespread and very important, as effective reaction media, in different geological environments including the aquifers apt to geological CO_2 sequestration. Any aqueous electrolyte solution comprises the solvent water, some ionic solutes and some neutral solutes. Examples of ionic solutes are both the free ions, such as Na^+ and Cl^-, and the ionic complexes, such as $CaHCO_3^+$ and $NaSO_4^-$. Examples of neutral solutes are both dissolved molecules, such as $CO_2(aq)$ and $SiO_2(aq)$, and neutral complexes, such as $NaHCO_3(aq)$ and $CaCO_3(aq)$.

Many features and details of aqueous electrolyte solutions (e.g. the thermodynamic properties of electrolyte solutes) are omitted from this presentation as the interested reader can find this information in many textbooks. However, we will recall the basic concepts on which are founded both our present understanding of aqueous electrolyte solutions and, consequently, the geochemical modelling of these intriguing and fascinating media. As discussed in Chapter 7, the geochemical modelling of aqueous electrolyte solutions is a quite complex exercise, consisting in solving the mass balance, charge balance and equilibrium relations involving the activities of relevant solutes and of the solvent water. A delicate part of this exercise is the correct computation of the activity coefficients of solutes and water activity. We recall that the activity coefficient of the ith solute component, γ_i, is related to the partial molar excess Gibbs free energy, $\overline{G_i}^{\text{ex}}$, by following the relationship (e.g. Denbigh, 1971; Prausnitz et al., 1999):

$$\left(\frac{\partial G^{\text{ex}}}{\partial n_i}\right)_{T,P,n_j} = \overline{G_i}^{\text{ex}} = RT \cdot \ln \gamma_i \qquad (4\text{-}1)$$

where n_i are the moles of the considered component and n_j are the moles of all the other components $j \neq i$. A similar relation holds for solvent water as well. G^{ex} is the difference between the Gibbs free energy and the Gibbs free energy of ideal mixing. The activity coefficients of solutes and water activity might deviate significantly from unity under the temperature, pressure conditions of interest. Although many excellent investigations were carried out so far, a considerable research effort is still needed to improve the overall understanding of this subject.

4.2. The Debye–Hückel theory

One of the major steps forward in the theoretical understanding of electrolyte solutions was done in 1923, when two scientific contributions by Peter Debye and Erich Hückel were published. They reported a technique for computing activity coefficients of electrolytes in dilute solutions.

Debye and Hückel (1923a,b) considered the solvent as an ideal dielectric fluid without any structure and the solute ions as spheres with their charges situated at their centres, and recognized that the most important interactions were long-range Coulombic forces. In contrast, the interactions acting between the neutral molecules of the solvent and non-electrolytes are much weaker van der Waals forces, which are felt over shorter distances. This difference in the range of the acting forces explains the large departures from ideality in electrolyte solutions even at low concentrations, where the particles of ionic solutes are far apart. The method used by Debye and Hückel to calculate the electrical contribution of long-range Coulombic forces to the chemical potential of the ionic solutes, which are responsible for such interactions, can be found in many books, such as Debye (1954) and Robinson and Stokes (1968).

4.2.1 The mean stoichiometric activity coefficient of a binary electrolyte

Here, it is important to recall the output of the Debye–Hückel method, which is conveniently summarized by the following relationship (Helgeson and Kirkham, 1974b), usually known as Debye–Hückel function:

$$\log \gamma_{\pm} = \frac{-A_{\gamma} \cdot |Z_{+} \cdot Z_{-}| \cdot \sqrt{\bar{I}}}{1 + \mathring{a} \cdot B_{\gamma} \cdot \sqrt{\bar{I}}} \tag{4-2}$$

where $\gamma_{\pm}$ is the mean *stoichiometric* activity coefficient of a completely dissociated binary electrolyte (the term *stoichiometric* indicates that ion dissociation is complete) consisting of cations and anions with charges Z_{+} and Z_{-}, respectively, $\bar{I}$ the true ionic strength, $\mathring{a}$ the ion-size parameter and A_{γ} and B_{γ} are given by

$$A_{\gamma} = \left(2000\pi N_{A}\rho\right)^{1/2} \cdot \frac{e^{3}}{\ln(10) \cdot \left(4\pi\varepsilon_{o}\varepsilon k_{B}T\right)^{3/2}} = \frac{1.82483 \cdot 10^{6} \cdot \rho^{1/2}}{\left(\varepsilon T\right)^{3/2}} \tag{4-3}$$

and

$$B_{\gamma} = \left(\frac{N_{\mathrm{A}}\rho \cdot e^{2}}{5\varepsilon_{\mathrm{o}}\varepsilon k_{\mathrm{B}}T}\right)^{1/2} = \frac{5.02916 \cdot 10^{9} \cdot \rho^{1/2}}{\left(\varepsilon T\right)^{1/2}}. \tag{4-4}$$

In these equations, N_A is the Avogadro's number ($N_A = 6.02214 \times 10^{23}$ mol^{-1}), k_B the Boltzmann's constant ($k_B = R/N_A = 1.38066 \times 10^{-23}$ J K^{-1}), e the absolute electronic charge ($e = 1.60218 \times 10^{-19}$ C), ε_o the permittivity of free space ($\varepsilon_o = 8.85419 \times 10^{-12}$ C^2 N^{-1} m^{-2}), ρ the density (in units of g cm^{-3}), ε the dielectric constant or relative permittivity of water and T the absolute temperature. The dielectric constant of water measures the effect of water in decreasing the force F acting between dissolved ionic solutes. For two particles of opposite charge situated at distance r, F is as follows:

$$F = \frac{Z_+ \cdot Z_-}{\varepsilon \cdot r^2}. \tag{4-5}$$

The dielectric constant is a dimensionless number, since it is obtained by determining the capacitance of water against that of a vacuum.

The true ionic strength of the aqueous solution is related to the molal concentrations of dissolved ionic solutes, obtained through speciation calculation, by the equation:

$$\bar{I} = \frac{1}{2}\sum_i m_i \cdot Z_i^2. \tag{4-6}$$

The overbar on the $\bar{I}$ symbol distinguishes the true ionic strength from the stoichiometric ionic strength I. The latter is computed assuming complete ion dissociation.

The parameters A_γ and B_γ at different temperatures and pressures are reported by Helgeson and Kirkham (1974b, pp. 1202, 1256 and 1257), whereas the electrostatic and thermodynamic properties of water, including ε and ρ, are given by Helgeson and Kirkham (1974a). At 298.15 K (25°C), with $\varepsilon = 78.47$ and $\rho = 0.9971$ g cm^{-3}, A_γ and B_γ result to be 0.5092 kg$^{1/2}$ mol$^{-1/2}$ and 0.3283×10^8 kg$^{1/2}$ mol$^{-1/2}$ cm^{-1}, respectively.

The Debye–Hückel function (4-2) can be applied up to ionic strengths of ~0.1 mol kg^{-1}. Below ionic strengths of ~0.01 mol kg^{-1}, the denominator of equation (4-2) is close to unity and the equation reduces to:

$$\log \gamma_\pm = -A_\gamma \cdot |Z_+ \cdot Z_-| \cdot \sqrt{\bar{I}}, \tag{4-7}$$

which is the so-called Debye–Hückel limiting law. Since this equation implies a linear relation between the logarithm of the mean activity coefficient and the square root of the ionic strength, it can be used to extrapolate the experimental measurements of log $\gamma_\pm$ against $\sqrt{\bar{I}}$, back to $\bar{I} = 0$.

Activity coefficients computed by means of equation (4-2) decrease continuously with increasing ionic strength, whereas measured $\gamma_\pm$ decrease with increasing ionic strength, attain a minimum at $\bar{I}$ of some tenths mol kg^{-1} to some mol kg^{-1} and increase afterwards at higher $\bar{I}$. Extension of equation (4-2) to high ionic strengths at low temperatures requires the introduction of additional terms in an ascending power function of $\bar{I}$, such as proposed by Lietzke and Stoughton (1962). The simplest extension, suggested by Hückel (1925), is as follows:

$$\log \gamma_\pm = \frac{-A_\gamma \cdot |Z_+ \cdot Z_-| \cdot \sqrt{\bar{I}}}{1 + \mathring{a} \cdot B_\gamma \cdot \sqrt{\bar{I}}} + \overset{\circ}{B} \cdot \bar{I}, \tag{4-8}$$

where $\overset{o}{B}$ (B-dot) is an empirical parameter typical of each electrolyte. A possible way to obtain values of the B-dot parameter is by defining the following deviation function:

$$\overset{o}{B} = \frac{1}{\bar{I}} \cdot \left[\log \gamma_{\pm} + \frac{A_{\gamma} \cdot |Z_{+} \cdot Z_{-}| \cdot \sqrt{\bar{I}}}{1 + \mathring{a} \cdot B_{\gamma} \cdot \sqrt{\bar{I}}} \right], \tag{4-9}$$

in which the experimental values of $\gamma_{\pm}$ are plugged in. This approach was suggested by Helgeson (1969), based on early researches by Scatchard (1936) and others. More recently, Helgeson et al. (1981) have split the B-dot parameter into a solvation parameter derived from the corrected Born equation and a second parameter accounting for short-range interactions between ions of opposite charges (see equation (4-14) and related discussion). In this way, values of $\gamma_{\pm}$ can be computed up to ionic strengths of 5 mol kg^{-1} for NaCl and KCl at 25°C, 1 bar (see Table 5 of Helgeson et al., 1981).

The shape of the functions considered so far, i.e., the Debye–Hückel function (4-2), the Debye–Hückel limiting law (4-7), the B-dot deviation function (4-9) and the extended Debye–Hückel function (4-8) for NaCl at 25°C, 1 bar are reported in Figs. 4.1a, b. These curves have been computed considering an ion-size $\mathring{a}$ of 3.72×10^{-8} cm (Helgeson et al., 1981), a B-dot parameter of 0.041 kg mol^{-1} (Helgeson, 1969) and other parameters as specified above. In Fig. 4.1a the true ionic strength, in logarithmic scale, is reported on the *x*-axis, whereas in Fig. 4.1b the square root of the true ionic strength, in linear scale, was preferred, as it appears in many of the considered functions. Note that, due to this choice, the Debye–Hückel limiting law becomes linear in Fig. 4.1b. Both figures show that the Debye–Hückel function does not turn upwards at high values of $\bar{I}$, and that this behaviour is suitably reproduced through addition of the B-dot term.

4.2.2 The activity coefficient of individual ions

A brief digression is needed here to underscore that it is impossible to determine experimentally the activity coefficients of individual ions, γ_j, despite some authors' claim that they have been able to do so. For instance, Vera and co-workers have published several papers (see references in Malatesta, 2005) on the determination of γ_j starting from the measurement of the electromotive force of potentiometric cells with liquid junctions and based on the assumption that liquid junction potentials E_L may be computed with enough precision. However, Malatesta (2005) has shown that the E_Ls are a function of the γ_j, i.e., of the unknown parameters to be determined or, in other words, that a vicious loop is present in the approach of Vera and co-workers.

Actually, this problem dates back at least to the beginning of the 20th century, when it was tackled by scientists of great renown, such as Lewis, Randall and Harned. For instance, Lewis and Randall (1921) wrote that, "*it would be of much theoretical interest if we could determine the actual activity of an ion in a solution of any concentration. This indeed might be accomplished if we had any general method of calculating the potential at a liquid junction. Such an attempt to estimate individual ion activities even in very concentrated solutions*

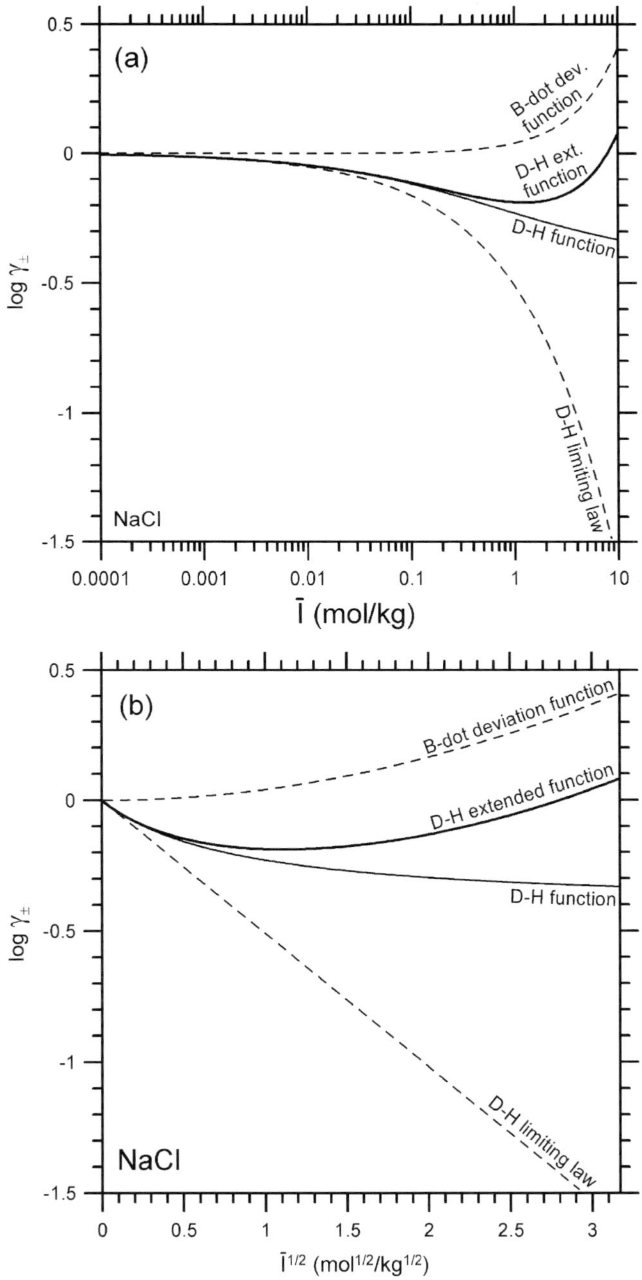

Figure 4.1. Values of log $\gamma_\pm$ (mean stoichiometric activity coefficient) of NaCl at 25°C, 1 bar computed by means of the Debye–Hückel function (4-2), the Debye–Hückel limiting law (4-7) and the extended Debye–Hückel function (4-8) against (a) the true ionic strength and (b) the square root of the true ionic strength. The B-dot deviation function (4-9) is also reported.

has been made by Harned [1920], but he was obliged to make certain assumptions regarding the elimination of liquid potentials which may be very far from valid.

Supposing that we consider the cell, H_2|dil. HCl|conc. HCl|H_2, the electromotive force may be expressed by the equation

$$E = \frac{RT}{F} \cdot \ln\left(\frac{a_+}{a'_+}\right) + E_L$$

where the activity of hydrogen ion in the dilute solution is a_+ and in the concentrated a'_+, while E_L is the potential at the liquid junction. Now, if a_+ were very small, so that it could be taken as equal to the molality, we could determine directly a'_+ if we knew E_L. This, unfortunately, we do not know. . . . Efforts to eliminate such a liquid potential by interposing, between the two solutions, some concentrated salt solutions, like potassium chloride, have doubtless served in many cases to reduce such a potential to a few millivolts. But by the method so far proposed the elimination almost certainly has never been complete, and the uncertainties increase with the concentration of the solutions, between which the liquid potential is to be estimated."

After treating the case of a liquid junction between two different solutions of the same concentration, Lewis and Randall (1921) wrote that, "*at the present time we must conclude that the determination of the absolute activity of the ions is an interesting problem, but one which is yet unsolved.*" Why Lewis and Randall did not write a very clear sentence on the impossibility to determine experimentally the activity coefficients of individual ions? I do not know, but that missing sentence would have certainly discouraged attempts to determine this sort of philosopher's stone.

Again, only the mean stoichiometric activity coefficient, $\gamma_\pm$, of a binary electrolyte can be measured experimentally. The $\gamma_\pm$, however, cannot be applied as such to a single ion in a complex aqueous solution. This is unfortunate, since activity coefficients of individual ions, γ_j, and their activities $a_j = m_j \cdot \gamma_j$ are the properties required in thermodynamic calculations of chemical equilibrium (see Chapter 2). How to obtain the γ_j from the $\gamma_\pm$?

First, for a binary electrolyte fully dissociating in ν_+ moles of cations and ν_- moles of anions, the mean stoichiometric activity coefficient is related to the activity coefficients of the single cation, γ_+, and of the single anion, γ_-, by the following equation:

$$\gamma_\pm = \left(\gamma_+^{\nu+} \cdot \gamma_-^{\nu-}\right)^{1/\nu}, \tag{4-10}$$

where $\nu = \nu_+ + \nu_-$. Second, it is assumed that $\gamma_{K^+} = \gamma_{Cl^-} = \gamma_{\pm,KCl}$. This assumption was proposed by MacInnes (1919) and is known as MacInnes' convention. It appears reasonable since at each concentration the two ions K^+ and Cl^- "*have nearly the same weight and mobility*" (MacInnes, 1919). Lewis and Randall (1921) wrote, . . . "*this convention which, at least in the concentrated solutions, must be regarded as purely arbitrary, we may adopt as well as any other for the sake of obtaining a table of individual activity coefficients for the ions.*" Although this observation is correct, the MacInnes' convention is exactly what we need together with equation (4-10) to obtain the γ_j from the $\gamma_\pm$. For instance, referring to a monovalent chloride MCl, whose mean stoichiometric activity coefficient, $\gamma_{\pm,\,MCl}$, has been determined experimentally, it is possible to compute the activity coefficient of the cation M^+, γ_{M^+}, by using the following equation:

$$\gamma_{\pm,MCl} = \left(\gamma_{M^+} \cdot \gamma_{Cl^-}\right)^{1/2} = \left(\gamma_{M^+} \cdot \gamma_{\pm,KCl}\right)^{1/2}, \tag{4-11}$$

which is solved with respect to γ_{M^+}. This technique, known as mean salt method, can be easily extended to anions and salts of different stoichiometry (Lewis and Randall, 1921, 1923).

Therefore, equation (4-8) can be adapted to individual ions in the following form:

$$\log \gamma_j = \frac{-A_\gamma \cdot Z_j^2 \cdot \sqrt{\bar{I}}}{1 + \mathring{a}_j \cdot B_\gamma \cdot \sqrt{\bar{I}}} + \overset{o}{B} \cdot \bar{I}, \tag{4-12}$$

where Z_j and $\mathring{a}_j$ are the charge and the ion-size parameter of the considered ion, respectively, whereas A_γ, B_γ, the B-dot parameter (and obviously $\bar{I}$) are the same as above, as they refer either to the dominant solute (B-dot) or to the solvent (A_γ, B_γ).

Sometimes the empirical Davies equation:

$$\log \gamma_j = -A_\gamma \cdot Z_j^2 \cdot \left(\frac{\sqrt{I}}{1+\sqrt{I}}\right) - 0.3 \cdot I \tag{4-13}$$

is preferred to equation (4-12). Because of the lack of the ion-size parameter, equation (4-13) gives the same γ_j for all ions of the same charge, at fixed I, and is less accurate than equation (4-12) at low values of I. However, it can be used up to ionic strengths of 0.1–0.7 mol kg^{-1}.

According to Helgeson et al. (1981) the individual ion activity coe fficient of the jth species is given by

$$\log \gamma_j = \frac{-A_\gamma \cdot Z_j^2 \cdot \sqrt{\bar{I}}}{1 + \mathring{a}_j \cdot B_\gamma \cdot \sqrt{\bar{I}}} + \Gamma_\gamma + \omega_j^{abs} \cdot \sum_k b_k \cdot y_k \cdot \bar{I} + \sum_l \frac{2 \cdot b_{jl} \cdot \bar{y}_l \cdot \bar{I}}{Z_l^2}. \tag{4-14}$$

The parameter ω_j^{abs} is the absolute Born coefficient of the jth species (see next section), which is defined as follows:

$$\omega_j^{abs} = \frac{N \cdot e^2 \cdot Z_j^2}{2 \cdot r_{e,j}} \tag{4-15}$$

where $r_{e,j}$ is the effective electrostatic radius of the considered species (see next section).

The terms y_k and $\bar{y}_l$ are the stoichiometric ionic strength fraction and the true ionic strength fraction of the subscripted species, respectively, which are defined as follows:

$$y_k = \frac{m_k \cdot Z_l^2}{2 \cdot I} \quad \text{and} \quad \bar{y}_l = \frac{m_l \cdot Z_l^2}{2 \cdot \bar{I}}. \tag{4-16}$$

The symbol b_k stands for the electrostatic solvation parameter for the kth electrolyte, whereas b_{jl} represents the short-range interaction parameter for the jth positively charged species interacting with the lth negatively charged species.

The term Γ_γ converts the rational activity coefficient into the molal scale through:

$$\Gamma_\gamma = -\log\left(1 + \frac{18.0153 \cdot m^*}{1000}\right) \tag{4-17}$$

where m^* is the sum of the molalities of all solute species and 18.0153 is the molecular weight of water.

Moreover, the $\mathring{a}_j$ parameter in the denominator of equation (4-14) is computed from the effective electrostatic radii of aqueous ions through the following relation:

$$\mathring{a}_j = 2 \cdot \sum_j^{\bar{j}} \frac{r_{e,j}}{\bar{j}} = \frac{\sum_k v_k \cdot 2 \cdot \sum_j \frac{v_{j,k} \cdot r_{e,j}}{v_k}}{\sum_k v_k} \tag{4-18}$$

Equation (4-14) can be considered as a modified version of the extended Debye–Hückel function. It embodies a solvation parameter derived from the Born equation and a second parameter taking into account short-range interactions. These substantial changes are based on the HKF model and revised HKF model, which is the subject of Section 4.3.

4.2.3 The activity coefficient of neutral dissolved species

The activity coefficients of neutral solutes, e.g., $CO_{2(aq)}$ and $SiO_{2(aq)}$, can be determined experimentally upon attainment of chemical equilibrium between the aqueous neutral solute of interest and the same pure substance, such as gaseous CO_2 or solid silica (e.g., quartz). By changing the composition of the aqueous solution, the activity coefficient of the neutral solute is determined under the conditions of interest.

For instance, we write again reaction (3-33):

$$CO_{2(aq)} \rightleftharpoons CO_{2(g)}, \tag{3-33}$$

whose thermodynamic equilibrium constant is as follows:

$$K_{CO_2} = \frac{a_{CO_{2(g)}}}{a_{CO_{2(aq)}}} = \frac{f_{CO_{2(g)}} / f^o_{CO_{2(g)}}}{m_{CO_{2(aq)}} \cdot \gamma_{CO_{2(aq)}}}. \tag{4-19}$$

Since the first dissociation constant of CO_2(aq) is less than 10^{-6} at any temperature between 0 and 300°C and saturation pressures, dissociation products (i.e. HCO_3^- and CO_3^{2-} ions) can be safely neglected if experimental conditions are suitably chosen. Since the standard state adopted for gases is unit fugacity of the hypothetical gas at 1 bar and any temperature, $f^o_{CO_2(g)}$ is equal to unity and

$$\gamma_{CO_{2(aq)}} = \frac{f_{CO_{2(aq)}}}{m_{CO_{2(aq)}} \cdot K_{CO_2}}. \tag{4-20}$$

Equation (4-20) indicates that the activity coefficient of $CO_{2(aq)}$ can be obtained by determining the fugacity of gaseous CO_2 and the molality of dissolved CO_2, provided that the thermodynamic equilibrium constant K_{CO_2} is accurately known. Also, the activity coefficient of $CO_{2(aq)}$ in a salt solution, $\gamma_{CO_2(aq),sol}$, is equal to the ratio between the Henry's law coefficients (which are not thermodynamic constants!) in the salt solution ($K_{H,CO_2,sol}$) and in pure water ($K_{H,CO_2,water}$):

$$\gamma_{CO_{2(aq)},sol} = \frac{K_{H,CO_2,sol}}{K_{H,CO_2,water}}. \tag{4-21}$$

This is the approach followed by Helgeson (1969) to compute the activity coefficient of $CO_{2(aq)}$ in NaCl solutions of different molalities based on the Henry's law coefficients of Ellis and Golding (1963).

In general, the dependence of the activity coefficient of a neutral solute N on the true ionic strength of a single electrolyte solution, $\bar{I}$, is expressed by the following equation (Helgeson et al., 1981):

$$\log\gamma_N = b_{\gamma,N} \cdot \bar{I} + \Gamma_\gamma, \tag{4-22}$$

where $b_{\gamma,N}$ is the short-range interaction parameter for the Nth neutral solute and Γ_γ the mole fraction to molality conversion factor (see equation (4-17)). Equation (4-22) is equivalent to the Setchénow equation (Setchénow, 1892), which has been used by many authors to describe the activity coefficients of neutral solutes in aqueous electrolyte solutions (see Oelkers and Helgeson, 1991 and references therein). Since $b_{\gamma,N} > 0$, γ_N increases with I. This behaviour is consistent with the so-called *salting-out effect*.

4.2.4 The activity of water

The activity of the solvent, a_{H_2O}, is related to the osmotic coefficient of the electrolyte solution, ϕ, by (Helgeson et al., 1981)

$$\ln a_{H_2O} = -\left(\frac{18.0153 \cdot m^*}{1000}\right)\phi \tag{4-23}$$

where, as already said, m^* is the sum of the molalities of all solute species.

The activity of the solvent can also be related to the molality-based excess Gibbs free energy of the solution, $G^{ex,m}$, by

$$\ln a_{H_2O} = -\frac{18.0153\, m^*}{1000} + \frac{1}{RT}\left(\frac{\partial G^{ex,m}}{\partial n_{H_2O}}\right). \tag{4-24}$$

If the activity coefficient of individual ions is computed by means of the Davies equation (4-12), then water activity is given by (Wolery, 1992)

$$\ln a_{\mathrm{H_2O}} = \frac{18.0153}{1000}\left(-\frac{m^*}{\ln(10)} + \frac{2}{3}A_\gamma I^{3/2}\,\sigma\left(I^{1/2}\right) - 2(0.2)A_\gamma I^2\right). \tag{4-25}$$

If the activity coefficient of individual ions is calculated by using the B-dot equation, then the activity of water is approximated by the following relation, which is quasi-consistent with the B-dot equation (Wolery, 1992):

$$\ln a_{\mathrm{H_2O}} = \frac{18.015}{1000}\left(-\frac{m^*}{\ln(10)} + \frac{2}{3}A_\gamma \bar{I}^{3/2}\,\sigma\left(\mathring{a}_j \cdot B_\gamma \cdot \bar{I}^{1/2}\right) - \overset{o}{B}\bar{I}^{2}\right). \tag{4-26}$$

In equations (4-25) and (4-26) the σ-parameter is given by

$$\sigma(x) = \frac{3}{x^3}\left[1 + x - \frac{1}{1+x} - 2\cdot\ln\left(1+x\right)\right] \tag{4-27}$$

and, again, $x = I^{1/2}$ in equation (4-25) but $x = \mathring{a}_j \cdot B_\gamma \cdot \bar{I}^{1/2}$ in equation (4-26).

4.3. The HKF model for aqueous electrolytes

A theoretical model for predicting the thermodynamic properties of electrolyte solutions at elevated temperatures and pressures was elaborated by Harold C. Helgeson and co-workers between 1974 and 1981 (Helgeson and Kirkham, 1974a,b; Helgeson and Kirkham, 1976; Helgeson et al., 1981) and later revised by Tanger and Helgeson (1988) and Shock et al. (1992). This model is frequently called HKF model or revised HKF model after the authors of the 1981 paper.

The Born (1920) equation, as corrected by Bjerrum (1929), represents a good starting point for presenting the HKF model. This equation describes the absolute standard molal Gibbs free energy of solvation of the jth ion:

$$\Delta\bar{G}^{o,\mathrm{abs}}_{s,j} = \frac{N\cdot e^2\cdot Z_j^2}{2\cdot r_{e,j}}\cdot\left(\frac{1}{\varepsilon}-1\right) = \omega_j^{\mathrm{abs}}\cdot\left(\frac{1}{\varepsilon}-1\right). \tag{4-28}$$

Although all the symbols given above have been defined already, it is worth to underscore that equation (4-28) defines the Gibbs free energy change associated with the displacement of an ion j of charge $e \times Z_j$ and effective electrostatic radius $r_{e,j}$ from a vacuum to a solvent of dielectric constant ε. When applying this equation we do not have to worry for the local collapse of the solvent structure (electrostriction collapse) as these effects cancel out in the two-way transfer process, i.e., to and from a vacuum. Moreover, it is assumed that effects related to void space in the solvation shell, dielectric saturation and the finite size of the solvent dipoles are incorporated in the $r_{e,j}$ term.

Helgeson et al. (1981) related $r_{e,j}$ to the crystallographic ionic radius ($r_{x,j}$) through the following simple equation:

$$r_{e,j} = r_{x,j} + k_{Z_\pm} \cdot |Z_j|, \tag{4-29}$$

where $k_{Z_\pm}$ is equal to 0.94 Å for cations and 0 Å for anions. Therefore, the $r_{e,j}$ term was considered to be independent of *P* and *T* in the HKF model. In the revised HKF model, instead, the previous equation is used to compute the $r_{e,j}$ term at the reference temperature and pressure, whereas the effective electrostatic radius of the *j*th ion at any *P*,*T* is as follows:

$$r_{e,j,P,T} = r_{e,j,Pr,Tr} + |Z_j| \cdot g, \tag{4-30}$$

where *g* is a solvent function dependent on *T* and *P*, derived through regression of heat capacity and volume data for aqueous NaCl (Tanger and Helgeson, 1988; Shock et al., 1992).

To complete the HKF model, the other effects brought about by addition of an ion to water (e.g. the non-electrical part of the electrostriction volume loss) are taken into account through addition of an empirical term with several adjustable parameters.

Two conventions are needed to derive thermodynamic properties. First, the conventional standard molal properties of the hydrogen ion are assumed to be zero, at any pressure and temperature; the corresponding properties of anions are equivalent to those of the associated acid electrolyte. For instance, this assumption allows one to obtain the conventional electrostatic Born coefficient as follows:

$$\omega_j = \omega_j^{\text{abs}} - Z_j \cdot \omega_{\text{H}^+}^{\text{abs}} = \frac{N \cdot e^2 \cdot Z_j^2}{2 \cdot r_{e,j}} - Z_j \cdot \left(\frac{N \cdot e^2 \cdot Z_{\text{H}^+}^2}{2 \cdot r_{e,\text{H}^+}} \right). \tag{4-31}$$

Then, the conventional standard partial molal Gibbs free energy of solvation of the *j*th ion is

$$\Delta \overline{G}_{s,j}^{\text{o}} = \omega_j \cdot \left(\frac{1}{\varepsilon} - 1 \right). \tag{4-32}$$

The second convention is the additivity rule relating any partial molal property, $\overline{\Xi}$, of individual ions to those of the corresponding electrolyte:

$$\overline{\Xi}_k^{\text{o}} = \sum_j \nu_{j,k} \cdot \overline{\Xi}_j^{\text{o}}, \tag{4-33}$$

where $\nu_{j,k}$ represents the stoichiometric number of moles of the *j*th ion in one mole of the *k*th electrolyte. Based on this rule, the Born function can be written also for the *k*th aqueous electrolyte as follows:

$$\Delta \overline{G}_{s,k}^{\text{o}} = \omega_k \cdot \left(\frac{1}{\varepsilon} - 1 \right), \tag{4-34}$$

where

$$\omega_k = \sum_j \nu_{j,k} \cdot \omega_j. \tag{4-35}$$

Then, the conventional entropy and volume of solvation of either the jth ion or the kth electrolyte are obtained through differentiation of either equation (4-32) or equation (4-34). To save space, subscripts j and k are omitted below:

$$\Delta\overline{S}_s^{\text{o}} = -\left(\frac{\partial \Delta\overline{G}_s^{\text{o}}}{\partial T}\right)_P = \omega \cdot Y - \left(\frac{1}{\varepsilon} - 1\right) \cdot \left(\frac{\partial \omega}{\partial T}\right)_P \tag{4-36}$$

and

$$\Delta\overline{V}_s^{\text{o}} = \left(\frac{\partial \Delta\overline{G}_s^{\text{o}}}{\partial P}\right)_T = -\omega \cdot Q + \left(\frac{1}{\varepsilon} - 1\right) \cdot \left(\frac{\partial \omega}{\partial P}\right)_T. \tag{4-37}$$

The standard heat capacity at constant pressure of solvation is then obtained as follows:

$$\Delta\overline{C}_{P,s}^{\text{o}} = T \cdot \left(\frac{\partial \Delta\overline{S}_s^{\text{o}}}{\partial T}\right)_P = \omega \cdot T \cdot X + 2 \cdot T \cdot Y \cdot \left(\frac{\partial \omega}{\partial T}\right)_P - T \cdot \left(\frac{1}{\varepsilon} - 1\right) \cdot \left(\frac{\partial^2 \omega}{\partial T^2}\right)_P. \tag{4-38}$$

Equations (4-36)–(4-38) requires values for the negative reciprocal of the dielectric constant of water, $Z = -1/\varepsilon$, and its partial derivatives, which are given by the following Born functions (Helgeson and Kirkham, 1974a; Helgeson et al., 1981):

$$Q = \left(\frac{\partial Z}{\partial P}\right)_T = \frac{1}{\varepsilon^2} \cdot \left(\frac{\partial \varepsilon}{\partial P}\right)_T \tag{4-39}$$

$$Y = \left(\frac{\partial Z}{\partial T}\right)_P = \frac{1}{\varepsilon^2} \cdot \left(\frac{\partial \varepsilon}{\partial T}\right)_P \tag{4-40}$$

$$X = \left(\frac{\partial Y}{\partial T}\right)_P = \left(\frac{\partial^2 Z}{\partial T^2}\right)_P = \frac{1}{\varepsilon^2} \cdot \left(\frac{\partial^2 \varepsilon}{\partial T^2}\right)_P - 2 \cdot \varepsilon \cdot Y^2. \tag{4-41}$$

Values of $\Delta\overline{V}_s^{\text{o}}$ and $\Delta\overline{C}_{P,s}^{\text{o}}$ computed by means of equations (4-37) and (4-38) were compared with experimental data of $\Delta\overline{V}^{\text{o}}$ and $\Delta\overline{C}_P^{\text{o}}$. Deviations between the two sets of data were fitted with the following functions:

$$\Delta\overline{V}_{ns}^{\text{o}} = a_1 + \frac{a_2}{\Psi + P} + \frac{a_3}{T - \Theta} + \frac{a_4}{(\Psi + P) \cdot (T - \Theta)} \tag{4-42}$$

and

$$\Delta\overline{C}^{\mathrm{o}}_{P,ns} = c_1 + \frac{c_2}{(T-\Theta)^2}, \tag{4-43}$$

where coefficients a_1, a_2, a_3, a_4, c_1 and c_2 have specific values for each solute. Equations (4-42) and (4-43) describe the contributions to the conventional standard partial molal volume and to the conventional standard partial molal heat capacity of the *j*th ion and of the *k*th electrolyte, which are evidently due to processes other than solvation. In the revised HKF model, Θ is equal to 228 K (−45°C), which is a sort of singular temperature for water, as suggested by investigations of super-cooled water carried out by Austen Angell and his collaborators (Speedy and Angell, 1976; Angell, 1982; Speedy, 1983), although things are not totally clear yet (Debenedetti and Stanley, 2003). In spite of this, many properties of water approach $\pm\infty$ when $T \rightarrow 228$ K. The Ψ parameter takes the value of 2,600 bar.

Based on equations (4-37) and (4-42), the conventional standard partial molal volume of the *j*th ion and of the *k*th electrolyte is given by

$$\overline{V}^{\mathrm{o}} = \Delta\overline{V}^{\mathrm{o}}_{s} + \Delta\overline{V}^{\mathrm{o}}_{ns} = -\omega\cdot Q + \left(\frac{1}{\varepsilon}-1\right)\cdot\left(\frac{\partial\omega}{\partial P}\right)_T + a_1 + \frac{a_2}{\Psi+P} + \frac{a_3}{T-\Theta} + \frac{a_4}{(\Psi+P)\cdot(T-\Theta)} \tag{4-44}$$

Taking into account equations (4-38) and (4-43), the conventional standard partial molal heat capacity of the *j*th ion and of the *k*th electrolyte is expressed as a function of temperature (but not of pressure !) through the following relation:

$$\overline{C}^{\mathrm{o}}_{P} = \Delta\overline{C}^{\mathrm{o}}_{P,s} + \Delta\overline{C}^{\mathrm{o}}_{P,ns} = \omega TX + 2TY\left(\frac{\partial\omega}{\partial T}\right)_P - T\left(\frac{1}{\varepsilon}-1\right)\cdot\left(\frac{\partial^2\omega}{\partial T^2}\right)_P + c_1 + \frac{c_2}{(T-\Theta)^2}. \tag{4-45}$$

To determine the pressure dependence of conventional standard partial molal heat capacities we have to consider the following expression:

$$\overline{C}^{\mathrm{o}}_{P,P} = \overline{C}^{\mathrm{o}}_{P,Pr} + \int_{Pr}^{P} -T\left(\frac{\partial^2\overline{V}^{\mathrm{o}}}{\partial T^2}\right)\mathrm{d}P, \tag{4-46}$$

which is integrated to give:

$$\overline{C}^{\mathrm{o}}_{P,P} = \overline{C}^{\mathrm{o}}_{P,Pr} - \frac{2T}{(T-\Theta)^3}\cdot\left[a_3\cdot(P-P_r) + a_4\cdot\ln\left(\frac{\Psi+P}{\Psi+P_r}\right)\right]. \tag{4-47}$$

Taking into account relations (4-38), (4-43), (4-45) and (4-47), the conventional standard partial molal heat capacity of the *j*th ion and of the *k*th electrolyte is expressed as a function of both temperature and pressure through the following relation:

$$\overline{C}^{\mathrm{o}}_{P,P,T} = \Delta\overline{C}^{\mathrm{o}}_{P,s} + \Delta\overline{C}^{\mathrm{o}}_{P,ns} + \frac{2T}{(T-\Theta)^3}\cdot\left[a_3\cdot(P-P_r)+a_4\cdot\ln\left(\frac{\Psi+P}{\Psi+P_r}\right)\right]$$

$$= \omega TX + 2TY\left(\frac{\partial\omega}{\partial T}\right)_P - T\left(\frac{1}{\varepsilon}-1\right)\left(\frac{\partial^2\omega}{\partial T^2}\right)_P \tag{4-48}$$

$$+c_1+\frac{c_2}{(T-\Theta)^2}-\frac{2T}{(T-\Theta)^3}\left[a_3(P-P_r)+a_4\ln\left(\frac{\Psi+P}{\Psi+P_r}\right)\right].$$

Starting from equations (4-44) and (4-45), expressing the pressure dependence of $\overline{V}^{\mathrm{o}}$ and the temperature dependence of $\overline{C}^{\mathrm{o}}_P$, respectively, the other conventional standard partial molal properties of interest, namely the entropy, enthalpy and Gibbs free energy of the *j*th ion and *k*th electrolyte, are obtained through integration. This exercise gives the following results:

$$\overline{S}^{\mathrm{o}}_{P,T} = \overline{S}^{\mathrm{o}}_{Pr,Tr} + \int_{Tr}^{T}\frac{\overline{C}^{\mathrm{o}}_P}{T}\mathrm{d}T - \int_{Pr}^{P}\left[\left(\frac{\partial\overline{V}^{\mathrm{o}}}{\partial T}\right)_P\right]_T\mathrm{d}P$$

$$= \overline{S}^{\mathrm{o}}_{Pr,Tr} + c_1\cdot\ln\frac{T}{T_r} - \frac{c_2}{\Theta}\cdot\left[\left(\frac{1}{T-\Theta}\right)-\left(\frac{1}{T_r-\Theta}\right)+\frac{1}{\Theta}\cdot\ln\left(\frac{T_r\cdot(T-\Theta)}{T\cdot(T_r-\Theta)}\right)\right] \tag{4-49}$$

$$+\left(\frac{1}{T-\Theta}\right)^2\cdot\left[a_3\cdot(P-P_r)+a_4\cdot\ln\left(\frac{\Psi+P}{\Psi+P_r}\right)\right]+\omega Y-\left(\frac{1}{\varepsilon}-1\right)\left(\frac{\partial\omega}{\partial T}\right)_P-\omega_{Pr,Tr}Y_{Pr,Tr}$$

$$\overline{H}^{\mathrm{o}}_{P,T} = \overline{H}^{\mathrm{o}}_{Pr,Tr} + \int_{Tr}^{T}\overline{C}^{\mathrm{o}}_P\,\mathrm{d}T + \int_{Pr}^{P}\left[\overline{V}^{\mathrm{o}}-T\left(\frac{\partial\overline{V}^{\mathrm{o}}}{\partial T}\right)_P\right]_T\mathrm{d}P$$

$$= \overline{H}^{\mathrm{o}}_{Pr,Tr} + c_1\cdot(T-T_r) - c_2\left[\left(\frac{1}{T-\Theta}\right)-\left(\frac{1}{T_r-\Theta}\right)\right]+a_1(P-P_r)+a_2\cdot\ln\left(\frac{\Psi+P}{\Psi+P_r}\right)$$

$$+\left(\frac{2T-\Theta}{(T-\Theta)^2}\right)\left[a_3(P-P_r)+a_4\cdot\ln\left(\frac{\Psi+P}{\Psi+P_r}\right)\right]+\omega\left(\frac{1}{\varepsilon}-1\right)+\omega TY \tag{4-50}$$

$$-T\left(\frac{1}{\varepsilon}-1\right)\left(\frac{\partial\omega}{\partial T}\right)_P-\omega_{Pr,Tr}\left(\frac{1}{\varepsilon_{Pr,Tr}}-1\right)-\omega_{Pr,Tr}T_rY_{Pr,Tr}$$

$$\overline{G}^{o}_{P,T} = \overline{G}^{o}_{Pr,Tr} - \overline{S}^{o}_{Pr,Tr}\left(T - T_r\right) + \int_{Tr}^{T} \overline{C}^{o}_{P}\,\mathrm{d}T - \int_{Tr}^{T} \frac{\overline{C}^{o}_{P}}{T}\,\mathrm{d}T + \int_{Pr}^{P} \overline{V}^{o}\,\mathrm{d}P =$$

$$= \overline{G}^{o}_{Pr,Tr} - \overline{S}^{o}_{Pr,Tr}(T - T_r) - c_1\left[T \cdot \ln\left(\frac{T}{T_r}\right) - T + T_r\right] + a_1\left(P - P_r\right) + a_2 \cdot \ln\left(\frac{\Psi + P}{\Psi + P_r}\right)$$

$$-c_2\left\{\left[\left(\frac{1}{T - \Theta}\right) - \left(\frac{1}{T_r - \Theta}\right)\right] \cdot \left(\frac{\Theta - T}{\Theta}\right) - \frac{T}{\Theta^2} \cdot \ln\left(\frac{T_r \cdot (T - \Theta)}{T \cdot (T_r - \Theta)}\right)\right\} \quad (4\text{-}51)$$

$$+\left(\frac{1}{T - \Theta}\right) \cdot \left[a_3\left(P - P_r\right) + a_4 \cdot \ln\left(\frac{\Psi + P}{\Psi + P_r}\right)\right] + \omega\left(\frac{1}{\varepsilon} - 1\right) - \omega_{Pr,Tr}\left(\frac{1}{\varepsilon_{Pr,Tr}} - 1\right)$$

$$-\omega_{Pr,Tr}\left(T - T_r\right) Y_{Pr,Tr}$$

Since the HKF model is specifically addressed to work at elevated temperatures and pressures, a central role is played by the $\overline{C}^{o}_{P}$ and $\overline{V}^{o}$ of ions and electrolytes. Both the conventional standard partial molal properties have a characteristic inverted U-shape when plotted against temperature, as shown in Fig. 4.2. It seems reasonable to attribute the shape of these functions to the existence of two singular temperatures for water: one is the critical temperature T_c (374.14°C), whereas the other one is situated at −45°C (228 K), as already recalled above.

According to the HKF model there are two contributions to $\overline{C}^{o}_{P}$ and $\overline{V}^{o}$ of solutes: one comes from solvation and is described by the Born–Bierrum theory, whereas the other is the non-solvation contribution and is accounted for empirically. In the revised HKF model the non-solvation contribution to $\overline{C}^{o}_{P}$ and $\overline{V}^{o}$ has been forced to dominate at high temperatures, whereas the solvation contribution has been constrained to prevail at low temperatures (and attains $-\infty$ when $T \rightarrow T_c$). Summation of the two contributions determines the expected maximum in the $\overline{C}^{o}_{P} - T$ and $\overline{V}^{o} - T$ functions.

Due to the shape of these two functions, the thermodynamic properties of ions and electrolytes approach $\pm\infty$ when temperature gets close to the two singular values (374.14°C and −45°C) and have either a maximum or a minimum or an inflection point between these two limiting temperatures.

4.4. The Pitzer model for aqueous electrolytes

A distinct theoretical model for aqueous electrolyte solutions was developed by Kenneth S. Pitzer and his co-workers, initially for solutions near room temperature (Pitzer, 1973, 1975; Pitzer and Kim, 1974). The Pitzer theory has been thoroughly reviewed by Pitzer (1979, 1987, 1992), to whom the reader is addressed for an exhaustive description of this topic. Only a brief summary is given here.

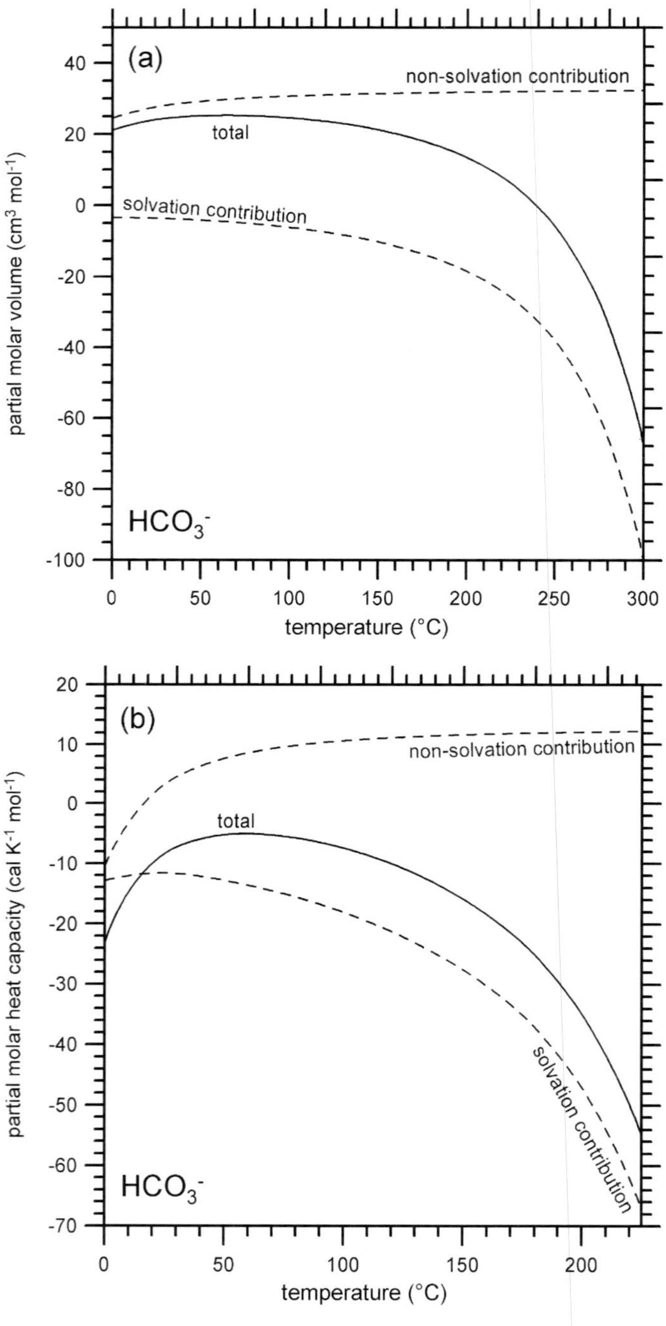

Figure 4.2. (a) Conventional standard partial molal volume and (b) heat capacity of bicarbonate ion according to the revised HKF model for aqueous electrolytes. The separate contributions of solvation and non-solvation processes are also shown. Data from Johnson et al. (1992).

4.4.1. The semi-empirical Pitzer's equations

The equations of Pitzer were derived from the following virial expansion (something similar to the virial equation of state for gases, see Section 3.3) of the excess Gibbs free energy, G^{ex}:

$$\frac{G^{ex}}{w_w RT} = f(I) + \sum_i \sum_j m_i \cdot m_j \cdot \lambda_{ij} + \sum_i \sum_j \sum_k m_i \cdot m_j \cdot m_k \cdot \mu_{ijk} + \cdots, \tag{4-52}$$

where w_w is the weight (in kg) of solvent water and m_i, m_j, ... are the molalities of the subscripted solutes. The λ_{ij} are second-order interaction (or virial) coefficients, and μ_{ijk} are third-order interaction (or virial) coefficients, representing the effects of short-range forces between two ions and three ions, respectively. The λ_{ij}'s depend on the ionic strength I, but the μ_{ijk}'s are independent of it. Both the λ_{ij}'s and the μ_{ijk}'s are symmetric, i.e., $\lambda_{ij} = \lambda_{ji}$ and $\mu_{ijk} = \mu_{ikj} = \mu_{jik}$.

Fourth or higher order interaction coefficients are required for extremely concentrated solutions only. For the interaction between ions and neutral molecules, these coefficients are generally assumed to be constant at a given temperature and independent of I.

The Debye–Hückel function $f(I)$ accounts for long-range electrostatic forces and depends not only on the ionic strength but also on temperature and solvent properties. The Debye–Hückel function $f(I)$ is somewhat different from the equations presented in Section 4.2, as it is

$$f(I) = -\left(\frac{4A_\phi I}{b}\right) \cdot \ln\left(1 + b\sqrt{I}\right). \tag{4-53}$$

The parameter A_ϕ is related to the parameter A_γ (see equation (4-3)) through:

$$A_\phi = \frac{\ln(10) \cdot A_\gamma}{3}. \tag{4-54}$$

At 25°C, $A_\gamma = 0.509$ kg$^{1/2}$ mol$^{-1/2}$ (see Section 4.2.1) and $A_\phi = 0.391$ kg$^{1/2}$ mol$^{-1/2}$. The parameter b is a constant equal to 1.2 kg$^{1/2}$ mol$^{-1/2}$ for all electrolytes. Therefore, differences in the ion-size parameter are not accounted for by the Debye–Hückel function of Pitzer.

Starting from equation (4-52), expressions for the activity coefficient of the ith species and the osmotic coefficient, ϕ, are obtained by taking the following partial derivatives:

$$\ln\gamma_i = \left[\frac{\left(\partial G^{ex}/w_w RT\right)}{\partial m_i}\right]_{w_w}$$

$$= \left(\frac{Z_i^2}{2}\right)\left(\frac{df(I)}{dI}\right) + 2 \cdot \sum_j \lambda_{ij} m_j + \left(\frac{Z_i^2}{2}\right) \cdot \sum_j \sum_k \left(\frac{d\lambda_{jk}}{dI}\right) \cdot m_j \cdot m_k \tag{4-55}$$

$$+ 3 \cdot \sum_j \sum_k \mu_{ijk} \cdot m_j \cdot m_k + \cdots.$$

and

$$\phi - 1 = -\frac{\left(\partial G^{\text{ex}} / \partial w_{\text{w}}\right)_{m_i}}{kT \cdot \sum_i m_i}$$

$$= \frac{1}{\sum_i m_i} \left\{ \left[I \cdot \left(\frac{\mathrm{d}f(I)}{\mathrm{d}I} \right) - f(I) \right] + \sum_i \sum_j \left[\lambda_{ij} + I \cdot \left(\frac{\mathrm{d}\lambda_{ij}}{\mathrm{d}I} \right) \right] \cdot m_i \cdot m_j \right. \tag{4-56}$$

$$\left. + 2 \cdot \sum_i \sum_j \sum_k \mu_{ijk} \cdot m_i \cdot m_j \cdot m_k + \cdots \right\}.$$

The activity of water is given by

$$\ln a_{\text{H}_2\text{O}} = -\frac{M_{\text{W,H}_2\text{O}} \sum m}{1000} - \frac{M_{\text{W,H}_2\text{O}}}{1000} \left[I \cdot \left(\frac{\mathrm{d}f(I)}{\mathrm{d}I} \right) - f(I) \right]$$

$$+ \frac{M_{\text{W,H}_2\text{O}}}{1000} \left\{ \sum_i \sum_j \left[\lambda_{ij} + I \cdot \left(\frac{\mathrm{d}\lambda_{ij}}{\mathrm{d}I} \right) \right] \cdot m_i \cdot m_j + 2 \cdot \sum_i \sum_j \sum_k \mu_{ijk} \cdot m_i \cdot m_j \cdot m_k \right\}. \tag{4-57}$$

Since the virial coefficients λ_{ij} and μ_{ijk} cannot be obtained from experimental data, it is convenient to rewrite equation (4-52) in terms of parameters that can be measured. For mixed electrolytes the appropriate relation is as follows:

$$\frac{G^{\text{ex}}}{w_{\text{w}} RT} = f(I) + \sum_c \sum_a m_c \cdot m_a \cdot \left[B_{ca} + C_{ca} \cdot \left(\sum_c m_c \cdot Z_c \right) \right]$$

$$+ \sum_c \sum_{<c'} m_c \cdot m_{c'} \left[2\Phi_{cc'} + \sum_a m_a \Psi_{cc'a} \right]$$

$$+ \sum_a \sum_{<a'} m_a \cdot m_{a'} \left[2\Phi_{aa'} + \sum_c m_c \Psi_{caa'} \right] + 2 \sum_n \sum_c m_n m_c \lambda_{nc} + 2 \sum_n \sum_a m_n m_a \lambda_{na} \tag{4-58}$$

$$+2 \sum_n \sum_{<n'} m_n m_{n'} \lambda_{nn'} + \sum_n m_n^2 \lambda_{nn} \cdots$$

in which the notations $\sum_c \sum_{<c'}$, $\sum_a \sum_{<a'}$ and $\sum_n \sum_{<n'}$ indicate that sums are carried out over pairs of different cations c, c', anions a, a', and neutral species n, n'. If neutral solutes are present in the aqueous solution, the terms including the virial coefficients λ_{ij} and μ_{ijk} from

equation (4-52) are maintained in equation (4-58), although only the second-order interaction coefficients are shown in (4-58) for simplicity.

New equations are also obtained for the osmotic coefficient, the individual activity coefficients of the cation $M_{\nu+}$ and the anion $X_{\nu-}$, and the mean ionic activity coefficient of the electrolyte $M_{\nu+}X_{\nu-}$:

$$\begin{aligned}\phi-1=\frac{2}{\sum_i m_i}\cdot\Bigg[&\frac{-A_\phi I^{3/2}}{1+bI^{1/2}}+\sum_c\sum_a m_c m_a\left(B_{ca}^\phi+C_{ca}\sum_i m_i|Z_i|\right)\\&+\sum_c\sum_{<c'} m_c m_{c'}\left(\Phi_{cc'}^\phi+\sum_a m_a\Psi_{cc'a}\right)\\&+\sum_a\sum_{<a'} m_a m_{a'}\left(\Phi_{aa'}^\phi+\sum_c m_c\Psi_{caa'}\right)+\sum_n\sum_c m_n m_c\lambda_{nc}+\sum_n\sum_a m_n m_a\lambda_{na}\\&+\sum_n\sum_{<n'} m_n m_{n'}\lambda_{nn'}+\frac{1}{2}\sum_n m_n^2\lambda_{nn}+\cdots\Bigg]\end{aligned}\qquad(4\text{-}59)$$

$$\begin{aligned}\ln\gamma_M=Z_M^2\Bigg\{&-A_\phi\left[\frac{I^{1/2}}{1+bI^{1/2}}+\frac{2}{b}\cdot\ln\left(1+bI^{1/2}\right)\right]\\&+\sum_c\sum_a m_c m_a B'_{ca}+\sum_c\sum_{<c'} m_c m_{c'}\Phi'_{cc'}+\sum_a\sum_{<a'} m_a m_{a'}\Phi'_{aa'}\Bigg\}\\&+\sum_a m_a\left(2B_{Ma}+C_{Ma}\sum_i m_i|Z_i|\right)+\sum_c m_c\left(2\Phi_{Mc}+\sum_a m_a\Psi_{Mca}\right)\\&+\sum_a\sum_{<a'} m_a m_{a'}\Psi_{Maa'}+|Z_M|\sum_c\sum_a m_c m_a C_{ca}+2\sum_n m_n\lambda_{nM}+\cdots\end{aligned}\qquad(4\text{-}60)$$

$$\begin{aligned}\ln\gamma_X=Z_X^2\Bigg\{&-A_\phi\left[\frac{I^{1/2}}{1+bI^{1/2}}+\frac{2}{b}\cdot\ln\left(1+bI^{1/2}\right)\right]\\&+\sum_c\sum_a m_c m_a B'_{ca}+\sum_c\sum_{<c'} m_c m_{c'}\Phi'_{cc'}+\sum_a\sum_{<a'} m_a m_{a'}\Phi'_{aa'}\Bigg\}\\&+\sum_c m_c\left(2B_{cX}+C_{cX}\sum_i m_i|Z_i|\right)+\sum_a m_a\left(2\Phi_{Xa}+\sum_c m_c\Psi_{cXa}\right)\\&+\sum_c\sum_{<c'} m_c m_{c'}\Psi_{cc'X}+|Z_X|\sum_c\sum_a m_c m_a C_{ca}+2\sum_n m_n\lambda_{nX}+\cdots\end{aligned}\qquad(4\text{-}61)$$

$$\ln \gamma_{-,\mathrm{MX}} = |Z_\mathrm{M} Z_\mathrm{X}| \left\{ -A_\phi \left[\frac{I^{1/2}}{1+bI^{1/2}} + \frac{2\ln\left(1+bI^{1/2}\right)}{b} \right] \right.$$

$$\left. + \sum_c \sum_a m_c m_a B'_{ca} + \sum_c \sum_{<c'} m_c m_{c'} \Phi'_{cc'} + \sum_a \sum_{<a'} m_a m_{a'} \Phi'_{aa'} \right\}$$

$$+ \frac{v_M}{v_\mathrm{MX}} \sum_a m_a \left(2B_{\mathrm{M}a} + C_{\mathrm{M}a} \sum_i m_i |Z_i| + 2\frac{v_\mathrm{X}}{v_\mathrm{M}} \Phi_{\mathrm{X}a} \right)$$

$$+ \frac{v_\mathrm{X}}{v_\mathrm{MX}} \sum_c m_c \left(2B_{c\mathrm{X}} + C_{c\mathrm{X}} \sum_i m_i |Z_i| + 2\frac{v_\mathrm{M}}{v_\mathrm{X}} \Phi_{\mathrm{M}c} \right)$$

$$+ \sum_c \sum_a m_c m_a v_\mathrm{MX}^{-1} \left(2v_\mathrm{M} Z_\mathrm{M} C_{ca} + v_\mathrm{M} \Psi_{\mathrm{M}ca} + v_\mathrm{X} \Psi_{caX} \right)$$

$$+ \sum_c \sum_{<c'} m_c m_{c'} \frac{v_\mathrm{X}}{v_\mathrm{MX}} \Psi_{cc'\mathrm{X}} + \sum_a \sum_{<a'} m_a m_{a'} \frac{v_\mathrm{M}}{v_\mathrm{MX}} \Psi_{\mathrm{M}aa'}$$

$$+ \frac{2}{v_\mathrm{MX}} \sum_n m_n \left(v_\mathrm{M} \lambda_{n\mathrm{M}} + v_\mathrm{X} \lambda_{n\mathrm{X}} \right) \tag{4-62}$$

where B' and Φ' are the derivatives of B and Φ with respect to I. Also:

$$v_\mathrm{MX} = v_\mathrm{M} + v_\mathrm{X} \tag{4-63}$$

$$B_{ca}^{\phi} = B_{ca} + I \cdot B'_{ca} \tag{4-64}$$

$$\Phi_{cc'}^{\phi} = \Phi_{cc'} + I \cdot \Phi'_{cc'} \tag{4-65}$$

and an analogous equations defines $\Phi_{aa'}^{\phi}$. Again, the third-order interaction coefficients for neutral species are not reported in equations (4-59)–(4-62).

In these new equations ((4-58)–(4-62)), the parameters B and C are obtained from data determined on aqueous solutions of single electrolytes, whereas the parameters Φ and Ψ are derived from experimental data for multi-electrolyte aqueous solutions sharing a common ion. With respect to the original equations, the parameters B and Φ correspond to combinations of second-order interaction coefficients λ_{ij}, whereas the parameters C and Ψ correspond to combinations of third-order interaction coefficients μ_{ijk}.

The parameter C_MX is related to the parameter C_MX^{φ}, which is usually reported, by the following relation:

$$C_\mathrm{MX} = \frac{C_\mathrm{MX}^{\phi}}{2|Z_\mathrm{M} Z_\mathrm{X}|^{1/2}}. \tag{4-66}$$

The parameters B^{ϕ}_{MX}, B_{MX} and B'_{MX} depend on the ionic strength as follows:

$$B^{\phi}_{\text{MX}} = \beta^{(0)}_{\text{MX}} + \beta^{(1)}_{\text{MX}} \cdot \exp\left(-\alpha_1 I^{1/2}\right) + \beta^{2}_{\text{MX}} \cdot \exp\left(-\alpha_2 I^{1/2}\right) \tag{4-67}$$

$$B_{\text{MX}} = \beta^{(0)}_{\text{MX}} + \beta^{(1)}_{\text{MX}} \cdot g\left(\alpha_1 I^{1/2}\right) + \beta^{(2)}_{\text{MX}} \cdot g\left(\alpha_2 I^{1/2}\right) \tag{4-68}$$

$$B'_{\text{MX}} = \frac{\beta^{(1)}_{\text{MX}} \cdot g'\left(\alpha_1 I^{1/2}\right) + \beta^{(2)}_{\text{MX}} \cdot g'\left(\alpha_2 I^{1/2}\right)}{I}. \tag{4-69}$$

The functions g and g' are defined as

$$g(x) = \frac{2\left[1-(1+x)\cdot e^{-x}\right]}{x^2} \tag{4-70}$$

and

$$g'(x) = \frac{-2\left[1-\left(1+x+\frac{x^2}{2}\right)\cdot e^{-x}\right]}{x^2}. \tag{4-71}$$

The coefficients $\beta^{(0)}_{\text{MX}}$, $\beta^{(1)}_{\text{MX}}$ and $\beta^{(2)}_{\text{MX}}$ in equations (4-67)–(4-69) are specific for each electrolyte. They are obtained through fitting of isothermal or isobaric data available for aqueous solutions of the considered electrolyte. The coefficient $\beta^{(2)}_{\text{MX}}$ is needed only for two-two (e.g. $MgSO_4$) or higher-valence electrolytes, in which formation of electrostatic ion pairs is important. For solutions containing at least an univalent ion, α_1 = 2.0 kg$^{1/2}$ mol$^{-1/2}$ and $\alpha_2 = 0$, but for two-two electrolytes at 25°C, α_1 = 1.4 kg$^{1/2}$ mol$^{-1/2}$ and α_2 = 1.2 kg$^{1/2}$ mol.$^{-1/2}$ These values are often assumed to be independent of T and P.

The Φ-parameters are defined as follows:

$$\Phi^{\phi}_{ij} = \theta_{ij} + {}^{E}\theta_{ij}(I) + I\,{}^{E}\theta'_{ij}(I) \tag{4-72}$$

$$\Phi_{ij} = \theta_{ij} + {}^{E}\theta_{ij}(I) \tag{4-73}$$

$$\Phi'_{ij} = {}^{E}\theta'_{ij}(I). \tag{4-74}$$

They exhibit an important dependence on I for unsymmetrical cases, (e.g. between Na^+ and Ca^{2+} and Cl^- and SO_4^{2-}) which derives from long-range electrostatic forces. The coefficients ${}^{E}\theta_{ij}(I)$ and ${}^{E}\theta'_{ij}(I)$ are a function of the ionic strength, the charges of the two cationic or anionic species, and the solvent properties density and dielectric constant (and consequently temperature and pressure). The θ_{ij} coefficient is usually taken as a constant at fixed temperature and pressure, although it might have a significant dependence on the ionic strength in some cases.

The parameters Ψ_{ijk} (third virial coefficients for mixing effects) are taken independent of the ionic strength and are defined either for a cation k and two different anions i,j or for an anion k and two different cations i,j.

Summing up, provided that the empirical parameters $\beta_{MX}^{(0)}$, $\beta_{MX}^{(1)}$, $\beta_{MX}^{(2)}$, C_{MX}^{ϕ}, θ_{ij} and Ψ_{ijk} are known and their dependence on T and P is eventually known as well, the solubilities in binary, ternary and higher order systems can be computed.

Contrary to what is shown above for charged solutes, for neutral solutes it is not necessary to rearrange the terms in λ_{ij} and μ_{ijk}. Therefore, the activity coefficient of a neutral solute N in an electrolyte solution is given by the following relation:

$$\ln \gamma_N = 2 \cdot \left(\sum_c m_c \lambda_{Nc} + \sum_a m_a \lambda_{Na} + \sum_n m_n \lambda_{Nn} \right), \tag{4-75}$$

which is written leaving the third virial terms (in μ_{ijk}) out of consideration. The equation complete of these terms is given by Clegg and Brimblecombe (1990). In an aqueous solution of pure neutral solute, its activity coefficient is as follows:

$$\ln \gamma_{N,H_2O} = 2m_N \lambda_{NN} + 3m_N^2 \mu_{NNN}, \tag{4-76}$$

whose interaction coefficients λ_{NN} and μ_{NNN} are directly observable. In general the μ_{NNN}-term is important only in very concentrated solutions, as found also for CO_2 in water (Barta and Bradley, 1985).

Other needed thermodynamic properties, such as excess volumes, heat capacities, entropies and enthalpies, are obtained by suitable differentiation of the excess Gibbs free energy given by equation (4-52) and are not shown here. These thermodynamic properties can also be derived from experimental data. Hence, it is possible to investigate the pressure and temperature dependence of the parameters involved in the Pitzer equations.

4.4.2. Applications of the Pitzer's model

The Pitzer equations were used by Harvie and Weare (1980) to implement a chemical model (HW model) for describing both solution properties and mineral–solution equilibria in the Na–K–Mg–Ca–Cl–SO_4–H_2O system at 25°C and high ionic strengths (up to 20 mol kg^{-1}). Then this model was extended by Harvie, Møller and Weare (1984), who included acid–base and carbonate equilibria. The HMW model predicts the solubilities of minerals in the Na–K–Mg–Ca–H–Cl–SO_4–OH–HCO_3–CO_3–CO_2–H_2O system at 25°C, high ionic strengths and variable P_{CO2}. These works and other investigations have proven that the Pitzer theory is particularly apt for concentrated, multi-component natural brines. For instance, the mineral parageneses produced through evaporation of seawater up to almost total dryness were successfully reproduced (Weare, 1987).

More recently, the Pitzer's model was extended to high temperature and applied to different systems. For instance, Pabalan and Pitzer (1987) proposed a model for the Na–K–Mg–Cl–SO_4–OH–H_2O system, which can be applied up to 200°C, and Pabalan and Pitzer (1988) elaborated a model for the Na–Cl–SO_4–OH–H_2O system, which can be used up to 300°C. The pressure dependence was investigated by Connaughton et al. (1989) and Monnin (1989).

Corti et al. (1990) applied the Pitzer's theory to aqueous solutions of (i) NaCl and CO_2 up to temperatures of 250°C and pressures of 600 bar and (ii) Na_2SO_4 and CO_2 at temperatures of 50–75°C and pressures up to 200 bar. It was shown that the semiempirical Pitzer's equations describe accurately the salt solubilities in these two ternary systems.

4.5. Implications for CO_2 solubility in concentrated aqueous solutions

In the previous sections of this chapter, it was shown that activity coefficients of ionic and neutral solutes in aqueous solutions can be computed by means of different approaches.

According to the simplest equations, such as the B-dot equation (4-12) and the Davies function (4-13), activity coefficients depend essentially on the ionic strength and are not explicit function of the concentrations of main dissolved species.

However, it was shown experimentally (e.g. Harned and Owen, 1958) that, above the dilute interval, activity coefficients depend not only on the ionic strength of the aqueous solution but also on the concentrations of major solutes. This dependence can be described by means of equations involving suitable interaction parameters, as in the Pitzer's model (see equations (4-55), (4-60), (4-61), (4-62), (4-75) and (4-76)).

As an example, we can use the program EQ3NR, version 7.2b, to model the aqueous solutions of single electrolytes (HCl, NaCl, KCl, $MgCl_2$ and $CaCl_2$) at 25°C, 1 bar, under a CO_2 fugacity of 1 bar, and variable ionic strength, up to 6 mol kg^{-1}. The version 7.2b of program EQ3NR computes activity coefficients by means of either the Davies equation or the B-dot equation or the Pitzer's equations, and a compatible thermodynamic database must be chosen in each case. Let us follow two different approaches to compute activity coefficients, namely (i) the B-dot equation, using the SUP thermodynamic database (Johnson et al., 1992) and (ii) the Pitzer's equations, using the HMW thermodynamic database (Harvie et al., 1984).

Among charged solutes species, the hydrogen ion plays a pivotal role due to its ubiquitary presence in aqueous solutions and its participation in both acid–base equilibria and hydrolysis reactions. Owing to the utmost interest of hydrogen ion, the dependence of the logarithm of its activity coefficient on the true ionic strength of the considered electrolyte solutions is plotted in Fig. 4.3a. As expected, the log γ_{H^+} values computed by means of the B-dot equation are the same for all the considered electrolytes, since they depend on the ionic strength only, and are independent of the concentrations of major solutes, at constant temperature, pressure conditions (as in this example). In contrast, the log γ_{H^+} values calculated by using the Pitzer's theory for a given aqueous electrolyte are very different from those of other electrolytes, thus defining a series of log $\gamma_{H^+} - \bar{I}$ functions in Fig. 4.3a. These functions have a sort of parabolic shape and converge towards $\gamma_{H^+} = 1$ for $I \rightarrow 0$ but diverge strongly for increasing $\bar{I}$.

Again, the B-dot equation gives the same value of $\gamma_{CO_2(aq)}$ for all the electrolyte solutions, whereas the Pitzer's equations give distinct values of $\gamma_{CO_2(aq)}$ for each

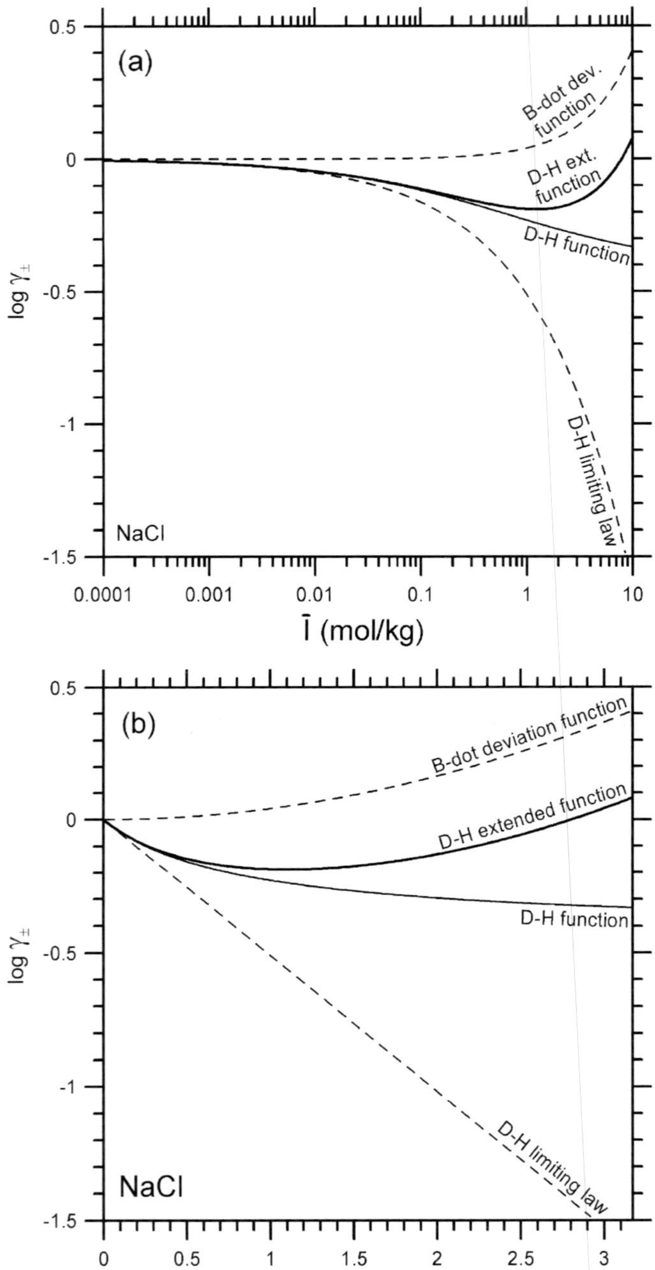

Figure 4.3. Plot of the logarithm of the activity coefficient of (a) hydrogen ion and (b) aqueous CO_2 versus the true ionic strength, for HCl, NaCl, KCl, $MgCl_2$ and $CaCl_2$ aqueous solutions at 25°C, 1 bar, under a CO_2 fugacity of 1 bar. Activity coefficients were computed by means of both the B-dot equation and the Pitzer's model using the EQ3NR program, version 7.2b.

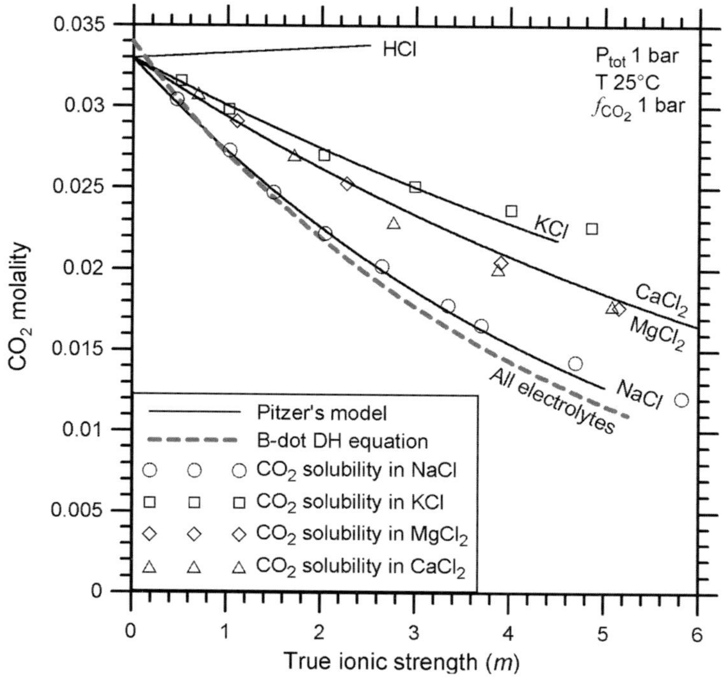

Figure 4.4. Plot of the solubility of aqueous CO_2 versus the true ionic strength, for HCl, NaCl, KCl, $MgCl_2$ and $CaCl_2$ aqueous solutions at 25°C, 1 bar, under a CO_2 fugacity of 1 bar. Activity coefficients were computed by means of both the B-dot equation and the Pitzer's model using the EQ3NR program, version 7.2b. Experimental data are from Yasunishi and Yoshida (1979). Reprinted from Harvie et al. (1984), Copyright (1984), with permission from Elsevier (modified).

electrolyte, as shown in Fig. 4.3b. Not surprisingly, the log $\gamma_{CO_2(aq)} - \bar{I}$ functions converge at $\gamma_{CO_2(aq)} = 1$ for $\bar{I} \rightarrow 0$ but diverge for increasing Ī. However, the log $\gamma_{CO_2(aq)}$ varies linearly with the ionic strength, in contrast to what observed above for hydrogen ion.

The obvious implication of the distinct activity coefficients of dissolved CO_2 for the different electrolyte solutions predicted by the Pitzer's model is CO_2 solubility varying from electrolyte to electrolyte (Fig. 4.4). It must be underscored that these predictions of the Pitzer's theory reproduce satisfactorily the experimental data, which were used to derive the Pitzer's parameters (see Harvie et al., 1984 and references therein). In contrast, the B-dot equation reproduces reasonably the experimental CO_2 solubilities in NaCl solutions which were used to derive the parameters of the B-dot equation (Helgeson, 1969), but it is at variance with the other experimental data.

Summing up, use of the B-dot equation is acceptable for dilute to slightly concentrated aqueous solutions of NaCl composition, whereas use of the Pitzer's model is compulsory for very concentrated electrolytes.

Chapter 5

The Product Solid Phases

Virtually permanent CO_2 sequestration in form of solid carbonates into relatively deep geological formations and the industrial process of mineral carbonation are different approaches to get rid of large amounts of anthropogenic CO_2 (e.g., Seifritz, 1990; Lackner et al., 1995). In spite of the differences in these two methodologies of CO_2 disposal, there is a very important point common to both. Indeed, both involve the dissolution of primary phases, mainly silicates and aluminium–silicates, and the precipitation of new phases, mainly carbonates, silica minerals and clay minerals.

In both methodologies of CO_2 disposal, the expected product carbonates are chiefly Ca-, Mg- and Ca–Mg-carbonates, which are therefore the main carbonate minerals of interest to us. Their mineralogical and chemical characteristics will be summarized and their thermodynamic stability will be outlined below. Nevertheless Fe(II)-bearing carbonates (e.g., siderite and ankerite) may also be important in some situations, as well as dawsonite [$NaAlCO_3(OH)_2$], which has recently received considerable attention (see Section 5.1.5).

This chapter is also devoted to briefly review the thermodynamic stability of the other expected product solid phases of geological CO_2 sequestration and mineral carbonation, i.e. the silica minerals and the clay minerals.

5.1. Major carbonate minerals

Calcite and dolomite are the two most important carbonate minerals and have been the subject of several mineralogical and geochemical investigations, recently reviewed by Goldsmith (1990), Reeder (1990) and Wenk et al. (1990), among the others.

The CO_3 group is the fundamental building block of carbonate minerals and its basic configuration is almost constant. The CO_3 group is a sort of equilateral triangle with a carbon atom in the centre and three oxygen atoms at the corners. The O–C–O angle was found to be 120° on the average, in 30 different carbonate minerals (Zemann, 1981), in perfect agreement with the ideal value. The average C–O bond length is 1.284 Å with a standard deviation of 0.004 Å. This short C–O bond length brings about a close approach for oxygen atoms within the CO_3 group, i.e. 2.22 Å only. Slightly shorter distances between oxygen atoms are found in $NaNO_3$ and stishovite (Table 5 in Shannon and Prewitt, 1969).

This means that the carbonates have a rather close-packed structure, which is a nice perspective for the sequestration of CO_2 through injection in deep reservoirs, since the generation of dense secondary minerals is a way to "save space" or, in other terms, to minimize the loss in rock porosity and permeability caused by precipitation of product solid phases.

The C–O bond is strongly covalent and remarkably stronger than other bonds between oxygen atoms and metal atoms. For example, in calcite, the C–O bond strength computed by means of the empirical technique of Brown and Shannon (1973) is ~4 times greater than for the Ca–O bond.

5.1.1. The structure of calcite and the $R\bar{3}c$ carbonates

The general structure of calcite was established by Bragg (1914), while its positional and thermal parameters were determined with good precision by Sass et al. (1957), Inkinen and Lahti (1964) and Chessin et al. (1965). The formal description of calcite structure is given in Table 5.1.

What can we do with these numbers? As an example, cell parameters can be used to compute the unit cell volume, from which molar volume and density are easily obtained. Let us do the exercise for calcite. For any mineral, independent on the crystal system, the volume of the unit cell, υ_0, is given by (Pauling, 1970):

$$\upsilon_0 = abc(1+2\cos\alpha\ \cos\beta\cos\gamma-\cos^2\alpha-\cos^2\beta-\cos^2\gamma)^{1/2}. \tag{5-1}$$

For the hexagonal crystal system: $\alpha = \beta = 90°$ and $\gamma = 120°$. Hence:

$$\begin{aligned}\upsilon_0 &= abc\times(1-\cos^2 120)^{1/2}\\ &= 4.9896\times 4.9896\times 17.0610\times(1-\cos^2 120)^{1/2} = 367.85\ \text{Å}^3.\end{aligned}$$

TABLE 5.1

Formal description of calcite and dolomite structure (from Reeder, 1990)

	Calcite	Dolomite
Class symmetry	Trigonal scalenohedral	Trigonal rhombohedral
Space group	$R\bar{3}c$ (No. 167)	$R\bar{3}$ (No. 148)
Unit cell	Hexagonal, rhombohedrally centred	Hexagonal, rhombohedrally centred
Cell contents	$6CaCO_3$	$3CaMg(CO_3)_2$
Cell parameters	$a = b = 4.9896$ Å	$a = b = 4.8069$ Å
	$c = 17.0610$ Å	$c = 16.0034$ Å
Atom positions	Ca in 6 (b): 0, 0, 0	Ca in 3 (a): 0, 0, 0
	C in 6 (a): 0, 0, 1/4	Mg in 3 (b): 0, 0, 1/2
	O in 18 (e): x, 0, 1/4	C in 6 (c): 0, 0, z_c
	$x = 0.2568$	O in 18 (f): x_o, y_o, z_o
		$z_c = 0.24282(4)$
		$x_o = 0.24776(7)$
		$y_o = -0.03525(7)$
		$z_o = 0.24404(2)$

Cell parameters and positional parameters of dolomite refer to the Lake Arthur specimen of composition $[Ca_{1.00}][Mg_{0.99}Ca_{0.01}](CO_3)_2$ (Reeder, 1990).

Knowing the unit cell volume, the molar volume v (cm^3 mol^{-1}) is computed as follows:

$$v = \frac{\upsilon_0 \times N_A \times 10^{-24}}{Z} = \frac{367.85 \times 6.02214 \cdot 10^{23} \times 10^{-24}}{6} = 36.92\,\text{cm}^3 \cdot \text{mol}^{-1}, \tag{5-2}$$

where Z [adimensional] is the number of molecules or formula units of the substance in the unit cell (which is 6 for calcite), N_A the Avogadro's number (6.02214×10^{23} mol^{-1}) and 10^{-24} the transformation factor from Å^3 to cm^3. Then, the density ρ is readily calculated:

$$\rho = \frac{\text{MW}}{v} = \frac{100.091}{36.92} = 2.71\,\text{g} \cdot \text{cm}^{-3}. \tag{5-3}$$

Isn't it amazing to find that cell parameters of minerals can be easily related to density, which is part of every day life or nearly so?

But let us go back to the calcite structure. In the calcite structure, Ca is coordinated to six oxygen atoms, each belonging to distinct CO_3 groups, thus forming slightly distorted CaO_6 octahedra (Fig. 5.1). The CO_3 groups are distributed in different layers and have like orientations within each layer but reversed orientations in successive layers. The shortest O–O distances between different CO_3 groups are 3.19 Å, across CO_3 layers, and 3.26 Å, in the same CO_3 layer. Within a given layer, the octahedra share neither edges nor corners and are, therefore, totally independent. However, each octahedron shares its corners with other octahedra, three from below and three from above.

The calcite structure is also exhibited by the carbonates of Mg (magnesite), Fe (siderite), Co (sphaerocobaltite or cobaltocalcite), Ni (gaspeite), Mn (rhodochrosite), Zn (smithsonite) and Cd (otavite). Otavite and gaspeite are rare minerals. All these metals are in the divalent state, as required by the electroneutrality condition. Indeed, the presence of divalent Fe in $FeCO_3$ and divalent Mn in $MnCO_3$ was confirmed by spectroscopic studies (Takashima and Ohashi, 1968; Wildeman, 1970; Urch and Wood, 1978). All these divalent

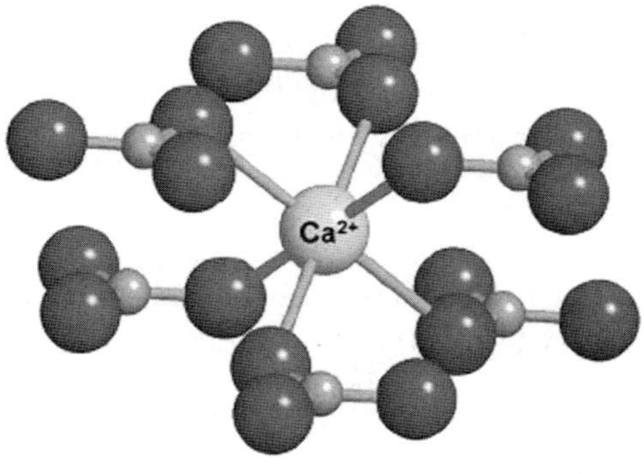

Figure 5.1. Schematic representation of the calcite structure. The large grey ball at the centre is a calcium ion in octahedral coordination, the large black balls at the corners of the octahedron are oxygen atoms and the small grey balls are carbon atoms. (Reprinted from Elzinga et al., 2002, Copyright 2002, with permission from Elsevier.)

cations are smaller than Ca^{2+}, whose size is considered to represent the limiting value for 6-fold coordination. Indeed, cations larger than Ca^{2+}, such Ba^{2+}, Sr^{2+} and Pb^{2+}, form carbonates with orthorhombic aragonite structure, in which the cation has 9-fold coordination. However, at high temperatures, the orthorhombic carbonates $BaCO_3$ and $SrCO_3$ invert to rhombohedral calcite-type structure (Chang, 1965). As commonly observed for isostructural series, cation size, M–O bond length, polyedral volume of the MO_6 octahedra and unit cell volume are strictly interrelated (Reeder, 1990). The plot of the unit cell volume vs. 6-fold coordination ionic radius (Fig. 5.2a) shows a strict linear correlation for the carbonates of Mg, Fe, Co, Ni, Mn, Zn and Cd, whereas the unit cell volume of calcite appears to be somewhat higher than expected. Again, in the plot of the M–O bond length vs. 6-fold coordination ionic radius (Fig. 5.2b), a strict linear correlation is displayed by the carbonates of Mg, Fe, Mn, Zn and Cd, whereas the Ca–O bond length in calcite is higher than expected (the data used to prepare these plots are given in Table 5.2). Apart from the calcite anomaly, the strict correlation between these parameters suggests that the $R\bar{3}c$ carbonates constitute a very well behaved isostructural series with cation size being the major governing parameter, but this is not totally true as shown by Reeder (1990).

Complete miscibility is observed between the carbonates of Fe and Mg, Ca and Cd, Mg and Co, Fe and Mn and Mg and Mn, as expected based on the small differences in cation radii ($\Delta^{VI}r \leq 0.11$ Å) whereas limited miscibility occurs between calcite and the carbonates of Ni ($\Delta^{VI}r = 0.31$ Å), Mg ($\Delta^{VI}r = 0.28$ Å), Co ($\Delta^{VI}r = 0.255$ Å), Fe ($\Delta^{VI}r = 0.22$ Å) and Mn ($\Delta^{VI}r = 0.17$ Å), as well as between the carbonates of Cd and Mg ($\Delta^{VI}r = 0.23$ Å) and Ni and Mg, although the $\Delta^{VI}r$ is 0.03 Å only. The anomalous behaviour of the Ni–Mg pair confirms that the miscibility between different carbonate minerals is not solely governed by cation size.

In spite of the limited miscibility of calcite with most $R\bar{3}c$ carbonates, precipitation of calcite is likely to scavenge Mg, Fe, Co, Ni, Mn, Zn and Cd from the aqueous solution, as discussed by Rimstidt et al. (1998) and Marini et al. (2001).

5.1.2. The structure of dolomite and the $R\bar{3}$ carbonates

Dolomite is the most abundant of the over fifteeen $R\bar{3}$ carbonates, five of which have the dolomite structure. In addition to dolomite [$CaMg(CO_3)_2$], they are: ankerite or ferroan dolomite [$Ca(Mg, Fe, Mn)(CO_3)_2$], kutnahorite [$CaMn(CO_3)_2$], minrecordite [$CaZn(CO_3)_2$] and synthetic cadmium dolomite [$CdMg(CO_3)_2$].

The crystallographic properties of dolomite are given in Table 5.1. The structure of dolomite is similar to that of calcite but has a somewhat lower symmetry (see below). The dolomite structure can be described either through combination of alternating calcite- and magnesite-like layers (Lippmann, 1973) or by substitution of Mg atoms for the Ca atoms in every other cation layer. Since the CO_3 layers remain practically undisturbed by Mg–Ca substitution, the unit cell of dolomite is comparable in size to that of calcite.

Also the C–O bond length in dolomite is virtually unchanged from that of calcite-like carbonate minerals. On the contrary, the Ca–O bond length, 2.38 Å, is much greater than the Mg–O bond length, 2.08 Å. Consequently, the oxygen atoms in 3-fold coordination result to be closer to Mg than Ca. This difference in Ca–O and Mg–O bond lengths is

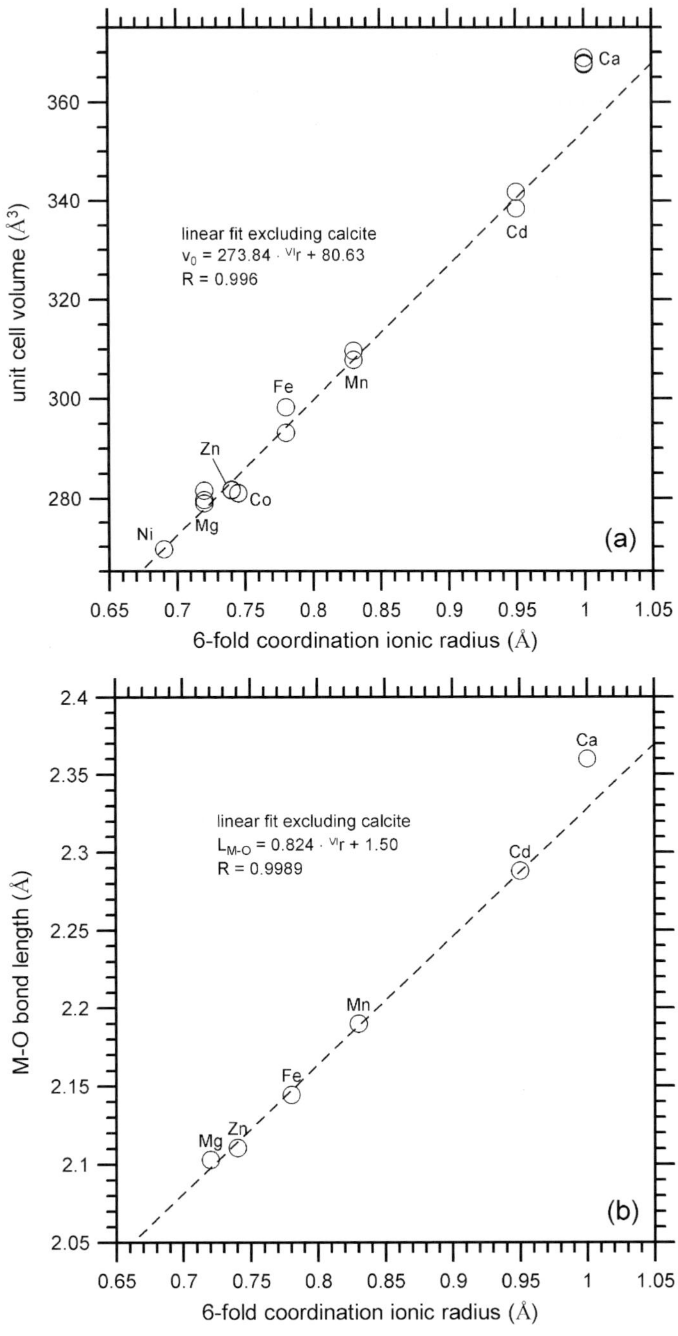

Figure 5.2. (a) Unit cell volume (Å^3) and (b) M–O bond length (Å) for the $R\bar{3}c$ carbonates vs. the 6-fold coordination ionic radius (Å) of the corresponding cations (data in Table 5.2).

TABLE 5.2

Selected crystallographic data of carbonates and 6-fold coordination ionic radius (^{VI}r) of relevant cations (HS = high spin)

Mineral	Unit cell volume (Å^3)	M–O bond length (Å)	Ref.	Cation	^{VI}r (Å)	Ref.
Calcite	367.85	2.3598	(1)	Ca^{2+}	1.00	(7)
	367.78	–	(2)			
	367.59	2.359	(4)			
	368.93	2.362	(5)			
Otavite	341.85	2.288	(5)	Cd^{2+}	0.95	(7)
	338.48	–	(2)			
Rhodochrosite	307.86	2.1900	(1)	Mn^{2+} (HS)	0.83	(7)
	309.68	–	(2)			
Siderite	293.17	2.1445	(1)	Fe^{2+} (HS)	0.78	(7)
	298.29	–	(2)			
Sphaerocobaltite	281.16	–	(2)	Co^{2+} (HS)	0.745	(7)
	281.07	–	(6)			
Smithsonite	281.68	2.1107	(1)	Zn^{2+}	0.74	(7)
	281.77	–	(2)			
Magnesite	279.05	2.1018	(1)	Mg^{2+}	0.72	(7)
	281.62	–	(2)			
	279.74	2.105	(3)			
Gaspeite	269.51	–	(6)	Ni^{2+}	0.69	(7)

References: (1) = Effenberger et al. (1981); (2) = http://webmineral.com; (3) = Oh et al. (1973); (4) = Finger (1975); (5) = Borodin et al. (1979); (6) = Graf (1961); (7) = Shannon (1976).

accommodated through a uniform rotation of all the CO_3 groups of a given layer around the 3-fold axis with respect to their position in calcite. Therefore, the CO_3 groups in successive layers are not related by a *c*-glide and the diad axes are also absent in dolomite. The CaO_6 coordination octahedra are weakly rotated with respect to their position in the calcite structure, but in an opposite sense with respect to that of the CO_3 groups. The MgO_6 octahedra are also rotated in the same sense and by the same extent as the CaO_6 octahedra.

The thermodynamic properties of dolomite are remarkably affected by the order/disorder, which is defined by the ordering parameter *s* (Helgeson et al., 1978):

$$s = 2X_{\mathrm{Ca,A}} - 1 \tag{5-4}$$

where $X_{\mathrm{Ca,A}}$ is the mole fraction of Ca atoms on the A site in dolomite. The ordering parameter *s* ranges from 1 if dolomite is completely ordered to 0 if it is totally disordered. Navrotsky and Loucks (1977) used the Bragg–Williams theory to calculate the temperature dependence of the ordering parameter *s*, which is given by:

$$T/T_c = s/(\tanh^{-1} \cdot s) \tag{5-5}$$

where T_c is the temperature where dolomite becomes totally disordered. For T_c = 1,200°C (Goldsmith and Heard, 1961), the temperature dependence of the Bragg–Williams ordering

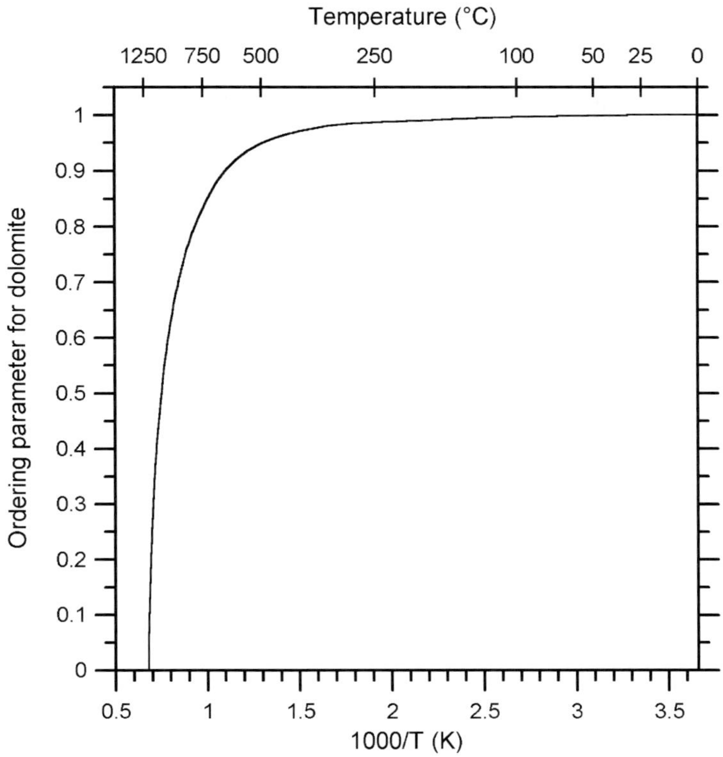

Figure 5.3. Temperature dependence of the Bragg–Williams ordering parameter of dolomite. Reprinted from Helgeson et al. (1978), modified, copyright (1978), with permission from American Journal of Science.

parameter of dolomite reported in Fig. 5.3 is obtained. Dolomites synthetized at the relatively low temperatures expected for the geological sequestration of CO_2 and mineral carbonation, therefore, should be completely ordered or almost so.

However, this expectation is not supported by available data on recent, low-temperature dolomites, which exhibit strongly attenuated superstructure reflections possibly suggesting some cation disorder (Wenk et al., 1990). According to Reeder (1990), there are many descriptions of "disordered" or "poorly-ordered" low-temperature dolomites in the literature, but there are no published determinations of site occupancies apart from the data for Lake Arthur dolomite (Table 5.1), which is a stoichiometric, ideally ordered dolomite. More crystal chemical data are needed to elucidate this matter.

5.1.3. The thermal expansion of carbonates

Experimental thermal expansion data of solid phases are usually fitted using the polynomial expression (Fei, 1995):

$$\alpha(T) = a_0 + a_1\, T + a_2\, T^{-2}, \tag{5-6}$$

which is analogous to the Maier–Kelley equation (2-36), generally used for fitting heat capacities data. The constants a_0, a_1 and a_2 are called isobaric volumetric thermal expansion coefficients. If they are known, the molar volume of the solid phase is computed through the expression:

$$v_T = v_{\mathrm{Tr}} \cdot \exp\left[\int_{\mathrm{Tr}}^{T} \alpha(T) \cdot \mathrm{d}T\right] = v_{\mathrm{Tr}} \cdot \exp\left[a_0 \cdot (T - \mathrm{Tr}) + \frac{a_1}{2} \cdot (T^2 - \mathrm{Tr}^2) - a_2 \cdot \left(\frac{1}{T} - \frac{1}{\mathrm{Tr}}\right)\right. \quad (5\text{-}7)$$

where v_{Tr} is the molar volume of the solid at the reference temperature Tr. The same equation is used to compute the cell parameters *a* and *c* at any temperature, starting from their values at the reference temperature and the isobaric linear thermal expansion coefficients. These are obtained by fitting experimental data to polynomial expressions analogous to equation (5-6).

Both volumetric and linear thermal expansion coefficients of a large number of solids have been collected and reviewed by Fei (1995). Those of carbonates are listed in Table 5.3.

By plugging the thermal expansion coefficients of calcite in equation (5-7), it turns out that the lattice of calcite changes with temperature in a rather unusual way, in that the *a* cell parameter decreases with increasing temperature up to 900°C (Markgraf and Reeder, 1985, and references in Reeder, 1990). However, the *c* cell parameter experiences an

TABLE 5.3

Linear and volumetric thermal expansion coefficients of carbonate minerals

Mineral	Cell parameters or volume	Thermal expansion coefficients			T range (K)	Ref.
		$a_0(10^{-4})$	$a_1(10^{-8})$	a_2		
Calcite	*a*	−0.0315	0	0	297–1173	(1)
	c	0.1922	2.5183	−1.214	297–1173	(1)
	υ	0.0713	3.3941	−1.214	297–1173	(1)
Otavite	*a*	−0.056	0	0	293-593	(2)
	c	0.227	0	0	293–593	(2)
	υ	0.115	0	0	293–593	(2)
Magnesite	*a*	0.0775	0.2934	−0.5809	297–773	(1)
	c	0.0037	4.2711	0	297–773	(1)
	υ	0.1686	4.7429	−1.1618	297–773	(1)
Rhodochrosite	*a*	0.018	0	0	297–773	(3)
	c	0.192	0	0	297–773	(3)
	υ	0.228	0	0	297–773	(3)
Siderite	*a*	0.054	0	0	297–773	(4)
	c	0.161	0	0	297–773	(4)
	υ	0.269	0	0	297–773	(4)
Dolomite	*a*	0.0271	0.6045	−0.1152	297–973	(5)
	c	0.1233	2.2286	−0.3089	297–973	(5)
	υ	0.1928	3.1703	−0.5393	297–973	(5)

References: (1) = Markgraf and Reeder (1985); (2) = Bayer (1971); (3) = Rao and Murthy (1970); (4) = Pfaff (1935); (5) = Reeder and Markgraf (1986)

increase with temperature greater than the decrease in *a*, in the same temperature range. Therefore, the overall effect of these changes is the increase in the unit cell volume of calcite with temperature.

Otavite has a behaviour similar to calcite, whereas the volumetric and linear thermal expansion coefficients of dolomite, magnesite, rhodochrosite and siderite are all positive, indicating that both the *a* and *c* cell parameters (and the unit cell volume as well) increase with temperature.

What else can we do with these numbers? When dealing with solid phases, their volume is usually assumed independent of P and T. Although this assumption is reasonable, it is worthwhile to give a look to the changes in the molar volume of calcite with temperature in the range of interest, from 25 to 125°C. To this purpose, we compute the molar volume of calcite in the two following ways: (a) directly considering the volumetric thermal expansion coefficients, and (b) computing first the cell parameters *a* and *c* at different temperatures, starting from the linear thermal expansion coefficients, and then calculating the molar volume by means of equations (5-1) and (5-2). Although results should be the same, there are small discrepancies, which increase with temperature, but are always ≤ 0.01 cm^3 mol^{-1} (Fig. 5.4). In spite of these uncertainties, it can be concluded that the

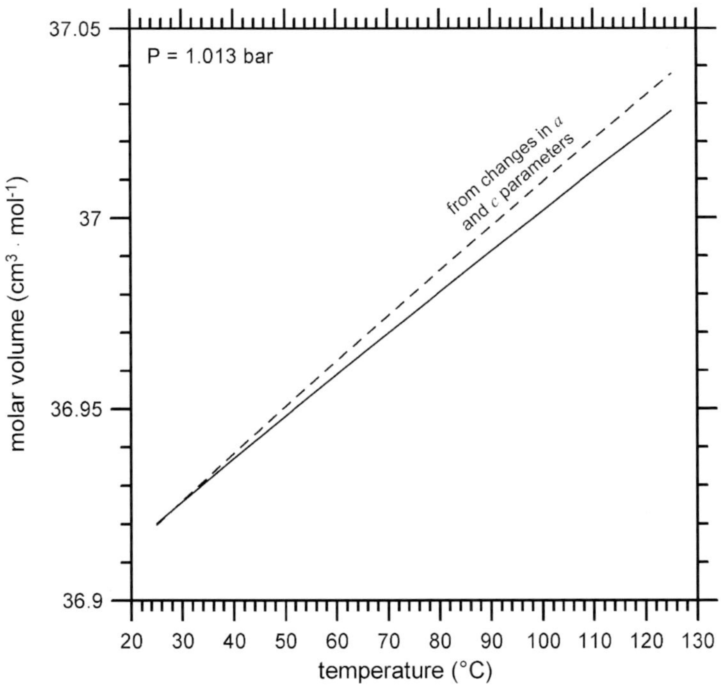

Figure 5.4. Computed molar volume of calcite at different temperatures from 25 to 125°C. The solid curve refers to the molar volume calculated considering the volumetric thermal expansion coefficients, whereas the dashed curve was obtained computing first the cell parameters *a* and *c* at different temperatures (based on the linear thermal expansion coefficients) and then calculating the molar volume using equations (5-1) and (5-2).

molar volume of calcite increases with temperature from 36.92 $cm^3\ mol^{-1}$ at 25°C to 37.03−37.04 $cm^3\ mol^{-1}$ at 125°C. As expected, the volume increase is relatively small, close to 0.3%, and can be neglected for the purposes of the present discussion.

5.1.4. The stability of the carbonate minerals of Ca and Mg

The subject of this section is the evaluation of the relative stability of carbonate minerals of calcium and magnesium. This evaluation will be carried out on a merely thermodynamic basis, referring first to the systems CaO–CO_2–H_2O and MgO–CO_2–H_2O, and then to the system CaO–MgO–CO_2–H_2O. Although these are very simple systems, they provide a reasonable approximation of several natural environments as well as useful insights into the expected product solid phases of carbonation reactions. Obviously, we will take into account only the Ca-, Mg- and Ca–Mg-minerals whose thermodynamic properties are known. Luckily, these minerals are quite numerous, as seen in Table 5.4, which reports their reactions of dissolution and the corresponding thermodynamic equilibrium constants at specified temperatures and pressures. All these data are taken from the thermodynamic database COM of the EQ3/6 software package.

The thermodynamic equilibrium constants in the system CO_2–H_2O are also given, in order to compute the log K for dissolution reactions written in other ways. For instance, the dissolution reaction for magnesite is given in Table 5.4 as:

$$\text{Magnesite} + H^+ = Mg^{2+} + HCO_3^- \tag{5-8}$$

and the corresponding log K at 25°C is 2.2936.

If we want to know the log K for the reaction:

$$\text{Magnesite} = Mg^{2+} + CO_3^{2-}, \tag{5-9}$$

i.e. the solubility product of magnesite, we first notice that reaction (5-9) is obtained by summation of reaction (5-8) plus the dissociation reaction of bicarbonate ion:

$$HCO_3^- = CO_3^{2-} + H^+, \tag{5-10}$$

whose log K is −10.3288 at 25°C. Then we simply sum the log K of reactions (5-8) and (5-10), obtaining 2.2936 − 10.3288 = −8.0352 for the log K_{sp} of magnesite.

It must be noted that the four decimal figures for the log K values are really too many, with respect to the uncertainties of these data. However, data are reported as such in the EQ3/6 database, to avoid rounding off errors.

5.1.4.1. The system CaO–CO_2–H_2O

The relevant mineral phases in the system CaO–CO_2–H_2O are calcite, aragonite, vaterite, monohydrocalcite and portlandite. Of the two natural polymorphs of $CaCO_3$ other than calcite, aragonite is stable under high pressures, >3.0 kbar at 25°C (Boettcher and

TABLE 5.4

Thermodynamic equilibrium constants for the dissolution reactions of Ca-, Mg- and Ca-Mg-carbonates and dawsonite at specified temperatures and total pressures

Mineral formula	System/reaction of dissolution	log K at T,P					
		0.01°C 1.013 bar	25°C 1.013 bar	60°C 1.013 bar	100°C 1.013 bar	150°C 4.7572 bar	150°C 500 bar
	$CaO–CO_2–H_2O$ system						
$Ca(OH)_2$	Portlandite + $2H^+ = Ca^{2+} + 2H_2O$	24.6242	22.5552	20.1960	18.0608	15.9702	–
$CaCO_3$	Calcite + $H^+ = Ca^{2+} + HCO_3^-$	2.2257	1.8487	1.3330	0.7743	0.0999	0.3145
$CaCO_3$	Aragonite + $H^+ = Ca^{2+} + HCO_3^-$	2.3715	1.9931	1.4762	0.9179	0.2467	0.4443
$CaCO_3{\cdot}H_2O$	Monohydrocalcite + $H^+ = Ca^{2+} + HCO_3^- + H_2O$	3.0114	2.6824	2.3085	1.9820	1.6827	–
	$MgO–CO_2–H_2O$ system						
$Mg(OH)_2$	Brucite + $2H^+ = Mg^{2+} + 2H_2O$	18.0898	16.2980	14.2674	12.4514	10.6978	10.7749
$MgCO_3$	Magnesite + $H^+ = Mg^{2+} + HCO_3^-$	2.9734	2.2936	1.4383	0.5875	−0.3520	−0.1556
$MgCO_3{\cdot}3H_2O$	Nesquehonite + $H^+ = Mg^{2+} + HCO_3^- + 3H_2O$	5.4163	4.9955	4.8188	4.7099	4.5592	4.6864
$MgCO_3{\cdot}5H_2O$	Lansfordite + $H^+ = Mg^{2+} + HCO_3^- + 5H_2O$	–	4.8409	–	–	–	–
$Mg_2CO_3(OH)_2{\cdot}3H_2O$	Artinite + $3H^+ = 2Mg^{2+} + HCO_3^- + 5H_2O$	21.7371	19.6560	17.2642	15.1017	12.9693	13.1582
$Mg_5(CO_3)_4(OH)_2{\cdot}4H_2O$	Hydromagnesite + $6H^+ = 5Mg^{2+} + 4HCO_3^- + 6H_2O$	35.5731	30.8539	25.2242	19.9126	14.3962	15.2261
	$CaO–MgO–CO_2–H_2O$ system						
$CaMg(CO_3)_2$ [ordered]	Ord-dolomite + $2H^+ = Ca^{2+} + Mg^{2+} + 2HCO_3^-$	3.4063	2.5135	1.3314	0.0944	−1.3493	−0.9419
$CaMg(CO_3)_2$ [disordered]	Dis-dolomite + $2H^+ = Ca^{2+} + Mg^{2+} + 2HCO_3^-$	5.1469	4.0579	2.6502	1.2070	−0.4400	−0.0325
$CaMg_3(CO_3)_4$	Huntite + $4H^+ = Ca^{2+} + 3Mg^{2+} + 4HCO_3^-$	12.9081	10.3010	7.0008	3.6895	0.0002	0.8157
	$Na_2O–Al_2O_3–CO_2–H_2O$ system						
$NaAl(CO_3)(OH)_2$	Dawsonite + $3H^+ = Na^+ + Al^{3+} + HCO_3^- + 2H_2O$	5.5675	4.3464	2.9528	1.7174	0.5411	–
	$CO_2–H_2O$ system						
	$CO_{2(g)} + H_2O = H^+ + HCO_3^-$	−7.6765	−7.8136	−8.0527	−8.3574	−8.7692	−8.7890
	$H_2CO_3{*} = H^+ + HCO_3^-$	−6.5804	−6.3447	−6.2684	−6.3882	−6.7235	−6.5261
	$HCO_3^- = CO_3^{2-} + H^+$	−10.6241	−10.3288	−10.1304	−10.0836	−10.2003	−9.9516

The thermodynamic equilibrium constants in the system $CO_2–H_2O$ are also given. All the data are from the thermodynamic database COM of the EQ3/6 software package.

Wyllie, 1968), whereas vaterite is restricted in natural occurrences to "repair tissues of fractured shells of certain gastropods, to gallstones, and to presence as an alteration product in hydrogel pseudomorphs after larnite" (Carlson, 1990). Moreover, vaterite is metastable with respect to calcite and aragonite under geological circumstances (Carlson, 1990). Both vaterite and aragonite are, therefore, left out of consideration in the following discussion.

Equilibrium coexistence of calcite and monohydrocalcite is expressed by the reaction:

$$CaCO_3 \cdot H_2O = CaCO_3 + H_2O \tag{5-11}$$

monohydrocalcite calcite

whose thermodynamic equilibrium constant, assuming that solid phases are pure, is:

$$K_{\text{monohydrocalcite-calcite}} = a_{H_2O}. \tag{5-12}$$

Consequently, the transformation of calcite into monohydrocalcite and vice versa depends solely on the activity of water. In particular, at 25°C and 1.013 bar, coexistence of calcite and monohydrocalcite occurs for a_{H_2O} of 6.82. This is an impossible value, likely due to uncertainties in the thermodynamic data of monohydrocalcite. Anyway, this phase can be safely neglected, for our purposes, since it has been identified, so far, in antarctic lakes only (Sonnenfeld, 1984). Neglecting monohydrocalcite, the remaining two phases of interest, calcite and portlandite, are involved into the following reaction:

$$CaCO_3 + H_2O = Ca(OH)_2 + CO_{2(g)} \tag{5-13}$$

calcite portlandite

whose thermodynamic equilibrium constant is:

$$K_{\text{calcite-portlandite}} = f_{CO_2}, \tag{5-14}$$

if both calcite and portlandite are pure and liquid water is also pure or relatively pure (a dilute aqueous solution is an acceptable approximation of this limiting condition). The values of f_{CO_2} constrained by calcite–portlandite equilibrium coexistence at variable temperature, in the range 0–150°C, are shown in Fig. 5.5. They range from ~10^{-14} bar at 10°C to ~10^{-8} bar at 120°C. Portlandite is stable under very low f_{CO_2} values, which are rarely attained in natural environments, whereas calcite is stable under temperature, f_{CO_2} conditions which are common in nature. Consistently with these indications, calcite is a widespread mineral, whereas portlandite is a very rare mineral. To appreciate the meaning of the low f_{CO_2} values characterising portlandite stability, it is useful to take into account the high-pH, Ca–OH waters interacting with ultramafic rocks variably affected by serpentinization (Bruni et al., 2001, 2002; Marini and Ottonello, 2002; Cipolli et al., 2004). These waters have temperatures of 11–23°C and P_{CO_2} values of 10^{-8}–10^{-11} bar, which are very

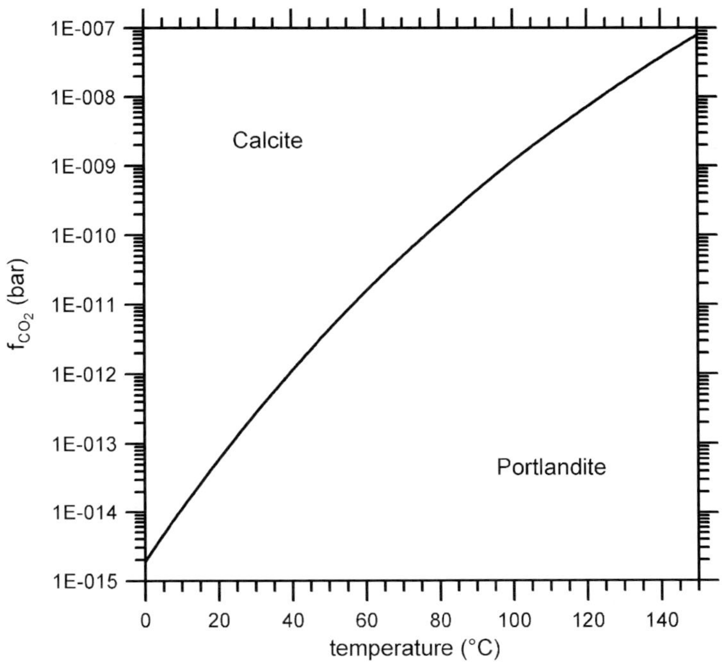

Figure 5.5. Stability diagram for the $CaO–CO_2–H_2O$ system from 0 to 150°C at 1.013 bar total pressure.

low P_{CO_2} values for natural systems. Nevertheless, they are at least two orders of magnitude higher than those dictated by calcite–portlandite coexistence.

Summing up, calcite is the expected product solid phase of carbonation reactions in the system $CaO–CO_2–H_2O$.

5.1.4.2. The system $MgO–CO_2–H_2O$

The mineral phases of interest in the system $MgO–CO_2–H_2O$ are more numerous than those in the system $CaO–CO_2–H_2O$ and include the Mg-hydroxide brucite, the Mg-carbonates magnesite, nesquehonite and lansfordite, and the Mg-hydroxy-carbonates artinite and hydromagnesite (Table 5.4).

Again, similarly to what discussed above for the hydrous carbonate of Ca, the stability of the hydrous carbonates of Mg is uniquely controlled by water activity. Both nesquehonite and lansfordite are typical of closed-basin lacustrine and marine environments, arid soils and evaporative cave pools (Sonnenfeld, 1984; Langmuir, 1997 and references therein), where brines form through evaporative concentration. Therefore, they are not considered anymore.

Magnesite is the most frequent Mg carbonate in different geological environments. Its stability with respect to the Mg-hydroxide and the Mg-hydroxy-carbonates is expected to depend on f_{CO_2} and deserves further consideration. Similarly to calcite and portlandite, magnesite and brucite are involved into the following reaction:

$$\underset{\text{magnesite}}{MgCO_3} + H_2O = \underset{\text{brucite}}{Mg(OH)_2} + CO_{2(g)} \tag{5-15}$$

whose thermodynamic equilibrium constant is:

$$K_{\text{magnesite}-\text{brucite}} = f_{CO_2}, \tag{5-16}$$

assuming that liquid water and the solid phases are pure. Figure 5.6 shows that the f_{CO_2} values constrained by magnesite–brucite coexistence varies from ~10^{-7} bar at 10°C to ~10^{-3} bar at 120°C and are therefore 1–4 orders of magnitude higher than those measured in high-pH, Ca–OH waters and 5–7 orders of magnitude greater than those fixed by calcite–portlandite coexistence (see above). Also shown in Fig. 5.6 are the stability relations between brucite, artinite and hydromagnesite. These are drawn by writing the equilibrium reactions in terms of $CO_{2(g)}$ and liquid H_2O, through summation of the reactions given in Table 5.4. For instance, for the hydromagnesite–artinite equilibrium we start to write the following three reactions:

$$2\text{ Hydromagnesite} + 12H^+ = 10Mg^{2+} + 8HCO_3^- + 12H_2O \tag{5-17}$$

$$10Mg^{2+} + 5HCO_3^- + 25H_2O = 5\text{Artinite} + 15H^+ \tag{5-18}$$

$$3H^+ + 3HCO_3^- = 3CO_{2(g)} + 3H_2O. \tag{5-19}$$

Reactions (5-17) and (5-18) are obtained by multiplying times 2 and 5, respectively, those reported in Table 5.4, to balance the number of Mg^{2+} ions on both sides. Reaction (5-19) is obtained by multiplying times 3 that of Table 5.4, to balance both H^+ and HCO_3^- ion. The summation of the three reactions leads to:

$$2\text{Hydromagnesite} + 10H_2O = 5\text{Artinite} + 3CO_{2(g)}, \tag{5-20}$$

which is exactly what we want. Again, if the solid phases are pure and liquid water is also pure, the log K of reaction (5-20) is simply equal to 3 log f_{CO_2}. Figure 5.6 shows that the two hydrated Mg-carbonates are metastable relative to magnesite, consistently with their relatively unfrequent occurrences in geological environments.

It must be underscored that the stability relations among solid phases in the system $MgO–CO_2–H_2O$ depicted in Fig. 5.6 are somewhat different from those shown in the same plot by Langmuir (1965). As recognized by Langmuir (1997), the reason for this is the solubility product of magnesite, which was considered to be −4.9 (at 25°C, 1.013 bar) by Langmuir (1965), whereas the value adopted here is −8.04 (from Helgeson et al., 1978). The latter value is also supported by the solubility measurements reported by Khodakovsky (Langmuir, 1997) and by the entropy and enthalpy data by Koziol and Newton (1995). These

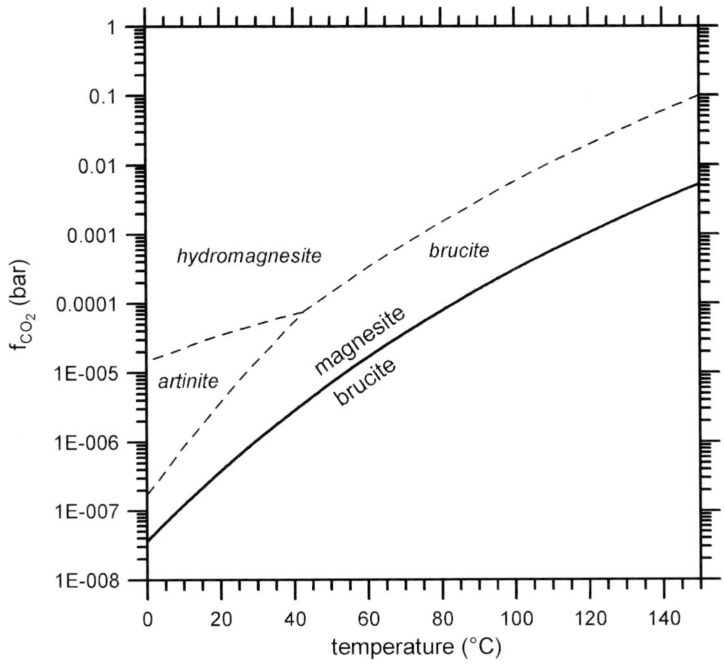

Figure 5.6. Stability diagram for the $MgO–CO_2–H_2O$ system from 0 to 150°C at 1.013 bar total pressure.

discrepant values in the geochemical literature reflect the difficulties in measuring the low-temperature solubility of magnesite due to its slow dissolution rate (Langmuir, 1965).

Summing up, the expected solid phase produced by carbonation reactions in the system $MgO–CO_2–H_2O$ is magnesite.

5.1.4.3. The system $CaO–MgO–CO_2–H_2O$

Based on the results of previous sections, for evaluating the relative stability of carbonate minerals in the system $CaO–MgO–CO_2–H_2O$, we may restrict consideration to: (i) calcite and magnesite, which are present under most geological circumstances and are the stable products of carbonation reactions, and (ii) the two hydroxides brucite and portlandite. In addition, we have to include the two Ca–Mg-carbonates dolomite and huntite.

Dolomite follows only calcite in the list of the most abundant carbonate minerals. As recalled in Section 5.1.2, dolomite is likely to be completely ordered or close to this condition at the relatively low temperatures of interest, although understanding of order/disorder in dolomites is still far from complete (see above). Due to these uncertainties, ordered-dolomite will be initially considered in the following discussion, whereas disordered-dolomite will be taken into account afterwards.

Huntite has been first described as late as 1953, but it is appearing to be not uncommon as an alteration or weathering product of dolomite- or magnesite-bearing rocks (Deer et al., 1992).

According to Langmuir (1984), the stability relations among Ca–Mg carbonates, at constant temperature and pressure, are conveniently displayed by a plot of the Ca/Mg molality ratio vs. CO_2 partial pressure, both in a logarithmic scale. We introduce some minor changes to this plot, constructing a log–log diagram of the Ca^{2+}/Mg^{2+} activity ratio vs. CO_2 fugacity. Differences between CO_2 partial pressures and CO_2 fugacities are expected to be minor, at least in the considered range. Also the $a_{Ca^{2+}}/a_{Mg^{2+}}$ ratio should be close to the $m_{Ca^{2+}}/m_{Mg^{2+}}$ ratio, as the individual ion activity coefficients of $Ca^{2+}$and Mg^{2+} ions tend to cancel out, at least if the aqueous solutions are not too concentrated. It is important to underscore, however, that the considered activities refer to the free ions Ca^{2+} and Mg^{2+} and, when dealing with aqueous solutions, they have to be properly computed by means of speciation calculations.

To construct the plot, we have first to write the equilibrium reactions among all Ca–Mg carbonates and hydroxides in terms of gaseous CO_2 and liquid water. In addition, Ca^{2+} and Mg^{2+} ions have to be on the opposite sides of the final reaction, with stoichiometric coefficient equal to unity. For instance, for the brucite–huntite equilibrium, we begin to write the reactions:

$$4\text{Brucite} + 8\text{H}^+ = 4\text{Mg}^{2+} + 8\text{H}_2\text{O} \tag{5-21}$$

$$\text{Ca}^{2+} + 3\text{Mg}^{2+} + 4\text{HCO}_3^- = \text{Huntite} + 4\text{H}^+ \tag{5-22}$$

$$4\text{CO}_{2(g)} + 4\text{H}_2\text{O} = 4\text{H}^+ + 4\text{HCO}_3^-. \tag{5-23}$$

Note that reaction (5-21) has been obtained by multiplying times 4 that of Table 5.4, in order to have one Mg^{2+} ion on the right side of the final reaction and one Ca^{2+} ion on its left side. Again, reaction (5-23) has been obtained by multiplying times 4 that of Table 5.4, to balance both H^+ and HCO_3^- ions as seen above for reaction (5-19). Then, we sum reactions (5-21) to (5-23) to obtain the following one:

$$4\text{Brucite} + \text{Ca}^{2+} + 4\text{CO}_{2(g)} = \text{Huntite} + \text{Mg}^{2+} + 4\text{H}_2\text{O}, \tag{5-24}$$

whose thermodynamic equilibrium constant can be rearranged as follows:

$$\log\frac{a_{\text{Ca}^{2+}}}{a_{\text{Mg}^{2+}}} = -\log K_{\text{brucite-huntite}} - 4\cdot\log f_{\text{CO}_2}. \tag{5-25}$$

This linear equation allows us to draw the brucite–huntite boundary in the plot under preparation. Repeating the same exercise for all the pairs of minerals of interest and inserting the log *K* values at 25°C and 1.013 bar, we obtain the linear relationships reported in Table 5.5. Therefore, we can draw a series of straight lines (each one expressing the equilibrium coexistence of a given pair of minerals) on the plot under construction (Fig. 5.7a). Note that, for simplicity, we have not computed the calcite–magnesite metastable boundary, because of the obvious presence of the dolomite field between the stability fields of calcite and magnesite. Also excluded is portlandite, as its stability field is limited by the brucite–portlandite

TABLE 5.5

Thermodynamic equilibrium constants, at specified temperatures and total pressures, of the chemical reactions used to construct the log–log diagram of Ca/Mg activity ratio vs. CO_2 fugacity

Reaction	$\log(a_{Ca^{2+}}/a_{Mg^{2+}}) - \log f_{CO_2}$ relation	log K values at T,P				
		25°C	60°C	100°C	150°C	150°C
		1.013 bar	1.013 bar	1.013 bar	4.7572 bar	500 bar
Portlandite + Mg^{2+} = brucite + Ca^{2+}	$\log(a_{Ca^{2+}}/a_{Mg^{2+}}) = \log K$	6.2572	5.9286	5.6094	5.2724	---
Calcite + Mg^{2+} + H_2O = brucite + Ca^{2+} + $CO_{2(g)}$	$\log(a_{Ca^{2+}}/a_{Mg^{2+}}) = \log K - \log f_{CO_2}$	−6.6357	−4.8817	−3.3197	−1.8287	−1.6714
Dolomite + Mg^{2+} + $2H_2O$ = 2brucite + Ca^{2+} + $2CO_{2(g)}$	$\log(a_{Ca^{2+}}/a_{Mg^{2+}}) = \log K - 2 \cdot \log f_{CO_2}$	−14.4553	−11.0980	−8.0936	−5.2065	−4.9137
Magnesite+ H_2O = brucite + $CO_{2(g)}$	$\log f_{CO_2} = \log K$	−6.1908	−4.7764	−3.5065	−2.2806	−2.1415
2Calcite + Mg^{2+} = dolomite + Ca^{2+}	$\log(a_{Ca^{2+}}/a_{Mg^{2+}}) = \log K$	1.1839	1.3346	1.4542	1.5491	1.5709
Dolomite + Mg^{2+} = 2magnesite + Ca^{2+}	$\log(a_{Ca^{2+}}/a_{Mg^{2+}}) = \log K$	−2.0737	−1.5452	−1.0806	−0.6453	−0.6307
Huntite + Mg^{2+} + $4H_2O$ = 4brucite + Ca^{2+} + $4CO_{2(g)}$	$\log(a_{Ca^{2+}}/a_{Mg^{2+}}) = \log K - 4 \cdot \log f_{CO_2}$	−23.6366	−17.8580	−12.6865	−7.7142	−7.1279
4/3Calcite + Mg^{2+} = 1/3huntite + Ca^{2+}	$\log(a_{Ca^{2+}}/a_{Mg^{2+}}) = \log K$	−0.9687	−0.5563	−0.1974	0.1331	0.1474
2Dolomite + Mg^{2+} = huntite + Ca^{2+}	$\log(a_{Ca^{2+}}/a_{Mg^{2+}}) = \log K$	−5.2740	−4.338	−3.5007	−2.6988	−2.6995
Huntite + Mg^{2+} = 4magnesite + Ca^{2+}	$\log(a_{Ca^{2+}}/a_{Mg^{2+}}) = \log K$	1.1266	1.2476	1.3395	1.4082	1.4381
Dis-dolomite + Mg^{2+} + $2H_2O$ = 2brucite + Ca^{2+} + $2CO_{2(g)}$	$\log(a_{Ca^{2+}}/a_{Mg^{2+}}) = \log K - 2 \cdot \log f_{CO_2}$	−12.9109	−9.7792	−6.981	−4.2972	−4.0043
2Calcite + Mg^{2+} = dis-dolomite + Ca^{2+}	$\log(a_{Ca^{2+}}/a_{Mg^{2+}}) = \log K$	−0.3605	0.0158	0.3416	0.6398	0.6615
Dis-dolomite + Mg^{2+} = 2magnesite + Ca^{2+}	$\log(a_{Ca^{2+}}/a_{Mg^{2+}}) = \log K$	−0.5293	−0.2264	0.032	0.264	0.2787
2dis-dolomite + Mg^{2+} = huntite + Ca^{2+}	$\log(a_{Ca^{2+}}/a_{Mg^{2+}}) = \log K$	−2.1852	−1.7004	−1.2755	−0.8802	−0.8807

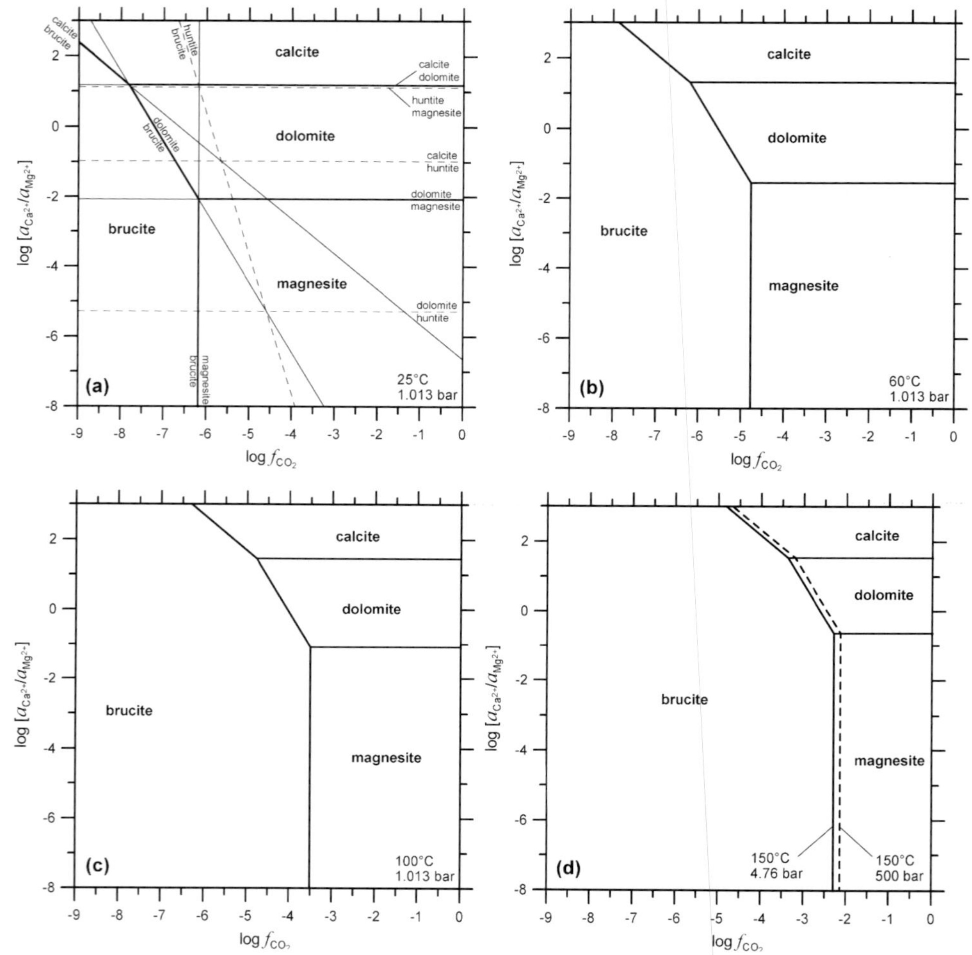

Figure 5.7. Log–log plots of the Ca^{2+}/Mg^{2+} activity ratio vs. CO_2 fugacity, showing the stability relations in the $CaO–MgO–CO_2–H_2O$ system at (a) 25°C, 1.013 bar total pressure, (b) 60°C, 1.013 bar, (c) 100°C, 1.013 bar and (d) 150°C and total pressure of either 4.76 bar (saturation pressure, solid line) or 500 bar (dashed line).

coexistence line, which corresponds to $\log(a_{Ca^{2+}}/a_{Mg^{2+}})$ values of 6.3–5.3, for temperatures of 25–150°C (see Table 5.5), i.e. well above the maximum of the *y* axis. Before going any further, let us also temporarily leave huntite out of consideration.

In order to define the stability fields of the considered solid phases (brucite, magnesite, dolomite and calcite), we have now to apply both the Schreinemakers' rule and some common sense. The Schreinemakers' rule states that *phase boundaries, when produced, must extend into fields with a higher number of phases* (Fisher, 1991) or, in other terms, that *no divariant assemblage can be stable within a sector that makes an angle of more than 180° measured between any two univariant lines in the same bundle* (Nordstrom and Munoz, 1985). For example, let us consider the intersection of the straight lines labelled dolomite–brucite, dolomite–magnesite and magnesite–brucite. Applying the Schreinemakers' rule we readily find that there are two possible solutions, one is that marked by the thick lines (this is the right solution, but we still do not know it) and the other is that marked by the thin lines. Other possibilities violates the Schreinemakers' rule and are rejected. In order to choose between the two possible solutions, we have to use some common sense. In this specific case, the stability fields of brucite and magnesite must correspond to regions of minimum f_{CO_2} and minimum $a_{Ca^{2+}}/a_{Mg^{2+}}$ ratio, respectively. It turns out that the only possible solution is shown by the thick lines.

We now consider the huntite–brucite, huntite–magnesite, huntite–dolomite and huntite–calcite coexistence lines. All these lines are situated in the stability fields of the other four solid phases, indicating that huntite is metastable with respect to them, at the considered temperature, pressure conditions.

The Ca^{2+}/Mg^{2+} activity ratio vs. CO_2 fugacity plot thus obtained shows that Ca–Mg-carbonates are stable, with respect to brucite, for f_{CO_2} greater than ~10^{-6}–10^{-9} bar. Among the Ca–Mg-carbonates, dolomite is stable in a very large range of the $a_{Ca^{2+}}/a_{Mg^{2+}}$ ratio. Dolomite is substituted by calcite for $a_{Ca^{2+}}/a_{Mg^{2+}} > 10$ and by magnesite for $a_{Ca^{2+}}/a_{Mg^{2+}} < 0.01$.

To investigate the effect of temperature on relevant multicomponent–multiphase equilibria, we have prepared the same diagram at different conditions, namely 60°C, 1.013 bar (Fig. 5.7b), 100°C, 1.013 bar (Fig. 5.7c) and 150°C, 4.76 bar (saturation pressure, Fig. 5.7d). In the latter plot also shown are the stability fields at the same temperature, but at the total pressure of 500 bar, to appreciate the pressure effect.

The temperature increase from 25 to 150°C brings about a considerable expansion of the stability field of brucite, a limited expansion of the magnesite field at dolomite's expense, and an even smaller expansion of the dolomite field at calcite's expense. In particular, dolomite is expected to be the prevailing product solid phase of carbonation reactions, in the temperature range 25–150°C, for Ca^{2+}/Mg^{2+} activity ratios larger than 0.01–0.2 and smaller than 10–35, i.e. in a rather wide range of aqueous solution compositions.

The pressure effect is totally negligible on the calcite–dolomite and dolomite–magnesite equilibria and very small on the brucite–carbonates reactions, also considering that the total pressure of 500 bar represents a very high value for our purposes.

So far, dolomite was considered to be completely ordered, i.e. the Bragg–Williams ordering parameter of dolomite, *s*, was taken equal to 1, as expected at relatively low temperatures. What happens if disordered dolomite ($s = 0$) is taken into account instead of ordered dolomite? To answer this question, a new series of log–log plots of the Ca^{2+}/Mg^{2+} activity ratio vs. CO_2 fugacity has been prepared (Fig. 5.8). They show the stability relations between disordered-dolomite, brucite, calcite and magnesite at the same pressure, temperature conditions explored in Fig. 5.7. Inspection of Fig. 5.8 shows that the stability field of disordered dolomite is very reduced, especially at the lowest temperatures, whereas calcite and magnesite are the prevailing phases under relatively high values of the $a_{Ca^{2+}}/a_{Mg^{2+}}$ ratio and relatively low values of the same ratio, respectively. Therefore, disordered-dolomite represents an unlikely product of carbonation reactions, apart from very peculiar conditions (i.e., special values of the $a_{Ca^{2+}}/a_{Mg^{2+}}$ ratio), as expected.

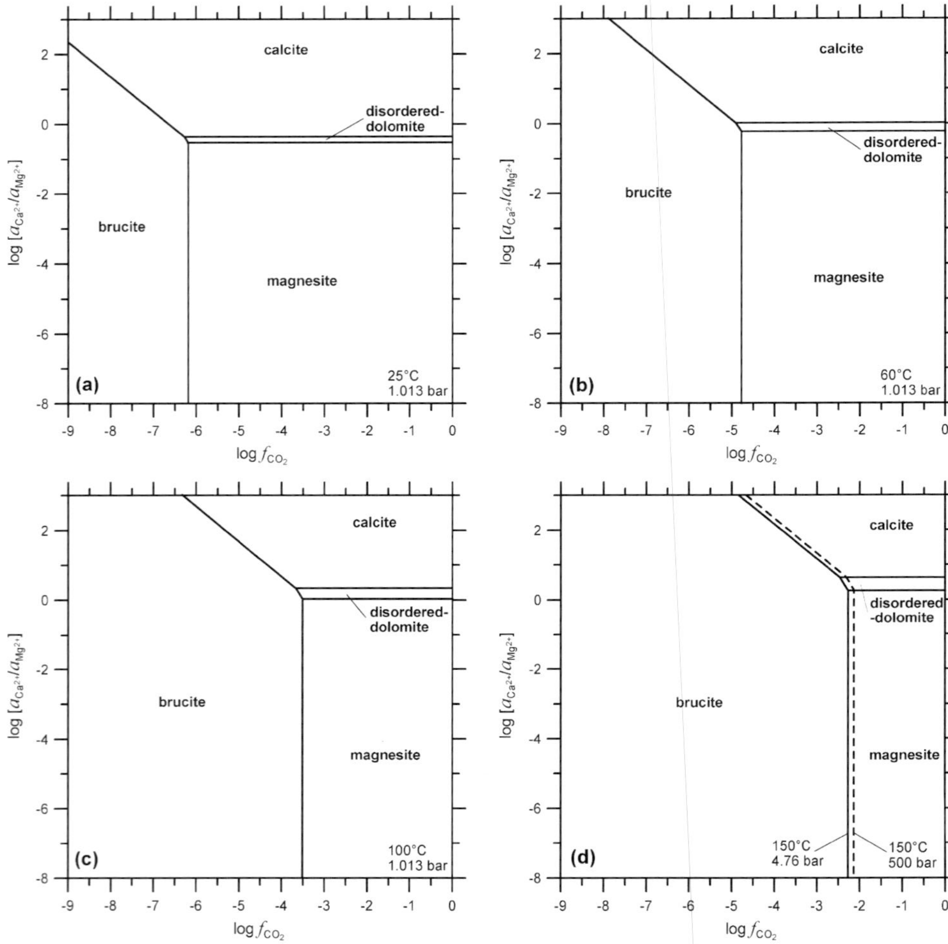

Figure 5.8. Log–log plots of the Ca^{2+}/Mg^{2+} activity ratio vs. CO_2 fugacity, showing the stability relations between brucite, calcite, disordered-dolomite and magnesite at (a) 25°C, 1.013 bar total pressure, (b) 60°C, 1.013 bar, (c) 100°C, 1.013 bar and (d) 150°C and total pressure of either 4.76 bar (saturation pressure, solid line) or 500 bar (dashed line).

All the outcomes of the previous discussion are based on purely thermodynamic considerations. Things might be different if the system does not attains stable chemical equilibrium, as assumed above. What happens if production of both dolomite and magnesite is hindered, e.g. for kinetic reasons? In this case, it is reasonable to assume that either huntite or calcite will form, depending on the Ca^{2+}/Mg^{2+} activity ratio. This possibility is quantitatively explored in the log–log plots of Ca^{2+}/Mg^{2+} activity ratio vs. CO_2 fugacity of Fig. 5.9, showing the stability relations between brucite, calcite and huntite at the same pressure, temperature conditions of Figs. 5.7 and 5.8. Temperature effects are rather

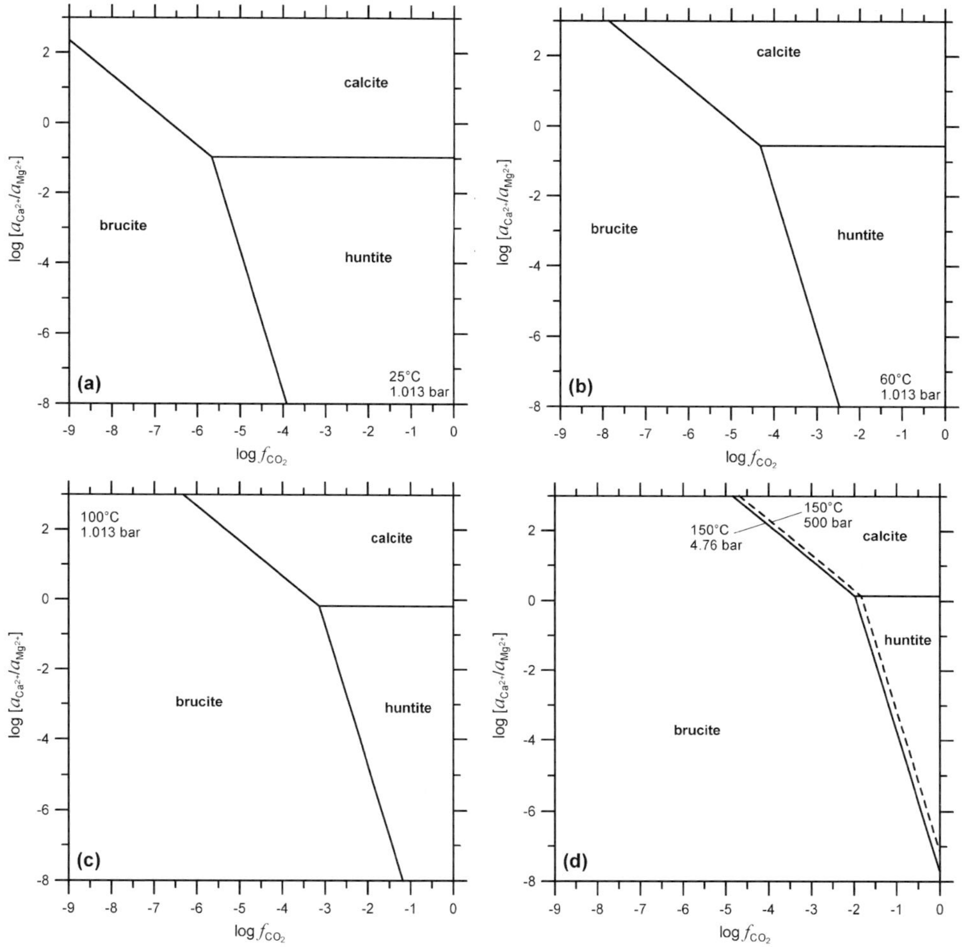

Figure 5.9. Log–log plots of the Ca^{2+}/Mg^{2+} activity ratio vs. CO_2 fugacity, showing the stability relations between brucite, calcite and huntite at (a) 25°C, 1.013 bar total pressure, (b) 60°C, 1.013 bar, (c) 100°C, 1.013 bar and (d) 150°C and total pressure of either 4.76 bar (saturation pressure, solid line) or 500 bar (dashed line).

important, as calcite prevails for $a_{Ca^{2+}}/a_{Mg^{2+}}$ ratios greater than 0.11 at 25°C, whereas this threshold becomes 0.28 at 60°C, 0.63 at 100°C and 1.4 at 150°C, irrespective of total pressure. It must be underscored that production of 1 mole of huntite [$CaMg_3(CO_3)_4$, $v^o_{huntite} = 122.9$ cm^3 mol^{-1}] instead of 1 mole of calcite ($v^o_{calcite} = 36.934$ cm^3 mol^{-1}) plus 3 moles of magnesite ($v^o_{magnesite} = 28.018$ cm^3 mol^{-1}) makes some difference, in terms of volume occupied by these solid phases. Indeed, the total volume occupied by calcite plus magnesite is $36.934 + 3 \times 28.018 = 120.988$ cm^3 mol^{-1}, whereas the corresponding huntite volume is only 1.9 cm^3 (or 1.6%) more.

5.1.5 Dawsonite: new perspectives for the geological sequestration of CO_2

Dawsonite is an hydroxycarbonate of sodium and aluminium [$NaAl(CO_3)(OH)_2$] named in 1874 after the Canadian geologist J.W. Dawson (1820–1899). Both K^+ and NH_4^+ ions are able to substitute for Na^+ ion in dawsonite (Zhang et al., 2004). However, Na-rich compositions are most common in nature.

The structure of dawsonite is orthorhombic–dipyramidal (space group *Imam*) and consists of distorted $NaO_4(OH)_2$ and $AlO_2(OH)_4$ octahedra and CO_3 groups (Corazza et al., 1977). Its cell dimensions are: $a = 6.73$ Å, $b = 10.36$ Å, $c = 5.575$ Å, $Z = 4$ (from the mineralogy database: http://webmineral.com). Based on these figures, a cell volume of 388.70 Å^3 is computed by means of equation (5-1) and a molar volume $v = 58.52$ cm^3 mol^{-1} is then obtained by using equation (5-2). This value is close to that compiled by Robie and Hemingway (1995), 59.30 cm^3 mol^{-1}. These figures indicate that the structure of dawsonite is not as close-packed as that of calcite, which is unfavourable for the sub-surface storage of CO_2.

Dawsonite is found either in colourless to white radially distributed acicular crystals or as aggregates with fibrous morphology. Owing to these characteristics, dawsonite is expected to block pores and vugs, thus decreasing rock porosity and permeability. This is also unfortunate for the geological sequestration of CO_2.

Dawsonite is produced through hydrothermal alteration of Na-bearing Al-silicates contained in igneous rocks, such as analcime, nepheline and albite; as such, it was found in different types of magmatic rocks, comprising alkaline, felsic, mafic and ultramafic. Besides, dawsonite was reported as a daughter mineral in hydrothermal fluid inclusions (Coveney and Kelly, 1971).

Dawsonite is also common in some sedimentary sequences, such as the Bowen-Gunnedah-Sydney (BGS) sedimentary basin in Australia (Goldberry and Loughnan, 1977; Baker et al., 1995; Golab, 2003) and the Upper Triassic turbidite sandstones of the Lam formation in Yemen (Worden, 2005). In the BGS basin, dawsonite is widely distributed in marine and non-marine rocks of Permo-Triassic age as a cement, a replacement of framework clasts, and pore-filling mineral. In the Lam formation, dawsonite occurs as a widespread cement and constitutes up to 8% by volume of the rocks. In both areas, dawsonite likely grew at the expense of detrital feldspars (and illitic clays in the BGS basin) under relatively high CO_2 fugacities, owing to the extensive influx of magmatic CO_2, as suggested by $\delta^{13}C$ and $\delta^{18}O$ values (Baker et al., 1995; Worden, 2005). Low temperatures of 25–35°C and 85–100°C were estimated for the generation of dawsonite in the BGS basin and the Lam formation, respectively, based on isotopic evidence.

Moore et al. (2005) documented the dissolution of both carbonate cement and detrital feldspar clasts and the production of dawsonite and kaolinite in the siltstones of the Permian Supai Formation, representing the productive part of the Springerville-St. Johns CO_2 field (Arizona and New Mexico, USA). Dawsonite constitutes 5–17 wt.% of these siltstones and represents the youngest pore-filling mineral together with kaolinite.

An other example of dawsonite occurrence in sedimentary rocks is represented by the lacustrine evaporites of the Green River Formation in Wyoming, Colorado and Utah (USA). This formation contains the largest known resource of trona ($Na_2CO_3 \cdot NaHCO_3 \cdot 2\,H_2O$) and the second largest known resource of nahcolite ($NaHCO_3$), as well as considerable amounts of dawsonite and other sodium carbonate minerals, namely eitelite, $Na_2CO_3 \cdot MgCO_3$ and shortite, $Na_2CO_3 \cdot 2\ CaCO_3$ (Dyne, 1996). Dawsonite occurs as microscopic crystals within the microcrystalline matrix of oil shales. Its generation, in this context, was attributed to reaction of clays with Na–HCO_3 brines and CO_2 of bacterial origin by several authors (see references in Worden, 2005).

As reported by Laubengayer and Weisz (1943), laboratory synthesis of dawsonite (mixed with prevailing alumina and some diaspore) was possibly achieved in 1894–1895 by Thugutt, who heated a solution of sodium aluminate, sodium hydroxide, and sodium carbonate at 184–190°C. Laubengayer and Weisz (1943) tried to repeat this experiment but obtained only poorly crystallized böhmite. More recently, dawsonite was synthesized by Jackson et al. (1972) at 175–200°C under a P_{CO_2} of 1.035 bars and high Na/Al molar ratio, close to 43. At room temperature, dawsonite was obtained from Na carbonate/bicarbonate aqueous solutions in Al-rich environments (e.g., Furmakova, 1981).

Thermodynamic data were produced by Ferrante et al. (1976) for temperatures up to 400°C at atmospheric pressure, whereas dawsonite stability at high pressure was never studied. Although dawsonite dissolves incongruently at atmospheric pressure, with concurrent precipitation of gibbsite (Furmakova, 1981), it is convenient to describe dawsonite dissolution through the congruent reaction reported in Table 5.4, whose thermodynamic equilibrium constant at different temperatures and pressures of interest (derived from the COM thermodynamic database of EQ3/6) is also given in the same table. Note that high temperature data were computed assuming constant enthalpy of reaction, i.e. by using equation 5-40 (see below), owing to the considerable uncertainty on the heat capacity of dawsonite. These data are used together with the log K_{sp} of albite, paragonite, kaolinite and quartz, and the log K of the $CO_{2(g)}$–HCO_3^- equilibrium reaction (also taken from the COM thermodynamic database of EQ3/6) to construct the stability plots of Fig. 5.10, in which the logarithm of CO_2 fugacity is plotted against the Na^+/H^+ activity ratio. To draw the phase boundaries in Fig. 5.10, both quartz saturation and Al conservation in the solid phases are assumed (see Section 5.3.3.3 for further details). Relevant reactions and their thermodynamic equilibrium constants are reported in Table 5.6. It must be pointed out that owing to both the uncertainties possibly affecting the thermodynamic data of dawsonite and (as already recalled) its unknown stability at high pressure, phase boundaries in the stability plots of Fig. 5.10 were not computed for total pressures higher than water saturation pressure.

Let us now inspect the stability plots of Fig. 5.10 to infer the conditions of dawsonite formation. First, if we choose a suitable value of Na^+ activity, we may deal with pH instead than with the Na^+/H^+ activity ratio. In particular, the $\log(a_{Na^+}/a_{H^+})$ becomes equal to pH for $a_{Na^+} = 1$, which pertains to a rather concentrated aqueous solution. This choice makes the use of these plots much simpler. For instance, at pH 4.8, kaolinite is expected to be converted to dawsonite + quartz under f_{CO_2} values of at least ~0.90 bar at 25°C, ~6.4 bar at 60°C, ~52 bar at 100°C and ~632 bar at 150°C. If we choose instead $a_{Na^+} = 0.158$ and again pH = 4.8, then $\log(a_{Na^+}/a_{H^+})$ is equal to 4 and kaolinite conversion into dawsonite + quartz is predicted to occur under f_{CO_2} values of at least ~5.7 bar at 25°C, ~41 bar at 60°C, ~331 bar at 100°C and ~4 kbar at 150°C. Therefore, below 100°C, the conversion

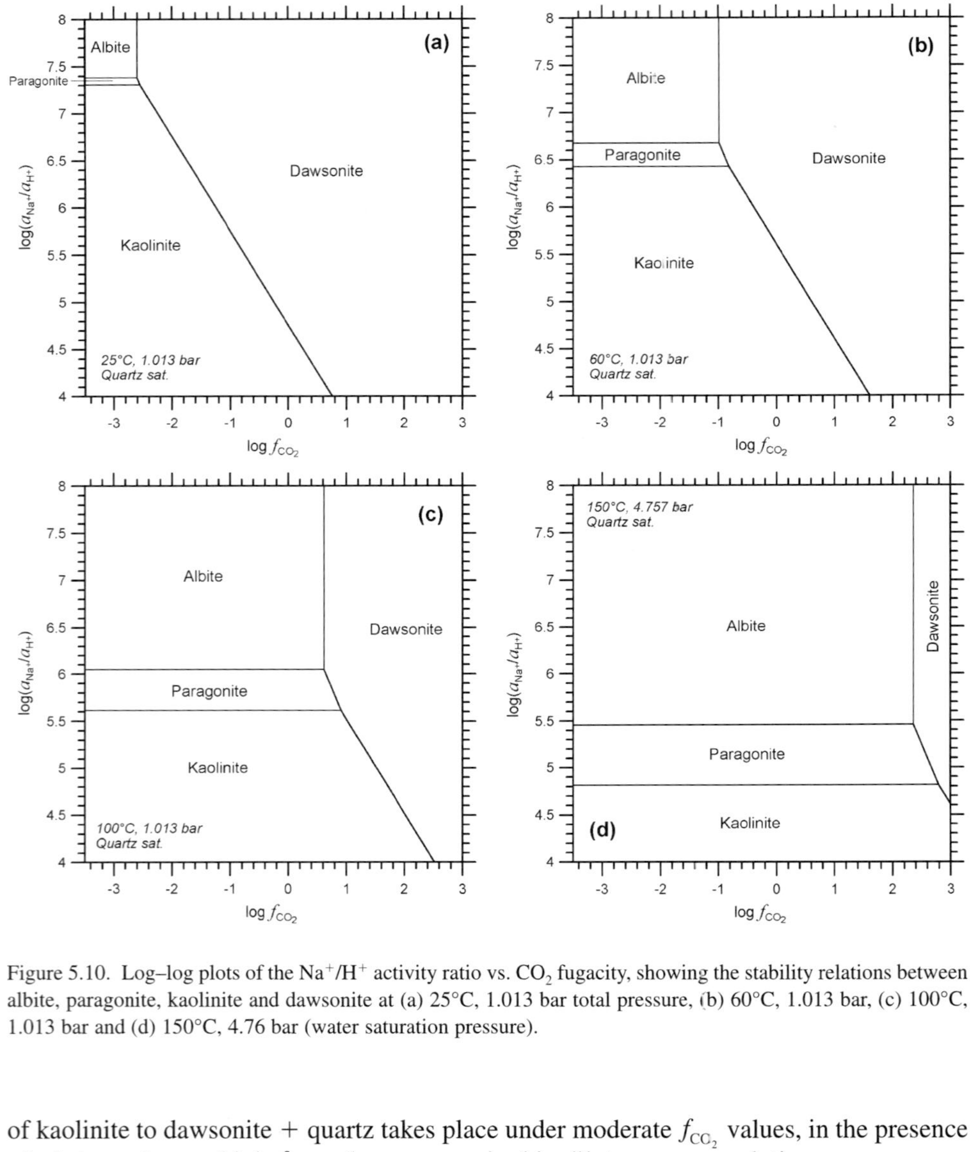

Figure 5.10. Log–log plots of the Na^+/H^+ activity ratio vs. CO_2 fugacity, showing the stability relations between albite, paragonite, kaolinite and dawsonite at (a) 25°C, 1.013 bar total pressure, (b) 60°C, 1.013 bar, (c) 100°C, 1.013 bar and (d) 150°C, 4.76 bar (water saturation pressure).

of kaolinite to dawsonite + quartz takes place under moderate f_{CO_2} values, in the presence of a brine, whereas high f_{CO_2} values are required in dilute aqueous solutions.

The plots of Fig. 5.10 also indicate that the conversion of albite to dawsonite + quartz is independent of the Na^+/H^+ activity ratio as it depends on f_{CO_2} only, and is expected to occur under f_{CO_2} values of at least ~0.0025 bar at 25°C, ~0.10 bar at 60°C, ~4.1 bar at 100°C and ~228 bar at 150°C, which are all lower than the corresponding minimum f_{CO_2} values for conversion of kaolinite to dawsonite + quartz.

Therefore, based on a purely thermodynamic point of view, the subsurface storage of CO_2 may bring about dawsonite formation especially if the reservoir rocks contain

TABLE 5.6

Thermodynamic equilibrium constants, at specified temperatures and total pressures, of the chemical reactions used to construct the log–log diagram of Na^+/H^+ activity ratio vs. CO_2 fugacity

Reaction	$\log(a_{Na^+}/a_{H^+}) - \log f_{CO_2}$ relation	log *K* at T,P			
		25°C, 1.013 bars	60°C, 1.013 bars	100°C, 1.013 bars	150°C, 4.7572 bar
3Albite + $2H^+$ = Paragonite + 6Quartz + $2Na^+$	$\log K_1 = 2\log(a_{Na^+}/a_{H^+})$	14.7673	13.3490	12.0961	10.9176
2Paragonite + $2H^+$ + $3H_2O$ = 3Kaolinite + $2Na^+$	$\log K_2 = 2\log(a_{Na^+}/a_{H^+})$	14.6137	12.8492	11.2255	9.6249
Albite + $CO_{2(g)}$ + H_2O = Dawsonite + 3Quartz	$\log K_3 = -\log f_{CO_2}$	2.6024	0.9825	−0.6166	−2.3572
Paragonite + $3CO_{2(g)}$ + $2Na^+$ + $3H_2O$ = 3Dawsonite + 3Quartz + $2H^+$	$\log K_4 = -2\log(a_{Na^+}/a_{H^+}) - 3\log f_{CO_2}$	−6.9601	−10.4015	−13.9459	−17.9892
Kaolinite + $2CO_{2(g)}$ + $2Na^+$ + H_2O = 2 Dawsonite + 2Quartz + 2 H^+	$\log K_5 = -2\log(a_{Na^+}/a_{H^+}) - 2\log f_{CO_2}$	−9.5113	−11.2174	−13.0391	−15.2011

abundant albite-rich feldspars and/or host high-salinity aqueous solutions, as already pointed out by Worden (2005).

5.2. The stability of silica minerals

One might ask why we have to be concerned about silica minerals. After all, we are interested in carbonation reactions, aren't we? Indeed, if we were interested in the carbonation of oxides and hydroxides, we could safely disregard silica minerals, as the carbonation reactions of interest would be:

$$MgO + CO_2 = MgCO_3, \tag{5-26}$$

for periclase, and

$$Mg(OH)_2 + CO_2 = MgCO_3 + H_2O \tag{5-27}$$

for brucite. Similar reactions apply to Ca oxide and Ca hydroxide. Unfortunately, periclase and brucite are rather uncommon minerals and lime [CaO] and portlandite [$Ca(OH)_2$] are even less frequent in natural environments.

Silicate minerals are the most abundant, naturally available reactants for carbonation reactions. Consequently, not only carbonates but also silica minerals are among the product solid phases. This is clearly indicated, as an example, by the following reactions between diopside and CO_2:

$$CaMgSi_2O_6 + 2CO_2 = CaMg(CO_3)_2 + 2SiO_2, \tag{5-28}$$

and serpentine and CO_2:

$$Mg_3Si_2O_5(OH)_4 + 3CO_2 = 3MgCO_3 + 2SiO_2 + 2H_2O, \tag{5-29}$$

as well as by the carbonation reactions of albite, paragonite and kaolinite, that we have already considered in the previous section (see Table 5.6). Note that the amounts of the product silica minerals, on a molar basis, can be even higher than those of carbonates (e.g., reaction (5-28)).

5.2.1. The crystallographic properties of silica minerals

At relatively low pressures, three crystalline minerals of silica are stable: quartz, tridymite and cristobalite. High-pressure solid phases, such as coesite and stishovite, can be safely disregarded for our purposes. Quartz and cristobalite have a low- and a high-temperature modification, called α- and β-, respectively (Deer et al., 1992). By contrast, seven different phases are reported for tridymite (Pryde and Dove, 1998 and references therein).

Another silica polymorph, moganite, first described by Flörke et al. (1976) has been the subject of several studies in recent years (see Gislason et al., 1997 and references therein). Evidence acquired so far suggests that chalcedony and chert are constituted by finely intergrown mixtures of moganite and microcrystalline quartz.

The crystal structure of all these silica polymorphs is made up of a three-dimensional framework of SiO_4 tetrahedra, which are linked together by sharing the corners with other tetrahedra. In these lattices, therefore, each oxygen atom has two silicon atoms as nearest neighbours and each silicon atom is surrounded by four oxygen atoms. In spite of these similarities, each silica polymorph has its own crystal structure.

The crystal structure of β-quartz is made up of SiO_4 tetrahedra arranged to form regular hexagonal and trigonal helices. In Fig. 5.11, these hexagonal helices are projected on [0001], i.e. looking down the vertical axis. Form this point of view, the silicon atoms occupy a two-dimensional hexagonal lattice. In the crystal structure of α-quartz, the SiO_4 tetrahedra are slightly displaced from their ideal positions and are distributed less regularly than in β-quartz. Consistently, the transition from β-quartz to α-quartz (which was discovered by Le Chatelier, 1889) involves minor atomic movements and no breakage of Si–O bonds.

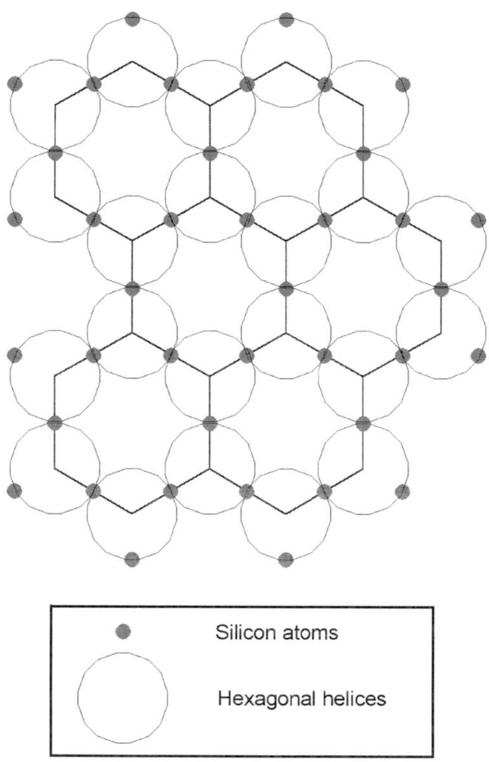

Figure 5.11. Crystal structure of β-quartz projected on [0001].

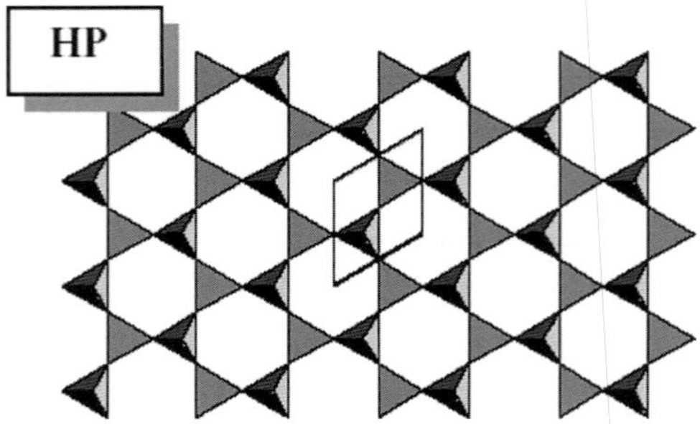

Figure 5.12. Crystal structure of a single layer of HP trydimite viewed down [001]. The unit cell is outlined. Reproduced from A. K. A. Pryde and M. T. Dove (1998) On the sequence of phase transitions in tridymite. Phys. Chem. Minerals, volume 26, pages 171–179, Figure 2 at page 173. Copyright 1998 by Springer-Verlag. With kind permission of Springer Science and Business Media.

In both forms of quartz, especially in α-quartz, SiO_4 tetrahedra are more densely packed than in the crystal structures of cristobalite and tridymite. These are formed by the same structural unit-layer, consisting of six-membered rings of corner-sharing SiO_4 tetrahedra, but they differ in the way of stacking together these sheets. In spite of this structural similarity, β-cristobalite experiences only one phase transition on cooling, whereas tridymite is characterized by a complex sequence of phase transitions (Fig. 5.12; Table 5.7).

The tridymite phase of highest temperature is the hexagonal HP phase (space group $P6_3$ *mmc*), which inverts into an other hexagonal phase on cooling, the LHP tridymite (space group $P6_322$). The LHP tridymite transforms into the *C*-centred orthorhombic tridymite (space group $C222_1$), which undergoes a transition to an other orthorhombic phase with incommensurate modulation, OS tridymite. This phase inverts either to orthorhombic primitive phase, OP tridymite (space group $P2_12_12_1$) or to the *C*-centred monoclinic phase, MC tridymite (space group *Cc*), depending on the experimental conditions. Finally, it is also possible to obtain a *C*-centred triclinic phase, MX-1 tridymite (space group *C*1), either by quenching to ~250 K or by grinding the MC phase (Pryde and Dove, 1998).

Again, the cell parameters of silica minerals can be used to compute the molar volume by means of equations (5-1) and (5-2). It can be seen that the transition from β-quartz to α-quartz takes place with a 4% reduction in volume, the transition from β-cristobalite to α-cristobalite determines a 6% decrease in volume, whereas the sequence of phase transitions in tridymite are characterized by small changes in volume, and probably in energy too.

5.2.2. The phase diagram for the unary system SiO_2 and the thermodynamic properties of silica minerals

The phase diagram for the unary system SiO_2 is reported in many papers and books but an unexpected disagreement is found on the slope of the tridymite–cristobalite phase boundary and, consequently, on the location of the triple point β-quartz–tridymite–cristobalite. According to Tuttle and Bowen (1958; this is a rather old study but it is quoted by

TABLE 5.7

Crystallographic properties of silica minerals

Mineral	Crystal system	Space group	Z	a(Å)	b(Å)	c(Å)	α(°)	β(°)	γ(°)	Cell volume (Å^3)	Molar volume (cm^3 mol^{-1})	Transition temperature[a] (°C)	Reference
α-Quartz	Trigonal	$P3_221$	3	4.9133	4.9133	5.4053	90	90	120	113.00	22.684	–	Mineralogy Database[c]
β-Quartz	Hexagonal	$P6_222$	3	4.9967	4.9967	5.4577	90	90	120	118.01	23.688	575	Carpenter et al. (1998)
MX-1-tridymite	Triclinic	*Cc* (average structure)	16	8.60	5.01	16.43	–	91.51	–	*707.90*[b]	*26.644*[b]	22	Löns and Hoffmann (1987); Graetsch and Topalovic-Dierdorf (1996)
MX-1-tridymite	Triclinic	*C*1 (modulated structure)	–	8.60	15.02	16.43	–	91.51	–	–	–	22	Löns and Hoffmann (1987); Graetsch and Topalovic-Dierdorf (1996)
MC tridymite	Monoclinic	*Cc*	48	18.52	5.00	23.81	90	105.82	90	2121.29	26.614	22	Dollase and Baur (1976)
OP tridymite	Orthorhombic	$P2_12_12_1$	24	26.171	4.99	8.20	90	90	90	1070.86	26.870	155	Kihara (1977)
OS tridymite	Orthorhombic	–	–	95–65[d]	5.02	8.18	90	90	90	–	–	150–190	Nukui et al. (1978)
OC tridymite	Orthorhombic	$C222_1$	8	8.74	5.04	8.24	90	90	90	362.97	27.323	220	Dollase (1967)
LHP tridymite	Hexagonal	$P6_322$	4	5.05	5.05	8.26	90	90	120	182.43	27.465	400	Cellai et al. (1994)
HP tridymite	Hexagonal	$P6_3mmc$	4	5.05	5.05	8.27	90	90	120	182.65	27.499	460	Kihara (1978)
α-cristobalite	Tetragonal	$P4_12_12$	4	4.971	4.971	6.918	90	90	90	170.95	25.737	–	Mineralogy Database[c]
β-cristobalite	Cubic	–	8	7.13	7.13	7.13	90	90	90	362.47	27.285	200–275	Deer et al. (1992)

[a]At 1 bar total pressure.

[b]Computed assuming $\alpha = \beta = \gamma = 90°$.

[c]http://webmineral.com.

[d]The range of lattice parameters is due to the existence of a series of OS phases.

many authors, including Hall, 1987, Deer et al., 1992 and Presnall, 1995), the tridymite–cristobalite phase boundary is virtually pressure-independent and, consequently, the triple point β-quartz/tridymite/cristobalite is situated at 1,470°C and close to 5 kb. On the contrary, Ernst (1976) presents the SiO_2 phase diagram of Boyd and England (1960) and underscores that "*since the molar volume of tridymite exceeds that of cristobalite, whereas tridymite has a smaller entropy than the high-temperature polymorph, the univariant curve along which both are stable has a negative slope*". Ernst (1976) makes reference to the well-known Clapeyron equation:

$$\frac{\mathrm{d}P}{\mathrm{d}T} = \frac{\Delta S}{\Delta V}. \tag{5-30}$$

Indeed, if $V_{\text{tridymite}} > V_{\text{cristobalite}}$ and $S_{\text{tridymite}} < S_{\text{cristobalite}}$, then the ratio $\Delta S/\Delta V$, which is equal to the slope of the tridymite–cristobalite phase boundary, is negative. If Ernst (1976) is correct, it follows that the triple point β-quartz/tridymite/cristobalite must be found at $T \ll 1{,}470°\text{C}$ and $P \ll 5$ kb.

In order to elucidate this matter we need the thermodynamic data of silica minerals. Helgeson et al. (1978) provide an extensive discussion of the α–β transition in quartz and give thermodynamic data for α-quartz, α-cristobalite and β-cristobalite, but neither for β-quartz (since it is unquenchable) nor for other silica polymorphs listed in Table 5.7, due to uncertainties in their experimental data. This is rather expected, based on the poor knowledge of the crystallographic properties of tridymite.

On the contrary, Berman (1988) provides thermodynamic data for all the three silica minerals of interest. These are accessible through *The MELTS Supplemental Calculator* (implemented by Mark S. Ghiorso) at the web-page http://decmelts.geology.washington.edu/meltsCALC/. Selecting temperature and pressure and choosing the phase of interest, *The MELTS Supplemental Calculator* returns its thermodynamic data, comprising Gibbs free energy, enthalpy, entropy, constant-pressure heat capacity and its first temperature derivative, molar volume and its first and second temperature and pressure derivative as well as the cross derivative $\partial^2 V/(\partial T\, \partial P)$.

In order to compute the slope of the tridymite–cristobalite phase boundary, we can position ourselves at the invariant point β-quartz/tridymite/cristobalite, i.e. at 1,207°C and 1,720 bar (see below). Here, the entropy is 146.143 J mol^{-1} K^{-1} for cristobalite and 146.067 J mol^{-1} K^{-1} for tridymite, whereas their molar volumes are 27.352 and 27.492 cm^3 mol^{-1}, respectively. Therefore:

$$\frac{\mathrm{d}P}{\mathrm{d}T} = \frac{(146.143 - 146.067)}{(27.352 - 27.492)} \times 9.998 = -5.43 \text{ bar K}^{-1},$$

where the factor 9.998 is needed for converting Joules to cm^3 bar.

With the data obtained through *The MELTS Supplemental Calculator* we can now construct the phase diagram for the unary system SiO_2. First, the Gibbs free energy data of the different minerals, quartz, cristobalite and tridymite, are computed at constant pressure and variable temperature, and are plotted vs. temperature. The point where two of these curves intersect defines the temperature, pressure condition of equilibrium coexistence of the two solid phases. Repeating this exercise at least twice (3 or 4 times is better, just to check for possible mistakes), the univariant equilibrium lines of interest can be drawn.

If you do this exercise you find that it is possible to determine the data needed to draw the univariant equilibrium lines β-quartz/cristobalite and β-quartz/tridymite, but it is impossible to establish the pressure, temperature conditions of tridymite/cristobalite equilibrium, due to numerical problems. However, since (i) the invariant point β-quartz/tridymite/cristobalite is defined by the intersection of the univariant curves β-quartz/cristobalite and β-quartz/tridymite, which occurs at 1,207°C and 1,720 bar, and (ii) the tridymite–cristobalite transition takes place at 1,470°C at 1 bar (Deer et al., 1992), it follows that the tridymite–cristobalite univariant curve can be obtained, to a first approximation, by connecting these two points (Fig. 5.13). In this way, the slope of this phase boundary results to be -6.54 bar K^{-1}, a value relatively close to that computed before, -5.43 bar K^{-1}. To make an eyeball comparison of the two dP/dT values, a dashed line of slope -5.43 bar $°C^{-1}$ is drawn in Fig. 5.13, starting from the tridymite–cristobalite transition temperature at atmospheric pressure. This exercise also suggests that the high uncertainty in the tridymite–cristobalite inversion is explained by the small changes in volume and especially in entropy, only 0.08 J mol^{-1} K^{-1}, which is a very small and highly uncertain value.

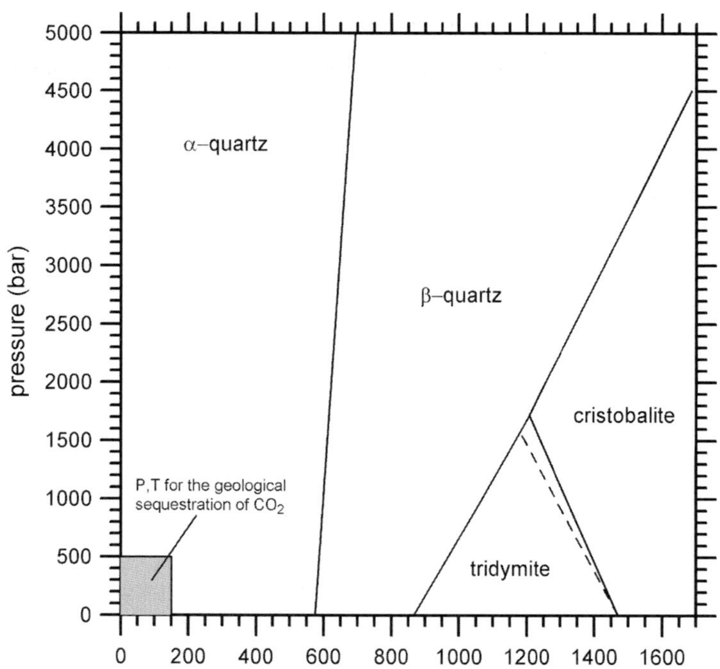

Figure 5.13. Phase diagram for the unary system SiO_2.

This also explains why the tridymite–cristobalite phase boundary is drawn in different ways, depending on the author of the phase diagram.

To complete our plot, we have still to draw the α-quartz/β-quartz phase boundary. This is an easy task, as the transition temperature is 575°C at 1 bar pressure and it increases linearly with pressure of 0.024°C bar^{-1}, based on the data of Berman (1988), or 0.026°C bar^{-1}, according to Helgeson et al. (1978).

Inspection of the phase diagram for silica (Fig. 5.13) shows that at the pressure and temperature values of interest for the geological sequestration of CO_2 and industrial carbonation, the thermodynamically stable phase of silica is α-quartz, whereas β-quartz, tridymite and cristobalite are stable at much higher temperature and pressure values. However, at atmospheric pressure, metastable tridymite may form at temperatures much lower than the β-quartz/tridymite transition temperature, 867°C, as indicated in Table 5.7. Moreover, metastable α-cristobalite can exist from ambient temperature to 200–275°C and metastable β-cristobalite can exist above these temperatures (Deer et al., 1992).

Further complications are brought about by the possible formation of: (i) chalcedony (cryptocrystalline quartz), which is constituted by tiny crystals of quartz with submicroscopic pores, (ii) opal, which is made up of cryptocrystalline cristobalite with sub-microscopic, water-containing pores, (iii) moganite, which is apparently widespread near the Earth surface (see above) and (iv) amorphous silica, which differs from silica glass (lechatelierite) for the incorporation of interstitial H_2O molecules. Indeed, deposition of amorphous silica is a major problem in the use and disposal of geothermal liquids for electrical production.

The thermodynamic properties of chalcedony and amorphous silica are reported by Helgeson et al. (1978), in addition to those of α-quartz, α-cristobalite and β-cristobalite. All these data are largely based on the results of previous researches by Walther and Helgeson (1977). The thermodynamic properties of moganite are given by Gislason et al. (1997). The temperature dependence of the Gibbs free energy of formation from the elements for all these solid phases, in the temperature range 0–150°C, are shown both (i) at pressures of 1.013 bar, below 100°C, or saturation pressure, above 100°C (Fig. 5.14a), and (ii) at constant total pressure of 500 bar (Fig. 5.14b). These plots were constructed by running SUPCRT92 (Johnson et al., 1992), which stores the thermodynamic data of Helgeson et al. (1978) for the silica minerals. The thermodynamic properties of moganite were added to the SUPCRT92 database.

In the second plot (Fig. 5.14b), the Gibbs free energy–temperature curves for all silica polymorphs are significantly shifted upwards with respect to Fig. 5.14a, indicating the relatively important pressure influence on the Gibbs free energy of silica minerals. We have already seen that quartz is the thermodynamically stable silica phase, at least for our purposes. Consistently, at any temperature and pressure, quartz is the silica mineral with the lowest Gibbs free energy and it is followed, in order of increasing Gibbs free energy values, by chalcedony, α-cristobalite, moganite, β-cristobalite and amorphous silica.

Also shown in Fig. 5.14 is the Gibbs free energy of an unspecified tridymite at 25°C, 1.013 bar (data from Wagman et al., 1982, reported in the database of the EQ3/6 software package), which is only 234 cal mol^{-1} higher than that of quartz and 136 cal mol^{-1} lower than that of chalcedony.

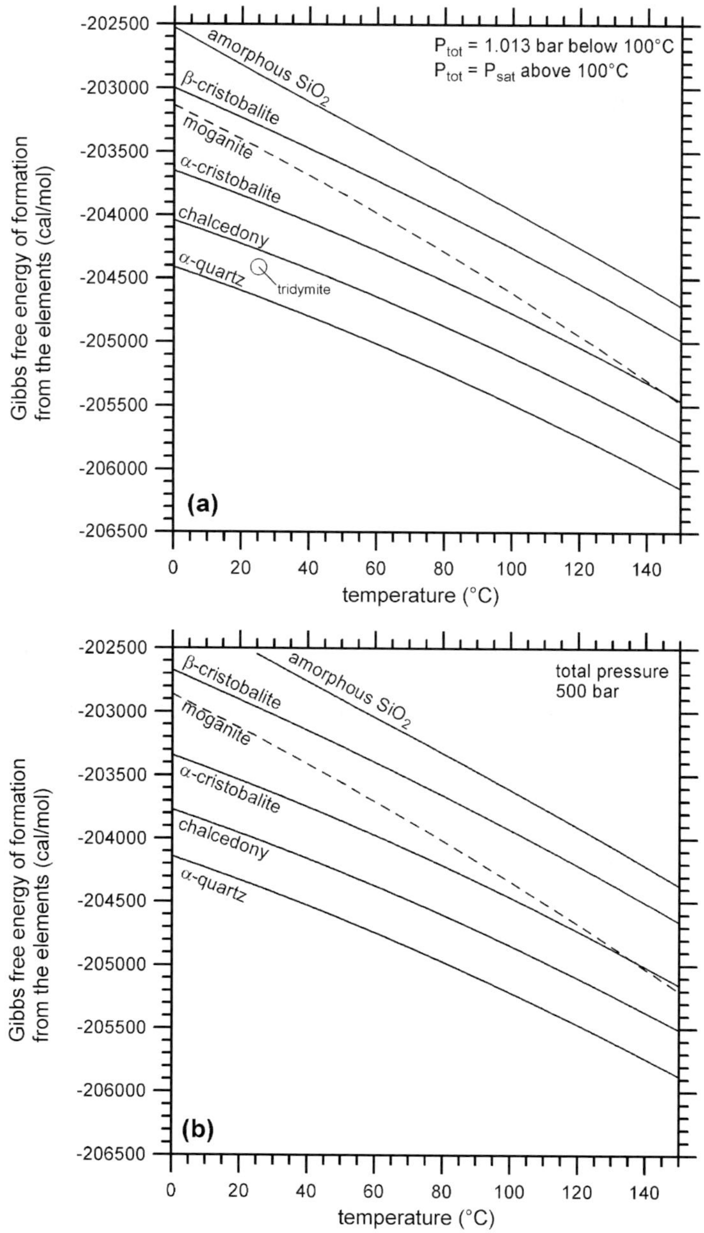

Figure 5.14. Temperature dependence of the Gibbs free energy of formation from the elements of α-quartz, chalcedony, α-cristobalite, β-cristobalite and amorphous silica (data from Helgeson et al., 1978) and moganite (data from Gislason et al., 1997), both (a) at pressures of 1.013 bar, below 100°C, or saturation pressure, above 100°C and (b) at constant total pressure of 500 bar (data computed by SUPCRT92). Also shown in (a) is the Gibbs free energy of an unspecified tridymite at 25°C, 1.013 bar (data from Wagman et al., 1982).

5.2.3. The solubilities of silica minerals

Let us now consider the dissolution reactions of silica minerals, which can be expressed as:

$$SiO_{2(s)} = SiO_{2(aq)} \qquad (5\text{-}31)$$

whose thermodynamic equilibrium constant is simply:

$$K = a_{SiO_{2(aq)}} \qquad (5\text{-}32)$$

if the considered solid phase is pure. It must be underscored that, since dissolved SiO_2 is a neutral species, its activity coefficient, $\gamma_{SiO_{2(aq)}}$, is unity or close to unity, even in relatively saline solutions. Consequently, the activity of dissolved silica is virtually equal to its molality, provided that the salinity of the aqueous solution is not too high, and, at saturation, it corresponds to the solubility of the considered mineral. Activities of dissolved silica at saturation with all the silica minerals of interest were readily computed by means of SUPCRT92, in the temperature range 0–150°C, both (i) at pressures of 1.013 bar, below 100°C, or saturation pressure, above 100°C, and (ii) at constant total pressure of 500 bar. The solubility of the unspecified tridymite of Wagman et al. (1982) and of moganite (Gislason et al., 1997) was also taken into account. All these data are plotted in the $a_{SiO_2\,(aq)}$ vs. temperature plot (Fig. 5.15), whose inspection shows that the thermodynamically most stable phase, quartz, is also the least soluble silica mineral and, again, it is followed, in order of increasing solubility, by chalcedony, α-cristobalite, moganite, β-cristobalite and amorphous silica. The solubility of the unspecified tridymite of Wagman et al. (1982) is similar to that of chalcedony, as expected due to the similar Gibbs free energy values.

Also evident from Fig. 5.15 is the remarkable pressure effect on the solubility of silica minerals, which is not surprising, due to the pressure influence on the Gibbs free energy of silica minerals (see above).

At this point of the discussion, it is useful to compare the solubilities of silica minerals computed by means of SUPCRT92, with those provided by the simple relations of Fournier (1973), which are valid from 20 to 250°C (SiO_2 indicates silica concentration in mg kg^{-1}; temperature is in Kelvin):

$$\log SiO_2 = 5.19 - \frac{1309}{T}, \quad \text{for quartz;} \qquad (5\text{-}33)$$

$$\log SiO_2 = 4.69 - \frac{1032}{T}, \quad \text{for chalcedony;} \qquad (5\text{-}34)$$

$$\log SiO_2 = 4.78 - \frac{1000}{T}, \quad \text{for } \alpha\text{-cristobalite;} \qquad (5\text{-}35)$$

$$\log SiO_2 = 4.51 - \frac{781}{T}, \quad \text{for opal-CT,} \qquad (5\text{-}36)$$

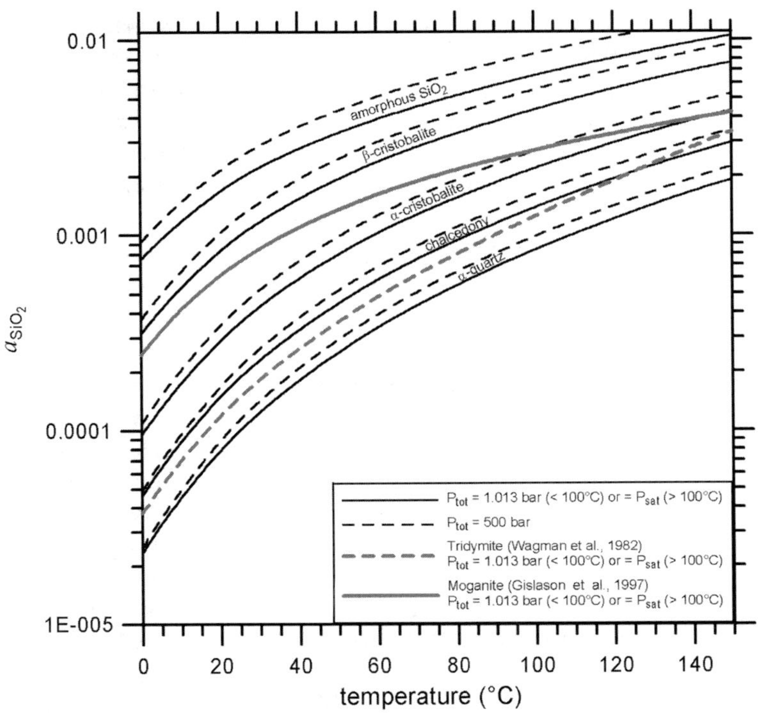

Figure 5.15. Temperature dependence of the solubility of α-quartz, chalcedony, α-cristobalite, moganite, β-cristobalite and amorphous silica at different pressures, as specified (data computed by SUPCRT92). Also shown is the temperature dependence of the solubility of an unspecified tridymite (data from Wagman et al., 1982).

(this was incorrectly identified as β-cristobalite by Fournier, 1973, as pointed out by Fournier, 1991).

$$\log SiO_2 = 4.52 - \frac{731}{T}, \quad \text{for amorphous silica.} \tag{5-37}$$

Results are shown in Table 5.8, which shows a satisfactory agreement between the two approaches, in spite of the strong contrast between the simplicity of the relations proposed by Fournier (1973), and the much more complex, thermodynamically rigorous approach by Helgeson and coworkers.

Assuming that the results provided by SUPCRT92 are right, the obvious question is: why the simple relations of Fournier (1973) works so well? The effectiveness of these relations, actually, is not surprising, since the dissolution reaction of silica mineral (5-31) is isocoulombic, i.e. it has the same number of ionic species having the same charge on each side (zero in reaction 5-31).

TABLE 5.8

Solubilities (mg kg^{-1}) of silica minerals at different temperatures, computed by means of the software code SUPCRT92 and by using the relationships of Fournier (1973)

SUPCRT92						Fournier (1973)				
T (°C)	Quartz	Chalcedony	α-cristobalite	β-cristobalite	SiO_2(am)	Quartz	Chalcedony	α-cristobalite	Opal-CT	SiO_2 (am)
25	6.0	11	21	59	116	6.3	17	27	78	117
60	20	35	61	142	236	18	39	60	146	212
100	50	83	131	263	393	48	84	126	261	364
150	115	178	260	458	630	125	178	261	462	620

It is known that the heat capacity of isocoulombic reactions, $\Delta_r C_P^o$, is close to zero (e.g., Anderson and Crerar, 1993 and Langmuir, 1997 and references therein). Consequently, the enthalpy of isocoulombic reactions is nearly constant at any temperature (at least within reasonable limits), since the following relation holds:

$$\left|\frac{\partial(\Delta_r H^o)}{\partial T}\right|_P = \Delta_r C_P^o. \tag{5-38}$$

To gain more insights into this matter let us recall equation (2-41):

$$\ln K_T = \ln K_{\mathrm{Tr}} - \frac{\Delta_r H^o_{\mathrm{Tr}}}{R}\left(\frac{1}{T} - \frac{1}{\mathrm{Tr}}\right) + \frac{\Delta a}{R}\left(\ln\frac{T}{\mathrm{Tr}} + \frac{\mathrm{Tr}}{T} - 1\right) + \frac{\Delta b}{2R}\left(T + \frac{\mathrm{Tr}^2}{T} - 2\cdot\mathrm{Tr}\right) \tag{5-39}$$

$$+ \frac{\Delta c}{2R}\left(\frac{-T^2 - \mathrm{Tr}^2 + 2\cdot T\cdot\mathrm{Tr}}{T\cdot\mathrm{Tr}}\right)$$

which reduces to

$$\log K_T = \log K_{\mathrm{Tr}} - \frac{\Delta_r H^o_{\mathrm{Tr}}}{2.303R}\left(\frac{1}{T} - \frac{1}{\mathrm{Tr}}\right), \tag{5-40}$$

for $\Delta_r C_P^o = 0$ (i.e., $\Delta a = \Delta b = \Delta c = 0$). Recalling equation (5-32), equation (5-40) can be rewritten as follows:

$$\begin{aligned}\log a_{\mathrm{SiO_2},T} &= \log a_{\mathrm{SiO_2},\mathrm{Tr}} - \frac{\Delta_r H^o_{\mathrm{Tr}}}{2.303R}\left(\frac{1}{T} - \frac{1}{\mathrm{Tr}}\right)\\ &= \left(\log a_{\mathrm{SiO_2},\mathrm{Tr}} + \frac{\Delta_r H^o_{\mathrm{Tr}}}{2.303R\,\mathrm{Tr}}\right) - \left(\frac{\Delta_r H^o_{\mathrm{Tr}}}{2.303R} \times \frac{1}{T}\right)\end{aligned} \tag{5-41}$$

Comparing equation (5-41) with the Fournier's equations, it is evident that the first and second terms on the right-hand side of the Fournier's equations correspond to the first and second terms on the right-hand side of equation (5-41), respectively. Therefore, it can be concluded that Fournier selected a proper function to express the temperature dependence of the solubility of silica minerals, as the enthalpy of their dissolution reaction is practically temperature-independent. This good choice explains why the Fournier's equations work so well.

5.2.4. The molar volumes of silica minerals

At the end of this excursus on silica minerals, we turn to consider their molar volumes, which were already taken into account at the beginning of this discussion. Again, the molar volumes of quartz, chalcedony, α-cristobalite, β-cristobalite and amorphous silica were extracted from SUPCRT92, in the temperature range 0–150°C, both (i) at pressures of 1.013 bar, below 100°C, or saturation pressure, above 100°C, and (ii) at constant total pressure of 500 bar. The molar volume of moganite is given by Heaney (1994). These data are plotted against temperature in Fig. 5.16.

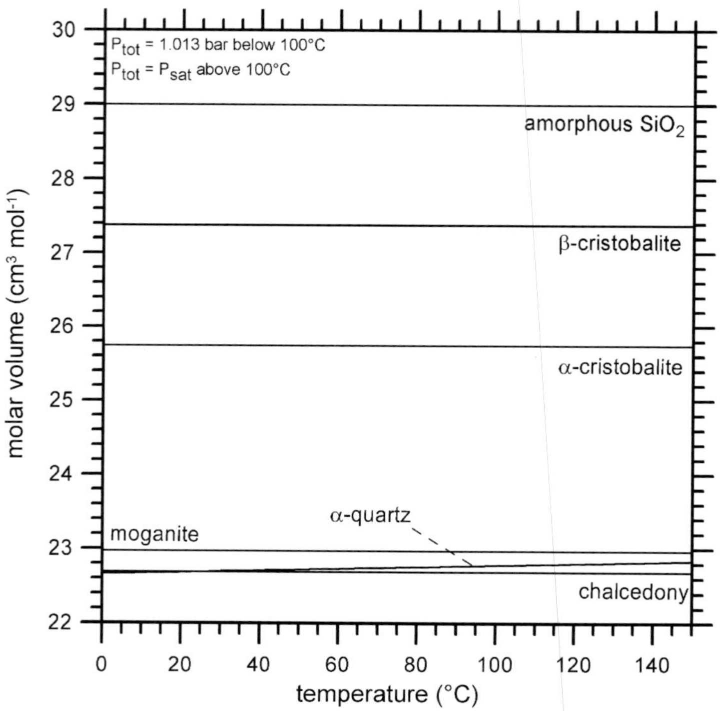

Figure 5.16. Temperature dependence of the molar volume of α-quartz, chalcedony, α-cristobalite, β-cristobalite and amorphous silica (data from Helgeson et al., 1978) and moganite (data from Heaney, 1994).

The molar volumes of chalcedony, α-cristobalite, β-cristobalite and amorphous silica are independent of pressure and temperature, as SUPCRT92 assumes $(\partial v°/\partial T)_P = 0$ and $(\partial v°/\partial P)_T = 0$. As recognized by Johnson et al. (1992), this assumption is acceptable for most minerals in the pressure, temperature ranges covered by SUPCRT92, i.e. 1–5,000 bar and 0–1,000°C. Nevertheless, there are two important exceptions, coesite and quartz, and the latter is of special importance for us. Indeed, quartz has a very high volumetric thermal expansion, which contrasts with the behaviour of other minerals (Skinner, 1966) and may control fluid circulation at depth in high-temperature geothermal areas (Marini and Manzella, 2005). The thermodynamic implications of $(\partial v°/\partial T)_P \neq 0$ are accounted for in SUPCRT92, through the incorporation of a series of equations given by Helgeson et al. (1978). Consequently, in Fig. 5.16, the molar volume of quartz changes with temperature, though to a limited extent, from 22.658 cm^3 mol^{-1} at 0°C to 22.838 cm^3 mol^{-1} at 150°C.

Apart from these temperature effects on the molar volume of quartz, Fig. 5.16 evidentiate the relatively small molar volumes of quartz, its cryptocrystalline variety (chalcedony) and moganite, contrasting with the comparatively large molar volume of amorphous silica, with α-cristobalite and β-cristobalite in intermediate positions. Since the molar volume of amorphous silica is 28% greater than that of quartz and chalcedony, the precipitation of the former instead of the latter phases makes a lot of difference during geological CO_2 sequestration, and it can determine a significant reduction in porosity and permeability of the aquifer formations, as pointed out by Cipolli et al. (2004).

5.3. Clay minerals and related solid phases

5.3.1. Clay minerals as by-products of carbonation reactions

Clay minerals (such as kaolinite, smectites and illites) are typical weathering products. Aluminium-silicates usually undergo incongruent dissolution. In other terms, their dissolution is accompanied by precipitation of some chemical components, which are chiefly incorporated into authigenic clay minerals and oxy-hydroxides. The formation of these secondary minerals is due to their low solubility and determines, to a first approximation, Al conservation in solid phases. In other words, almost all the Al released by Al-silicate dissolution is incorporated in secondary minerals, whereas the amount present in the aqueous solution is nil to negligible, at least under the typical pH values of natural waters, 5–8.5 approximately.

Since both the geological sequestration of CO_2 and mineral carbonation can be considered as peculiar types of weathering, clay minerals are expected to form during these processes, at least when Al is released through dissolution of primary Al-silicates.

For example, possible carbonation reactions of anorthite, accompanied by formation of clay minerals, are:

$$\underset{\text{anorthite}}{CaAl_2Si_2O_8} + CO_{2(g)} + 2H_2O = \underset{\text{calcite}}{CaCO_3} + \underset{\text{kaolinite}}{Al_2Si_2O_5(OH)_4} \tag{5-42}$$

$$\begin{aligned} &1.165\underset{\text{anorthite}}{CaAl_2Si_2O_8} + 1.34SiO_{2(aq)} + CO_{2(g)} + H_2O \\ &= \underset{\text{calcite}}{CaCO_3} + \underset{\text{Ca-beidellite}}{Ca_{0.165}Al_{2.33}Si_{3.67}O_{10}(OH)_2} \end{aligned} \tag{5-43}$$

$$\begin{aligned} &0.835\underset{\text{anorthite}}{CaAl_2Si_2O_8} + 2.33SiO_{2(aq)} + 0.33Mg^{2+} + 0.34CO_{2(g)} + H_2O \\ &= 0.33Ca^{2+} + 0.34\underset{\text{calcite}}{CaCO_3} + \underset{\text{Ca-montmorillonite}}{Ca_{0.165}Mg_{0.33}Al_{1.67}Si_4O_{10}(OH)_2}\,. \end{aligned} \tag{5-44}$$

In reaction (5-42), all the Ca released through anorthite dissolution is incorporated into the calcite lattice and all the Al and Si provided by anorthite are used to produce kaolinite. The chemical components used in the generation of the authigenic minerals, calcite and kaolinite, come entirely from the consumed primary mineral, anorthite.

In contrast, in reactions (5-43) and (5-44), the chemical components supplied by anorthite are not sufficient to make up the secondary phases: some additional dissolved SiO_2 is needed to produce Ca-beidellite, whereas both additional SiO_2 and Mg are required to generate Ca-montmorillonite.

From these simple examples it is evident that clay minerals are rather complex from the chemical point of view. They are also rather complex from the structural point of view, which represents the object of the next section. For further detailed information on clay minerals, the reader is referred to Deer et al. (1992) and to the monographs edited by Brindley and Brown (1980), Newman (1987) and Bailey (1988).

5.3.2. The crystal structure of clay minerals

Structurally, the clay minerals belong to the family of sheet silicates or phyllosilicates. The last name derives from the word *phyllon*, which is Greek for leaf. Phyllosilicates are constituted by the superposition of layers of cations coordinated with oxygen ions and/or hydroxyl ions. In addition, cations and water molecules can be sandwiched between these layers. In particular, two different types of layers are present in clay minerals, namely the so-called *octahedral* and *tetrahedral* layers.

A single tetrahedral layer is mainly constituted by Si^{4+} ions, and less frequent Al^{3+} ions, in 4-fold (tetrahedral) coordination with oxygen ions and/or hydroxyl ions. This kind of coordination is possible due to the small size of these two cations, whose ionic radii, in 4-fold coordination are 0.26 Å for Si^{4+} and 0.39 Å for Al^{3+} (Shannon, 1976). These small ionic radii contrast with those of oxygen ion, 1.40 Å, and hydroxyl ion, 1.37 Å.

To understand the characteristics of the octahedral layer we have to take into account the crystal structures of brucite and gibbsite (Fig. 5.17), although they are hydroxide minerals, not clay minerals. Both brucite and gibbsite have a layered structure, with

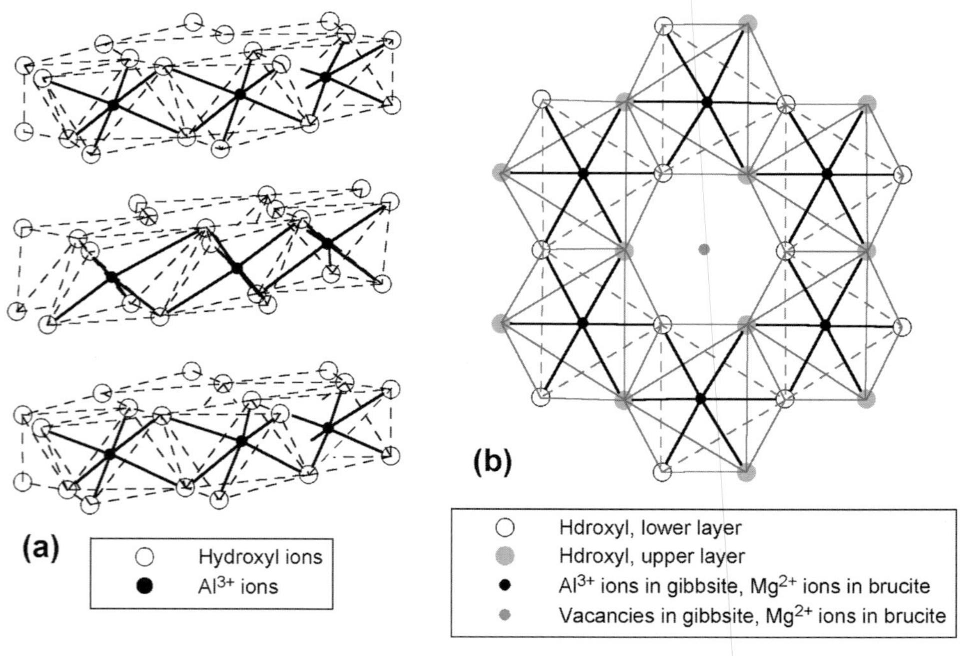

Figure 5.17. (a) Schematic representation of the crystal structure of gibbsite. (b) View of a sheet of the gibbsite (brucite) lattice from above.

two sheets of hydroxyl ions in hexagonal close-packing and a sandwiched sheet of either Mg^{2+} ions (in brucite) or Al^{3+} ions (in gibbsite), in 6-fold (octahedral) coordination.

In the crystal structure of brucite, all the octahedral sites are occupied by Mg^{2+} ions. Consequently, each Mg^{2+} ion has six OH^- ions as nearest neighbours, whereas each OH^- ion is surrounded by three Mg^{2+} ions, on one side, and by three OH^- ions of the next layer, on the other side. The corresponding stoichiometric formula of brucite is $Mg_3(OH)_6$ or, more simply, $Mg(OH)_2$. The phyllosilicates structurally based on this type of octahedral sheet are named trioctahedral.

In the gibbsite structure, only two out of three octahedral sites are occupied by Al^{3+} ions, in order to satisfy the electroneutrality condition. If all these sites were occupied, there would be an excess of positive charges, which is not possible. To verify this impossible condition, reference can be made to the chemical formula of brucite, substituting Al^{3+} for Mg^{2+}: the positive charges are $3 \times 3 = 9$, whereas the negative charges are only $6 \times 1 = 6$. With only two out of three octahedral sites occupied by Al^{3+} ions, the stoichiometric formula of gibbsite is $Al_2(OH)_6$ or, more simply, $Al(OH)_3$. The phyllosilicates with this kind of octahedral sheet are called dioctahedral.

The main clay minerals groups are:

(1) The kandite group, which comprises kaolinite, and its two polymorphs dickite, and nacrite, as well as halloysite.
(2) The illite group, which includes illite, hydro-micas, phengite, glauconite and celadonite.
(3) The smectite group, including montmorillonite, beidellite, saponite and nontronite.
(4) Vermiculite, a peculiar mineral whose name derives from the word *vermiculus*, which is Latin for little worm (see below).
(5) The palygorskite group, comprising palygorskite and sepiolite.

Kandites are T-O (or 1:1 or diphormic) clay minerals, as they are made up of tetrahedral (T) and octahedral (O) layers in equal proportions. Moreover, all kandites have a characteristics basal spacing close to 7 Å.

Illites, smectites and vermiculites are T-O-T (or 2:1 or triphormic) clay minerals, since they are all based on a structural unit consisting of two tetrahedral layers with one octahedral layer sandwiched in between. Their typical basal spacings are 10 Å for illites, 15 Å for smectites and 14.5 Å for vermiculite, although layer separation is variable in halloysite, smectites and vermiculite, due to either swelling, upon entrance of liquids (water or organic substances) in the structure, or shrinkage upon loss of these substances. The minerals of the palygorskite group are also 2:1 clay minerals, but their T-O-T layers form fibres instead of continuous sheets.

Chlorites have a separated octahedral layer in addition to the T-O-T sequence, with basal spacings typically close to 14 Å. They are defined as 2:1 + 1 or T-O-T + O phyllosilicates.

5.3.2.1. Kaolinite and other T-O phyllosilicates

Kaolinite. Kaolinite is one of the simplest clay minerals and it is also very common. In the structure of kaolinite (Fig. 5.18), the SiO_4 tetrahedra are arranged in a hexagonal array, with their bases situated on the same plane and all their vertices pointing in one direction. These apical oxygens substitute for some hydroxyl groups at the base of the octahedral,

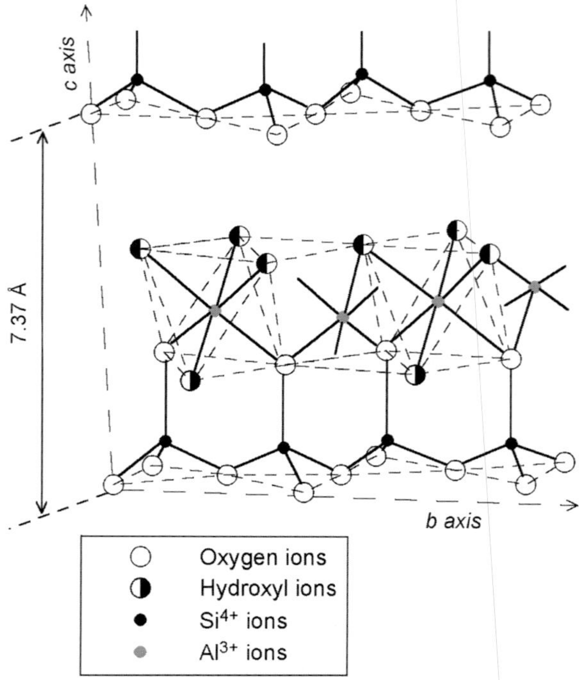

Figure 5.18. Schematic representation of the kaolinite crystal structure. (Reprinted from Hu et al., 2003, Copyright 2003 with permission from Elsevier, adapted.)

gibbsite-type layer. Since oxygen ions and hydroxyl ions are very similar in size (see above), the substitution of O^{2-} for OH^- and vice versa makes little difference from the steric point of view, provided that the positive and negative charges are balanced in the overall crystal structure.

It must be noted that the horizontal spacings in the gibbsite layers are slightly smaller than those in the tetrahedral, silica layers. This mismatch is accommodated either through distorsion of the tetrahedral layer, as in kaolinite and its polymorphs dickite and nacrite, or by making a curvature in the layers, as in halloysite. Halloysite has, therefore, a tubular structure. It also differs from kaolinite as a layer of H_2O molecules is interposed between the kaolinite-type sheets.

The unit cell of kaolinite is triclinic, with a cell volume of 327.34 Å^3, based on the cell parameters reported in Table 5.9, and equation (5-1). The corresponding molar volume is 98.56 cm^3 mol^{-1} (eq. (5-2)), which is 1% lower than the value reported by Helgeson et al. (1978), 99.52 cm^3 mol^{-1}. Table 5.9 also shows that dickite and nacrite have molar volume equal to and slightly higher than that of kaolinite, respectively, whereas the molar volume of halloysite is even higher, consistently with its peculiar structure.

The chemical composition of kaolinite usually departs little from the stoichiometric composition $Al_2Si_2O_5(OH)_4$. Although the tri-octahedral analogues of kaolinite occurs (e.g.,

TABLE 5.9

Crystallographic data of clay minerals and related minerals (from the mineralogy database: http://webmineral.com)

Mineral name	Chemical formula	Crystal system	Space group	Z	a(Å)	b(Å)	c(Å)	α(°)	β(°)	γ(°)	Cell volume (Å³)	Molar volume (cm³ mol⁻¹)
Kaolinite	$Al_2Si_2O_5(OH)_4$	Triclinic	*P*1	2	5.14	8.93	7.37	91.8	104.5	90.016	327.34	98.56
Dickite	$Al_2Si_2O_5(OH)_4$	Monoclinic	*Cc*	4	5.138	8.918	14.39	90	96.74	90	654.68	98.56
Nacrite	$Al_2Si_2O_5(OH)_4$	Monoclinic	*Cc*	4	5.146	8.909	15.7	90	113.7	90	658.95	99.21
Halloysite	$Al_2Si_2O_5(OH)_4$	Monoclinic	*Cc*	4	5.14	8.9	14.9	90	101.9	90	666.97	100.41
Chrysotile	$Mg_3Si_2O_5(OH)_4$	Monoclinic	*A*2/*m*	4	5.313	9.12	14.64	90	93.167	90	708.15	106.61
Lizardite	$Mg_3Si_2O_5(OH)_4$	Hexagonal	*P*31*m*	1	5.325	5.325	7.259	90	90	120	178.26	107.35
Antigorite	$Mg_{2.25}Fe^{2+}{}_{0.75}(Si_2O_5)(OH)_4$	Monoclinic	*Cm*	16	43.53	9.259	7.263	90	91.633	90	2926.12	110.13
Amesite	$Mg_2Al_2SiO_5(OH)_4$	Triclinic	*C*1	4	5.31	9.212	14.4	102.11	90.2	90.1	688.75	103.69
Greenalite	$Fe^{2+}{}_{2.3}Fe^{3+}{}_{0.5}Si_{2.2}O_5(OH)_{3.3}$	Monoclinic	*unknown*	2	5.54	9.55	7.44	90	104.2	90	381.60	114.90
Cronstedtite	$Fe^{2+}{}_2Fe^{3+}{}_2SiO_5(OH)_4$	Trigonal	$P3_1m$	1	5.486	5.486	7.095	90	90	120	184.92	111.36
Pyrophyllite	$Al_2Si_4O_{10}(OH)_2$	Triclinic	$P\bar{1}$	2	5.161	8.957	9.351	91.03	100.37	89.75	425.14	128.01
Talc	$Mg_3Si_4O_{10}(OH)_2$	Monoclinic	*C* 2/*c*	4	5.27	9.12	18.85	90	100.016	90	892.17	134.32
Montmorillonite	$Na_{0.2}Ca_{0.1}Al_2Si_4O_{10}(OH)_2(H_2O)_{10}$	Monoclinic	*C* 2/*m*	1	5.17	8.94	9.95	90	99.9	90	453.04	272.82
Beidellite	$Na_{0.5}Al_{2.5}Si_{3.5}O_{10}(OH)_2{\cdot}(H_2O)$	Monoclinic	*C* 2/*m*	2	5.14	8.93	15	90	99.54	90	678.98	204.44
Nontronite	$Na_{0.3}Fe^{3+}{}_2Si_3AlO_{10}(OH)_2{\cdot}4(H_2O)$	Monoclinic	*C* 2/*m*	2	5.23	9.11	15.25	90	96	90	722.61	217.58
Saponite	$Ca_{0.1}Na_{0.1}Mg_{2.25}Fe^{2+}{}_{0.75}Si_3AlO_{10}(OH)_2{\cdot}4(H_2O)$	Monoclinic	*C* 2/*m*	2	5.3	9.16	12.4	90	96.5	90	598.13	180.10
Illite	$K_{0.6}(H_3O)_{0.4}Al_{1.3}Mg_{0.3}Fe^{2+}{}_{0.1}Si_{3.5}O_{10}(OH)_2{\cdot}(H_2O)$	Monoclinic	*C* 2/*m*	2	5.18	8.98	10.32	90	101.83	90	469.85	141.47
Glauconite	$K_{0.6}Na_{0.05}Fe^{3+}{}_{1.3}Mg_{0.4}Fe^{2+}{}_{0.2}Al_{0.3}Si_{3.8}O_{10}(OH)_2$	Monoclinic	*C* 2/*m*	2	5.234	9.066	10.16	90	100.5	90	474.03	142.73
Celadonite	$KMg_{0.8}Fe^{2+}{}_{0.2}Fe^{3+}{}_{0.9}Al_{0.1}Si_4O_{10}(OH)_2$	Monoclinic	*C* 2/*m*	2	5.223	9.047	10.197	90	100.433	90	473.87	142.68
Brammallite	$Na_{0.66}(H_3O)_{0.33}AlMg_{0.6}Fe^{2+}{}_{0.1}Si_3AlO_{10}(OH)_2{\cdot}(H_2O)$	Monoclinic	*C* 2/*c*	4	5.12	8.91	19.26	90	95.83	90	874.08	131.59
Vermiculite	$Mg_{1.8}Fe^{2+}{}_{0.9}Al_{4.3}SiO_{10}(OH)_2{\cdot}4(H_2O)$	Monoclinic	*C* 2/*m*	2	5.26	9.23	14.97	90	96.82	90	721.65	217.29
Clinochlore	$Mg_{3.75}Fe^{2+}{}_{1.25}Si_3Al_2O_{10}(OH)_8$	Monoclinic	*C* 2/*m*	2	5.3	9.3	14.3	90	97	90	699.59	210.65
Sepiolite	$Mg_4Si_6O_{15}{\cdot}6(H_2O)$	Orthorhombic	*Pnan*	4	13.43	26.88	5.281	90	90	90	1906.43	287.02
Palygorskite	$Mg_{1.5}Al_{0.5}Si_4O_{10}(OH){\cdot}4(H_2O)$	Monoclinic	*C* 2/*m*	4	12.33	17.89	5.24	90	105.2	90	1115.42	167.93

serpentine and greenalite, see below) substitution of Mg^{2+} and Fe^{2+} for Al^{3+} in the gibbsite-type layer (with or without substitution of Al^{3+} for Si^{4+} in the tetrahedral layer) is negligible.

Serpentine. The minerals of the serpentine group (chrysotile, lizardite and antigorite) are not clay minerals. Nevertheless, they are considered here as they are tri-octahedral analogues of kaolinite. Their ideal formula is $Mg_3Si_2O_5(OH)_4$. Like kaolinite, also the structure of serpentine is made up of a silica layer, with the SiO_4 tetrahedra arranged in a hexagonal planar network. This, contrary to kaolinite, is joined to a brucite-type layer in which, on one side, two out of three hydroxyls are substituted by the apical oxygens of the SiO_4 tetrahedra. The repeat distance between these composite sheets is ~7.3 Å.

Due to differences in the horizontal spacings between the brucite-type layer and the silica layer, considerable mismatching is brought about by joining these two distinct layers. This mismatch can be accommodated in several ways and this fact determines the peculiar crystallographic and morphological features of the serpentine minerals.

Both chrysotile and antigorite belong to the monoclinic crystal system, whereas lizardite is triclinic. Inserting the cell parameters (Table 5.9) into the proper equations (5-1 and 5-2), molar volumes of 106.61, 110.13 and 104.67 $cm^3\ mol^{-1}$ are computed for chrysotile, antigorite and lizardite, respectively. The molar volume of chrysotile is 1.7% lower than that reported by Helgeson et al. (1978), 108.5 $cm^3\ mol^{-1}$. These authors give for antigorite the chemical formula $Mg_{48}Si_{34}O_{85}(OH)_{62}$ and the molar volume of 1749.13 $cm^3\ mol^{-1}$, which corresponds to 109.32 $cm^3\ mol^{-1}$ on a three-Mg basis. This value differs by less than 1% from that computed above.

Similar to kaolinite, also the chemical composition of serpentine minerals does not deviate substantially from the ideal composition, with little replacement of Si^{4+} by Al^{3+} and of Mg^{2+} by Al^{3+}, Fe^{2+} and Fe^{3+}. Also the Ni^{2+} content is usually very small, in spite of the existence of a nickel-sepentine, garnierite.

Septechlorites. Also the minerals of this group amesite [$Mg_2Al(AlSiO_5)(OH)_4$], chamosite [$Fe_2Al(AlSiO_5)(OH)_4$], greenalite [$Fe_3Si_2O_5(OH)_4$] and cronstedtite [$Fe_2Fe(FeSiO_5)(OH)_4$] are not clay minerals. Again, they are considered here because they are tri-octahedral analogues of kaolinite, like the serpentine minerals. They have unit layer thickness close to 7 Å, like kaolinite and the serpentine minerals. However, they differ from serpentine minerals because significant amounts of Fe^{2+}, Fe^{3+} and Al^{3+} ions substitute for Mg^{2+} in the brucite-type layer, and some Al^{3+} and/or Fe^{3+} substitute for Si^{4+} ion in the silica layer. Differences in the arrangement of layer stacking bring about differences in unit cells.

The molar volumes computed based on the cell parameters (Table 5.9) are 103.69 $cm^3\ mol^{-1}$ for amesite, 114.90 $cm^3\ mol^{-1}$ for greenalite and 111.36 $cm^3\ mol^{-1}$ for cronstedtite. These figures are comparable with those reported by Helgeson et al. (1978): 103.00 $cm^3\ mol^{-1}$ for amesite, 115.0 $cm^3\ mol^{-1}$ for greenalite and 110.9 $cm^3\ mol^{-1}$ for cronstedtite.

5.3.2.2. T-O-T phyllosilicates

Many minerals are included under this heading, such as those of the mica group, the smectite (montmorillonite) group, the chlorite group, pyrophyllite, talc and several mixed-layer minerals. Pyrophyllite and talc are the simplest phyllosilicates of this type.

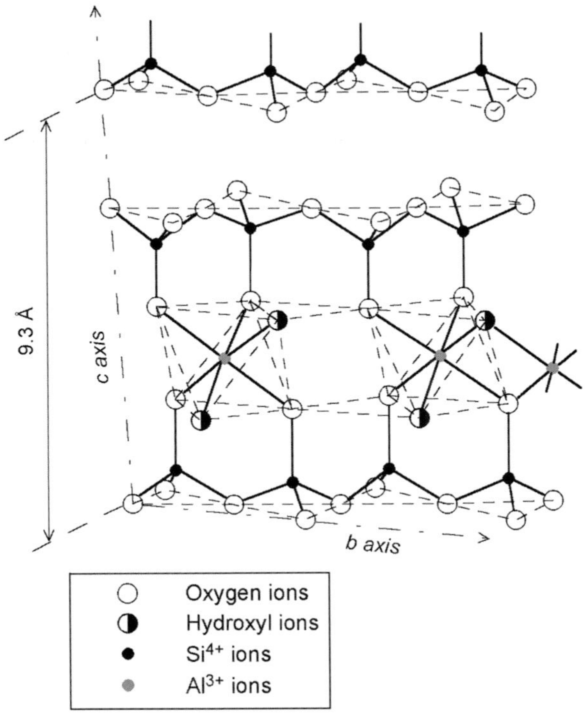

Figure 5.19. The crystal structure of pyrophyllite. (Reprinted from Hu et al., 2003, Copyright 2003, with permission from Elsevier, modified.)

Pyrophyllite has a layered crystal structure made up of a sheet of Al^{3+} ions in octahedral coordination, which is interposed between two inward pointing sheets of SiO_4 tetrahedra (Fig. 5.19). Besides, pyrophyllite is a di-octahedral layered mineral since only the 2/3 of the available octahedral sites are occupied by Al^{3+} ions, as in gibbsite, while the other sites are vacant.

Talc has a structure similar to that of pyrophyllite, but the octahedral sites in talc are all occupied by Mg^{2+} ions, as in brucite, and none is empty. Therefore, talc is a tri-octahedral layered mineral. Since the overall structure of both pyrophyllite and talc is electrically neutral, no interlayer cations are accommodates in these minerals, which are therefore called neutral-layer minerals.

All the other T-O-T phyllosilicates are related to either talc or pyrophyllite by these processes:

(1) Either substitution of Al^{3+} ion for Si^{4+} ion in the tetrahedral layers or substitution of Mg^{2+} and/or Fe^{2+} ions for Al^{3+} ion in the octahedral, gibbsite-type layers. Since these substitutions bring about a deficiency of positive charges on the composite sheet, interlayer cations are present.

(2) Substitution of Fe^{3+} and Al^{3+} for Mg^{2+} in the octahedral, brucite-type layer. This substitution causes an excess of positive charges on the composite sheet. In this

case, the electroneutrality condition is satisfied by replacement of monovalent cations for divalent interlayer cations or by presence of vacant sites in the octahedral layers.

Smectite (montmorillonite) group. Smectites are clay minerals characterized by the occurrence of replacements in both octahedral and tetrahedral sites and by the consequent presence of interlayer cations, chiefly Ca^{2+} and Na^{+}, as well as by water molecules in interlayer positions. Typically, 0.33 monovalent interlayer cations (or an equivalent amount of cations of different charge) are present per formula unit on a $O_{10}(OH)_2$-basis. Due to the presence of a small number of interlayer cations, smectites are also called low-charge clay minerals.

In *montmorillonite sensu stricto* substitution of Mg^{2+} for Al^{3+} in the octahedral layer of pyrophyllite leads to the average formula: $(Ca_{0.5},Na)_{0.33}(Al_{1.67}Mg_{0.33})Si_4O_{10}(OH)_2$.

Beidellite is characterized by replacement of Al^{3+} for Si^{4+} in the tetrahedral layer of pyrophyllite, which leads to the average formula: $(Ca_{0.5},Na)_{0.33}Al_2(Si_{3.67}Al_{0.33})O_{10}(OH)_2$.

The relevant features of *nontronite* are the substitution of Fe^{3+} for Al^{3+} in the octahedral layer of pyrophyllite, accompanied again by replacement of Al^{3+} for Si^{4+} in the tetrahedral layer of pyrophyllite (as seen above for beidellite). The average formula of nontronite is: $(Ca_{0.5},Na)_{0.33}Fe_2(Si_{3.67}Al_{0.33})O_{10}(OH)_2$.

Replacement of Al^{3+} for Si^{4+}, but in the tetrahedral layer of talc (instead of pyrophyllite) occurs in *saponite*, whose average formula is $(Ca_{0.5},Na)_{0.33}Mg_3(Si_{3.67}Al_{0.33})O_{10}(OH)_2$.

Montmorillonite, beidellite and nontronite are di-octahedral minerals and form a continuous mineral series, whereas saponite is tri-octahedral. Complete solid solution between di- and tri-octahedral smectites appears to be unlikely.

Further compositional changes are caused by the presence, in the octahedral sites, of other cations such as Fe^{2+}, Mn^{2+}, Ni^{2+}, Zn^{2+} (e.g., in sauconite) and Li^{+} (e.g., in hectorite).

For what concerns the interlayer cations, in addition to Ca^{2+} and Na^{+}, which are the most frequent ones, as already recalled, H^{+}, K^{+}, Cs^{+}, Mg^{2+}, Sr^{2+} and other cations can also be present. These are exchangeable to varying extents.

The number of layers of water molecules in the smectite structure depends on the type of smectite, the physical conditions, and the kind of interlayer cations. For instance, Ca-montmorillonite has usually two layers per cell, whereas Na-montmorillonite has a variable amount, i.e. two, three, or even more layers per cell.

Based on the cell parameters, the computed molar volumes of smectites are: 272.82 $cm^3\ mol^{-1}$ for montmorillonite, 204.44 $cm^3\ mol^{-1}$ for beidellite, 217.58 $cm^3\ mol^{-1}$ for nontronite and 180.10 $cm^3\ mol^{-1}$ for saponite (Table 5.9). Since these minerals are not considered by Helgeson et al. (1978), their molar volumes are compared with those reported in the EQ3/6 database, which are derived by Wolery (1978). Beidellites have molar volumes ranging from 123.19 to 133.7 $cm^3\ mol^{-1}$, whereas saponites have molar volumes from 133.66 to 139.85 $cm^3\ mol^{-1}$, with the lowest values referring to the Mg-endmembers and the highest to the K-endmembers. These values are significantly lower than those computed on the basis of cell parameters, as they refer to anhydrous compositions. Molar volumes of montmorillonites and nontronites are not given in the EQ3/6 database.

Illites and related minerals. From the structural point of view, illites are very similar to micas. Illites are mostly di-octahedral like muscovite [$KAl_2(AlSi_3)O_{10}(OH)_2$], but some are tri-octahedral like biotite, whose most common end-member components are phlogopite [$KMg_3(AlSi_3)O_{10}(OH)_2$] and annite [$KFe_3(AlSi_3)O_{10}(OH)_2$]. The typical chemical formula of illites, $K_{1-x}Al_2(Al_{1-x}Si_{3+x})O_{10}(OH)_2$, is closely related to that of muscovite, but illites differ from muscovite being poorer in K and Al and richer in Si.

A similarity in crystal structure, between illites and muscovite, corresponds to the similarity in chemical composition. Let us take into account first the crystal structure of muscovite. This can be derived from that of pyrophyllite through replacement of one Al^{3+} ion for one Si^{4+} ion of the tetrahedral sites, in such a way that one quarter of these sites are occupied by Al and three quarters by Si. Consequent introduction of a K^+ ion in the 12-fold coordinated interlayer positions occurs to fulfill the electroneutrality condition. The structure of muscovite (and other micas) is held together by the strong electrostatic forces between the positive K^+ ions and the negatively charged Al-silicate layers. In illite (Fig. 5.20), due to

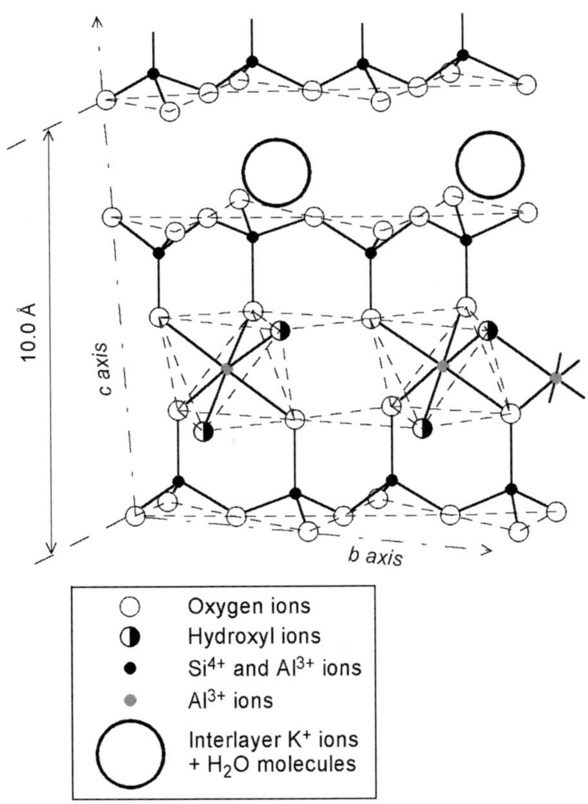

Figure 5.20. The crystal structure of illite. Note the presence of cations and water molecules in the interlayer positions, which is typical of T-O-T minerals. (Reprinted from Hu et al., 2003, Copyright 2003, with permission from Elsevier, modified.)

the presence of less interlayer cations than in muscovite, electrostatic forces are weaker. Also layer stacking is less regular in illites than in muscovite. The content of interlayer water in illites is usually small or nil.

The number of interlayer cations in illites is greater than in smectites. Therefore, illites and related minerals are also called high-charge clay minerals. Illites are the prevailing clay minerals in shales and clayey rocks, but they are also present in other sedimentary deposits, such as in limestones.

Brammallite is an uncommon mineral similar to illite, in which sodium is present as interlayer cation.

Glauconite is characterized by an excess of interlayer cations (K^+, Na^+, Ca^{2+}) with respect to muscovite, which is balanced by the presence of divalent cations (Fe^{2+} and Mg^{2+}) instead of trivalent cations in the octahedral sites. In addition, Fe^{3+} prevails over Al^{3+}. Glauconite is present almost exclusively in marine sedimentary deposits.

Celadonite is characterized by a very small substitution of Al for Si, ferric iron as main octahedral cation, and high contents of interlayer cations. Its typical chemical formula is $K_{0.85}(Fe^{3+}{}_{0.9}Fe^{2+}{}_{0.25}Mg_{0.6}Al_{0.25})(Al_{0.05}Si_{3.95})O_{10}(OH)_2$ (Odin et al., 1988).

The molar volumes reported in Table 5.9 for illite, glauconite and celadonite are very similar, as they range from 141.47 to 142.73 $cm^3\ mol^{-1}$, whereas that of brammallite is somewhat lower, with 131.59 $cm^3\ mol^{-1}$. Wolery (1978) reports a molar volume of 157.10 $cm^3\ mol^{-1}$ for a celadonite of idealized formula $KMgAlSi_4O_{10}(OH)_2$.

Vermiculites. Vermiculites are structurally and chemically similar to tri-octahedral smectites. Both are made up of talc-like sheets, whose deficiency of positive charge is balanced by the presence of cations in the interlayer positions. The charge deficiency in vermiculites is usually greater than in smectites and is mainly brought about by replacement of Al^{3+}, Fe^{3+}, Mg^{2+} and Fe^{2+} for Si^{4+} ions in the tetrahedral layer. The interlayer positions are generally occupied by 0.35 divalent cations (or an equivalent amount) per formula unit [$O_{10}(OH)_2$], chiefly Mg^{2+} and less frequently Ca^{2+} and Na^+. Vermiculites, similar to smectites, also contain water molecules in the interlayer spaces, but the amount of water in vermiculites is more constant (and close to two layers) than in smectites. All these compositional variabilities are suitably accounted for by the formula:

$$(Mg,Ca)_{0.35-0.5}\underbrace{Mg_{1.75-2.5}(Fe^{3+},Al)_{1.25-0.5}}_{3.0}\underbrace{Al_{1.0-1.25}Si_{3-2.75}}_{4.0}O_{10}(OH)_2\cdot(3.5-4.5)H_2O.$$

Vermiculites show, upon sudden heating to ~300°C, a peculiar exfoliation, which is due to quick production of steam, whose escape deforms and separates the structural layers.

In spite of the similarity with tri-octahedral smectites, the molar volume computed for vermiculite, 217.29 $cm^3\ mol^{-1}$, is significantly higher than that of saponite, 180.10 $cm^3\ mol^{-1}$ (Table 5.9).

Sepiolite and palygorskite. In these two minerals, the T-O-T layers form fibres, instead of sheets, comprising six SiO_4 tetrahedra in sepiolite [$Mg_4Si_6O_{15}(OH)_2 \cdot 6H_2O$] and four

SiO_4 tetrahedra in palygorskite, also called attapulgite [$(Mg,Al,Fe)_4Si_8O_{20}(OH)_2nH_2O$]. Some exchangeable cations and some Al are usually present in both minerals.

The molar volume computed for sepiolite based on cell parameters, 287.02 cm^3 mol^{-1}, agrees with that reported by Helgeson et al. (1978), whereas neither these authors nor the EQ3/6 database include palygorskite.

5.3.2.3. Chlorites

As recalled above chlorites are T-O-T+O phyllosilicates. Since chlorites are other possible product minerals of carbonation reactions their structural and chemical characteristics are briefly summarized here, even though they are not clay minerals. Nevertheless, chlorites are structurally similar to clay minerals. Indeed, a regular alternation of talc-like and brucite-like layers makes up their structure. The resulting basal spacing, ~14 Å, is very close to those of smectites and vermiculites.

The chemical characteristics of chlorites can be related to the end-member hypothetical composition $Mg_3Si_4O_{10}(OH)_2 + Mg_3(OH)_6$, which refers to an equal number of talc and brucite sheets. Silicon ion is substituted by Al^{3+} up to the limit Si_2Al_2, whereas Mg^{2+} ion is replaced by Al^{3+}, in both layers, up to the limit Mg_4Al_2. Concurrent occurrence of both substitutions prevents the unbalance of electrical charges. Taking into account these two replacements only, the composition of chlorites ranges from $Mg_6Si_4O_{10}(OH)_8$ to $(Mg_4Al_2)(Si_2Al_2)O_{10}(OH)_8$. These are the stoichiometric formulae of serpentine and amesite, respectively, both of which are T–O phyllosilicates, and have basal spacing close to 7 Å (see above).

However, replacement of Fe^{2+} for Mg^{2+} is also very important in chlorites and occurs without any restriction, so that the ratio $Fe^{2+}/(Fe^{2+} + Mg^{2+})$ varies from 0 to 1. Presence of ferric iron is also usual and it is accompanied by substitution of Al for Si. Small amounts of transition metals, such as Mn, Cr, Ni and Ti, can also be present.

Interestingly, the molar volume computed for a clinochlore of composition $Mg_{3.75}Fe^{2+}{}_{1.25}Si_3Al_2O_{10}(OH)_8$, 210.65 cm^3 mol^{-1} (Table 5.9), is intermediate between those reported by Helgeson et al. (1978) for 14Å-clinochlore [$Mg_5Al(AlSi_3)O_{10}(OH)_8$], 207.11 cm^3 mol^{-1} and 14Å-daphnite [$Fe_5Al(AlSi_3)O_{10}(OH)_8$], 213.42 cm^3 mol^{-1}.

Chlorites are widespread in low-grade metamorphic rocks and in hydrothermally altered igneous rocks, but they are also produced through weathering of clayey rocks and Fe-rich sediments.

5.3.2.4. Mixed-layer minerals

Since the crystal structures of T-O-T and T-O-T+O phyllosilicates are very similar, mixed-layer minerals may form. They are made up by layers typical of distinct minerals, such as illite–smectite, chlorite–smectite, etc.

5.3.3. The thermodynamic data of clay minerals and related solid phases

Different methodologies have been applied in the geochemical literature to evaluate the Gibbs free energy of formation and other thermodynamic properties of clay minerals and other minerals characterized by non-stoichiometric, highly variable chemical composition.

Helgeson (1969) computed the ΔG_f^o of illite, Mg-chlorite, Ca-, Na-, K- and Mg-montmorillonite assuming that the composition of given natural waters is representative of equilibrium with these solid phases.

Other authors (e.g., Tardy and Garrels, 1974; Nriagu, 1975; Mattigod and Sposito, 1978) obtained the ΔG_f^o of layer silicate minerals through summation of those of either hydroxide or oxide components suitably chosen to form the desired mineral.

A similar approach was applied by others, who used other components, namely beidellites (Helgeson and MacKenzie, 1970), illites (Stoessell, 1981), clay mineral end-members (Tardy and Fritz, 1981) and fictive components (Robinson and Haas, 1983).

Aagaard and Helgeson (1983) suggested to compute the activities of the thermodynamic end-member components of montmorillonites, illites and mixed-layer clay minerals from site mixing approximations. The activities of these thermodynamic end-member components were then correlated with the compositions of the clay minerals and used to delineate the stability boundaries in activity diagrams. Following Aagaard and Helgeson (1983), the activities of the end-member components pyrophyllite and alkali mica in potassium clay minerals are calculated by means of the equations:

$$a_{\text{pyrophyllite}} = k_1 \cdot X_{v,\text{A}} \cdot X^2_{\text{Al,M(2)}} \cdot X^4_{\text{Si,T}} \tag{5-45}$$

and

$$a_{\text{alkali-mica}} = k_2 \cdot X_{j,\text{A}} \cdot X^2_{\text{Al,M(2)}} \cdot X_{\text{Al,T}} \cdot X^3_{\text{Si,T}} \tag{5-46}$$

where k_1 and k_2 are proportionality constants used to normalize the activity of the end-members components to 1; $X_{j,\text{A}}$ and $X_{\text{v,A}}$ are the mole fractions of K atoms and vacancies on the interlayer positions, respectively; $X_{\text{Si,T}}$, $X_{\text{Al,T}}$ and $X_{\text{Al,(M2)}}$ are the mole fractions of Si and Al on the tetrahedral [T] and octahedral [M(2)] sites, respectively.

It must be noted that equations (5-45) and (5-46) represent special cases of the general equation relating site occupancy in solid mixtures to the activity of the *i*th thermodynamic component, a_i, under the assumptions of random mixing and equal interaction of atoms on energetically equivalent sites. This general equation is (Helgeson et al., 1978)

$$a_i = k_i \prod_s \prod_j X_{j,s}^{v_{s,j,i}} \tag{5-47}$$

where, again, k_i is a proportionality constant determined by the stoichiometry of the *i*th component, $v_{s,j,I}$ represents the stoichiometric number of energetically equivalent sth sites occupied by the jth species in one mole of the ith component and $X_{j,s}$ is the mole fraction of the jth species on the sth sites in the solid mixtures. This, in turn, is given by

$$X_{j,s} = \frac{n_{j,s}}{\sum_j n_{j,s}}, \tag{5-48}$$

where $n_{j,s}$ is the number of moles of the jth species on the sth sites. The proportionality constant k_i is defined as follows:

$$k_i = \prod_s \prod_j \left(\frac{\nu_{s,j,i}}{\nu_{s,i}} \right)^{-\nu_{s,j,i}}. \tag{5-49}$$

Consequently, $\lim_{X_i \to 1} a_i = 1$, where X_i represents the mole fraction of the ith thermodynamic end-member component of the solid mixture and $\nu_{s,i}$ is the stoichiometric number of energetically equivalent sth sites in one mole of the ith component. The condition $\nu_{s,i} = \nu_{s,j,I}$ is attained if the sth sites are occupied by one kind of atom only. Otherwise,

$$\nu_{s,i} = \sum_j \nu_{s,j,i}. \tag{5-50}$$

For the same reason, $k_i = 1$ if each kind of energetically equivalent sites is occupied by one kind of atom only.

Equation (5-47) can be used only for solid mixtures whose thermodynamic end-member components have equal standard molal volumes or, to a first approximation, if these deviate by 5–10% of each other. Equation (5-47) is consistent with the following expression for the molal Gibbs free energy of mixing (Helgeson et al., 1978):

$$G_{\text{mixing}} = RT \cdot \sum_i X_i \cdot \left(\ln k_i + \sum_s \sum_j \nu_{s,j,i} \cdot \ln X_{j,s} \right). \tag{5-51}$$

The approach of Aagaard and Helgeson (1983) was adopted by Giggenbach (1985) to construct activity diagrams involving di-octahedral potassium clay minerals, such as illites, beidellites, phengites, montmorillonites and celadonites.

These theoretical efforts for evaluating the thermodynamic stability of clay minerals are justified by the difficulties encountered in the experimental determination of their solubility, which could represent (if it were known) a good starting point for evaluating their Gibbs free energy of formation.

For instance, May et al. (1986) tried to determine the solubilities of five smectite minerals with mineral suspensions suitably adjusted to approach equilibrium from over- and undersaturation conditions, at pH values of 5–8. Unfortunately, solubilities could not be measured for these smectites as the chemistry of the aqueous phase resulted to be controlled by authigenic gibbsite or amorphous aluminium hydroxide. May et al. (1986), however, were able to measure the aqueous solubility of a kaolinite sample (from Dry Branch, Georgia) at pH 4 and 25°C. The mineral-solution equilibrium condition was attained after 1,237 days and indicates a Gibbs free energy of formation for the kaolinite sample of $-3{,}789.51 \pm 6.60$ kJ mol^{-1}. This value coincides, within the experimental uncertainty, with that proposed by Helgeson et al. (1978), –3789.09 kJ mol^{-1}, which is incorporated in the thermodynamic database of the software package EQ3/6.

In addition to kaolinite, this database contains the hydrolysis constants of other clay minerals, namely the Na-, K-, Mg-, Ca- and H-idealized endmembers of montmorillonites, beidellites, nontronites and saponites, as well as an illite and a celadonite of idealized compositions (Table 5.10). All these data were obtained by Wolery (1978) and are listed in Table 5.10, together with the hydrolysis constants of solid phases related to clay minerals.

Table 5.10 shows that the endmember components of montmorillonites, beidellites, nontronites and saponites differ only for the cation in the interlayer position. However, some compositional changes in the tetrahedral and octahedral layers can also be reproduced by mixing together these endmembers.

Although Wolery (1978) did not follow the approach of Aagaard and Helgeson (1983), which represents the most rigorous way to describe the thermodynamic stability of clay minerals and other minerals with variable chemical composition, we will refer to EQ3/6 thermodynamic database, and consequently to the Wolery's data on clay minerals, in the following discussion. The revision of the EQ3/6 thermodynamic database goes well beyond the aims of this book.

The activity diagrams represent the best tool to evaluate the thermodynamic stability of clay minerals. These diagrams were proposed by Helgeson (1968), who gave a rigorous description of their features. Bowers et al. (1984) presented a large number of activity diagrams, for total pressures from 1 bar to 5 kbar and temperatures ranging from 25 to 600°C.

It is convenient to start with the simple systems $MgO–SiO_2–H_2O$ and $Al_2O_3–SiO_2–H_2O$, since they comprise the building blocks of clay minerals, i.e. brucite and talc and gibbsite and pyrophyllite, respectively.

5.3.3.1. The system $MgO–SiO_2–H_2O$

The thermodynamic equilibrium constants of the dissolution reactions for the solid phases of interest in this system (Table 5.10) can be arranged as follows, assuming $a_{H_2O} = 1$ and unit activity for all the solid phases:

$$\log\left(\frac{a_{Mg^{2+}}}{a^2_{H^+}}\right) = \log K_{brucite} \tag{5-52}$$

$$\log\left(\frac{a_{Mg^{2+}}}{a^2_{H^+}}\right) = \frac{1}{3}\log K_{chrysotile} - \frac{2}{3}\log a_{SiO_{2(aq)}} \tag{5-53}$$

$$\log\left(\frac{a_{Mg^{2+}}}{a^2_{H^+}}\right) = \frac{1}{48}\log K_{antigorite} - \frac{34}{48}\log a_{SiO_{2(aq)}} \tag{5-54}$$

$$\log\left(\frac{a_{Mg^{2+}}}{a^2_{H^+}}\right) = \frac{1}{3}\log K_{talc} - \frac{4}{3}\log a_{SiO_{2(aq)}} \tag{5-55}$$

$$\log\left(\frac{a_{Mg^{2+}}}{a^2_{H^+}}\right) = \frac{1}{4}\log K_{sepiolite} - \frac{6}{4}\log a_{SiO_{2(aq)}} \tag{5-56}$$

TABLE 5.10

Hydrolysis constants of clay minerals and related minerals (from the COM thermodynamic database of the EQ3/6 software package)

Mineral formula	Reaction of dissolution	log K at T,P					
		0.01°C 1.013 bar	25°C 1.013 bar	60°C 1.013 bar	100°C 1.013 bar	150°C 4.7572 bar	150°C 500 bar
$Mg(OH)_2$	Brucite + $2H^+ = Mg^{2+} + 2H_2O$	18.0898	16.298	14.2674	12.4514	10.6978	10.7749
$Mg_3Si_4O_{10}(OH)_2$	Talc + $6H^+ = 3Mg^{2+} + 4SiO_{2(aq)} + 4H_2O$	23.1104	21.1383	18.1118	15.085	12.0522	12.6129
$Mg_3Si_2O_5(OH)_4$	Chrysotile + $6H^+ = 3Mg^{2+} + 2SiO_{2(aq)} + 5H_2O$	34.4271	31.1254	26.9983	23.1597	19.409	19.826
$Mg_{48}Si_{34}O_{85}(OH)_{62}$	Antigorite + $96H^+ = 48Mg^{2+} + 34SiO_{2(aq)} + 79H_2O$	527.8152	477.1943	413.2746	353.596	295.2109	301.9369
$Mg_4Si_6O_{15}(OH)_2 \cdot 6H_2O$	Sepiolite + $8H^+ = 4Mg^{2+} + 6SiO_{2(aq)} + 11H_2O$	32.3876	30.4439	27.171	23.8968	20.7229	21.3869
$Al(OH)_3$	Gibbsite + $3H^+ = Al^{3+} + 3H_2O$	9.3787	7.756	5.8286	3.9979	2.0853	2.2758
$Al_2Si_2O_5(OH)_4$	Kaolinite + $6H^+ = 2Al^{3+} + 2SiO_{2(aq)} + 5H_2O$	9.0182	6.8101	3.8468	0.9541	−2.0187	−1.4457
$Al_2Si_4O_{10}(OH)_2$	Pyrophyllite + $6H^+ = 2Al^{3+} + 4SiO_{2(aq)} + 4H_2O$	1.6336	0.4397	−1.7853	−4.197	−6.7775	−6.0649
$Mg_4Al_4Si_2O_{10}(OH)_8$	Amesite-14Å $+20H^+ = 4Mg^{2+} + 4Al^{3+} + 2SiO_{2(aq)} + 14H_2O$	87.992	75.4571	60.6239	46.9345	33.3083	–
$Ca_{0.165}Al_{2.33}Si_{3.67}O_{10}(OH)_2$	Beidellite-Ca $+7.32H^+ = 0.165Ca^{2+} + 2.33Al^{3+} + 3.67SiO_{2(aq)} + 4.66H_2O$	7.7832	5.5914	2.2752	−1.1147	−4.6679	–
$H_{0.33}Al_{2.33}Si_{3.67}O_{10}(OH)_2$	Beidellite-H $+6.99H^+ = 2.33Al^{3+} + 3.67SiO_{2(aq)} + 4.66H_2O$	6.7025	4.6335	1.4615	−1.7934	−5.2084	–
$K_{0.33}Al_{2.33}Si_{3.67}O_{10}(OH)_2$	Beidellite-K $+7.32H^+ = 0.33K^+ + 2.33Al^{3+} + 3.67SiO_{2(aq)} + 4.66H_2O$	7.3152	5.3088	2.2062	−0.9869	−4.3427	–
$Mg_{0.165}Al_{2.33}Si_{3.67}O_{10}(OH)_2$	Beidellite-Mg $+7.32H^+ = 0.165Mg^{2+} + 2.33Al^{3+} + 3.67SiO_{2(aq)} + 4.66H_2O$	7.7945	5.5537	2.1824	−1.2549	−4.8515	–
$Na_{0.33}Al_{2.33}Si_{3.67}O_{10}(OH)_2$	Beidellite-Na $+7.32H^+ = 0.33Na^+ + 2.33Al^{3+} + 3.67SiO_{2(aq)} + 4.66H_2O$	7.7364	5.6473	2.4564	−0.8088	−4.2257	–
$KMgAlSi_4O_{10}(OH)_2$	Celadonite $+6H^+ = K^+ + Mg^{2+} + Al^{3+} + 4SiO_{2(aq)} + 4H_2O$	8.2378	7.4575	5.7846	3.9696	2.0983	–
$Fe_2Al_2SiO_5(OH)_4$	Chamosite-7 Å $+10H^+ = 2Fe^{2+} + 2Al^{3+} + SiO_{2(aq)} + 7H_2O$	38.5531	32.8416	26.0443	19.737	13.4183	–
$Mg_5Al_2Si_3O_{10}(OH)_8$	Clinochlore-14 Å $+16H^+ = 5Mg^{2+} + 2Al^{3+} + 3SiO_{2(aq)} + 12H_2O$	76.7345	67.2391	55.7725	45.152	34.6423	35.7886

TABLE 5.10 (*Continued*)

Mineral formula	Reaction of dissolution	log *K* at T,P					
		0.01°C 1.013 bar	25°C 1.013 bar	60°C 1.013 bar	100°C 1.013 bar	150°C 4.7572 bar	150°C 500 bar
$Mg_5Al_2Si_3O_{10}(OH)_8$	Clinochlore-7 Å $+16H^+ = 5Mg^{2+} + 2Al^{3+} + 3SiO_{2(aq)} + 12H_2O$	80.327	70.6124	58.9053	48.0755	37.3689	38.5421
$Fe_5Al_2Si_3O_{10}(OH)_8$	Daphnite-14 Å $+16H^+ = 5Fe^{2+} + 2Al^{3+} + 3SiO_{2(aq)} + 12H_2O$	60.2452	52.2821	42.5135	33.3581	24.1876	–
$Fe_5Al_2Si_3O_{10}(OH)_8$	Daphnite-7 Å $+16H^+ = 5Fe^{2+} + 2Al^{3+} + 3SiO_{2(aq)} + 12H_2O$	63.8597	55.6554	45.621	36.2333	26.8434	–
$K_{0.6}Mg_{0.25}Al_{2.3}Si_{3.5}O_{10}(OH)_2$	Illite $+8H^+ = 0.6K^+ + 0.25Mg^{2+} + 2.3Al^{3+} + 3.5SiO_{2(aq)} + 5H_2O$	11.386	9.026	5.5551	2.0473	−1.5896	–
$Ca_{0.165}Mg_{0.33}Al_{1.67}Si_4O_{10}(OH)_2$	Montmor-Ca $+ 6H^+ = 0.165Ca^{2+} + 0.33Mg^{2+} + 1.67Al^{3+} + 4SiO_{2(aq)} + 4H_2O$	3.6657	2.4952	0.3114	−2.0464	−4.535	–
$K_{0.33}Mg_{0.33}Al_{1.67}Si_4O_{10}(OH)_2$	Montmor-K $+ 6H^+ = 0.33K^+ + 0.33Mg^{2+} + 1.67Al^{3+} + 4SiO_{2(aq)} + 4H_2O$	3.1183	2.1423	0.1755	−1.9903	−4.2957	–
$Mg_{0.495}Al_{1.67}Si_4O_{10}(OH)_2$	Montmor-Mg $+ 6H^+ = 0.495Mg^{2+} + 1.67Al^{3+} + 4SiO_{2(aq)} + 4H_2O$	3.5977	2.3879	0.1599	−2.2353	−4.7571	–
$Na_{0.33}Mg_{0.33}Al_{1.67}Si_4O_{10}(OH)_2$	Montmor-Na $+ 6H^+ = 0.33Na^+ + 0.33Mg^{2+} + 1.67Al^{3+} + 4SiO_{2(aq)} + 4H_2O$	3.546	2.4844	0.4328	−1.7939	−4.1399	–
$Ca_{0.165}Fe_2Al_{0.33}Si_{3.67}O_{10}(OH)_2$	Nontronite-Ca $+ 7.32H^+ = 0.165Ca^{2+} + 0.33Al^{3+} + 2Fe^{3+} + 3.67SiO_{2(aq)} + 4.66H_2O$	−11.3915	−11.5822	−12.6234	−13.9486	−15.4751	–
$H_{0.33}Fe_2Al_{0.33}Si_{3.67}O_{10}(OH)_2$	Nontronite-H $+ 6.99H^+ = 0.33Al^{3+} + 2Fe^{3+} + 3.67SiO_{2(aq)} + 4.66H_2O$	−12.4711	−12.5401	−13.4382	−14.6295	−16.0018	–
$K_{0.33}Fe_2Al_{0.33}Si_{3.67}O_{10}(OH)_2$	Nontronite-K $+ 7.32H^+ = 0.33K^+ + 0.33Al^{3+} + 2Fe^{3+} + 3.67SiO_{2(aq)} + 4.66H_2O$	−11.8593	−11.8648	−12.6926	−13.8213	−15.1505	–
$Mg_{0.165}Fe_2Al_{0.33}Si_{3.67}O_{10}(OH)_2$	Nontronite-Mg $+ 7.32H^+ = 0.165Mg^{2+} + 0.33Al^{3+} + 2Fe^{3+} + 3.67SiO_{2(aq)} + 4.66H_2O$	−11.3804	−11.62	−12.716	−14.0884	−15.6578	–

$Na_{0.33}Fe_2Al_{0.33}Si_{3.67}O_{10}(OH)_2$	Nontronite-Na + $7.32H^+ = 0.33Na^+ + 0.33Al^{3+} + 2Fe^{3+} + 3.67SiO_{2(aq)} + 4.66H_2O$	−11.4385	−11.5263	−12.442	−13.6424	−15.0323	–
$Ca_{0.165}Mg_3Al_{0.33}Si_{3.67}O_{10}(OH)_2$	Saponite-Ca + $7.32H^+ = 0.165Ca^{2+} + 3Mg^{2+} + 0.33Al^{3+} + 3.67SiO_{2(aq)} + 4.66H_2O$	29.244	26.29	22.1907	18.2025	14.2134	–
$H_{0.33}Mg_3Al_{0.33}Si_{3.67}O_{10}(OH)_2$	Saponite-H + $6.99H^+ = 3Mg^{2+} + 0.33Al^{3+} + 3.67SiO_{2(aq)} + 4.66H_2O$	28.1635	25.3321	21.3768	17.5233	13.6722	–
$K_{0.33}Mg_3Al_{0.33}Si_{3.67}O_{10}(OH)_2$	Saponite-K + $7.32H^+ = 0.33K^+ + 3Mg^{2+} + 0.33Al^{3+} + 3.67SiO_{2(aq)} + 4.66H_2O$	28.776	26.0075	22.1217	18.3303	14.5386	–
$Mg_{3.165}Al_{0.33}Si_{3.67}O_{10}(OH)_2$	Saponite-Mg $+7.32H^+ = 3.165Mg^{2+} + 0.33Al^{3+} + 3.67SiO_{2(aq)} + 4.66H_2O$	29.2521	26.2523	22.1016	18.0692	14.0402	–
$Na_{0.33}Mg_3Al_{0.33}Si_{3.67}O_{10}(OH)_2$	Saponite-Na $+7.32H^+ = 0.33Na^+ + 3Mg^{2+} + 0.33Al^{3+} + 3.67SiO_{2(aq)} + 4.66H_2O$	29.197	26.3459	22.3721	18.5087	14.6562	–

All these solubility relations can be suitably represented as straight lines on the activity plot with log ($a_{Mg^{2+}}/a^2_{H^+}$) as ordinate and $\log a_{SiO_{2(aq)}}$ as abscissa. It must be underscored that:

(1) The slopes of these straight lines are equal to the ratios between the stoichiometric coefficients of $SiO_{2(aq)}$ and Mg^{2+} ion in the dissolution reaction. These slopes are all negative, apart from that of the brucite line, as both $SiO_{2(aq)}$ and Mg^{2+} are on the same (right-hand) side of the dissolution reaction, whereas the slopes would be positive if chemical components were on the opposite sides of the reaction. The slope is 0 for brucite as no Si is present in this solid phase.

(2) The intercept of each linear solubility relation is equal to the logarithm of the thermodynamic equilibrium constant divided by the stoichiometric coefficient of Mg^{2+} ion.

Therefore, in general terms, equations (5-52)–(5-56) can be written as follows:

$$\log\left(\frac{a_{Mg^{2+}}}{a^2_{H^+}}\right) = \frac{1}{v_{Mg^{2+}}}\log K_{mineral} - \frac{v_{SiO_{2(aq)}}}{v_{Mg^{2+}}}\log a_{SiO_{2(aq)}}, \quad (5\text{-}57)$$

where v_i represents the stoichiometric coefficient of the ith component in the dissolution reaction.

Activity plots of log ($a_{Mg^{2+}}/a^2_{H^+}$) vs. $\log a_{SiO_{2(aq)}}$ have been prepared for different conditions, namely for temperatures of 25, 60 and 100°C and pressure of 1.013 bar, and for 150°C and both 4.76 bar (saturation pressure) and 500 bar (Fig. 5.21).

In each activity plot, the heavy line limits the stability field of the aqueous solution, which is defined by saturation (equilibrium) with respect to the least soluble, thermodynamically stable solid phases. These are brucite, antigorite, talc and quartz, for increasing $a_{SiO_{2(aq)}}$ values. Also shown is the solubility line for the most soluble silica mineral, i.e. amorphous silica.

Figure 5.21 also indicates that chrysotile may form only from aqueous solutions oversaturated with brucite at low $a_{SiO_{2(aq)}}$ values, whereas sepiolite is expected to precipitate from strongly oversaturated aqueous solutions at any $a_{SiO_{2(aq)}}$ values. Talc is the least soluble phase not only at saturation with quartz but also at higher $a_{SiO_{2(aq)}}$, up to saturation with amorphous silica and even above it, although oversaturation with respect to amorphous silica is uncommon in natural systems, due to its fast precipitation kinetics.

Although temperature changes determine significant shifts in solubility relations, their relative positions do not vary significantly. Finally, the 100-fold increase in pressure at 150°C (Fig. 5.21d) has small effects on solubility relations in the considered system.

5.3.3.2. The system Al_2O_3–SiO_2–H_2O

Again, assuming unit activity for H_2O and all the relevant minerals, the thermodynamic equilibrium constants of their dissolution reactions (Table 5.10), in general terms, can be written as

$$\log\left(\frac{a_{Al^{3+}}}{a^3_{H^+}}\right) = \frac{1}{v_{Al^{3+}}}\log K_{mineral} - \frac{v_{SiO_{2(aq)}}}{v_{Al^{3+}}}\log a_{SiO_{2(aq)}}. \quad (5\text{-}58)$$

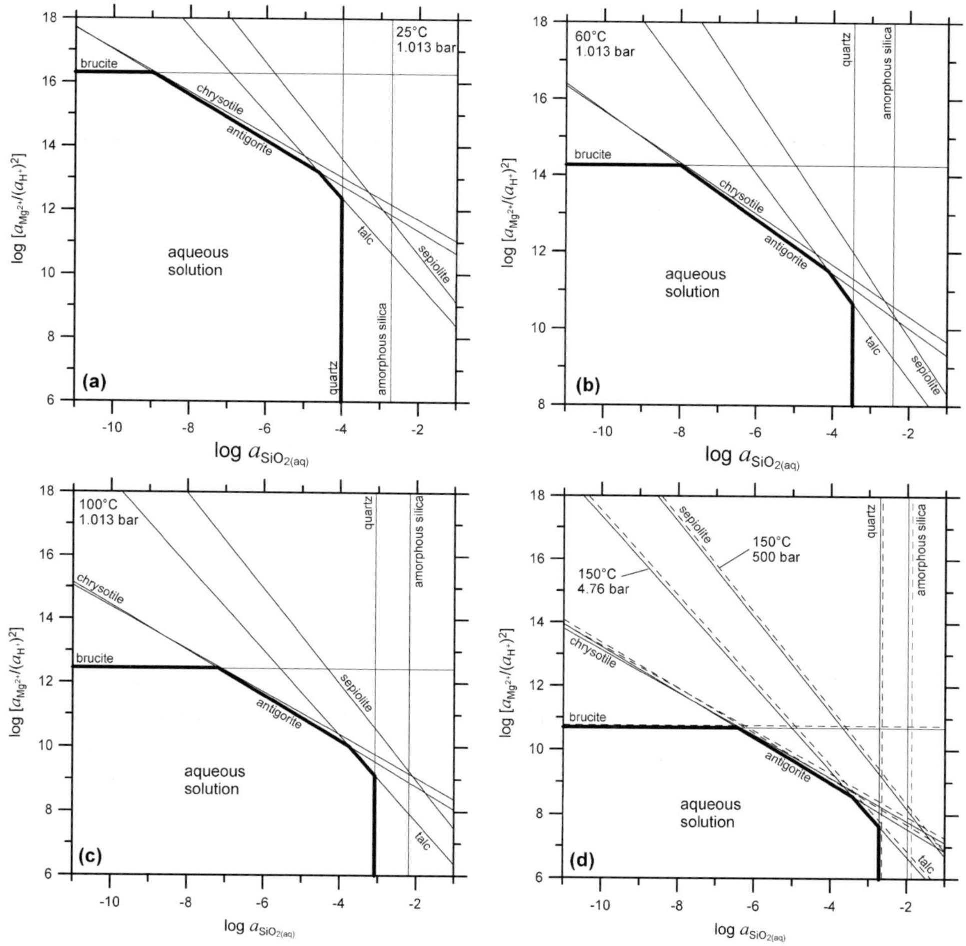

Figure 5.21. Activity plots of log $(a_{Mg^{2+}}/a^2_{H^+})$ vs. log $a_{SiO_{2(aq)}}$ for the system MgO–SiO_2–H_2O at (a) 25°C and 1.013 bar, (b) 60°C and 1.013 bar, (c) 100°C and 1.013 bar and (d) 150°C and both 4.76 bar (saturation pressure) and 500 bar. The thick line refers to the aqueous solution in equilibrium with the least soluble, thermodynamically stable minerals, namely brucite, antigorite, talc and quartz.

This corresponds to write

$$\log\left(\frac{a_{Al^{3+}}}{a^3_{H^+}}\right) = \log K_{\text{gibbsite}} \qquad \text{for gibbsite} \tag{5-59}$$

$$\log\left(\frac{a_{Al^{3+}}}{a^3_{H^+}}\right) = \frac{1}{2}\log K_{\text{kaolinite}} - \log a_{SiO_{2(aq)}} \qquad \text{for kaolinite} \tag{5-60}$$

$$\log\left(\frac{a_{Al^{3+}}}{a_{H^+}^3}\right) = \frac{1}{2}\log K_{pyrophyllite} - 2\log a_{SiO_{2(aq)}} \qquad \text{for pyrophyllite.} \tag{5-61}$$

Straight lines representative of the three solubility relations (5-59)–(5-61) are drawn on the activity plots of $\log\left(a_{Al^{3+}}/a_{H^+}^3\right)$ vs. $a_{SiO_{2(aq)}}$ (Fig. 5.22), which have been elaborated for the same temperature, pressure conditions of the previous figure.

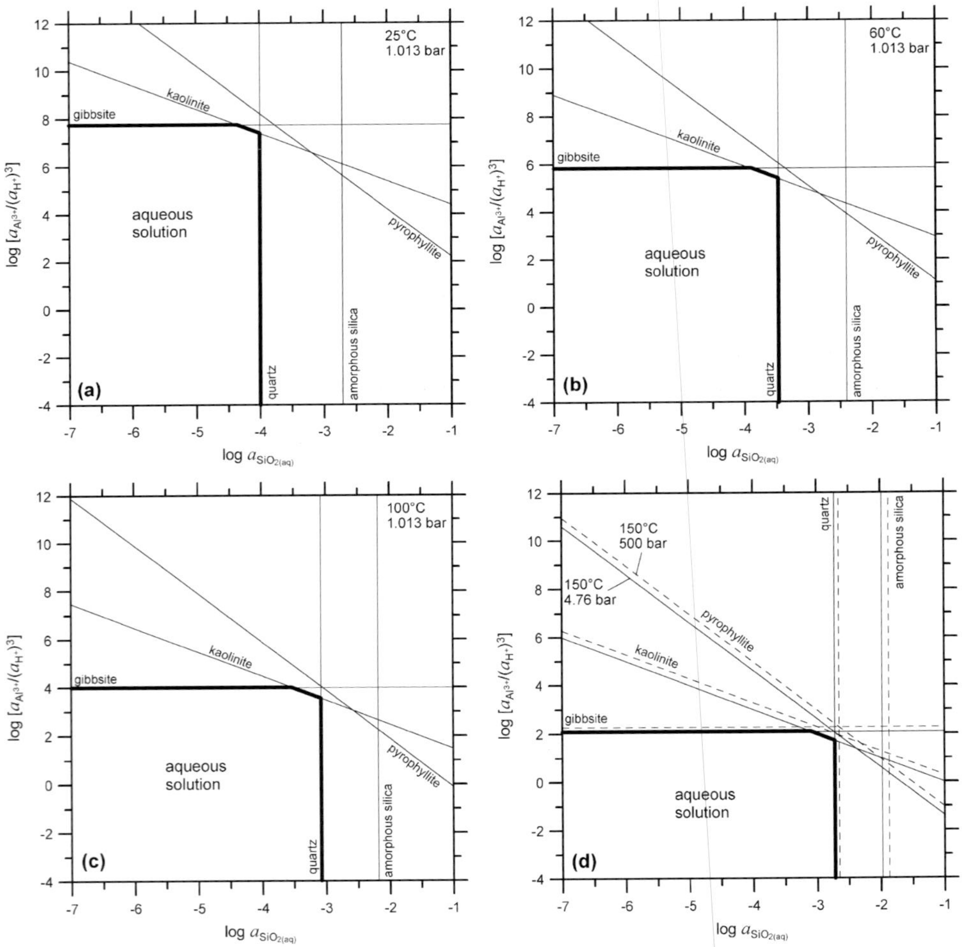

Figure 5.22. Activity plots of $\log\left(a_{Al^{3+}}/a_{H^+}^3\right)$ vs. $\log a_{SiO_{2(aq)}}$ for the system Al_2O_3–SiO_2–H_2O at (a) 25°C and 1.013 bar, (b) 60°C and 1.013 bar, (c) 100°C and 1.013 bar and (d) 150°C and both 4.76 bar (saturation pressure) and 500 bar. The thick line refers to the aqueous solution in equilibrium with the least soluble, thermodynamically stable minerals, namely gibbsite, kaolinite and quartz.

In this case, the stability field of the aqueous solution is limited by the solubility line of gibbsite at low $a_{SiO_{2(aq)}}$ values and by the solubility lines of kaolinite and quartz at higher $a_{SiO_{2(aq)}}$ values. These three minerals are the least soluble, thermodynamically stable solid phases in the system Al_2O_3–SiO_2–H_2O. Figure 5.22 also suggests that pyrophyllite may precipitate from oversaturated aqueous solutions at any $a_{SiO_{2(aq)}}$ values.

Similar to what observed for the system MgO–SiO_2–H_2O, variations in temperature bring about considerable shifts in the solubility relations of gibbsite, kaolinite and pyrophyllite, but their relative positions exhibit little changes. Also in this case, the pressure increase from 1 to 500 bar at 150°C (Fig. 5.22d) determines small shifts in the solubility lines of relevant solid phases.

5.3.3.3. The system MgO–Al_2O_3–SiO_2–H_2O

In order to draw the stability relations for this four-component system on two-dimensional plots, it is necessary to reduce the number of variables. This is usually accomplished assuming that aluminium is entirely conserved in the solid phases, which is a reasonable hypothesis, since the amount of dissolved Al is very small, at least in the pH range of most natural waters (as already recalled). This assumption corresponds to consider Al_2O_3 as an inert component in the sense of Thompson (1955). It should be noted that we are now talking about stability relations instead of *solubility* relations (as we did in the previous sections), because we are now neglecting Al concentration and without this information the solubility of Al-bearing minerals cannot be evaluated.

Operatively, we have to write dissolution–precipitation reactions involving pairs of minerals, in such a way that all the aluminium remains in the solid phases. For the kaolinite–gibbsite pair we sum the dissolution reaction of kaolinite (given in Table 5.10) plus the precipitation reaction of gibbsite (which is the reverse of its dissolution reaction given in Table 5.10) multiplied times 2. This is done to cancel out Al^{3+} ion. In this way, the following kaolinite–gibbsite reaction is obtained:

$$\text{Kaolinite} + H_2O = 2\text{Gibbsite} + 2SiO_{2(aq)}, \tag{5-62}$$

whose thermodynamic equilibrium constant reduces to:

$$\log a_{SiO_{2(aq)}} = \frac{\log K_{\text{Kaolinite}} - 2 \cdot \log K_{\text{Gibbsite}}}{2}, \tag{5-63}$$

assuming unit activity for H_2O and for the two solid phases. In the activity plot of $\log\,(a_{Mg^{2+}}/a^2_{H^+})$ vs. $\log a_{SiO_{2(aq)}}$, equation (5-63) corresponds to a vertical straight line whose position is constrained by the logarithm of the thermodynamic equilibrium constant of reaction (5-62) divided by the stoichiometric coefficient of $SiO_{2(aq)}$.

Similarly, for the pyrophyllite–kaolinite pair we get:

$$\text{Pyrophyllite} + H_2O = \text{Kaolinite} + 2SiO_{2(aq)}, \tag{5-64}$$

whose thermodynamic equilibrium constant is

$$\log a_{SiO_{2(aq)}} = \frac{\log K_{Pyrophyllite} - \log K_{Kaolinite}}{2}, \tag{5-65}$$

which is also represented by a vertical straight line in the activity plot of log $(a_{Mg^{2+}}/a^2_{H^+})$ vs. $\log a_{SiO_{2(aq)}}$. The position of this new straight line, which is representative of pyrophyllite–kaolinite coexistence, is fixed by the logarithm of the thermodynamic equilibrium constant of reaction (5-64) divided by the stoichiometric coefficient of $SiO_{2(aq)}$.

The equilibrium condition between pyrophyllite, Mg-montmorillonite and the aqueous solution is described by the reaction

$$\begin{aligned} &\text{Mg-montmorillonite} + 0.99\text{H}^+ \\ &= (1.67/2)\text{Pyrophyllite} + 0.495\text{Mg}^{2+} + 0.66\text{SiO}_{2(aq)} + 0.66\text{H}_2\text{O} \end{aligned} \tag{5-66}$$

whose thermodynamic equilibrium constant is

$$\begin{aligned} \log\left(\frac{a_{Mg^{2+}}}{a^2_{H^+}}\right) &= -\frac{0.66}{0.495}\cdot \log a_{SiO_{2(aq)}} \\ &+ \frac{\log K_{Mg\text{-}montmorillonite} - (1.67/2)\cdot \log K_{Phyrophyllite}}{0.495}, \end{aligned} \tag{5-67}$$

where 1.67 and 2 are the stoichiometric coefficients of Al^{3+} ion in the dissolution reaction of Mg-montmorillonite and pyrophyllite, respectively (see Table 5.10).

For the kaolinite–Mg-montmorillonite–aqueous solution equilibrium coexistence, we write:

$$\begin{aligned} &\text{Mg-montmorillonite} + 0.99\text{H}^+ + 0.175\text{H}_2\text{O} \\ &= 0.835\text{Kaolinite} + 0.495\text{Mg}^{2+} + 2.33\text{SiO}_{2(aq)} \end{aligned} \tag{5-68}$$

and

$$\begin{aligned} \log\left(\frac{a_{Mg^{2+}}}{a^2_{H^+}}\right) &= -\frac{2.33}{0.495}\cdot \log a_{SiO_{2(aq)}} \\ &+ \frac{\log K_{Mg\text{-}montmorillonite} - 0.835\cdot \log K_{Kaolinite}}{0.495}, \end{aligned} \tag{5-69}$$

The equilibrium coexistence of Mg-saponite, Mg-montmorillonite and the aqueous solution is described by the reaction:

$$\begin{aligned} &\text{Mg-saponite} + 6.1344\ \text{H}^+ \\ &= 0.1976\text{Mg - montmorillonite} + 3.0672\text{Mg}^{2+} + 2.8796\text{SiO}_{2(aq)} + 3.8696\text{H}_2\text{O} \end{aligned} \tag{5-70}$$

whose thermodynamic equilibrium constant is

$$\log\left(\frac{a_{Mg^{2+}}}{a_{H^+}^2}\right) = -\frac{2.8796}{3.0672}\cdot\log a_{SiO_{2(aq)}} + \frac{\log K_{Mg-saponite} - 0.1976\cdot\log K_{Mg-montmorillonite}}{3.0672}, \tag{5-71}$$

For the Mg–saponite–kaolinite pair we write the reaction:

$$\text{Mg–saponite} + 6.33H^+ = 0.165\text{Kaolinite} + 3.165Mg^{2+} + 3.34SiO_{2(aq)} + 3.835H_2O \tag{5-72}$$

and its thermodynamic equilibrium constant:

$$\log\left(\frac{a_{Mg^{2+}}}{a_{H^+}^2}\right) = -\frac{3.34}{3.165}\cdot\log a_{SiO_{2(aq)}} + \frac{\log K_{Mg-saponite} - 0.165\cdot\log K_{kaolinite}}{3.165}. \tag{5-73}$$

For the Mg–saponite–gibbsite pair we obtain the reaction:

$$\text{Mg–saponite} + 6.33H^+ = 0.33\text{Gibbsite} + 3.165Mg^{2+} + 3.67SiO_{2(aq)} + 3.67H_2O \tag{5-74}$$

whose thermodynamic equilibrium constant is

$$\log\left(\frac{a_{Mg^{2+}}}{a_{H^+}^2}\right) = -\frac{3.67}{3.165}\cdot\log a_{SiO_{2(aq)}} + \frac{\log K_{Mg-saponite} - 0.33\cdot\log K_{Gibbsite}}{3.165}. \tag{5-75}$$

We have also to consider the pairs Mg–saponite–14 Å-clinochlore and 14 Å-clinochlore–gibbsite. For the first one we get

$$\text{Mg–saponite} + 4.68H^+ = 0.165\text{Clinochlore} + 2.34Mg^{2+} + 3.175SiO_{2(aq)} + 2.68H_2O \tag{5-76}$$

and

$$\log\left(\frac{a_{Mg^{2+}}}{a_{H^+}^2}\right) = -\frac{3.175}{2.34}\cdot\log a_{SiO_{2(aq)}} + \frac{\log K_{Mg-saponite} - 0.165\cdot\log K_{Clinochlore}}{2.34}, \tag{5-77}$$

For the second pair we have

$$\begin{aligned}&\text{Clinochlore} + 10\text{H}^{+}\\&= 2\text{Gibbsite} + 5\text{Mg}^{2+} + 3\text{SiO}_{2(\text{aq})} + 6\text{H}_2\text{O}\end{aligned} \tag{5-78}$$

and

$$\log\left(\frac{a_{\text{Mg}^{2+}}}{a^2_{\text{H}^+}}\right) = -\frac{3}{5}\cdot\log a_{\text{SiO}_{2(\text{aq})}} + \frac{\log K_{\text{Clinochlore}} - 2\cdot\log K_{\text{Gibbsite}}}{5}. \tag{5-79}$$

Similar to what observed above for the systems $MgO–SiO_2–H_2O$ and $Al_2O_3–SiO_2–H_2O$, a general relation can be obtained also for the $MgO–Al_2O_3–SiO_2–H_2O$ system, under the assumption of Al conservation in the solid phases. This general relation is

$$\log\left(\frac{a_{\text{Mg}^{2+}}}{a^2_{\text{H}^+}}\right) = -\frac{v_{\text{SiO}_2,\text{aq}}}{v_{\text{Mg}^{2+}}}\cdot\log a_{\text{SiO}_{2(\text{aq})}} + \frac{\log K_{\text{solid,react}} - \dfrac{v_{\text{solid,prod}}}{v_{\text{solid,react}}}\cdot\log K_{\text{solid,prod}}}{v_{\text{Mg}^{2+}}}. \tag{5-80}$$

where $K_{\text{solid,react}}$ and $K_{\text{solid,prod}}$ are the thermodynamic equilibrium constants of the dissolution reactions of the mineral phases appearing as solid reactant and as solid product in the overall reaction, respectively, and the stoichiometric coefficients v_i refer to this overall reaction. Note that $v_{\text{solid,prod}}$ is equal to $v_{\text{Al}^{3+}}$ in the dissolution reaction and $v_{\text{solid,react}}$ is equal to $v_{\text{Al}^{3+}}$ in the precipitation reaction.

In addition to the reactions listed above and their thermodynamic equilibrium constants, other reactions and stability relations could be obtained among the considered minerals and other minerals could be taken into account in the considered system, such as amesite and Mg-beidellite. These additional stability relationships are disregarded as they are metastable.

We have already seen that equations (5-63) and (5-65) are used to draw vertical straight lines (of slope = 1) in the activity plot of log ($a_{\text{Mg}^{2+}}/a^2_{\text{H}^+}$) vs. $\log a_{\text{SiO}_{2(\text{aq})}}$. All the other thermodynamic equilibrium constants, i.e. expressions (5-67), (5-69), (5-71), (5-73), (5-75), (5-77) and (5-79) are represented as straight lines of variable, negative slope on the same activity plot. As already noted above, all these slopes are negative, since both Mg^{2+} and $SiO_{2(aq)}$ are on the same side of the corresponding reactions, i.e. (5-66), (5-68), (5-70), (5-72), (5-74), (5-76) and (5-78).

Using the log *K* values reported in Table 5.10, activity plots of log ($a_{\text{Mg}^{2+}}/a^2_{\text{H}^+}$) vs. $\log a_{\text{SiO}_{2(\text{aq})}}$ have been drawn for temperatures of 25, 60 and 100°C and total pressure of 1.013 bar, and for 150°C and both 4.76 bar (saturation pressure) and 500 bar (Fig. 5.23).

Also shown in these activity diagrams are the saturation (equilibrium) conditions with respect to quartz and amorphous silica, which have the same meaning of Figs. 5.21 and 5.22 and are representative of the minimum and maximum concentrations of dissolved SiO_2 in most natural waters. In principle, all the aqueous solutions with dissolved SiO_2 concentration higher than that fixed by quartz saturation are metastable, but in practice dissolution and precipitation of quartz proceed very slowly at low temperatures (Rimstidt and

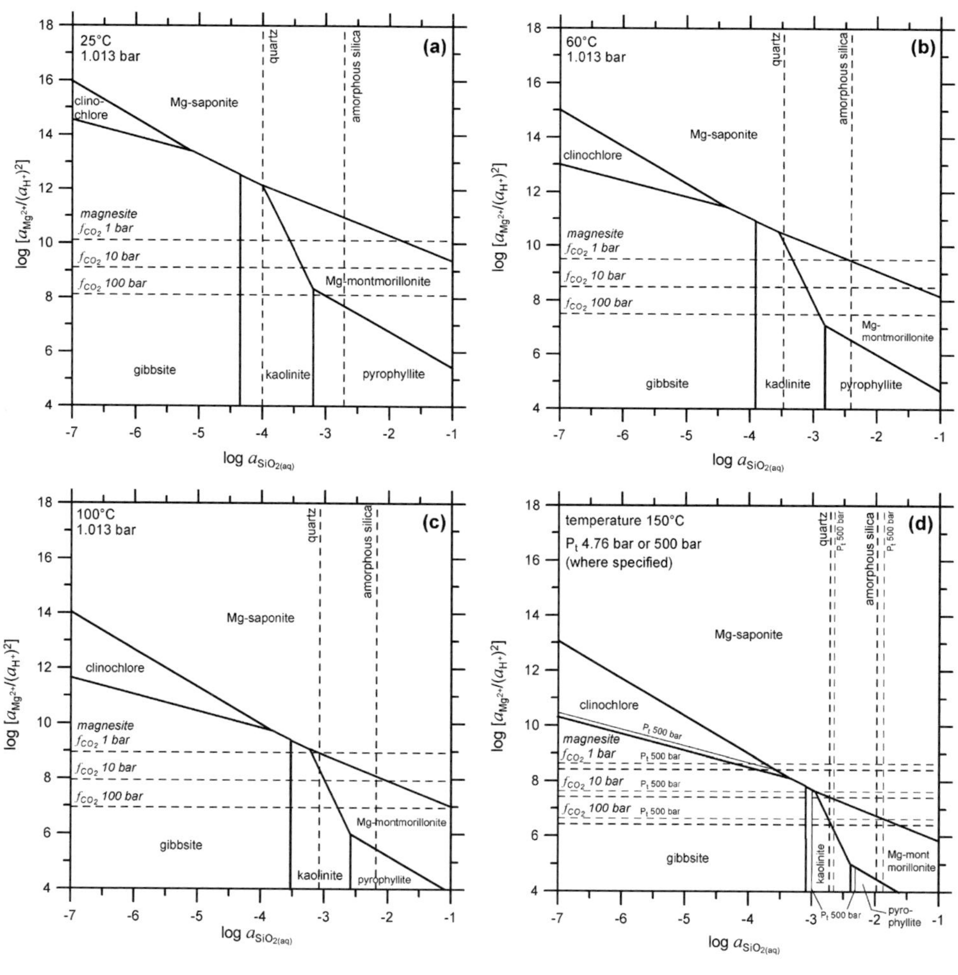

Figure 5.23. Activity plots of log ($a_{Mg^{2+}}/a^2_{H^+}$) vs. log $a_{SiO_{2(aq)}}$ for the system MgO–Al_2O_3–SiO_2–H_2O at (a) 25°C and 1.013 bar, (b) 60°C and 1.013 bar, (c) 100°C and 1.013 bar and (d) 150°C and both 4.76 bar (saturation pressure) and 500 bar. Also shown are the saturation lines for magnesite, at CO2 fugacities of 1, 10 and 100 bar, as well as for quartz and amorphous silica.

Barnes, 1980). Consequently, saturation with respect to quartz is attained only at relatively high temperatures, typically above 180°C in the geothermal reservoirs, after relatively long water–rock interaction (Fournier, 1991). On the contrary, dissolution and precipitation of amorphous silica occur at relatively fast rates and, consequently, amorphous silica solubility represents the upper threshold for aqueous SiO_2 in natural waters.

Also shown in the activity plots of Fig. 5.23 are three horizontal straight lines (i.e., characterized by $\log (a_{Mg^{2+}}/a^2_{H^+})$ = constant), which are representative of magnesite saturation for CO_2 fugacities of 1, 10 and 100 bar. These lines have been obtained referring to the dissolution reaction of magnesite:

$$\text{Magnesite} + 2H^+ = Mg^{2+} + CO_{2(g)} + H_2O \tag{5-81}$$

whose thermodynamic equilibrium constant can be rearranged as follows:

$$\log\left(\frac{a_{Mg^{2+}}}{a^2_{H^+}}\right) = \log K_{\text{magnesite}} - \log K_{CO_{2(g)}-HCO_3^-} \\ -\log f_{CO_2} = \log K'_{\text{magnesite}} - \log f_{CO_2} \tag{5-82}$$

(note that both $\log K_{\text{magnesite}}$ and $\log K_{CO_{2(g)}-HCO_3^-}$ refer to the reactions given in Table 5.4, whose stoichiometry is different from that of reaction (5-81)). Consequently, the $\log (a_{Mg^{2+}}/a^2_{H^+})$ ratio is a constant for a given f_{CO_2} value at fixed temperature and pressure. Of course, it is impossible to have a f_{CO_2} of 100 bar and a total pressure of 1.013 or 4.76 bar only. However, since the phase boundaries between solid phases are affected to a negligible extent by a change in total pressure from few bars to 500 bars (see above), it is acceptable to use these activity plots, in spite of this inconsistency between total pressure and f_{CO_2}.

The magnesite saturation lines are of special interest for the geological sequestration of CO_2. For instance, let us assume that CO_2 is injected, at a constant fugacity of 100 bar, into a deep reservoir, where magnesite precipitation takes place and where the activity of aqueous SiO_2 is fixed by equilibrium with quartz. The intersection of the magnesite equilibrium line with the quartz equilibrium line in the activity plots of Fig. 5.23 indicates the expected product solid phase(s), which is kaolinite, at temperatures of 25–100°C, and either kaolinite or Mg-montmorillonite or both at 150°C. The effect of a pressure increase from 1 to 500 bar is depicted in the plot drawn for the temperature of 150°C: it results to be very small. It must be underscored that these indications, concerning the expected product solid phase(s), refer to the relatively simple system $MgO–Al_2O_3–SiO_2–H_2O$. Kaolinite could be unstable with respect to other secondary minerals in more complex systems involving other chemical components, such as CaO, Na_2O and K_2O.

5.3.3.4. The system $CaO–Al_2O_3–SiO_2–H_2O$

The activity plots for the system $MgO–Al_2O_3–SiO_2–H_2O$ are a good starting point for constructing the activity plots for the system $CaO–Al_2O_3–SiO_2–H_2O$. Of course, no stability field corresponding to that of clinochlore is expected to be present in the diagram of

$\log\left(a_{Ca^{2+}}/a^2_{H^+}\right)$ vs. $\log a_{SiO_{2(aq)}}$, as no Ca-analogue of clinochlore is known to exist. The stability fields of gibbsite, kaolinite and pyrophyllite are the same for both sets of activity plots, as these minerals are free of both Mg and Ca. In contrast, the stability fields of the Mg-endmembers of both saponite and montmorillonite are expected to be substituted by the corresponding Ca-endmembers. Calcium-saponite and Ca-montmorillonite, however, contain not only Ca but also Mg, implying that both Ca^{2+} and Mg^{2+} ions are present among the products of their dissolution, in addition to Al^{3+} ion and $SiO_{2(aq)}$. Therefore, we have to find a way to get rid not only of Al^{3+} ion, but also of Mg^{2+} ion, in order to construct the activity plot of $\log\left(a_{Ca^{2+}}/a^2_{H^+}\right)$ vs. $\log a_{SiO_{2(aq)}}$. Again, Al^{3+} ion is eliminated assuming its conservation in the solid phases (see Section 5.3.3.3), whereas a possible way to get rid of Mg^{2+} ion is by involving both Ca- and Mg-saponite and Ca- and Mg-montmorillonite in the dissolution/precipitation reactions of interest. Besides it is assumed that the Ca- and Mg-components have equal activities both in the saponite solid mixture and in the montmorillonite solid mixture. First, the following reactions among Ca- and Mg-saponite and Ca- and Mg-montmorillonite are written assuming Mg conservation in the solid phases:

$$\begin{aligned}&1.5\text{Ca-montmorillonite} + 3\text{H}^+\\&= \text{Mg-montmorillonite} + 0.2475\text{Ca}^{2+} + 0.835\text{Al}^{3+} + 2\text{SiO}_{2(aq)} + 2\text{H}_2\text{O}\end{aligned} \tag{5-83}$$

and

$$\begin{aligned}&1.055\text{Ca-saponite} + 0.4026\text{H}^+\\&= \text{Mg-saponite} + 0.1741\text{Ca}^{2+} + 0.01815\text{Al}^{3+} + 0.20185\text{SiO}_{2(aq)} + 0.2563\text{H}_2\text{O}.\end{aligned} \tag{5-84}$$

The montmorillonite–saponite phase boundary is obtained by summation of reactions (5-83) and (5-84), assuming now Al conservation in the solid phases:

$$\begin{aligned}&1.055\text{Ca-saponite} + 0.02174\text{Mg-montmorillonite} + 0.3374\text{H}^+\\&= \text{Mg-saponite} + 0.03261\text{Ca-montmorillonite} + 0.1687\text{Ca}^{2+}\\&\qquad + 0.1584\text{SiO}_{2(aq)} + 0.2128\text{H}_2\text{O}.\end{aligned} \tag{5-85}$$

The thermodynamic equilibrium constant of reaction (5-85) is

$$\begin{aligned}\log\left(\frac{a_{Ca^{2+}}}{a^2_{H^+}}\right) &= -\frac{0.1584}{0.1687}\cdot\log a_{SiO_{2(aq)}}\\&+\frac{1.055\cdot\log K_{\text{Ca-saponite}} + 0.02174\cdot\log K_{\text{Mg-montmorillonite}} - \log K_{\text{Mg-saponite}} - 0.03261\cdot\log K_{\text{Ca-montmorillonite}}}{0.1687}\end{aligned} \tag{5-86}$$

Note that since the Ca- and Mg-components are assumed to have equal activities both in the saponite solid mixture and in the montmorillonite solid mixture, these activity terms cancel out and do not appear in equation (5-86). The same consideration holds for the subsequent relations involving either montmorillonite or saponite.

Then, the other phase boundaries of interest are found by summing the dissolution reactions of pyrophyllite, kaolinite and gibbsite with either reaction (5-83) or (5-84), again assuming Al conservation in the solid phases. For instance, the following reaction is obtained for the pyrophyllite–montmorillonite phase boundary:

$$\begin{aligned} &1.5\text{Ca - montmorillonite} + 0.495\text{H}^+ \\ &= 0.4175\text{Pyrophyllite} + \text{Mg-montmorillonite} \\ &\quad + 0.2475\text{Ca}^{2+} + 0.33\text{SiO}_{2(aq)} + 0.33\text{H}_2\text{O} \end{aligned} \tag{5-87}$$

whose thermodynamic equilibrium constant is

$$\begin{aligned} \log\left(\frac{a_{\text{Ca}^{2+}}}{a^2_{\text{H}^+}}\right) &= -\frac{0.33}{0.2475}\cdot\log a_{\text{SiO}_{2(aq)}} \\ &+\frac{1.5\cdot\log K_{\text{Ca-mont}} - 0.4175\cdot\log K_{\text{Pyro}} - \log K_{\text{Mg-mont}}}{0.2475}. \end{aligned} \tag{5-88}$$

Similarly, the following reaction is found for the kaolinite–montmorillonite phase boundary:

$$\begin{aligned} &1.5\text{Ca - montmorillonite} + 0.495\text{H}^+ + 0.0875\text{H}_2\text{O} \\ &= 0.4175\text{Kaolinite} + \text{Mg - montmorillonite} + 0.2475\text{Ca}^{2+} + 1.165\text{SiO}_{2(aq)} \end{aligned} \tag{5-89}$$

whose thermodynamic equilibrium constant is

$$\begin{aligned} \log\left(\frac{a_{\text{Ca}^{2+}}}{a^2_{\text{H}^+}}\right) &= -\frac{1.165}{0.2475}\cdot\log a_{\text{SiO}_{2(aq)}} \\ &+\frac{1.5\cdot\log K_{\text{Ca-mont}} - 0.4175\cdot\log K_{\text{Kaol}} - \log K_{\text{Mg-mont}}}{0.2475}. \end{aligned} \tag{5-90}$$

This reaction describes the saponite–kaolinite phase boundary:

$$\begin{aligned} &1.055\text{Ca - saponite} + 0.3482\text{H}^+ \\ &= \text{Mg - saponite} + 0.009075\text{Kaolinite} + 0.1741\text{Ca}^{2+} + 0.1837\text{SiO}_{2(aq)} + 0.2109\text{H}_2\text{O} \end{aligned} \tag{5-91}$$

whose thermodynamic equilibrium constant is:

$$\begin{aligned} \log\left(\frac{a_{\text{Ca}^{2+}}}{a^2_{\text{H}^+}}\right) &= -\frac{0.184}{0.174}\cdot\log a_{\text{SiO}_{2(aq)}} \\ &+\frac{1.055\cdot\log K_{\text{Ca-sapo}} - 0.00908\cdot\log K_{\text{Kaol}} - \log K_{\text{Mg-sapo}}}{0.1741}. \end{aligned} \tag{5-92}$$

Finally, the reaction describing the saponite–gibbsite phase boundary is

$$1.055\text{Ca saponite} + 0.3482\text{H}^+ \\ = \text{Mg saponite} + 0.01815\text{Gibbsite} + 0.1741\text{Ca}^{2+} + 0.2019\text{SiO}_{2(aq)} + 0.2019\text{H}_2\text{O} \quad (5\text{-}93)$$

whose thermodynamic equilibrium constant is

$$\log\left(\frac{a_{Ca^{2+}}}{a^2_{H^+}}\right) = -\frac{0.2019}{0.1741} \cdot \log a_{SiO_{2(aq)}} \\ + \frac{1.055 \cdot \log K_{\text{Ca-sapo}} - 0.01815 \cdot \log K_{\text{Gibbsite}} - \log K_{\text{Mg-sapo}}}{0.1741} \quad (5\text{-}94)$$

Based on the log K values reported in Table 5.10, the stability boundaries in the activity plots of $\log\left(a_{Ca^{2+}}/a^2_{H^+}\right)$ vs. $\log a_{SiO_{2(aq)}}$ have been readily drawn for the temperatures of 25, 60 and 100°C and total pressure of 1.013 bar, and for 150°C at 4.76 bar, i.e. saturation pressure (Fig. 5.24). Again, in these activity plots are also reported the saturation (equilibrium) lines for quartz and amorphous silica, as well as those for calcite at CO_2 fugacities of 1, 10 and 100 bar. The latter three saturation lines have been computed referring to the dissolution reaction of calcite:

$$\text{Calcite} + 2\text{H}^+ = \text{Ca}^{2+} + \text{CO}_{2(g)} + \text{H}_2\text{O} \quad (5\text{-}95)$$

whose thermodynamic equilibrium constant can be rearranged as follows:

$$\log\left(\frac{a_{Ca^{2+}}}{a^2_{H^+}}\right) = \log K_{\text{calcite}} - \log K_{CO_{2(g)}-HCO_3^-} - \log f_{CO_2}. \quad (5\text{-}96)$$

(again, note that both $\log K_{\text{calcite}}$ and $\log K_{CO_{2(g)}-HCO_3^-}$ refer to the reactions given in Table 5.4, whose stoichiometry is different from that of reaction (5-95)). Therefore, the $\log\left(a_{Ca^{2+}}/a^2_{H^+}\right)$ ratio is a constant if f_{CO_2}, temperature, and pressure are all fixed. Similar to what observed in the previous section for the magnesite saturation lines, also the calcite saturation lines are of utmost importance for the geological disposal of CO_2. Again, let us assume CO_2 injection, at a constant fugacity of 100 bar, into a deep reservoir, where calcite precipitation occurs and where $SiO_{2(aq)}$ activity is constrained by quartz saturation. The intersection of the calcite saturation line with the quartz saturation line in the activity plots of Fig. 5.24 suggests that the expected product solid phase is kaolinite, at all the considered temperatures, i.e. 25, 60, 100, and 150°C. Some words of caution are needed also here, as these inferences on the expected product solid phase hold true for the relatively simple system $CaO–Al_2O_3–SiO_2–H_2O$. Again, other secondary minerals could stably form instead of kaolinite in more complex systems involving other chemical components, such as the alkali oxides Na_2O and K_2O.

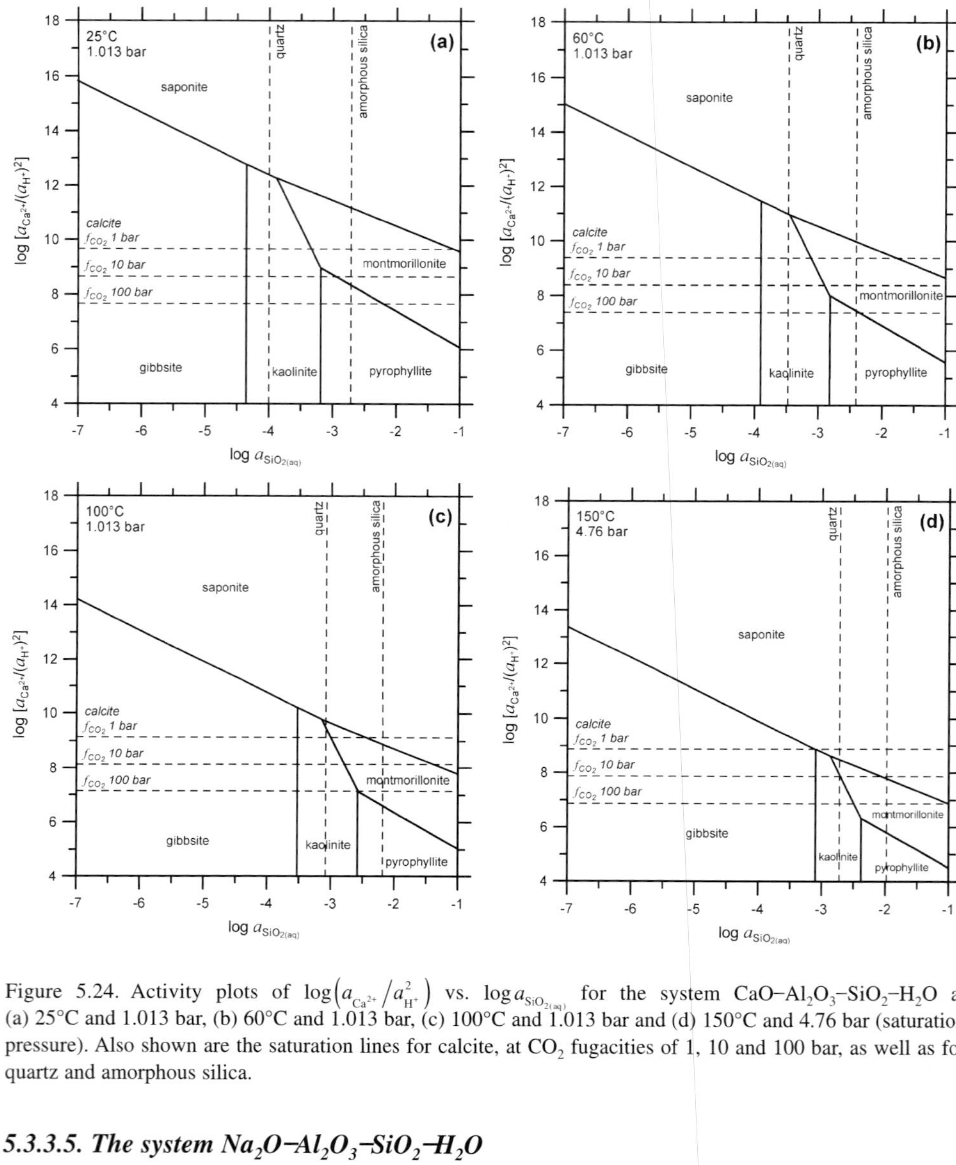

Figure 5.24. Activity plots of $\log\left(a_{Ca^{2+}}/a_{H^+}^2\right)$ vs. $\log a_{SiO_{2(aq)}}$ for the system $CaO–Al_2O_3–SiO_2–H_2O$ at (a) 25°C and 1.013 bar, (b) 60°C and 1.013 bar, (c) 100°C and 1.013 bar and (d) 150°C and 4.76 bar (saturation pressure). Also shown are the saturation lines for calcite, at CO_2 fugacities of 1, 10 and 100 bar, as well as for quartz and amorphous silica.

5.3.3.5. The system $Na_2O–Al_2O_3–SiO_2–H_2O$

The same approach used above for the system $CaO–Al_2O_3–SiO_2–H_2O$, can be applied to prepare the activity plots for the system $Na_2O–Al_2O_3–SiO_2–H_2O$. Therefore, there is no need to go through all the steps of this exercise and we can jump directly to the final results, which are depicted in Fig. 5.25 for the temperatures of 25, 60 and 100°C and total pressure of 1.013 bar, and for 150°C at 4.76 bar, i.e. saturation pressure. The saturation lines for quartz and amorphous silica and the equilibrium lines for dawsonite at CO_2 fugacities of 1, 10 and 100 bar are also shown in these activity plots. Since the dissolution reaction of dawsonite:

$$\text{Dawsonite} + 4H^+ = Na^+ + Al^{3+} + CO_{2(g)} + 3H_2O \tag{5-97}$$

involves not only Na^+ ion but also Al^{3+} ion, the equilibrium lines for dawsonite are independent of dissolved SiO_2 activity only in the gibbsite field. In fact, the thermodynamic equilibrium constant of the dawsonite–gibbsite reaction:

$$\text{Dawsonite} + H^+ = \text{Gibbsite} + Na^+ + CO_{2(g)} \tag{5-98}$$

can be written as

$$\log\left(\frac{a_{Na^+}}{a_{H^+}}\right) = \log K_{\text{dawsonite}} - \log K_{\text{gibbsite}} - \log K_{CO_{2(g)}-HCO_3^-} - \log f_{CO_2}. \tag{5-99}$$

In contrast, the thermodynamic equilibrium constant of the dawsonite–kaolinite reaction:

$$\text{Dawsonite} + H^+ + SiO_2 = 1/2\ \text{Kaolinite} + Na^+ + CO_{2(g)} + 1/2\ H_2O \tag{5-100}$$

can be rearranged as follows:

$$\log\left(\frac{a_{Na^+}}{a_{H^+}}\right) = \log K_{\text{dawsonite}} - \frac{1}{2}\log K_{\text{kaolinite}} - \log K_{CO_{2(g)}-HCO_3^-} - \log f_{CO_2} + \log a_{SiO_{2(aq)}}. \tag{5-101}$$

(again, note that both log $K_{\text{dawsonite}}$ and $\log K_{CO_{2(g)}-HCO_3^-}$ refer to the reactions given in Tables 5.4, whereas hydrolysis of kaolinite and its log K are reported in Table 5.10). Therefore, at constant f_{CO_2}, temperature, and pressure, the $\log(a_{Na^+}/a_{H^+})$ ratio fixed by dawsonite–kaolinite equilibrium coexistence increase with $\log a_{SiO_{2(aq)}}$ along a straight line of slope +1 (Fig. 5.25). Relations similar to equation (5-101) are readily obtained for equilibrium coexistence of dawsonite and pyrophyllite, dawsonite and montmorillonite and dawsonite and saponite. These allow to complete the equilibrium lines for dawsonite at CO_2 fugacities of 1, 10 and 100 bar in the activity plots of Fig. 5.25. To check the result of these calculations, $\log(a_{Na^+}/a_{H^+})$ and $\log a_{SiO_{2(aq)}}$ values at 25°C, 1.013 bar and f_{CO_2} of 1 bar were computed for coexistence of dawsonite with either gibbsite, or kaolinite, or saponite by means of EQ3NR, following the approach outlined in Section 7.1.2. EQ3NR results (squares in Fig. 5.25a) coincides with the $\log(a_{Na^+}/a_{H^+})$ and $\log a_{SiO_{2(aq)}}$ values computed above, confirming previous inferences.

Similar to what was observed in Section 5.3.3.4, the equilibrium lines for dawsonite suggest that different clay minerals are expected to coexist stably with dawsonite depending on temperature, f_{CO_2}, and the composition of the aqueous solution, depicted in term of the $\log(a_{Na^+}/a_{H^+})$ and $\log a_{SiO_{2(aq)}}$ parameters. Again, some words of caution are needed since reference is made to a simple system and several assumptions have been introduced to prepare the activity plots of Fig. 5.25.

5.3.3.6. The system K_2O–MgO–Al_2O_3–SiO_2–H_2O at magnesite saturation

In principle, it is possible to prepare the activity plots for the systems K_2O–Al_2O_3–SiO_2–H_2O along the same lines used for the system CaO–Al_2O_3–SiO_2–H_2O

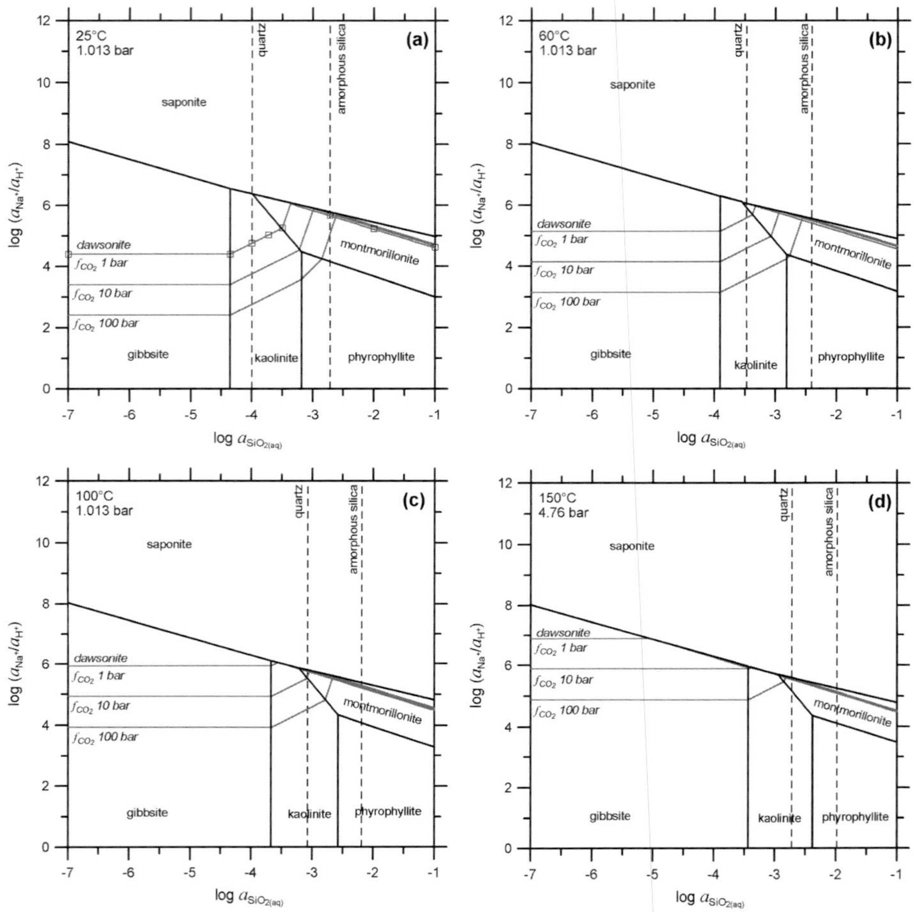

Figure 5.25. Activity plots of $\log(a_{Na^+}/a_{H^+})$ vs. $\log a_{SiO_{2(aq)}}$ for the system $Na_2O–Al_2O_3–SiO_2–H_2O$ at (a) 25°C and 1.013 bar, (b) 60°C and 1.013 bar, (c) 100°C and 1.013 bar and (d) 150°C and 4.76 bar (saturation pressure). Also shown are the saturation lines for dawsonite, at CO2 fugacities of 1, 10 and 100 bar, as well as for quartz and amorphous silica. The squares in Fig. 5.26a refer to the results of speciation calculations performed by means of the code EQ3NR.

and $Na_2O–Al_2O_3–SiO_2–H_2O$. However, this is a tedious exercise at this point. Therefore, a different approach is proposed here for adding K_2O to the system $MgO–Al_2O_3–SiO_2–H_2O$, whose activity plots have been explored in Section 5.3.3.3. Suppose we are chiefly interested to understand what happens when magnesite precipitates, it is worthwhile: (i) first, to write chemical reactions assuming Al conservation in the solid phases (as usually done in the preparation of activity plots), leaving the $\log(a_{Mg^{2+}}/a^2_{H^+})$ ratio as a variable in the corresponding thermodynamic constant expressions, and (ii) to insert then the $\log(a_{Mg^{2+}}/a^2_{H^+})$ ratio fixed by suitably chosen values of temperature, pressure and f_{CO_2} (see eq. (5-82)) in these thermodynamic constant expressions. This corresponds to sum the dissolution reaction of magnesite (5-81) to the considered chemical reactions.

Again, there is nothing new for what concerns the solid phases without K and Mg, i.e. gibbsite, kaolinite and pyrophyllite, whose phase boundaries are located as discussed above. The news are represented by the relationships involving the K–Mg-bearing solid phases, i.e. K-montmorillonite, illite and celadonite.

The K-montmorillonite–pyrophyllite phase boundary is found by referring to this reaction:

$$\begin{aligned}&\text{K-montmorillonite} + 0.99\text{H}^+\\&\quad = 0.835\text{Pyrophyllite} + 0.33\text{K}^+ + 0.33\text{Mg}^{2+} + 0.66\text{SiO}_{2(aq)} + 0.66\text{H}_2\text{O}\end{aligned} \tag{5-102}$$

whose thermodynamic equilibrium constant is

$$\begin{aligned}\log\left(\frac{a_{\text{K}^+}}{a_{\text{H}^+}}\right) &= -\frac{0.66}{0.33}\cdot\log a_{\text{SiO}_{2(aq)}} - \log\left(\frac{a_{\text{Mg}^{2+}}}{a^2_{\text{H}^+}}\right)\\&\quad + \frac{\log K_{\text{Mg-montmorillonite}} - 0.835\cdot\log K_{\text{Pyrophyllite}}}{0.33}.\end{aligned} \tag{5-103}$$

At magnesite saturation, equation (5-82) is also valid and the log $(a_{\text{Mg}^{2+}}/a^2_{\text{H}^+})$ ratio in equation (5-103) is substituted by $\log K'_{\text{magnesite}} - \log f_{\text{CO}_2}$, obtaining:

$$\begin{aligned}\log\left(\frac{a_{\text{K}^+}}{a_{\text{H}^+}}\right) &= -2\cdot\log a_{\text{SiO}_{2(aq)}} - \log \text{K}'_{\text{magnesite}}\\&\quad + \log f_{\text{CO}_2} + \frac{\log K_{\text{Mg-mont}} - 0.835\cdot\log K_{\text{Pyro}}}{0.33}.\end{aligned} \tag{5-104}$$

To constrain the K-montmorillonite–kaolinite phase boundary we begin to write the reaction:

$$\begin{aligned}&\text{K-montmorillonite} + 0.99\text{H}^+ + 0.175\text{H}_2\text{O}\\&\quad = 0.835\text{Kaolinite} + 0.33\text{K}^+ + 0.33\text{Mg}^{2+} + 2.33\text{SiO}_{2(aq)}\end{aligned} \tag{5-105}$$

whose thermodynamic equilibrium constant is

$$\begin{aligned}\log\left(\frac{a_{\text{K}^+}}{a_{\text{H}^+}}\right) &= -\frac{2.33}{0.33}\cdot\log a_{\text{SiO}_{2(aq)}} - \log\left(\frac{a_{\text{Mg}^{2+}}}{a^2_{\text{H}^+}}\right)\\&\quad + \frac{\log K_{\text{Mg-montmorillonite}} - 0.835\cdot\log K_{\text{Kaolinite}}}{0.33}.\end{aligned} \tag{5-106}$$

Then, assuming that the log $(a_{Mg^{2+}}/a^2_{H^+})$ ratio is controlled by magnesite saturation, it is replaced by $\log K'_{magnesite} - \log f_{CO_2}$, obtaining:

$$\log\left(\frac{a_{K^+}}{a_{H^+}}\right) = -\frac{2.33}{0.33}\cdot\log a_{SiO_{2(aq)}} - \log K'_{magnesite} + \log f_{CO_2} + \frac{\log K_{Mg-mont} - 0.835\cdot\log K_{Kaol}}{0.33}. \quad (5\text{-}107)$$

The line of coexistence of illite and kaolinite is determined by writing, first, the reaction:

$$\text{Illite} + 1.1H^+ + 0.75H_2O = 1.15\text{Kaolinite} + 0.6K^+ + 0.25Mg^{2+} + 1.2SiO_{2(aq)} \quad (5\text{-}108)$$

whose thermodynamic equilibrium constant is

$$\log\left(\frac{a_{K^+}}{a_{H^+}}\right) = -\frac{1.2}{0.6}\cdot\log a_{SiO_{2(aq)}} - \frac{0.25}{0.6}\log\left(\frac{a_{Mg^{2+}}}{a^2_{H^+}}\right) + \frac{\log K_{Illite} - 1.15\cdot\log K_{Kaolinite}}{0.6}. \quad (5\text{-}109)$$

Then, at magnesite saturation, we have

$$\log\left(\frac{a_{K^+}}{a_{H^+}}\right) = -\frac{1.2}{0.6}\cdot\log a_{SiO_{2(aq)}} - \frac{0.25}{0.6}\log K'_{magn} + \frac{0.25}{0.6}\log f_{CO_2} + \frac{\log K_{Illi} - 1.15\cdot\log K_{Kaol}}{0.6}. \quad (5\text{-}110)$$

Proceeding in the same way, we have then to deal with the following reactions, whose thermodynamic equilibrium constants are directly given below for the condition of magnesite saturation:

$$\text{Illite} + 1.1H^+ + 1.9H_2O = 2.3\text{Gibbsite} + 0.6K^+ + 0.25Mg^{2+} + 3.5SiO_{2(aq)} \quad (5\text{-}111)$$

$$\log\left(\frac{a_{K^+}}{a_{H^+}}\right) = -\frac{0.35}{0.6}\cdot\log a_{SiO_{2(aq)}} - \frac{0.25}{0.6}\log K'_{magn} + \frac{0.25}{0.6}\log f_{CO_2} + \frac{\log K_{Illi} - 2.3\cdot\log K_{Gibbs}}{0.6}. \quad (5\text{-}112)$$

$$1.3772\text{K-montmorillonite} + 0.2635H^+ + 0.1455K^+ = \text{Illite} + 0.2045Mg^{2+} + 2.009SiO_{2(aq)} + 0.509H_2O \quad (5\text{-}113)$$

$$\log\left(\frac{a_{K^+}}{a_{H^+}}\right) = \frac{2.009}{0.1455}\cdot \log a_{SiO_{2(aq)}} - \frac{0.2045}{0.1455}\log K'_{magn} + \frac{0.2045}{0.1455}\log f_{CO_2} + \frac{\log K_{Illi} - 1.3772\cdot \log K_{K-mont}}{0.1455} \tag{5-114}$$

$$2.3\text{Celadonite} + 5.8\text{H}^+ = \text{Illite} + 1.7\text{K}^+ + 2.05\text{Mg}^{2+} + 5.7\text{SiO}_{2(aq)} + 4.2\text{H}_2\text{O} \tag{5-115}$$

$$\log\left(\frac{a_{K^+}}{a_{H^+}}\right) = -\frac{5.7}{1.7}\cdot \log a_{SiO_{2(aq)}} - \frac{2.05}{1.7}\log \text{K}'_{magn} + \frac{2.05}{1.7}\log f_{CO_2} + \frac{2.3\cdot \log K_{Cela} - \log K_{Illi}}{1.7}. \tag{5-116}$$

$$1.67\text{Celadonite} + 4.02\text{H}^+ = \text{K-montmorillonite} + 1.34\text{K}^+ + 1.34\text{Mg}^{2+} + 2.68\text{SiO}_{2(aq)} + 2.68\text{H}_2\text{O} \tag{5-117}$$

$$\log\left(\frac{a_{K^+}}{a_{H^+}}\right) = -\frac{2.68}{1.34}\cdot \log a_{SiO_{2(aq)}} - \log K'_{magn} + \log f_{CO_2} + \frac{1.67\cdot \log K_{Cela} - \log K_{K\text{-mont}}}{1.34}. \tag{5-118}$$

Inserting in previous equations the log K values given in Table 5.10 and assuming that magnesite saturation occurs at f_{CO_2} of 100 bar, the activity plots of $\log\left(a_{K^+}/a_{H^+}\right)$ vs. $\log a_{SiO_{2(aq)}}$ have been readily obtained for 25°C and 1.013 bar, 60°C and 1.013 bar, 100°C and 1.013 bar and 150°C and 4.76 bar (saturation pressure) (Fig. 5.26). Again, the saturation (equilibrium) lines for quartz and amorphous silica are also drawn in these activity plots.

Accepting that the thermodynamically stable mineral of silica, i.e. quartz, is produced together with magnesite in the course of carbonation reactions, either kaolinite or illite or celadonite will form depending on the a_{K^+}/a_{H^+} ratio, as indicated by the activity plots of Fig. 5.26. Although the a_{K^+}/a_{H^+} ratio is the parameter fixed by mineral-solution equilibria, these two activities lumped together are much less informative than the two separate activities of K^+ and H^+.

As explained in Section 7.1, these two variables can be computed by means of a speciation saturation software code, such as EQ3NR (Wolery, 1992), for an aqueous solution in equilibrium with magnesite, quartz, kaolinite and illite, i.e. the aqueous phase corresponding to the intersection of the illite–kaolinite phase boundary with the quartz saturation line in the activity plots of Fig. 5.26.

At 150°C and f_{CO_2} of 100 bar, it turns out that the chemical speciation of the aqueous solution is dominated by HCO_3^-, among the anions, and by K^+ and $MgHCO_3{}^+$, among

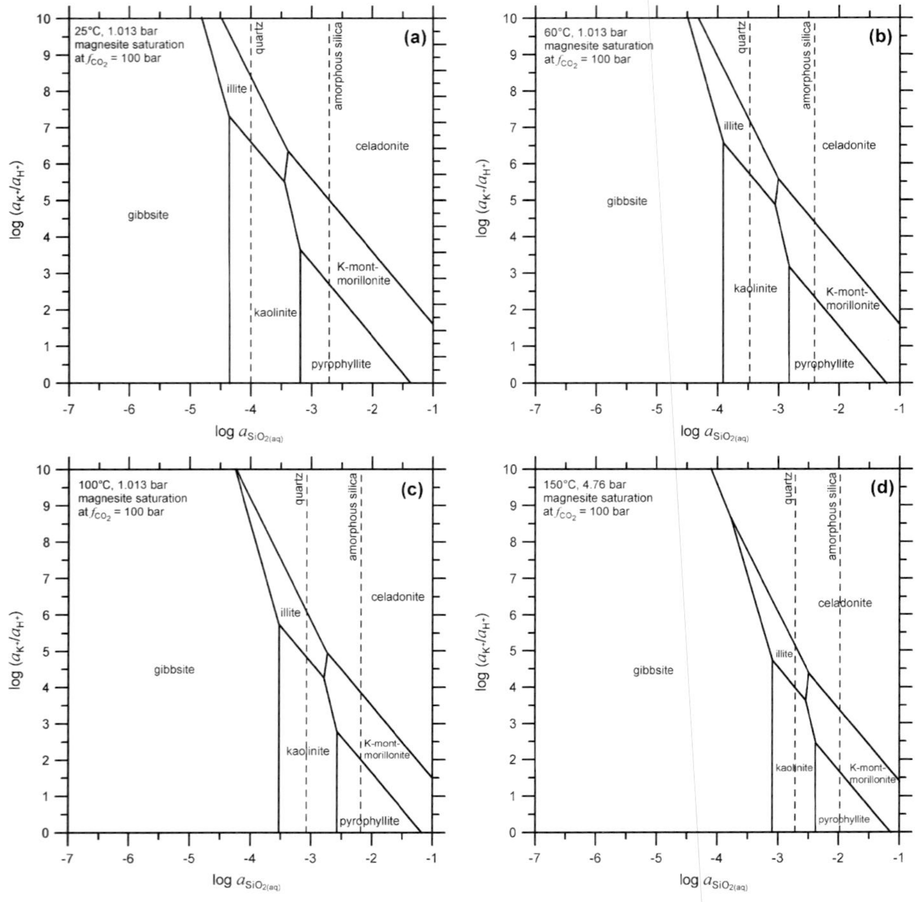

Figure 5.26. Activity plots of $\log(a_{K^+}/a_{H^+})$ vs. $\log a_{SiO_{2(aq)}}$ for the system K_2O–MgO–Al_2O_3–SiO_2–H_2O at magnesite saturation for CO_2 fugacity of 100 bar and at the following conditions: (a) 25°C and 1.013 bar, (b) 60°C and 1.013 bar, (c) 100°C and 1.013 bar and (d) 150°C and 4.76 bar (saturation pressure). Also shown are the saturation lines for quartz and amorphous silica.

the cations, whereas Mg^{2+} ion comes after these two species in order of decreasing importance, as listed below.

Species	Molality	Log molality	Log gamma	Log activity
CO_2(aq)	8.8426E–01	−0.0534	0.0077	−0.0457
HCO_3^-	5.2363E–02	−1.2810	−0.1092	−1.3901
K^+	5.1958E–02	−1.2843	−0.1310	−1.4154
SiO_2(aq)	1.9094E–03	−2.7191	0.0000	−2.7191
$MgHCO_3^+$	1.8405E–04	−3.7351	−0.1233	−3.8584
Mg^{2+}	1.1083E–04	−3.9553	−0.3933	−4.3486
H^+	5.1143E–06	−5.2912	−0.0955	−5.3868

The $\log\left(a_{K^+}/a_{H^+}\right)$ is 3.97 (as indicated also by Fig. 5.26d), pH is 5.39 and $\log a_{K^+}$ is −1.42. The free K^+ ion represents by far the prevailing K species and its concentration, 51.96 mmol kg^{-1}, coincides with total K molality, which corresponds to 2,032 mg kg^{-1}. Since this relatively high potassium concentration constitutes the lower threshold for illite generation, kaolinite (rather than illite) is expected to be the most likely phase during magnesite-producing carbonation reactions in the system K_2O–MgO–Al_2O_3–SiO_2–H_2O at 150°C and f_{CO_2} of 100 bar.

The minimum K concentration for illite generation increases with decreasing temperatures, i.e. 9,960 mg kg^{-1} at 100°C and 39,500 mg kg^{-1} at 60°C (no mathematical solution is found at 25°C), suggesting that illite production is even less likely at these lower temperatures.

This fact has important practical implications, as kaolinite forms booklet-like structures, which affect moderately permeability, whereas illite forms fibrous aggregates, which obstruct pores and reduce significantly permeability. It was shown (Blatt et al., 1980) that illite-cemented sandstones have permeability lower of at least one order of magnitude than the same kaolinite-cemented formation, for the same porosity value.

This example also shows the importance to consider complete rather than simplified chemical systems, which on the other hand are very useful for didactic purposes.

5.4. The thermodynamics of gas–solid carbonation reactions

The thermodynamics of gas–solid carbonation reactions has been recently reviewed by Lackner et al. (1995). Since this subject is strictly related to aqueous mineral carbonation and geological CO_2 sequestration, it is of extreme interest for us.

Lackner et al. (1995) underscored that Ca and Mg are the two most commonly available chemical elements that form stable, poorly soluble carbonate minerals, i.e. calcite, magnesite and dolomite. Iron might also be added to the list of such elements, since divalent Fe constitutes siderite, which has also low solubility in water. A substantial fraction of Fe is present in reactant minerals in the trivalent form which, however, cannot be carbonated as such. Since the reduction of Fe(III) to Fe(II) needs the involvement of one or more suitable reducing agents, Fe(III)-minerals are neglected in the following discussion. Besides, also Na and Al can be carbonated together producing dawsonite, as already discussed above (see Sections 5.1.5 and 5.3.3.5).

Minerals containing considerable amounts of MgO, CaO, FeO and $Na_2O + Al_2O_3$ (the latter two oxides may also be present in separate solid phases) in the presence of CO_2 can experience carbonation reactions, depending on the external conditions acting on the considered system. These reactions are initially exemplified referring to the oxides of Ca, Mg, Fe(II), Na_2O and Al_2O_3:

$$CaO + CO_{2(g)} \rightarrow CaCO_3 \tag{5-119}$$

$$MgO + CO_{2(g)} \rightarrow MgCO_3 \tag{5-120}$$

$$FeO + CO_{2(g)} \rightarrow FeCO_3 \qquad (5\text{-}121)$$

$$\tfrac{1}{2}Na_2O + \tfrac{1}{2}Al_2O_3 + CO_{2(g)} + H_2O_{(g)} \rightarrow NaAlCO_3(OH)_2. \qquad (5\text{-}122)$$

The Gibbs free energy of these reactions under standard state conditions, ΔG_r^o (i.e., f_{CO_2} of 1 bar and pure solid phases), at the temperature of 25°C and total pressure of 1 bar are −31,260, −15,318, −8,063 and −43,567 cal mol^{-1}, respectively, based on SUPCRT92 data (Johnson et al., 1992). Since these ΔG_r^o are negative, reactions (5-119) to (5-122) proceed spontaneously as they are written, i.e. consuming the Ca, Mg, Fe(II), Na and Al oxides and producing the corresponding carbonates.

Are these reactions spontaneous at any temperature and pressure? What is the influence of varying CO_2 fugacity? What happens if silicate minerals are carbonated instead of the pure oxides? All these important questions were answered by Lackner et al. (1995), who worked with the Gibbs free energies of formation from the oxides. Here the same subject is taken into account by using the Gibbs free energies of formation from the elements for the minerals of interest, mostly taken from the SUPCRT92 database and in some cases (almandine, pyrope, $CaSiO_3$ glass, $CaMgSi_2O_6$ glass and $CaAl_2Si_2O_8$ glass) from Robie and Hemingway (1995). Besides, Fe-silicates and Na–Al-silicates, which were neglected by Lackner et al. (1995), are also considered in the following discussion.

The Gibbs free energies of carbonation reactions for CO_2 fugacities higher than 1 bar, ΔG_r, were computed based on the fundamental relationship between the chemical potential and thermodynamic activity:

$$\mu_i = \mu_i^o + RT \cdot \ln a_i \qquad (2\text{-}9)$$

and recalling that the activity of a gas is equal to the ratio between its fugacity at the conditions of interest and its fugacity at standard state, which for gases is the hypothetical perfect gas at 1 bar and any specified temperature (this is the convention adopted in SUPCRT92). Therefore, equation (2-9) can be rewritten as follows:

$$\mu_i = \mu_i^o + RT \cdot \ln \frac{f_i}{f_i^o} = \mu_i^o + RT \cdot \ln \frac{f_i}{1}. \qquad (5\text{-}123)$$

Consequently, the ΔG_r for any f_{CO_2} value higher than 1 bar, is obtained by simply subtracting the term RT·ln f_{CO_2} (as CO_2 is a reactant in carbonation reactions) from the Gibbs free energies of carbonation for f_{CO_2} of 1 bar, ΔG_r^o.

The results of these simple calculations for the oxides involved in reactions (5-119) to (5-122), i.e. lime, periclase, ferrous oxide and sodium oxide plus corundum are shown in Fig. 5.27. The effect of total pressure is practically negligible for all the four reactions, as shown by the virtual superposition of the two ΔG_r–T lines, both of which were drawn for the same f_{CO_2} of 1 bar, but for different total pressure of either 500 bar or saturation pressure (which is a temperature function and varies from 1.013 bar at 100°C to 220.9 bar

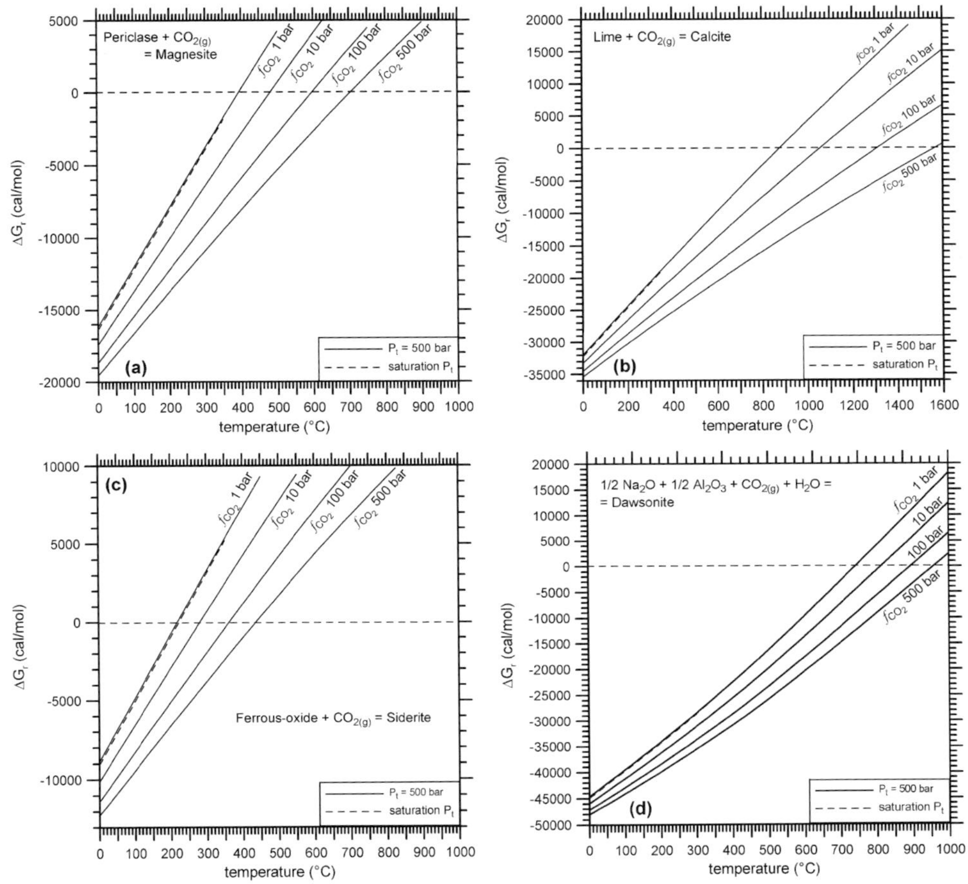

Figure 5.27. Temperature dependence of the Gibbs free energy of carbonation of (a) periclase, (b) lime, (c) ferrous oxide and (d) sodium oxide plus corundum at saturation pressure and CO_2 fugacity of 1 bar as well as at total pressure of 500 bar and CO_2 fugacities of 1, 10, 100 and 500 bar.

at the critical point of water, 374.14°C). We anticipate that a similarly negligible effect of total pressure is also observed for the carbonation of silicate minerals.

Figure 5.27a and Table 5.11 show that the carbonation of periclase proceeds spontaneously at temperatures lower than 393°C under f_{CO_2} of 1 bar (since $\Delta G_r < 0$) and that the maximum temperature of carbonation, T_{max}, increases with CO_2 fugacity, attaining 480°C for f_{CO_2} of 10 bar, 596°C for f_{CO_2} of 100 bar and 703°C for f_{CO_2} of 500 bar. Again, similar increases in T_{max} with f_{CO_2} are also observed for the carbonation of silicate minerals.

Lime and sodium oxide plus corundum carbonate spontaneously up to temperatures much higher than those of periclase carbonation, whereas the opposite occurs for ferrous oxide (Fig. 5.27 and Table 5.11).

TABLE 5.11

Maximum temperature of carbonation (°C), at different f_{CO_2} values for several minerals of interest and total pressure of 500 bar

Mineral	Chemical formula	$T_{max, f_{CO_2} = 1 bar}$	$T_{max, f_{CO_2} = 10 bar}$	$T_{max, f_{CO_2} = 100 bar}$	$T_{max, f_{CO_2} = 500 bar}$
Periclase	MgO	393	480	596	703
Lime	CaO	878	1,057	1,311	1,567
Ferrous-oxide	FeO	218	281	362	436
Sodium-oxide + Corundum	$Na_2O + Al_2O_3$	741	812	893	957
Forsterite	Mg_2SiO_4	211	272	351	423
Forsterite[a]	Mg_2SiO_4	195	255	335	413
Fayalite	Fe_2SiO_4	158	216	293	364
Pyrope	$Mg_3Al_2Si_3O_{12}$	241	293	356	409
Grossular	$Ca_3Al_2Si_3O_{12}$	237	300	378	446
Almandine	$Fe_3Al_2Si_3O_{12}$	166	213	270	319
Enstatite	$MgSiO_3$	191	250	327	397
Wollastonite	$CaSiO_3$	309	391	503	610
Ferrosilite	$FeSiO_3$	157	216	295	368
Diopside	$CaMgSi_2O_6$	206	271	356	437
Hedenbergite	$CaFeSi_2O_6$	205	272	362	447
Jadeite	$NaAlSi_2O_6$	135	176	224	263
Anthophyllite	$Mg_7Si_8O_{22}(OH)_2$	157	215	291	362
Tremolite	$(Ca_2Mg_5)Si_8O_{22}(OH)_2$	154	213	293	368
Chrysotile	$Mg_3Si_2O_5(OH)_4$	135	201	296	399
Antigorite	$Mg_{48}Si_{34}O_{85}(OH)_{62}$	124	187	276	371
Talc	$Mg_3Si_4O_{10}(OH)_2$	103	157	230	302
Paragonite	$NaAl_3Si_3O_{10}(OH)_2$	78	105	137	162
Margarite	$CaAl_4Si_2O_{10}(OH)_2$	214	246	282	310
Albite	$NaAlSi_3O_8$	85	116	153	182
Anorthite	$CaAl_2Si_2O_8$	232	267	308	340
Nepheline	$NaAlSiO_4$	156	192	233	266
Analcime	$NaAlSi_2O_6 \cdot H_2O$	91	124	162	193
Natrolite	$Na_2Al_2Si_3O_{10} \cdot 2H_2O$	99	132	172	204
Scolecite	$CaAl_2Si_3O_{10} \cdot 3H_2O$	138	190	257	316
Laumontite	$CaAl_2Si_4O_{12} \cdot 4H_2O$	197	266	370	558
$CaSiO_3$ glass	$CaSiO_3$	373	449	546	632
$CaMgSi_2O_6$ glass	$CaMgSi_2O_6$	310	383	479	568
$CaAl_2Si_2O_8$ glass	$CaAl_2Si_2O_8$	345	383	424	455

[a]Assuming production of amorphous silica instead of quartz.

Similar plots have been prepared also for several silicates and Al-silicates, writing the carbonation reactions on the basis of 1 mole of CO_2 (see Figs. 5.27–5.32 and Table 5.12), in order to obtain comparable results, and assuming that SiO_2 and Al_2O_3 released by primary minerals generally produce secondary quartz and kaolinite, respectively (apart from the Al stored in secondary dawsonite). Production of corundum instead of kaolinite was assumed by Lackner et al. (1995). However, considering that kaolinite decomposes at approximately 400°C, in a wide range of total pressure, and that corundum is stable above about 450°C (Deer et al., 1992), kaolinite is considered to be more representative than corundum for the purpose of this discussion.

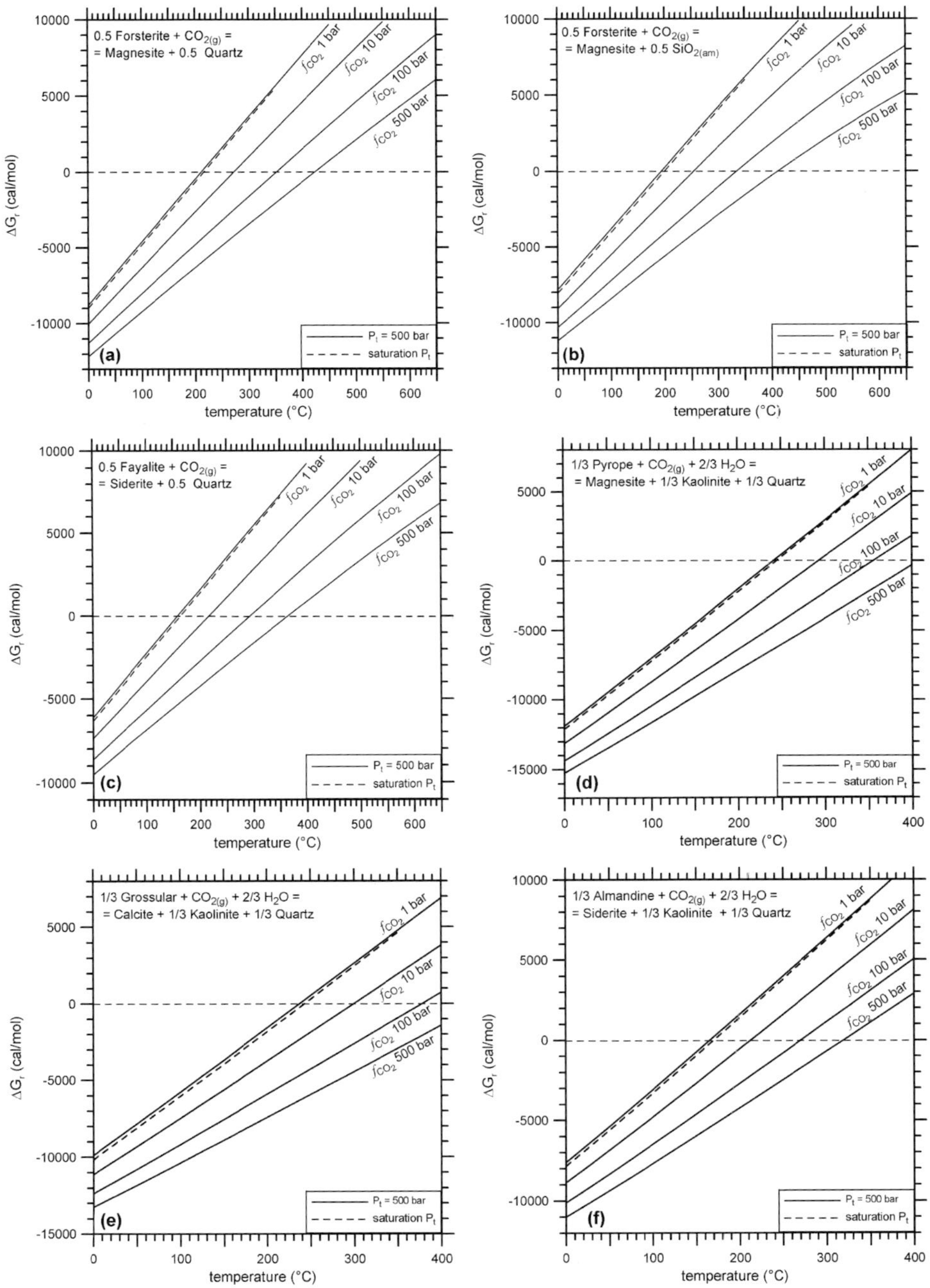

Figure 5.28. Temperature dependence of the Gibbs free energy of carbonation of (a) forsterite with production of quartz, (b) forsterite with production of amorphous silica, (c) fayalite, (d) pyrope, (e) grossular and (f) almandine at saturation pressure and CO_2 fugacity of 1 bar as well as at total pressure of 500 bar and CO_2 fugacities of 1, 10, 100 and 500 bar.

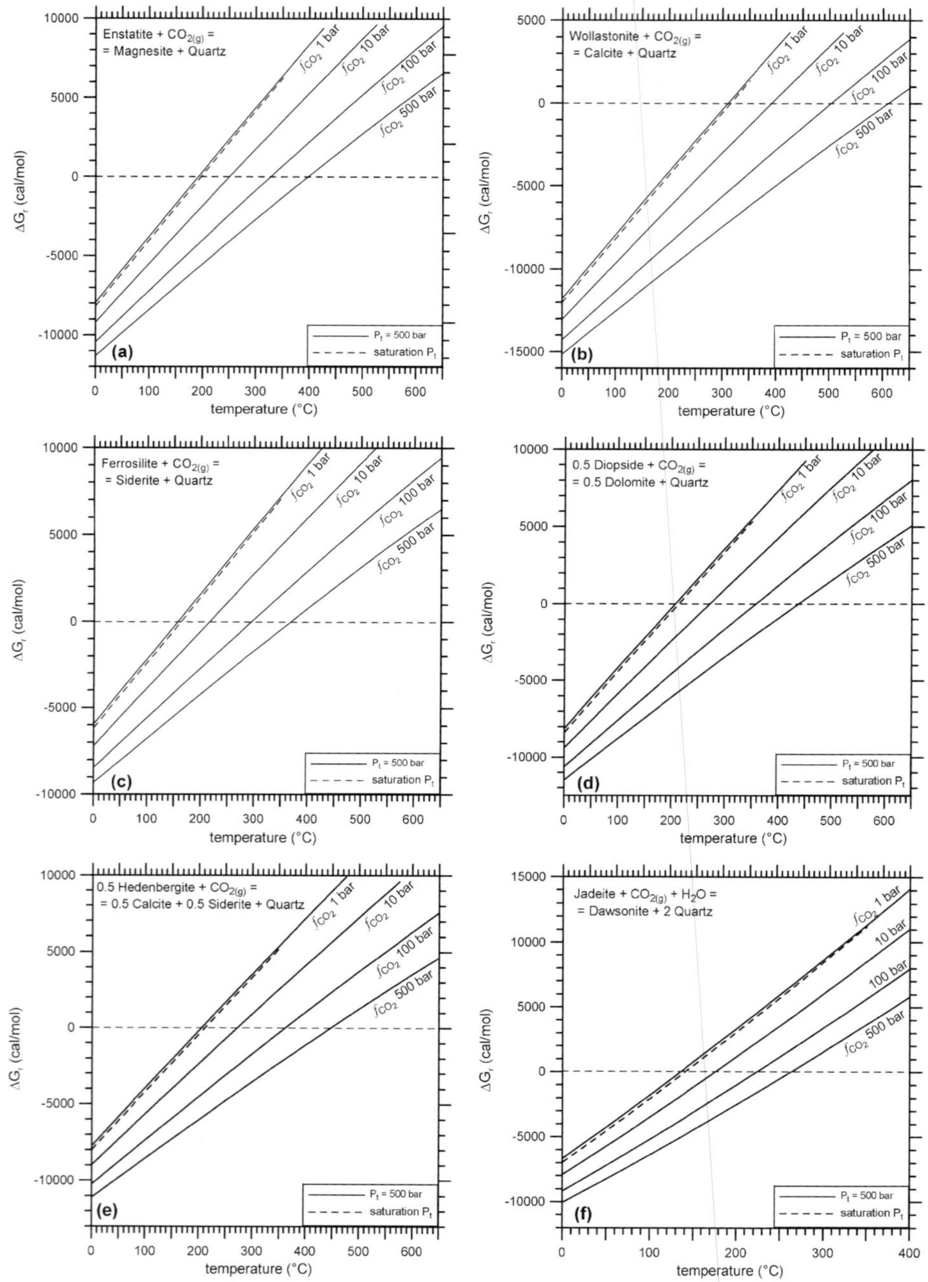

Figure 5.29. Temperature dependence of the Gibbs free energy of carbonation of (a) enstatite, (b) wollastonite, (c) ferrosilite, (d) diopside, (e) hedenbergite and (f) jadeite at saturation pressure and CO_2 fugacity of 1 bar as well as at total pressure of 500 bar and CO_2 fugacities of 1, 10, 100 and 500 bar.

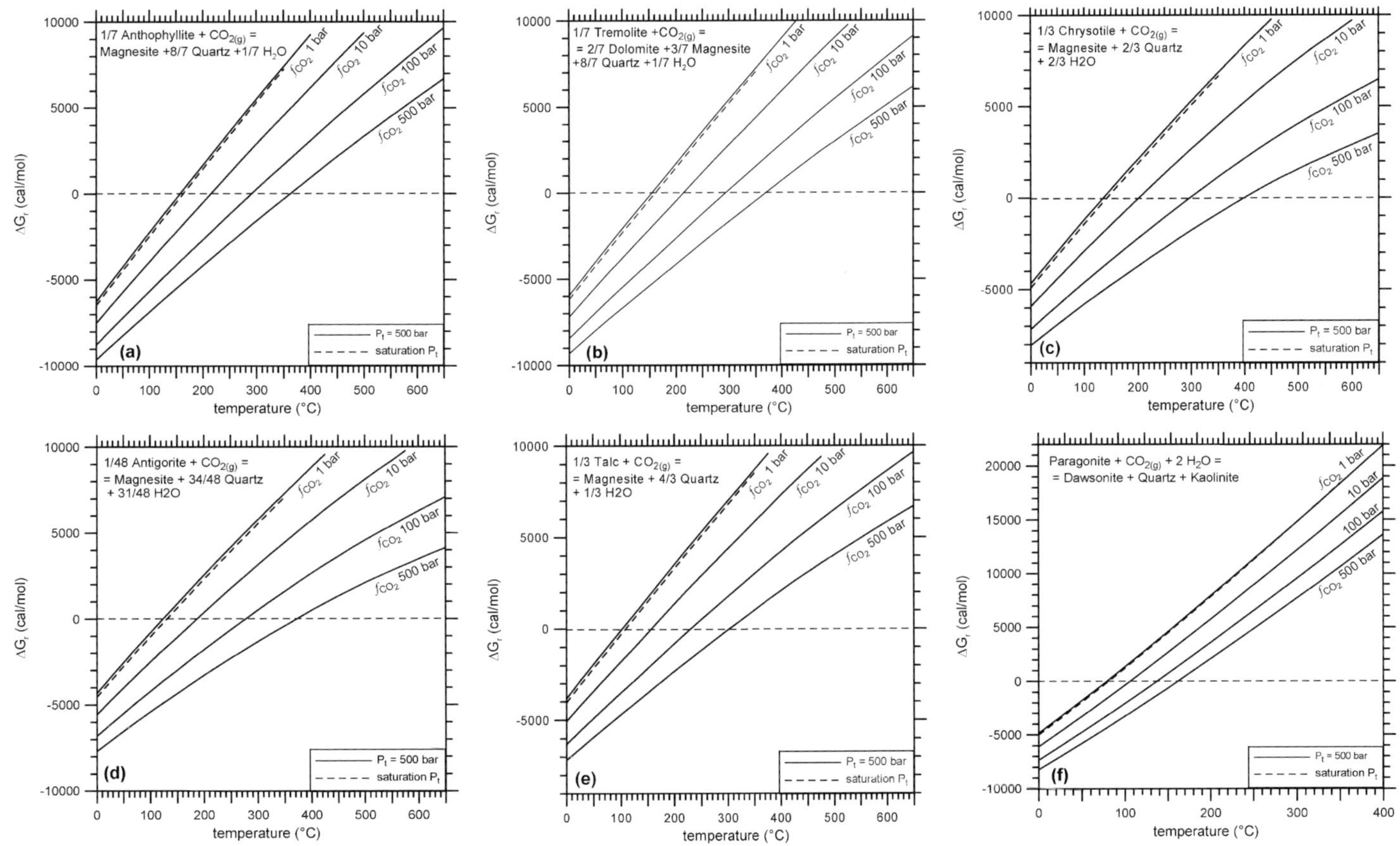

Figure 5.30. Temperature dependence of the Gibbs free energy of carbonation of (a) anthophyllite, (b) tremolite, (c) chrysotile, (d) antigorite, (e) talc and (f) paragonite at saturation pressure and CO_2 fugacity of 1 bar as well as at total pressure of 500 bar and CO_2 fugacities of 1, 10, 100 and 500 bar.

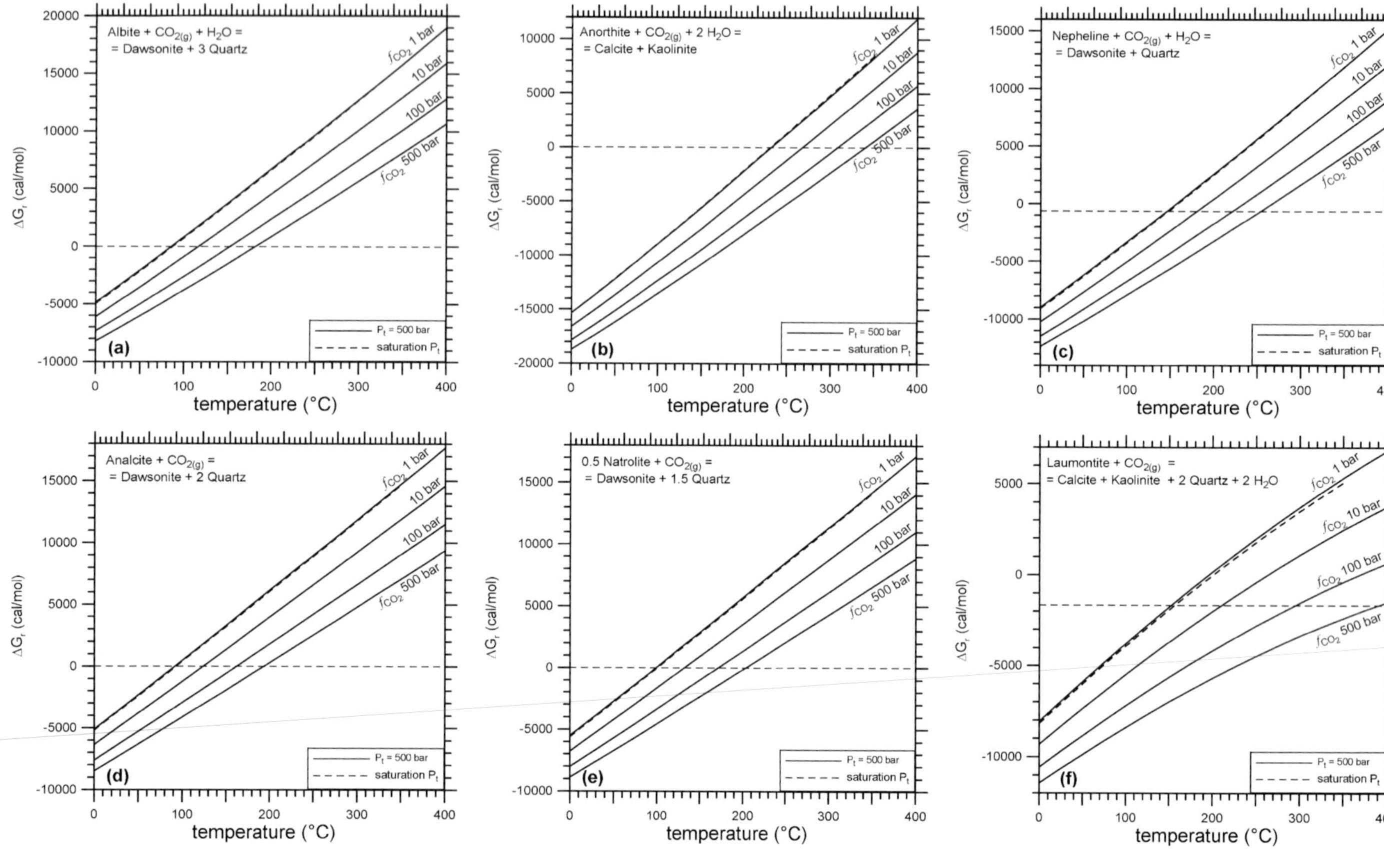

Figure 5.31. Temperature dependence of the Gibbs free energy of carbonation of (a) albite, (b) anorthite, (c) nepheline, (d) analcite, (e) natrolite and (f) laumontite at saturation pressure and CO_2 fugacity of 1 bar as well as at total pressure of 500 bar and CO_2 fugacities of 1, 10, 100 and 500 bar.

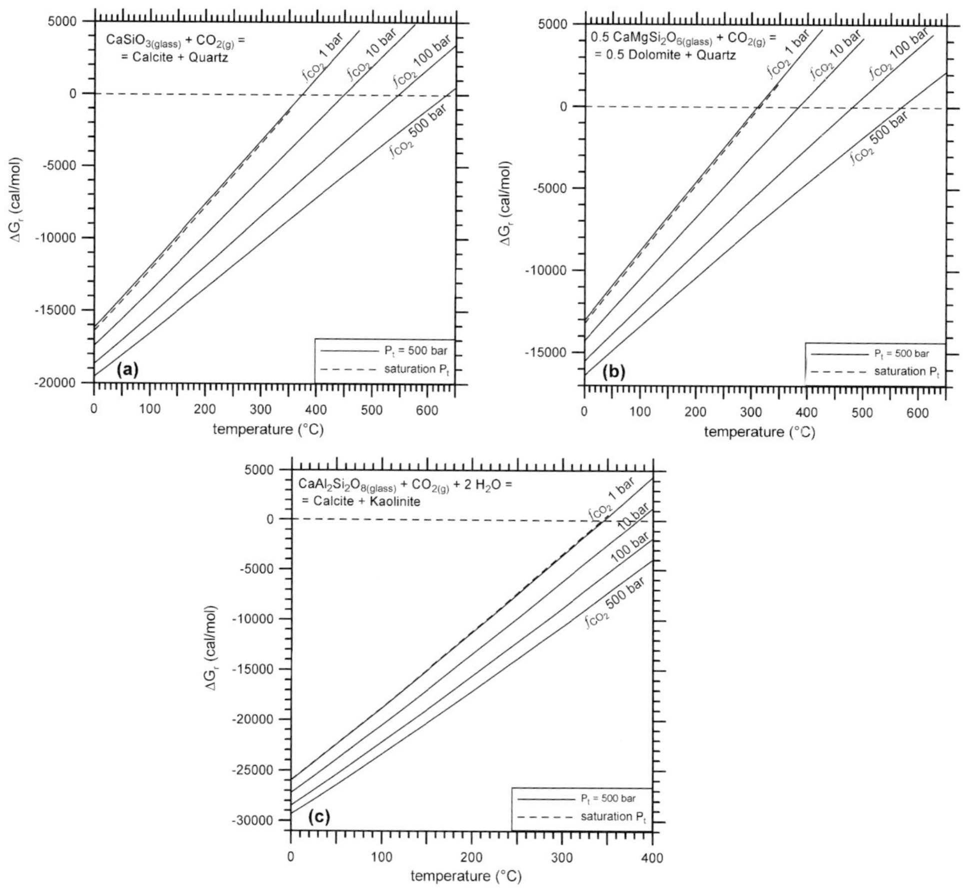

Figure 5.32. Temperature dependence of the Gibbs free energy of carbonation of (a) $CaSiO_3$ glass, (b) $CaMgSi_2O_6$ glass and (c) $CaAl_2Si_2O_8$ glass at saturation pressure and CO_2 fugacity of 1 bar as well as at total pressure of 500 bar and CO_2 fugacities of 1, 10, 100 and 500 bar.

Under f_{CO_2} of 1 bar, forsterite, the Mg-endmember of olivine, undergoes spontaneous carbonation up to 211°C, whereas the T_{max} for the Fe(II)-component of olivine, fayalite, is somewhat lower, 158°C, as expected on the basis of the behaviour of corresponding oxides (Fig. 5.28a,c; Table 5.11). Also note that the T_{max} of forsterite experiences limited changes (10–17°C) if amorphous silica is generated instead of quartz.

Among the garnets, the Mg-component, pyrope and the Ca-endmember, grossular, under f_{CO_2} of 1 bar, have similar T_{max} of 241 and 237°C, respectively, whereas the Fe(II)-bearing garnet, almandine, exhibits a markedly lower T_{max}, 166°C only (Fig. 5.28d–f; Table 5.11).

In Fig. 5.29 are presented the ΔG_r–T plots for different pyroxenes, namely enstatite, wollastonite, ferrosilite, diopside, hedenbergite and jadeite, referring to the reactions

TABLE 5.12

Standard molar volumes, $v°$, of relevant solid phases at 25°C and 1 bar, volume changes of solid phases (volumes of fluids are neglected) for their carbonation reactions, ΔV_r, and percentual change in the volume of solid phases caused by the carbonation reaction, $\Delta V/V$

Mineral	Carbonation reaction	$v°$ (cm^3 mol^{-1})	ΔV_r (cm^3)	$\Delta V/V$ (%)
Periclase	$MgO + CO_2 = MgCO_3$	11.25	16.8	149.1
Lime	$CaO + CO_2 = CaCO_3$	16.76	20.2	120.3
Ferrous-oxide	$FeO + CO_2 = FeCO_3$	12.00	17.4	144.8
Sodium-oxide		25.00		
Corundum	$1/2Na_2O + 1/2Al_2O_3 + CO_2 + H_2O = NaAlCO_3(OH)_2$	25.58	33.2	131.4
Forsterite	$0.5Mg_2SiO_4 + CO_2 = MgCO_3 + 0.5SiO_2$	43.79	17.5	79.8
Forsterite[a]	$0.5Mg_2SiO_4 + CO_2 = MgCO_3 + 0.5SiO_2$	43.79	20.6	94.2
Fayalite	$0.5Fe_2SiO_4 + CO_2 = FeCO_3 + 0.5SiO_2$	46.39	17.5	75.6
Pyrope	$1/3Mg_3Al_2Si_3O_{12} + CO_2 + 2/3H_2O =$ $MgCO_3 + 1/3Al_2Si_2O_5(OH)_4 + 1/3SiO_2$	113.27	31.0	82.1
Grossular	$1/3Ca_3Al_2Si_3O_{12} + CO_2 + 2/3H_2O =$ $CaCO_3 + 1/3Al_2Si_2O_5(OH)_4 + 1/3SiO_2$	125.30	35.9	86.0
Almandine	$1/3Fe_3Al_2Si_3O_{12} + CO_2 + 2/3H_2O =$ $FeCO_3 + 1/3Al_2Si_2O_5(OH)_4 + 1/3SiO_2$	115.28	31.7	82.5
Enstatite	$MgSiO_3 + CO_2 = MgCO_3 + SiO_2$	31.28	19.4	62.1
Wollastonite	$CaSiO_3 + CO_2 = CaCO_3 + SiO_2$	39.93	19.7	49.3
Ferrosilite	$FeSiO_3 + CO_2 = FeCO_3 + SiO_2$	32.95	19.1	58.0
Diopside	$0.5CaMgSi_2O_6 + CO_2 = 0.5CaMg(CO_3)_2 + SiO_2$	66.09	21.8	66.0
Hedenbergite	$0.5CaFeSi_2O_6 + CO_2 = 0.5CaCO_3 + 0.5FeCO_3 + SiO_2$	68.27	21.7	63.6
Jadeite	$NaAlSi_2O_6 + CO_2 + H_2O = NaAlCO_3(OH)_2 + 2SiO_2$	60.4	43.5	72.0
Anthophyllite	$1/7Mg_7Si_8O_{22}(OH)_2 + CO_2 =$ $MgCO_3 + 8/7SiO_2 + 1/7H_2O$	264.40	16.2	42.8
Tremolite	$1/7(Ca_2Mg_5)Si_8O_{22}(OH)_2 + CO_2 =$ $2/7CaMg(CO_3)_2 + 3/7MgCO_3 + 8/7SiO_2 + 1/7H_2O$	272.92	17.3	44.5
Chrysotile	$1/3Mg_3Si_2O_5(OH)_4 + CO_2 =$ $MgCO_3 + 2/3SiO_2 + 2/3H_2O$	108.5	7.0	19.3
Antigorite	$1/48Mg_{48}Si_{34}O_{85}(OH)_{62} + CO_2 =$ $MgCO_3 + 34/48SiO_2 + 31/48H_2O$	1749.13	7.6	21.0
Talc	$1/3Mg_3Si_4O_{10}(OH)_2 + CO_2 =$ $MgCO_3 + 4/3SiO_2 + 1/3H_2O$	136.25	12.9	28.3
Paragonite	$NaAl_3Si_3O_{10}(OH)_2 + CO_2 + 2H_2O =$ $NaAlCO_3(OH)_2 + Al_2Si_2O_5(OH)_4 + SiO_2$	132.53	48.2	36.4
Margarite	$CaAl_4Si_2O_{10}(OH)_2 + 2SiO_2 + CO_2 + 3H_2O =$ $CaCO_3 + 2Al_2Si_2O_5(OH)_4$	129.40	61.2	35.0
Albite	$NaAlSi_3O_8 + CO_2 + H_2O = NaAlCO_3(OH)_2 + 3SiO_2$	100.07	26.5	26.5
Anorthite	$CaAl_2Si_2O_8 + CO_2 + 2H_2O = CaCO_3 + Al_2Si_2O_5(OH)_4$	100.79	35.7	35.4
Nepheline	$NaAlSiO_4 + CO_2 + H_2O = NaAlCO_3(OH)_2 + SiO_2$	54.16	27.0	49.9
Analcime	$NaAlSi_2O_6 \cdot H_2O + CO_2 = NaAlCO_3(OH)_2 + 2SiO_2$	97.10	6.8	7.0
Natrolite	$1/2Na_2Al_2Si_3O_{10} \cdot 2H_2O + CO_2 =$ $NaAlCO_3(OH)_2 + 3/2SiO_2$	169.72	7.7	9.1
Scolecite	$CaAl_2Si_3O_{10} \cdot 3H_2O + CO_2 =$ $CaCO_3 + Al_2Si_2O_5(OH)_4 + SiO_2 + H_2O$	172.3	−13.1	−7.6
Laumontite	$CaAl_2Si_4O_{12} \cdot 4H_2O + CO_2 =$ $CaCO_3 + Al_2Si_2O_5(OH)_4 + 2SiO_2 + 2H_2O$	207.55	−25.7	−12.4

TABLE 5.12 (*Continued*)

Mineral	Carbonation reaction	$v°$ (cm^3 mol^{-1})	ΔV_r (cm^3)	$\Delta V/V$ (%)
$CaSiO_3$ glass	$CaSiO_3 + CO_2 = CaCO_3 + SiO_2$	40.13	19.5	48.6
$CaMgSi_2O_6$ glass	$0.5CaMgSi_2O_6 + CO_2 = 0.5CaMg(CO_3)_2 + SiO_2$	76.09	16.8	44.2
$CaAl_2Si_2O_8$ glass	$CaAl_2Si_2O_8 + CO_2 + 2H_2O =$ $CaCO_3 + Al_2Si_2O_5(OH)_4$	103.00	33.5	32.5

[a]Assuming production of amorphous silica instead of quartz.

reported in the same figures. In particular, production of dolomite was assumed in the diopside case whereas production of both calcite and siderite, instead of ferrodolomite [$CaFe(CO_3)_2$], was hypothesized in the hedenbergite case, as the thermodynamic data of ferrodolomite are unknown and no natural occurrence of this mineral has been recorded. Apart from the jadeite case, the T_{max}, under f_{CO_2} of 1 bar, is highest for wollastonite (309°C), lowest for ferrosilite (157°C) and intermediate for enstatite (191°C). Besides, the T_{max} values for diopside and hedenbergite are very similar, 206 and 205°C, respectively. This spread of T_{max} values is similar to that observed above for the oxides. In contrast with the oxides' behaviour, jadeite has maximum carbonation temperatures even lower than those of ferrosilite, indicating that the inferences drawn above for oxide minerals cannot be applied *tout court* to silicate minerals.

Amphiboles (i.e., anthophyllite and tremolite) and most phyllosilicates (i.e., chrysotile, antigorite, talc and paragonite) considered here have low T_{max} values (78–157°C) for f_{CO_2} of 1 bar, but the rather uncommon Ca-bearing mica, margarite, has a very high T_{max} of 214°C, under a f_{CO_2} of 1 bar (Fig. 5.30; Table 5.11).

Interestingly, there is a marked contrast between the T_{max} of the two plagioclase end-members, albite and anorthite, 85 and 232°C, respectively, under a f_{CO_2} of 1 bar (Fig. 5.31a,b; Table 5.11).

Also most of the other framework silicates taken into account, namely nepheline, analcime and two zeolites (natrolite and scolecite), have low T_{max} values (Fig. 5.31c–f; Table 5.11). Again, the sodic zeolite natrolite [$Na_2Al_2Si_3O_{10}{\cdot}2H_2O$] has a T_{max} of 99°C only, under a f_{CO_2} of 1 bar, which is markedly lower than the T_{max} of the corresponding calcic zeolite scolecite [$CaAl_2Si_3O_{10}{\cdot}3H_2O$], 138°C. The other calcic zeolite laumontite [$CaAl_2Si_4O_{12}{\cdot}4H_2O$] has a rather unexpected behaviour in that it carbonates spontaneously up to a temperature of 197°C, under a f_{CO_2} of 1 bar, and its T_{max} increases significantly with increasing f_{CO_2}.

Carbonation of $CaSiO_3$-, $CaMgSi_2O_6$- and $CaAl_2Si_2O_8$-glasses occurs spontaneously up to temperatures of 373, 310 and 345°C, respectively, under f_{CO_2} of 1 bar (Fig. 5.32). These values are significantly higher than those calculated for the corresponding crystalline counterparts wollastonite, diopside and anorthite, respectively, with increases in T_{max} of 64, 104 and 113°C, respectively. Therefore glasses represent very interesting materials, since they can be carbonated at relatively high temperatures where fast reaction rates are probably attained (Lackner et al., 1995).

It must be underscored that partial carbonation reactions instead of complete carbonation reactions may occur in both natural systems and chemical reactors (Lackner et al., 1995). For instance, carbonation of forsterite might proceed with production of enstatite and magnesite:

$$\underset{\text{Forsterite}}{Mg_2SiO_4} + CO_2 = \underset{\text{Enstatite}}{MgSiO_3} + \underset{\text{Magnesite}}{MgCO_3}, \tag{5-124}$$

rather than according to the reaction considered in Fig. 5.28 and Table 5.12. Further complications can be caused by occurrence of separate dehydration reactions at temperatures either above or below the maximum carbonation temperature. Although these processes are still poorly understood from the kinetic point of view, release of chemically bound water from the lattice of serpentine minerals leaves behind a highly reactive poorly crystalline or amorphous material (see Section 7.7.2).

Although these first theoretical calculations represent only a very preliminary step to understand the course of carbonation reactions of silicate minerals, some general rules can be derived, namely:

(1) the influence of total pressure on Gibbs free energy–temperature relations is negligible;
(2) on the contrary, the increase in f_{CO_2} brings about a significant increase in maximum carbonation temperatures;
(3) differences in maximum carbonation temperatures between Ca-, Mg-, Fe(II)- and Na–Al-silicates are much smaller than those between lime, periclase, ferrous oxide and sodium oxide plus corundum; generally, Fe(II)- and Na–Al-silicate minerals show the lowest values of maximum carbonation temperature;
(4) carbonation of all Na–Al-silicate minerals with production of dawsonite takes place spontaneously at temperatures much lower than 300°C (even at f_{CO_2} of 500 bar), which is the temperature at which dawsonite decomposes with release of CO_2 (Huggins and Green, 1973);
(5) carbonation of glasses occurs spontaneously up to temperatures significantly higher than those of the corresponding crystalline counterparts.

5.5. The volume changes of carbonation reactions

The volume changes of solid phases for the carbonation reactions considered in the previous section (all involving 1 mole of CO_2 to obtain comparable results) at variable temperature are shown in Fig. 5.33. The volumes of fluids are intentionally neglected in these computations (carried out by means of SUPCRT92) to focus on solid reactants and products. Apart from two zeolites (i.e., scolecite and laumontite), the volume of product solid phases is larger than that of solid reactants for all the considered carbonation reactions, i.e. the computed reaction volume for solids, ΔV_r, is always positive, although significant differences are observed from case to case.

This volume increase must be taken into account when mineral carbonation is carried out in a suitable reactor under controlled conditions. The implications of this positive volume

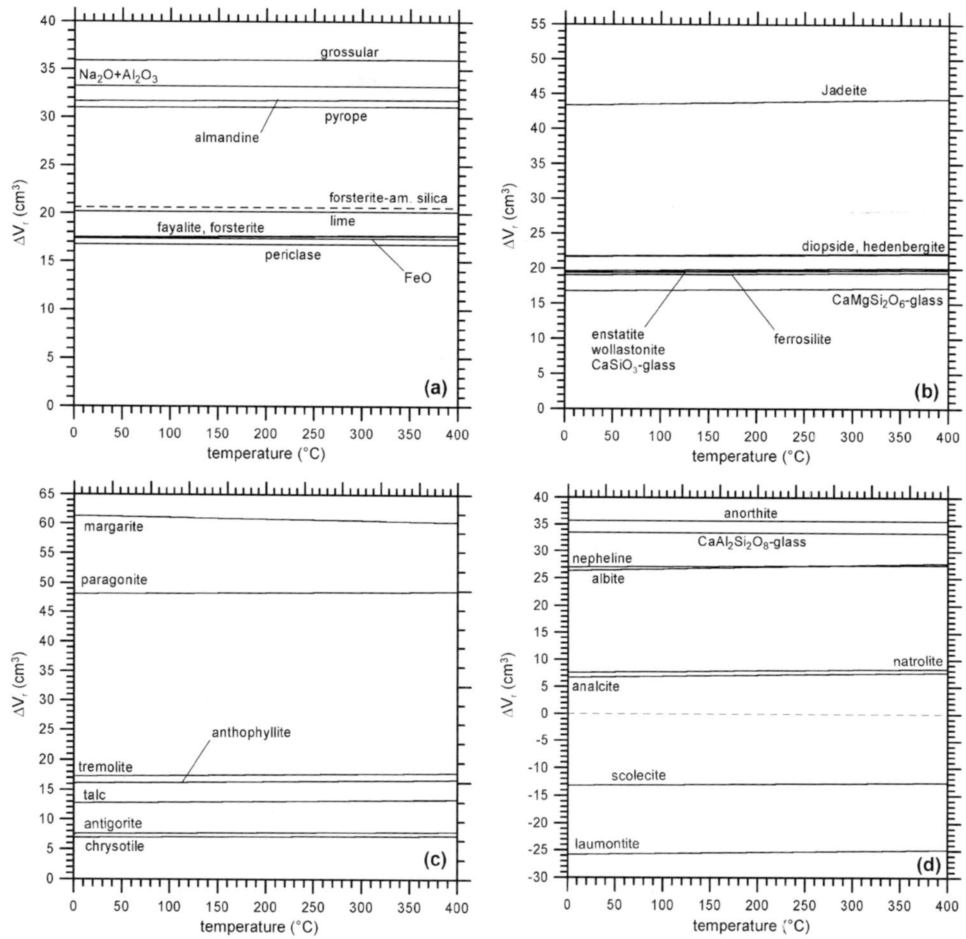

Figure 5.33. Volume changes of solid phases for different carbonation reactions, all written on the basis of 1 mole of CO_2, at variable temperature.

changes of solid phases are even more important for geological CO_2 sequestration, as they are expected to bring about either clogging of pores and fractures, which could hinder further CO_2 injection, or other unwanted processes such as fracturing and microearthquakes or changes in ground elevation, depending on the rheologic behaviour of rocks. In any case, low positive volume changes are desirable whereas high positive volume changes are worrying.

Figure 5.33 also shows that the ΔV_r is almost temperature independent. In fact, volume changes with temperature are only observed for the carbonation reactions involving quartz, since the molar volumes of all solid phases, except quartz, are considered constant at any temperature by SUPCRT92. Indeed, it is correct to treat quartz as an exception due to its very high volumetric thermal expansion with respect to all other minerals (Skinner, 1966; see Section 5.2.4), but again ΔV_r does not change significantly with temperature.

Consequently, the volumes of carbonation reactions can be investigated taking into account the standard molar volumes, $v°$, of relevant minerals at reference conditions, i.e. 25°C and 1 bar (Table 5.12).

Oxides of Mg, Ca and Fe(II) and most Al-free silicates have relatively low ΔV_r values, from 16 to 22 cm^3, with the notable exceptions of talc ($\Delta V_r = 12.9\ cm^3$) and the serpentine minerals chrysotile ($\Delta V_r = 7.0\ cm^3$) and antigorite ($\Delta V_r = 7.6\ cm^3$), owing to their relatively high $v°$, with respect to other Al-free silicates.

All water-free Al-silicates exhibit high ΔV_r values, from 26.5 cm^3 (in the case of albite) to 61.2 cm^3 (in the case of margarite). This is due to the high $v°$ of secondary kaolinite (99.5 $cm^3\ mol^{-1}$) and the relatively high $v°$ of dawsonite (58.5 $cm^3\ mol^{-1}$), in spite of the high $v°$ of some primary Al-silicates, as is the case of anorthite (100.8 $cm^3\ mol^{-1}$), albite (100.1 $cm^3\ mol^{-1}$) and $CaAl_2Si_2O_8$ glass (103 $cm^3\ mol^{-1}$).

Interestingly, water-bearing Al-silicates, namely analcime and the zeolite minerals (natrolite, scolecite and laumontite) have exceptionally low ΔV_r values, which are negative for scolecite and laumontite, as already noted above. These anomalously low ΔV_r values are evidently due to the loose structure of these framework silicates.

As anticipated, carbonation of forsterite was computed assuming production of either quartz or amorphous silica. In the first case the reaction volume is ~17.5 cm^3, whereas in the second case the reaction volume is significantly higher, ~20.6 cm^3, since the molar volume of amorphous silica is much greater than that of quartz (see Section 5.2.4).

Most glasses have molar volumes higher than the corresponding crystalline counterparts: for instance $CaAl_2Si_2O_8$ glass and $CaMgSi_2O_6$ glass are significantly less dense than anorthite and diopside, respectively. However, the difference between the molar volumes of $CaSiO_3$ glass and wollastonite is small. As expected, these differences reflect in reaction volumes.

Following Cipolli et al. (2004), it is useful to take into account also the percentual change in the volume of solid phases caused by the carbonation reaction, which is defined by the following equation:

$$\Delta V/V = \frac{\sum_P \nu_P \cdot v_P^\circ - \sum_R \nu_R \cdot v_R^\circ}{\sum_R \nu_R \cdot v_R^\circ} \cdot 100, \quad (5\text{-}125)$$

where v_i^o and ν_i represent the standard molar volume and the stoichiometric coefficient of the *i*th solid phase in the considered carbonation reaction, respectively, and subscripts P and R indicate products and reactants, respectively.

In Fig. 5.34, the $\Delta V/V$ parameter is plotted against the standard molar volume of each primary phase, multiplied times its stoichiometric coefficient in the carbonation reaction, ν. Close-packed oxide minerals have very high values of the $\Delta V/V$ parameter, between 120 and 149%. All the Al-free silicates distribute along a unique alignment, also comprising lime, periclase and ferrous oxide, and exhibit $\Delta V/V$ usually between 43 and 80%, apart from talc

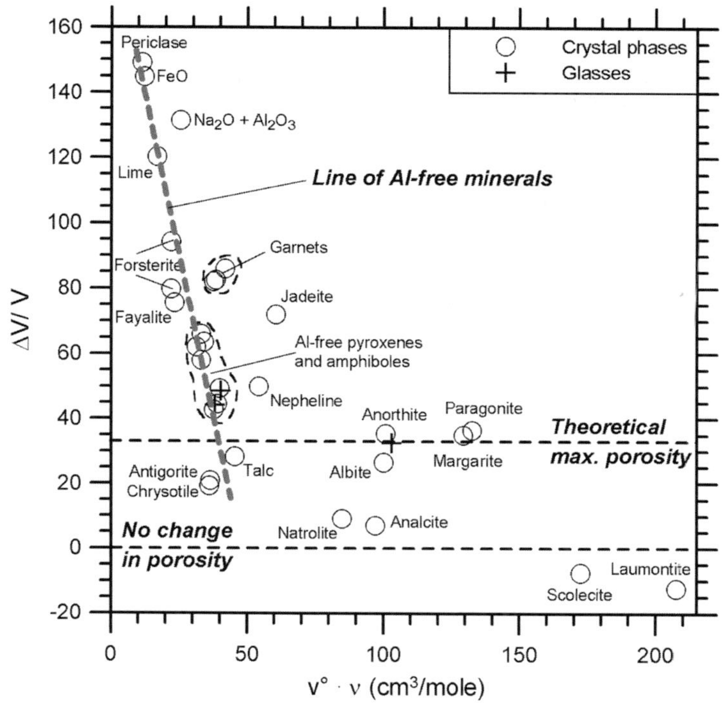

Figure 5.34. The standard molar volume of the solid reactant multiplied times its stoichiometric coefficient in the carbonation reaction, $v° \cdot v$, is reported on the abscissa against the percentual change in the volume of solid phases caused by different carbonation reactions, $\Delta V/V$.

(28%) and the serpentine minerals chrysotile (19%) and antigorite (21%). The oxides' pair $Na_2O + Al_2O_3$ and all the Al-silicates are situated to the right of the alignment of Al-free silicates, as the former phases have higher values of the product $v° \cdot v$ than the latter minerals. The $\Delta V/V$ parameter ranges between 26.5 and 86% in most cases. However, analcime and the considered zeolites have exceptionally low values of the $\Delta V/V$ parameter, which is negative for scolecite and laumontite, as expected on the basis of previous considerations.

The $\Delta V/V$ parameter can be compared directly with the porosity of a hypothetical monomineralic rock. If $\Delta V/V$ is larger than porosity, rock porosity is deemed to become zero, under closed system, before completion of carbonation reaction. Since the maximum theoretical porosity of an aggregate of uniform spherical grains is 33%, completion of carbonation is expected to be the exception rather than the rule, based on computed $\Delta V/V$ values, apart from the favourable cases represented by zeolites minerals and analcime, and to a lesser extent by serpentine minerals, talc and albite. However, this preliminary conclusion is funded on several simplifying assumptions (monomineralic rock, closed system, etc.) and the discussion on volume changes of carbonation reactions must be resumed by means of more complete geochemical models (see Chapter 7).

Chapter 6

The Kinetics of Mineral Carbonation

Chemical equilibrium models can be used to describe properly reversible and fast reactions only. In contrast, a kinetic approach becomes necessary when dealing with irreversible heterogeneous reactions like those of mineral carbonation, involving dissolved CO_2 and silicate minerals as reactants, and carbonate minerals and Si-bearing minerals as products. In the last few decades, an increasing number of laboratory experiments have been carried out, aimed at discriminating the numerous factors affecting the kinetics of dissolution and precipitation of several minerals. Unfortunately, very few of these experiments have been performed under CO_2 partial pressures of interest in geological CO_2 sequestration. Furthermore, most of the obtained laboratory rates are difficult to be compared with one another because of their dependence on either the adopted experimental apparatus (see Section 6.5.1) or the chosen boundary conditions (i.e. pH and temperature ranges, composition and ionic strength of the solution, structural and compositional variability of the minerals, wetted surface areas during dissolution, etc.). In spite of the above-mentioned heterogeneities, this wide dataset must be treated on a common theoretical basis to allow comparison and interpretation of experimental results and their use in geochemical modelling of the processes of interest. Most available data on the kinetics of dissolution and precipitation of silicates, oxides, hydroxides and carbonates will be summarized in this chapter, after a brief review of the theoretical basis, which is chiefly represented by the transition state theory (TST).

6.1. Fundamental concepts and relations

Let us consider a general reaction of stoichiometry:

$$aA + bB = cC. \tag{6-1}$$

Referring to the product substance C, the rate of the reaction could be expressed as the increase in the amount (moles) of C, dn_C/dt. Based on the reactant substances A and B, the rate of the reaction could be defined as the decrease in their amounts, either dn_A/dt or dn_B/dt, respectively. However, it is convenient that the rate of the reaction be the same for all the substances involved in it. To this purpose, the dn_i/dt ratio for each substance is divided by the corresponding stoichiometric coefficient. In this way, the rate of reaction (6-1) is defined as

$$-\frac{1}{a}\frac{dn_A}{dt} = -\frac{1}{b}\frac{dn_B}{dt} = \frac{1}{c}\frac{dn_C}{dt} = \frac{d\xi}{dt}, \tag{6-2}$$

where ξ is the "reaction progress" or "extent of reaction" (Prigogine, 1955 and references therein; Denbigh, 1971).

Note that the stoichiometric coefficients are conventionally negative for reactants and positive for products. Since the ratios dn_A/dt and dn_B/dt are negative and the ratio dn_C/dt is positive, all the terms in equation (6-2) are positive.

Experiments show that the reaction rate is often proportional to either small integral powers or simple fractional powers (e.g. ½) of the concentrations of reacting substances, as indicated by the equation:

$$\frac{1}{V}\cdot\frac{1}{c}\cdot\frac{dn_C}{dt} = k\cdot C_A^{\alpha}\cdot C_B^{\beta}, \tag{6-3}$$

where V is the volume of the considered system (which is not necessarily constant); C_A and C_B are the concentrations of A and B, respectively; α and β are small integers or simple fractions, not necessarily equal to the stoichiometric coefficients a and b; and k is called rate constant as it is independent of the concentrations. Note that the independence of the rate constant from composition is its characteristic property (e.g. Denbigh, 1971), although some authors call this parameter rate coefficient and recognize that it may be a function not only of temperature and pressure but also of composition (e.g. Sposito, 1994). The rate constant is also called the specific rate of the reaction, since k is equal to the rate if all the concentrations are equal to 1, in suitable units.

The so-called order of the reaction is defined by the numbers α and β. For instance, if $\alpha = 1$ and $\beta = 2$, the reaction is termed first order with respect to A and second order with respect to B, and the total reaction order is $\alpha + \beta$, which is 3 in the considered example.

If V is constant with time, than the left-hand side of equation (6-3) can be expressed as

$$\frac{1}{V}\cdot\frac{1}{c}\cdot\frac{dn_C}{dt} = \frac{1}{c}\cdot\frac{d(n_C/V)}{dt} = \frac{1}{c}\cdot\frac{dC_C}{dt}, \tag{6-4}$$

and equation (6-3) can be written as

$$\frac{1}{c}\cdot\frac{dC_C}{dt} = k\cdot C_A^{\alpha}\cdot C_B^{\beta}, \tag{6-5}$$

which can be used only if V is constant!

It is worth underscoring that the concentrations of chemical species, usually expressed in moles/unit volume, are generally used in rate laws instead of their activities, which are used in thermodynamic relations. This derives from the assumption that any gas reaction occurs through the collision of the reacting gas molecules. Indeed, according to the kinetic theory of gases, the number of collisions of any two molecules A and B, per unit time and volume, is proportional to the product of their volume concentrations, $C_A \times C_B$.

For several reactions that are relatively close to equilibrium, experiments show that the reaction rate can be written as a difference of two terms, one of which involves the concentrations of reactants and the other the concentrations of products. For instance, the rate of reaction (6-1), assuming that it is close to equilibrium, can be expressed as

$$\frac{1}{c}\cdot\frac{\mathrm{d}C_\mathrm{C}}{\mathrm{d}t} = k_+\cdot C_\mathrm{A}^{\alpha}\cdot C_\mathrm{B}^{\beta} - k_-\cdot C_\mathrm{C}^{\gamma}, \tag{6-6}$$

where k_+ and k_- are both constants.

As noted by Denbigh (1971), there is no thermodynamically based reason as to why the experimentally measured reaction rates have to be written as a difference of two terms as in equation (6-6). The only thermodynamic requirements are that (1) the reaction rate is positive in the direction in which the Gibbs free energy decreases and (2) the reaction rate becomes zero at equilibrium. The presence of two terms in equations such as (6-6) is usually explained from a merely kinetic point of view, assuming that the measured reaction rate is equal to the difference of the rates in the forward and backward directions, implicitly admitting that two opposite processes occur concurrently at the molecular scale. Again, only the net reaction rate is determined experimentally, whereas the forward and backward rates are matters of interpretation only.

At equilibrium, equation (6-6) becomes

$$k_+\cdot C_\mathrm{A}^{\alpha}\cdot C_\mathrm{B}^{\beta} - k_-\cdot C_\mathrm{C}^{\gamma} = 0, \tag{6-7}$$

which can be rearranged as

$$\frac{C_\mathrm{C}^{\gamma}}{C_\mathrm{A}^{\alpha}\cdot C_\mathrm{B}^{\beta}} = \frac{k_+}{k_-}. \tag{6-8}$$

The left-hand side of equation (6-8) is a sort of equilibrium constant defined in terms of concentrations, K_C, not to be confused with the corresponding thermodynamic equilibrium constant K involving the activities. However, the true K_C is

$$K_C = \frac{C_\mathrm{C}^{c}}{C_\mathrm{A}^{a}\cdot C_\mathrm{B}^{b}}, \tag{6-9}$$

in which concentrations are raised to the stoichiometric coefficients a, b and c. The left-hand term of equation (6-8) can be related to the right-hand term of equation (6-9) through a still unspecified function:

$$C_\mathrm{A}^{-\alpha}\cdot C_\mathrm{B}^{-\beta}\cdot C_\mathrm{C}^{\gamma} = f\left(C_\mathrm{A}^{-a}\cdot C_\mathrm{B}^{-b}\cdot C_\mathrm{C}^{c}\right). \tag{6-10}$$

Assuming that this function is a power function, equation (6-10) can be rewritten as

$$C_\mathrm{A}^{-\alpha}\cdot C_\mathrm{B}^{-\beta}\cdot C_\mathrm{C}^{\gamma} = \left(C_\mathrm{A}^{-a}\cdot C_\mathrm{B}^{-b}\cdot C_\mathrm{C}^{c}\right)^{m}. \tag{6-11}$$

From equation (6-11), it results that α, β and γ must be the same multiple m of the corresponding stoichiometric coefficients a, b and c:

$$\frac{-\alpha}{-a} = \frac{-\beta}{-b} = \frac{\gamma}{c} = m, \tag{6-12}$$

where m can be any positive number, including fractions, but it cannot be negative, as this would bring about a positive reaction rate in the direction in which the Gibbs free energy increases, which is against one of the two thermodynamic requirements given above.

Based on these considerations, equations (6-8) and (6-9) lead to

$$\frac{k_+}{k_-} = K_C^m. \tag{6-13}$$

Note that m is customarily assumed equal to 1 by most authors. Based on this assumption, equation (6-13) is written as follows:

$$\frac{k_+}{k_-} = \frac{C_C^c}{C_A^a \cdot C_B^b} = K \cdot \frac{\gamma_A^a \cdot \gamma_B^b}{\gamma_C^c}, \tag{6-14}$$

for the general reaction (6-1) at equilibrium, where the γ s are appropriate activity coefficients.

This relation, known as the *principle of detailed balancing*, is considered to represent the link between thermodynamics and kinetics. According to some authors (e.g. Lasaga, 1983a, 1998), it is a very powerful principle as it allows one to obtain the rate constant of the reverse reaction, knowing the rate constant of the forward reaction (or vice versa) and the thermodynamic equilibrium constant. However, some caution is necessary in its use, considering that equation (6-14) rests on the simplifying assumption of $m = 1$, as shown above (see also Manes et al., 1950; Hollingsworth, 1952; Blum and Luus, 1964).

It must be underscored that the *principle of detailed balancing* can be applied to elementary reactions (i.e. reactions occurring at the molecular level exactly as they are written) and not to overall reactions, resulting from a sequence of elementary reactions.

An example of application of this principle is given in the investigation of Rimstidt and Barnes (1980) on the rate of the dissolution and precipitation of quartz and amorphous silica minerals:

$$SiO_{2(s)} + 2H_2O \underset{k_-}{\overset{k_+}{\rightleftharpoons}} H_4SiO_4^o. \tag{6-15}$$

These are *heterogeneous* reactions, like all the reactions of dissolution and precipitation of solid phases, which are of utmost interest to us. If one solid phase is involved in the considered reaction, the rate depends on the area of its exposed surface per unit volume (or mass) of the aqueous solution. For instance, the following expression describes the rate of precipitation of silica minerals (modified from Rimstidt and Barnes, 1980):

$$\frac{dm_{H_4SiO_4^o}}{dt} = \frac{A_{SiO_2(s)}}{M_{H_2O}} \cdot k_- \cdot m_{H_4SiO_4^o}, \tag{6-16}$$

indicating that the rate law for the precipitation of silica minerals is first order with respect to the molality of $H_4SiO_4^\circ$. In Section 6.1.2, we will see more on the meaning of this sentence. Here it is important to underscore the presence of the exposed surface area of the silica mineral, $A_{SiO_{2(s)}}$ — in equation (6-16) — which has to be determined either directly by means of the BET method or indirectly by evaluating grain shape and size (see Section 6.5.2).

6.1.1. The rate expressions of zero-order reactions

The zero-order irreversible reactions of the type ν_R R→ ν_P P represent the simplest case and obey the rate equations:

$$\frac{1}{\nu_R}\cdot\frac{dC_R}{dt} = k \quad \text{or} \quad \frac{1}{\nu_P}\cdot\frac{dC_P}{dt} = k. \tag{6-17}$$

The concentration of a generic reactant R is considered in the first equation, whereas the second involves the concentration of a generic product P. Integration of equations (6-17) gives

$$C_R = C_{R,0} + \nu_R \cdot k \cdot t \quad \text{and} \quad C_P = C_{P,0} + \nu_P \cdot k \cdot t, \tag{6-18}$$

where $C_{R,0}$ and $C_{P,0}$ represent the initial concentration, at time t = 0, of R and P, respectively. Since ν_R is negative, C_R decreases with time, whereas since ν_P is positive, C_P increases with time, as expected. In a plot of concentration vs. time, zero-order reactions give straight lines and, consequently, rates of this kind are sometimes indicated as linear, although this term might be misleading.

Today, it is firmly established that the dissolution of most silicates in natural water–rock environments, from the low temperatures typical of weathering up to at least 200°C, follows zero-order kinetics (see Section 6.3.1).

6.1.2. The rate expressions of first-order reactions

For the generic *irreversible* reaction ν_R R→ ν_P P obeying first-order kinetics, the following equations apply:

$$\frac{1}{\nu_R}\cdot\frac{dC_R}{dt} = k\cdot C_R \quad \text{and} \quad \frac{1}{\nu_P}\cdot\frac{dC_P}{dt} = k\cdot C_P, \tag{6-19}$$

which can be integrated to obtain

$$C_R = C_{R,0}\cdot\exp(\nu_R \cdot k \cdot t) \quad \text{and} \quad C_P = C_{P,0}\cdot\exp(\nu_P \cdot k \cdot t) \tag{6-20}$$

or

$$\ln C_R = \ln C_{R,0} + \nu_R \cdot k \cdot t \quad \text{and} \quad \ln C_P = \ln C_{P,0} + \nu_P \cdot k \cdot t. \tag{6-21}$$

Therefore, a first-order reaction gives a straight line of slope either $\nu_R k$ (which is negative since ν_R is negative) in a plot of log C_R vs. time or $\nu_P k$ (which is positive since ν_P is positive) in a plot of log C_P vs. time. The rate constants of first-order reactions have units of time^{-1}. The time needed to consume half of the original mass of the reactant R is called the half-life or half-time of the reaction, $t_{1/2}$. It is obtained by inserting $C_R = ½C_{R,0}$ in the first equation (6-20) or (6-21):

$$t_{1/2} = -\frac{\ln 2}{\nu_R \cdot k} = -\frac{0.693}{\nu_R \cdot k}. \tag{6-22}$$

Radioactive decay is an example of first-order reactions. Reactions of radioactive decay are irreversible and have only a forward rate and a forward rate constant.

Let us now consider the case of a first-order forward reaction ν_R R→ ν_P P associated with a first-order backward reaction ν_P P → ν_R R. Let us suppose that the rate of both reactions is fast enough so that equilibrium or steady state is attained. The rate law is as follows:

$$\frac{1}{\nu_P} \cdot \frac{dC_P}{dt} = k \cdot (C_{P,S} - C_P), \tag{6-23}$$

where $C_{P,S}$ is the equilibrium or steady-state concentration of product P. Integration of equation (6-23) gives

$$\ln \frac{C_{P,S} - C_P}{C_{P,S} - C_{P,0}} = \nu_P \cdot k \cdot t. \tag{6-24}$$

Therefore, if $C_{P,0} = 0$, a straight line with slope $\nu_P k$ (which is positive since ν_P is positive) is obtained on a plot of $\ln\left[\left(C_{P,S} - C_P\right)/C_{P,S}\right]$ vs. t.

The rate expressions for second-order reactions are presented in several books (e.g. Castellan, 1971; Lasaga, 1998), whereas those of order higher than second are rarely important.

6.1.3. The temperature dependence of rate constants

The rates of chemical reactions increase significantly with temperature, since reaction rate constants exhibit an exponential dependence on temperature. This is frequently expressed by means of the Arrhenius equation (Arrhenius, 1889):

$$k = A_a \cdot \exp\left(-\frac{E_a}{RT}\right) \tag{6-25}$$

or

$$\ln k = \ln A_a - \frac{E_a}{RT}, \tag{6-26}$$

where A_a is the so-called pre-exponential factor, E_a the *activation energy* term (also called apparent activation energy or experimental activation energy), R the universal gas constant and T the absolute temperature. Although both A_a and E_a may depend on temperature (Lasaga, 1983a, 1998), it is instructive to differentiate equation (6-26) with respect to T

and $1/T$, assuming that both A_a and E_a are constants, thus obtaining

$$\frac{\mathrm{d}\ln k}{\mathrm{d}T} = \frac{E_a}{RT^2} \quad \text{and} \quad \frac{\mathrm{d}\ln k}{\mathrm{d}(1/T)} = -\frac{E_a}{R}. \tag{6-27}$$

Consequently, under these assumptions, a straight line with slope $-E_a/R$ should be obtained on a plot of ln$k$ vs. the inverse of the absolute temperature, thus allowing the estimation of the activation energy term of the chemical reaction of interest. In contrast, a non-linear trend in this plot indicates that E_a is temperature-dependent.

The importance of the activation energy goes beyond the temperature dependence of reaction rate, as its magnitude reflects the mechanism of the chemical reaction (Lasaga, 1983a, 1998). As discussed in Section 6.2, reactions of dissolution and precipitation can be governed by either transport or surface reactions or by a combination of these two mechanisms. Interestingly, transport-controlled reactions in solution have relatively low activation energies, typically lower than 20 kJ mol^{-1}, whereas surface-controlled reactions have higher activation energies, usually between 40 and 85 kJ mol^{-1} (Lasaga, 1984). All these values apply in the absence of catalysis, which can greatly change reaction rates.

6.1.4. The transition state theory

The TST was first proposed by Henry Eyring (Eyring, 1935a,b). Basically, it represents a way of computing reaction kinetics by means of statistical mechanics. Here we do not intend to enter this rather complex field, but simply to elucidate the fundamental principles on which TST is based and to show some of its implications. Further details are given by Castellan (1971) and Lasaga (1983b, 1998), among the others.

In any elementary reaction, both reactants and products are stable or metastable species and represent lows in a potential energy surface, separated by a potential energy barrier (Fig. 6.1). A *sine qua non* condition for the occurrence of the considered reaction is overcoming this potential energy barrier. A basic point in TST is the molecular configuration of the so-called *activated complex* (a concept first introduced by Arrhenius), which is situated at the top of this barrier. TST considers the activated complex as a true chemical species and hypothesizes that it is always in equilibrium with the reactants. The reaction rate is obtained by computing the number of possible activated complexes and the rate of their decomposition to yield products, by using statistical mechanics.

Let us take into account the general irreversible reaction

$$\mathrm{A+B \rightarrow AB}, \tag{6-28}$$

which is considered to occur through the formation of the activated complex $AB^{\ddagger}$ as a sort of intermediate species. This is described through the sequential reaction

$$\mathrm{A + B \leftrightarrows AB^{\ddagger} \rightarrow AB} \tag{6-29}$$

Note that the whole reaction (6-28) has been splitted into two steps, namely (1) a first, fast step resulting in the formation of the activated complex from the reactants and (2) a

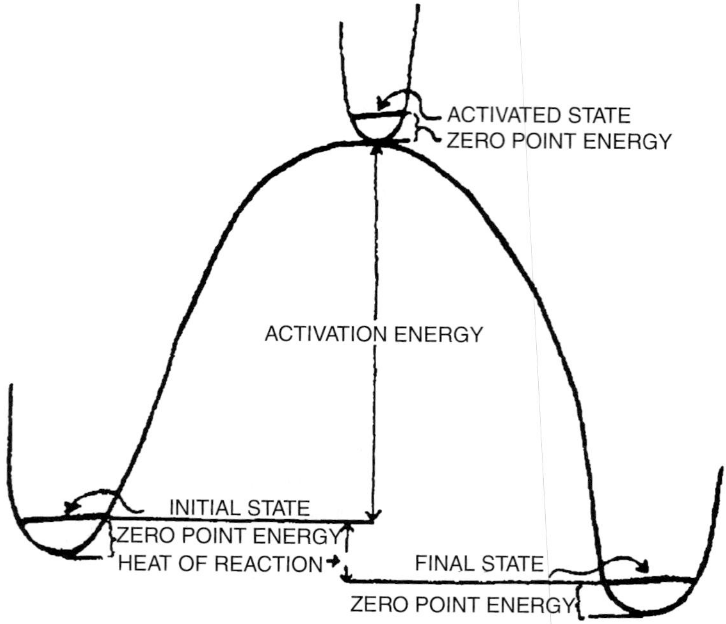

Figure 6.1. Cross-section showing the path of a point representing a generic reacting system moving along the potential energy surface. (Reprinted with permission from Eyring, 1935b, Copyright 1935 American Chemical Society.)

second, slow step resulting in the decomposition of the activated complex, which generates the reaction products. This second step is therefore rate-controlling.

Since the activated complex is assumed to be in equilibrium with the reactants, the first, fast step can be described by the following thermodynamic equilibrium constant:

$$K^{\ddagger} = \frac{a_{AB^{\ddagger}}}{a_A \cdot a_B} = \frac{C_{AB^{\ddagger}}}{C_A \cdot C_B} \frac{\gamma_{AB^{\ddagger}}}{\gamma_A \cdot \gamma_B}, \tag{6-30}$$

where C_i represents the concentration of the ith species in unspecified units and γ_i is its appropriate activity coefficient. Note that $K^{\ddagger}$ is different from the thermodynamic equilibrium constant of the whole reaction (6-28), which involves the activities of products and reactants as usual.

Following Eyring (1935b), the rate r of the whole reaction (6-28) is given by

$$r = \kappa \cdot C_{AB^{\ddagger}} \frac{k_B T}{h} \tag{6-31}$$

where k_B and h are the Boltzmann's constant and the Planck's constant, respectively, and κ is the probability of non-returning in crossing the potential energy barrier separating reactants and products. For many reactions, κ is close to unity.

The concentration of the activated complex $C_{AB^{\ddagger}}$ is obtained from reation (6-30) and inserted in equation (6-31), which yields

$$r = \kappa \cdot \frac{k_B T}{h} \cdot K^{\ddagger} \cdot \frac{\gamma_A \cdot \gamma_B}{\gamma_{AB^*}} \cdot C_A \cdot C_B. \tag{6-32}$$

As underscored by Eyring (1935b), the rate constant is given by the products of the first three terms of the right-hand member of equation (6-32), namely

$$k = \kappa \cdot \frac{k_B T}{h} \cdot K^{\ddagger}. \tag{6-33}$$

Note that the rate constant depends on both the so-called decay frequency of the activated complex $f = k_B T/h$, which is related to the second step, and the thermodynamic constant of the first step, $K^{\ddagger}$.

Recalling that $G = H - TS$ and based on equation (2-31), $K^{\ddagger}$ can be related to the differences in standard Gibbs free energy, enthalpy and entropy between the reactants and the activated complex ($\Delta G^{\ddagger o}$,$\Delta H^{\ddagger o}$ and $\Delta S^{\ddagger o}$, also called standard Gibbs free energy, enthalpy and entropy of activation, respectively) through the equation,

$$K^{\ddagger} = \exp\left(-\frac{\Delta G^{\ddagger o}}{RT}\right) = \exp\left(-\frac{\Delta H^{\ddagger o}}{RT}\right) \cdot \exp\left(\frac{\Delta S^{\ddagger o}}{R}\right). \tag{6-34}$$

Therefore, equation (6-33) can be rewritten as follows:

$$k = \kappa \cdot \frac{k_B T}{h} \cdot \exp\left(-\frac{\Delta H^{\ddagger o}}{RT}\right) \cdot \exp\left(\frac{\Delta S^{\ddagger o}}{R}\right) \tag{6-35}$$

or

$$\ln k = \ln T - \frac{\Delta H^{\ddagger o}}{RT} + \frac{\Delta S^{\ddagger o}}{R} + \ln \frac{\kappa \cdot k_B}{h}. \tag{6-36}$$

Equation (6-36) can be differentiated with respect to temperature, assuming that $\Delta H^{\ddagger o}$ and $\Delta S^{\ddagger o}$ are temperature-independent (which means to hypothesize that $\Delta C_p^{\,o} = 0$). Thus, the following relation is obtained:

$$\frac{\mathrm{d} \ln k}{\mathrm{d}T} = \frac{1}{T} + \frac{\Delta H^{\ddagger o}}{RT^2} = \frac{RT + \Delta H^{\ddagger o}}{RT^2}. \tag{6-37}$$

Considering the equation obtained through differentiation of the Arrhenius equation (6-26) with respect to temperature, assuming that both A_a and E_a are constant (see the first of equation (6-27)), it turns out that

$$E_a = RT + \Delta H^{\ddagger o}. \tag{6-38}$$

Since $R = 8.314$ J mol^{-1} K^{-1}, the product RT is about 2.5 kJ mol^{-1} at 25°C and is close to 6.4 kJ mol^{-1} at 500°C. Therefore, at not too high temperatures it can be neglected and it can be assumed that

$$E_a \cong \Delta H^{\ddagger o}. \tag{6-39}$$

Based on this approximated relationship, comparison of equations (6-25) and (6-35) shows that the pre-exponential factor A_a in the Arrhenius equation is

$$A_a \cong \frac{\kappa \cdot k_B T}{h} \exp\left(\frac{\Delta S^{\ddagger o}}{R}\right). \tag{6-40}$$

The standard entropy of activation, which is controlled by the configuration of the activated complex, can be calculated from the partition functions (Castellan, 1971; Lasaga, 1998) and the pre-exponential factor A_a can then be obtained by means of equation (6-40). If the structure of the activated complex is similar to that of the product molecule, the standard entropy of activation is close to the standard entropy of the whole reaction.

For variable pressure conditions and assuming a standard state of 1 bar, equation (6-35) must be rewritten to include an additional term, as follows:

$$k = \frac{\kappa \cdot k_B T}{h} \cdot \exp\left(-\frac{\Delta H^{\ddagger o}}{RT}\right) \cdot \exp\left(\frac{\Delta S^{\ddagger o}}{R}\right) \cdot \exp\left(-\frac{(P-1) \cdot \Delta V^{\ddagger o}}{RT}\right), \tag{6-41}$$

where $\Delta V^{\ddagger o}$ is the volume variation associated with the formation of the activated complex. This is assumed to be pressure-independent in equation (6-41) and, consequently,

$$E_a = RT + \Delta H^{\ddagger o} + (P-1) \cdot \Delta V^{\ddagger o}. \tag{6-42}$$

In most cases, however, $\Delta V^{\ddagger o}$ is negligible.

6.1.4.1. The dependence of the rate on the ionic strength: transition state theory predictions

The TST equation (6-32) can be used to investigate the dependence of the rate on the ionic strength for elementary reactions occurring in aqueous solutions and involving ionic species. This is the so-called primary salt effect, which was known before TST development (e.g. Brönsted, 1928; La Mer, 1932 and references therein). Let us collect together all the terms of equation (6-32), which do not depend on the properties of the solvent, defining the parameter r_o as follows:

$$r_0 = \kappa \cdot \frac{\mathrm{k_B} T}{\mathrm{h}} \cdot K^{\ddagger} \cdot C_A \cdot C_B \tag{6-43}$$

Therefore, equation (6-32) can be rewritten as,

$$r = r_o \cdot \frac{\gamma_A \cdot \gamma_B}{\gamma_{AB^{\ddagger}}} \tag{6-44}$$

or

$$\log r = \log r_0 + \log \gamma_A + \log \gamma_B - \log \gamma_{AB^\ddagger}. \tag{6-45}$$

Let us suppose that A and B are two ions of charge Z_A and Z_B, respectively, and that the activated complex has charge $Z_A + Z_B$. Below ionic strengths of ~0.01 mol kg^{-1}, the activity coefficients of individual ions can be computed by means of the Debye–Hückel limiting law (see Chapter 4) adapted to individual ions:

$$\log \gamma_i = -A_\gamma \cdot Z_i^2 \cdot \sqrt{I}. \tag{6-46}$$

Therefore, equation (6-45) can be rewritten as follows:

$$\log r = \log r_o - A_\gamma \cdot \sqrt{I} \cdot \left[Z_A^2 + Z_B^2 - (Z_A + Z_B)^2 \right] = \log r_o + 2 \cdot Z_A \cdot Z_B \cdot A_\gamma \cdot \sqrt{I}. \tag{6-47}$$

Based on equation (6-47), the logarithm of the rate of an elementary reaction between ions is a function of the square root of the ionic strength of the aqueous solution. In a plot of $\log r/r_0$ vs. $\sqrt{I}$, linear trends are expected, whose slope depends on the product of the charges of the ions A and B. In particular, if A and B have the same charge the slope is positive, whereas if A and B have opposite charges the slope is negative. The agreement between these theoretical predictions and experimental data is surprising (Fig. 6.2).

Note that in Fig. 6.2 (from La Mer, 1932), the difference, $\log k - \log k_o$, is plotted instead of the difference $\log r - \log r_0$. The reason for this is that La Mer (1932) obtained the relation

$$\log k = \log k_o + \log \frac{\gamma_A \cdot \gamma_B}{\gamma_{AB}}, \tag{6-48}$$

through integration of the velocity formula for a generic bimolecular reaction (Brönsted, 1928)

$$r = k_o \cdot C_A \cdot C_B \cdot \frac{\gamma_A \cdot \gamma_B}{\gamma_{AB}} \tag{6-49}$$

(TST was not yet developed at that time), under the assumption that the activity coefficients do not vary during the considered chemical reaction. However, La Mer (1932) also wrote the following relation:

$$r = k_o \cdot C_A \cdot C_B \cdot 10^{Z_A \cdot Z_B \cdot \sqrt{I}}, \tag{6-50}$$

and correctly concluded that "*the velocity is consequently an exponential function of the square root of the ionic strength.*" In our opinion, rate constants have to be constant (excuse me for the tautology), i.e. independent of concentrations and related properties of the aqueous solution such as the ionic strength. Indeed, based on TST it can be shown that rates, rather than rate constants, depend on the square root of the ionic strength (see above). The ionic strength effect has to be properly incorporated in rate laws, such as in equation (6-56). Unfortunately, this point of view is not shared by many people and

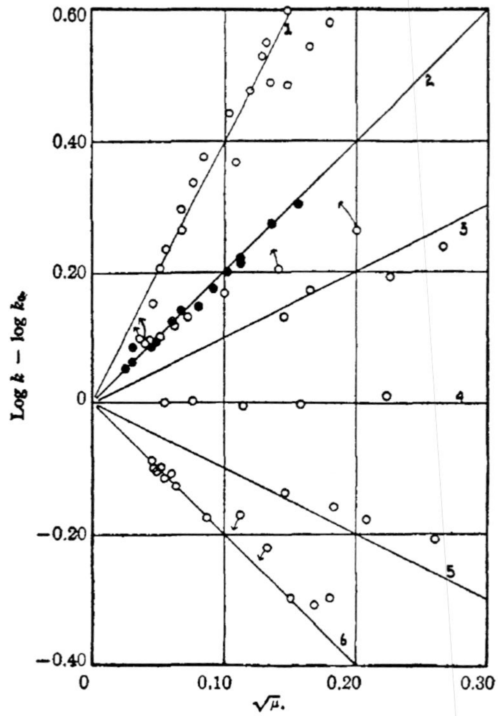

THE INFLUENCE OF IONIC STRENGTH ON THE VELOCITY OF THE IONIC REACTIONS

1. 2 $[Co(NH_3)_5Br]^{++} + Hg^{++} + 2\ H_2O \rightarrow 2\ [Co(NH_3)_5H_2O]^{+++} + HgBr_2$ (Bimolecular). No foreign salt added. Brönsted and Livingston: J. Am. Chem. Soc. **49,** 435 (1927).

2. Circles. $CH_2BrCOO^- + S_2O_3^{--} \rightarrow CH_2S_2O_3COO^- + Br^-$ as the sodium salt. No foreign salt added. La Mer: J. Am. Chem. Soc. **51,** 334 (1929).

Dots. $S_2O_8^{--} + I^- \rightarrow I_2 + SO_4^{--}$ as $Na_2S_2O_8$ and KI. King and Jacobs: J. Am. Chem. Soc. **53,** 1704 (1931).

3. Saponification of nitrourethane ion by hydroxyl ion. $[NO_2 = N - COOC_2H_5]^{--} + OH^- \rightarrow N_2O + CO_3^{--} + C_2H_5OH$. Brönsted and Delbanco: Z. anorg. Chem. **144,** 248 (1925).

4. $C_{12}H_{22}O_{11}^0 + OH^- \rightarrow$ Invert Sugar. Arrhenius: Z. physik. Chem. **1,** 111 (1887).

5. $H_2O_2 + H^+ + Br^- \rightarrow H_2O + Br_2$. Livingston: J. Am. Chem. Soc. **48,** 53 (1926).

6. $[Co(NH_3)_5Br]^{++} + OH^- \rightarrow [Co(NH_3)_5OH]^{++} + Br^-$. Brönsted and Livingston: J. Am. Chem. Soc. **49,** 435 (1927).

Figure 6.2. Dependence of the rate of ionic reactions (ordinate) on the ionic strength of the aqueous solution (abscissa). (Reprinted with permission from La Mer, 1932, Copyright 1932 American Chemical Society.)

sentences like ". . . the logarithm of the rate constant of an elementary reaction involving ions depends on the square root of the ionic strength of the solution" are commonly found in textbooks and scientific papers.

6.2. The kinetics of precipitation and dissolution of solid phases

The fundamental concepts on the kinetics of precipitation and dissolution reactions have been summarized by Berner (1983) and references therein, on which this section is largely based. Further details are given by Stumm and Morgan (1996), Drever (1982) and Lasaga (1998), among the others.

6.2.1. Nucleation and crystal growth

The precipitation of a solid phase comprises two distinct stages: nucleation, first, and crystal growth, afterwards. Nucleation is called homogeneous if the nuclei form in the bulk solution, whereas it is termed heterogeneous if they form on a solid surface. To describe the process, it is instructive to consider the variation of the free energy of precipitation, ΔG_n, as a function of the radius of a hypothetical single crystal (Fig. 6.3).

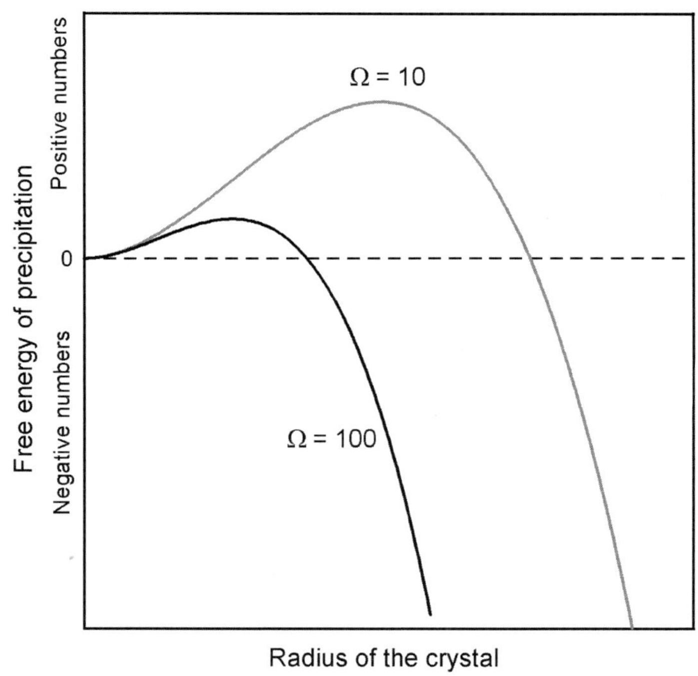

Figure 6.3. Schematic diagram of the free energy of precipitation, ΔG_n, vs. the radius, r, of a hypothetical single crystal. Note that ΔG_n increases up to a maximum during nucleation and decreases afterwards during crystal growth. The effect of the degree of oversaturation, Ω, is also shown. Reprinted from Stumm and Morgan (1996), modified, copyright (1996), with permission of Wiley.

The free energy of precipitation depends on the balance of two terms:

(i) The interfacial free energy, which represents the energy associated with the formation of the interface between the growing crystal and the aqueous solution or, in other terms, the work required to generate this surface; this term is positive and depends on the square of the radius for a spherical crystal.

(ii) The bulk free energy, which is negative and depends on the cube of the radius for a spherical crystal; besides, it is a function of the degree of oversaturation Ω (see Section 6.4 for the definition of Ω).

During nucleation (i.e. during the formation of the so-called *crystal embryo*), the interfacial free energy is greater than the bulk term and, consequently, the system experiences a net free energy increase. This increase occurs up to a maximum, corresponding to the balance between the interfacial and bulk terms. The growing solid phase, at this point, is called *critical nucleus*.

The further increase in the size of the critical nucleus is accompanied by a net decrease in free energy and the process becomes spontaneous for $\Delta G_n < 0$. This stage is known as crystal growth and the growing solid phase is referred to as a crystal. Of course, the transfer of material from the aqueous solution to the crystal goes on till the attainment of saturation (equilibrium).

The increase in Ω brings about both a decrease in the size of the critical nucleus and a decrease in the maximum of the ΔG_n–radius curve (Fig. 6.3). Since, during nucleation, the rate is an exponential function of the free energy of the process, nucleation proceeds fast for high values of Ω and vice versa.

In the considered system, nucleation and crystal growth compete for the dissolved material. For high values of Ω, nucleation may proceed so fast that most dissolved material is used to constitute critical nuclei, whereas little is available for crystal growth. If so, a very fine-grained precipitate is formed. On the contrary, for low values of Ω, nucleation may proceed so slow that most dissolved material is consumed by crystal growth, involving a small number of critical nuclei. If so, a coarser mineral is produced.

In general, crystal growth comprises several mechanisms and its rate is limited by the slowest one. In particular, the rate of crystal growth may be governed by either

(i) the transport of solute particles from the bulk aqueous solution to the crystal surface and, in this case, it is termed *transport-controlled*, or

(ii) one of the numerous reactions taking place at the surface of the growing crystal and, if so, it is called *surface-reaction controlled* or

(iii) by a combination of these two different mechanisms.

The change in the concentration of the relevant solute(s) moving away from the crystal surface towards the bulk aqueous solution is different, depending on the mechanism controlling crystal growth, as shown schematically in Fig. 6.4.

In transport-controlled crystal growth (Fig. 6.4a), addition of solute particles to the solid surface (i.e. the surface reaction under way) is so quick that solute transport in the aqueous phase cannot keep up with it. Therefore, solute concentration in the aqueous solution near the solid surface decreases sharply and gets close to the saturation (equilibrium) value. Solute transport in the aqueous solution takes place through either advection, which is the fastest way, or diffusion, which is slowest way. Transport-controlled crystal growth depends on hydrodynamic conditions and stirring accelerates it.

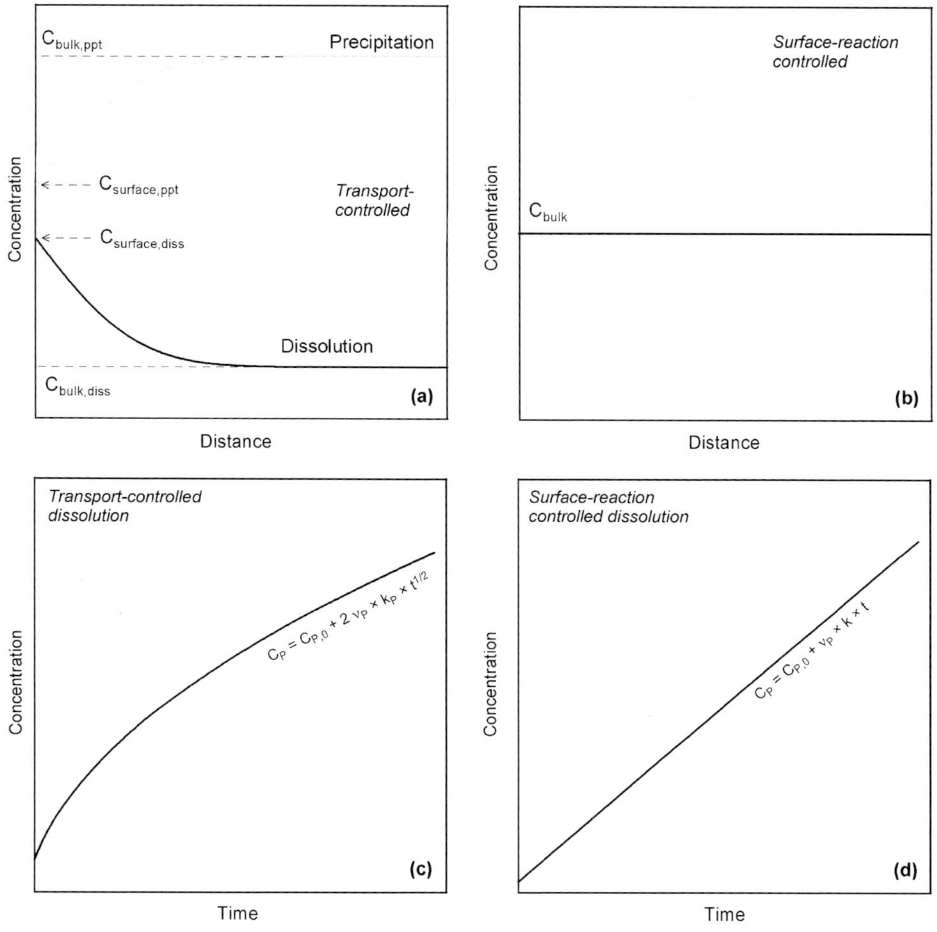

Figure 6.4. Plots (a) and (b) show schematically the change in the concentration of a relevant solute as a function of the distance from the crystal surface for (a) transport-controlled dissolution and precipitation and (b) surface-reaction-controlled processes. Plots (c) and (d) represent the change in concentration as a function of time in a generic batch experiment for (a) transport-controlled and (b) surface-reaction-controlled dissolution. Reprinted from Stumm and Morgan (1996), modified, copyright (1996), with permission from Wiley.

In surface-reaction controlled crystal growth (Fig. 6.4b), addition of solute particles to the solid surface is so slow that transport processes, even diffusion, are able to supply new solute particles near the growing crystal surface. In this case, there is little change in solute concentration between the aqueous layer close to the solid surface and the bulk solution, and crystal growth is virtually independent of hydrodynamic conditions.

Intermediate situations are possible depending on the relative speed of solute addition to the crystal surface and solute transport in the aqueous phase. Both mechanisms govern the rate of crystal growth.

To identify the mechanism controlling the rate of crystal growth, experimentally determined rates are compared with those computed for the slowest type of aqueous transport, i.e. molecular (ionic) diffusion. Measured rates faster than diffusion-controlled rates are evidently explained by advective transport, whereas slower measured rates suggest that crystal growth is governed by reactions occurring at the solid surface.

Following Nielsen (1964), the diffusion-controlled rate is computed by means of the following equation:

$$\frac{\mathrm{d}r_\mathrm{c}}{\mathrm{d}t} = \frac{v \cdot D_\mathrm{s} \cdot (C_\mathrm{B} - C_\mathrm{S})}{r_\mathrm{c}}, \tag{6-51}$$

where r_c is the mean radius of the crystals; v the molar volume of the precipitating substance; D_s the diffusion coefficient of solute particles in the aqueous solution; C_B and C_S are the concentrations of the solute in the bulk aqueous solution and close to the crystal surface, respectively; and t time. Assuming C_B to be constant, integration of equation (6-51) gives

$$r_\mathrm{c} = \left[r_{\mathrm{c},t=0}^2 + 2 \cdot v \cdot D_\mathrm{s} \cdot (C_\mathrm{B} - C_\mathrm{S}) \cdot t \right]^{1/2}, \tag{6-52}$$

where $r_{\mathrm{c,t=0}}$ is the average radius of the crystals at time “zero,” i.e. at the beginning of crystal growth, i.e. at the end of the nucleation step. Equations (6-51) and (6-52) apply to equidimensional crystals of regular shape (e.g. spheres, cubes, etc.) separated by at least five diameters.

Alternatively, crystal growth is experimentally carried out at different temperatures and the temperature dependence of the rate is established to infer the controlling mechanism (see Section 6.1.3). Besides, dependence on hydrodynamic conditions suggests transport control and vice versa, as already mentioned.

6.2.2. Dissolution

Dissolution is different from precipitation in that the dissolving crystals are already present in the system and therefore no initial step (corresponding to nucleation in precipitation) is needed.

Similar to precipitation, dissolution of solids also can be governed by either transport or surface reactions. For transport-controlled dissolution, solute concentration near the solid surface is higher than in the bulk aqueous solution (Fig. 6.4a), whereas in surface-reaction-controlled dissolution, again, solute concentration adjacent to the surface is nearly equal to that in the bulk solution (Fig. 6.4b).

The rate of diffusion-governed dissolution can be computed by means of equation (6-51). In this case, $\mathrm{d}r_\mathrm{c}/\mathrm{d}t < 0$, since $C_\mathrm{S} > C_\mathrm{B}$. Again, diffusion-controlled rates computed by means of equation (6-51) can be used to infer the mechanism controlling the rate of crystal dissolution, together with the temperature dependence of the rate and hydrodynamic effects (see above). In addition, electron microscopy on partially dissolved crystals further represents a tool for establishing the rate-governing mechanism. In surface-reaction-controlled dissolution, attack of the solid surface is slow and takes place preferentially in the sites of excess energy, causing the formation of etch pits. In contrast, in transport-controlled dissolution, surface attack is fast and non-specific, determining general rounding.

It is important to underscore that transport-controlled dissolution is governed by the so-called parabolic rate law, which can be expressed as

$$\frac{1}{\nu_P}\cdot\frac{dC_P}{dt} = k_P \cdot t^{-1/2}, \tag{6-53}$$

where the subscript P stands for a generic product as in Sections 6.1.1 and 6.1.2. Integration of equation (6-53) gives

$$C_P = C_{P,0} + 2\nu_P \cdot k_P \cdot t^{1/2}. \tag{6-54}$$

Equation (6-54) shows that, in transport-controlled dissolution, the concentration of a relevant solute is expected to increase with the square root of time (Fig. 6.4c). See Drever (1982) for further details.

In contrast, surface-controlled dissolution is described by the zero-order rate law (see equations (6-17) and (6-18)), and solute concentration is expected to increase linearly with time (Fig. 6.4d).

6.2.3. The effect of the chemical bond type

At molecular level, the kinetics of precipitation and dissolution of solid phases depends on the type of chemical bonds in the mineral. In solid phases such as fluorite (CaF_2) and silvite (KCl), where highly ionic bonds are present, ions are easily solvated by water molecules and detached from the mineral lattice.

In contrast, in aluminosilicates and oxides, the chemical bonds between oxygen and the framework ions (e.g. Si^{4+} and Al^{3+}) have a large degree of covalence and a low degree of ionicity. For instance, Si–O bonds are approximately 49% covalent and 51% ionic, whereas Al–O bonds are about 37% covalent and 63% ionic. Consequently, these framework ions are not easily solvated by water molecules and detached from the mineral structure. To be broken, these chemical bonds must be strongly perturbed by the attack of highly polarizing species, such as hydrogen ion and several inorganic and organic ligands.

We recall that the degree (or fraction) of ionicity, f_{ion}, is given by the following relation (Pauling, 1940):

$$f_{ion} = 1 - \exp\left[\frac{-(\chi_A - \chi_B)^2}{4}\right] \tag{6-55}$$

where χ_i is the electronegativity of the *i*th element, which is 1.0 for Li, 0.9 for Na, 0.8 for K, 1.5 for Be, 1.2 for Mg, 1.0 for Ca, 1.5 for Al, 1.8 for Si and Fe, 2.5 for C and S, 3.0 for Cl, 3.5 for O and 4.0 for F (Pauling, 1940).

6.3. The kinetics of chemical weathering

6.3.1. An historical perspective

It was already underscored in previous sections that both the geological sequestration of CO_2 and mineral carbonation can be considered as a sort of chemical weathering. Therefore, the kinetics of chemical weathering is of utmost interest for us.

Although the experimental investigations of mineral dissolution began during the second half of the nineteenth century (Daubrée, 1867; Beyer, 1871), the first important studies are probably those of Correns and von Engelhardt (1938) and Correns (1940), who showed that (i) the chemical components of silicates are released to the aqueous phase as true solute species (ions and molecules) and not as colloids and (ii) silicate dissolution is a non-stoichiometric process. Starting from these conclusions, the formation of a protective residual surface layer was hypothesized to explain the compositional data of aqueous solutions.

This hypothesis was refused by Marshall (1962) who suggested the occurrence of ion exchange reactions during feldspar dissolution, consistently with its surface structure. In particular, he proposed that exchange of H^+ for K^+ and Na^+ occurs on the feldspar surface during the first step of feldspar dissolution. Involvement of H_3O^+ instead of H^+ in the exchange site was later suggested by Bondam (1967).

In spite of these results, many models of silicate dissolution assumed the presence of a protective surface layer made up of either an authigenic (secondary) mineral phase (consistently with the incongruent dissolution of most silicates, i.e. dissolution of the primary silicate accompanied by precipitation of a secondary phase) or a non-stoichiometric residuum of the dissolving mineral (e.g. Helgeson, 1971; Busenberg and Clemency, 1976).

In particular, Wollast (1967) carried out experimental studies of dissolution of K-feldspar grains in a buffered aqueous solution, by monitoring the dissolved SiO_2 concentration, and he found that the rate of SiO_2 addition to the aqueous phase decreased with time. Since the concentration of solutes released by K-feldspar dissolution resulted to be proportional to the square root of time, the process was considered to follow the so-called "parabolic kinetics" or a kinetics of order 1/2. The formation of a protective Al-rich layer of an unspecified phase growing onto the surface of K-feldspar grains was suggested by Wollast (1967) to explain these experimental results, and diffusion of silica and cations through the protective surface layer was considered to be the process limiting the rate of K-feldspar dissolution. Again, the rate of feldspar dissolution was suggested to be governed by diffusion of cations through a surface-leached layer by Luce et al. (1972) and Pačes (1973).

However, the presence of an armouring, continuous surface layer, able to slow down the release of chemical components from the dissolving solid to the aqueous solution, did not find experimental confirmation, at least through the first attempted measurements of leached-layer thickness. For instance, the existence of this layer was not supported by transmission electron microscope and electron diffractometer analyses of the surface of reacted feldspar (Tchoubar and Oberlin, 1963; Wyart et al., 1963; Tchoubar, 1965) as well as by scanning electron micrographs and X-ray photoelectron spectroscopy analysis of reacted feldspar grains both from laboratory attacks and from soils carried out by Berner and co-workers (Petrović, 1976; Petrović et al., 1976; Berner and Holdren, 1977, 1979; Berner, 1978; Holdren and Berner, 1979). The surface of these K-feldspar grains was found totally unaltered and no protective surface layer (depleted in cations and enriched in Al) thicker than 15–20 Å was found.

These authors also obtained a linear (zero-th order) kinetics instead of a parabolic kinetics by treating feldspar grains with hydrogen fluoride to remove ultrafine grains and strained particles. These highly reactive particles, produced by grinding, dissolve faster than larger grains, thus causing the apparent parabolic kinetics. Thus, it was finally demonstrated that parabolic kinetics was an artefact due to improper sample preparation.

Based on these results, Berner and co-workers suggested that the rate-limiting mechanism in feldspar dissolution is the kinetics of reaction at activated sites on its surface and not diffusional transfer of chemical components through a continuous surface coating on the dissolving grains of the primary silicate. These achievements agree with earlier findings by Lagache et al. (1961) and Lagache (1965, 1976). The preferential occurrence of dissolution reaction at activated surface sites is also confirmed by the presence of typical etch pits on the surface of the partially dissolved feldspar grains.

Berner et al. (1980) and Schott et al. (1981), by means of both laboratory and soil studies, found that the dissolution of pyroxene and amphibole also follows a zero-th order kinetics and is controlled by surface reactions, in spite of limited cation depletion with respect to Si on the pyroxene surface.

Although parabolic kinetics was completely discredited, the formation of a Na- and Al-leached layer on the dissolving albite surface, under acidic conditions at 25°C, was suggested again by Chou and Wollast (1984) and Wollast and Chou (1985), based on solution chemistry data. In addition, new data obtained through secondary ion mass spectrometry, X-ray photoelectron spectroscopy, Auger electron spectroscopy, and transmission electron microscopy (Nesbitt and Muir, 1988; Muir et al., 1989, 1990; Casey and Bunker, 1990; Hochella, 1990; Shotyk and Nesbitt, 1990) support the occurrence of Al, Na, K and Ca leaching from the feldspar surface and the formation of a silica-rich leached layer under acidic conditions.

6.3.2. The present understanding

Today, based on the results of a large number of experimental studies carried out mostly in the last 25 yr, silicate dissolution can be considered to include two basic steps: (1) a first, provisional step of non-stoichiometric dissolution, due to the formation of a leached layer, followed by (2) a second step of steady-state, stoichiometric dissolution (Oelkers, 2001a).

It must be underscored that non-stoichiometric dissolution due to the formation of a leached layer is different from non-stoichiometry due to incongruent dissolution.

According to Oelkers et al. (1994) and Oelkers (1996, 2001a), silicates and Al-silicates can be considered multi-oxides and their dissolution often requires the breaking of more than one type of metal–oxygen bond and, for this reason, it differs from the dissolution of simple oxides and hydroxides. This represents the basis of the so-called *multi-oxide dissolution model.*

To understand silicate dissolution, it is advisable, therefore, to consider the dissolution of oxides and hydroxides first. This process typically takes place through metal for proton exchange reactions, i.e. by breaking the original metal–oxygen bonds of the (hydr)oxide structure and forming new proton–oxygen bonds (Stumm, 1992). These exchange reactions involve a number of protons, which is equal to the valence of the metal (Casey and Ludwig, 1996). In general, the higher the degree of ionicity of the M–O bond (see Section 6.2.3), the faster is the dissolution rate at constant temperature and pH, as shown in Fig. 6.5. This relationship is especially evident for the oxides of divalent alkali-earth metals BeO, MgO and CaO. In addition, the hydroxides of univalent alkali metals, such as NaOH and KOH, dissolve much faster than those plotted in Fig. 6.5, since the Na–O and K–O bonds have a very high degree of ionicity, and consequently, the Na^+, K^+ and OH^- ions are easily solvated and detached from these solids.

Among the silicates, the nesosilicates have the simplest structure, which is made up of isolated SiO_4 tetrahedra (i.e. a covalent silicate framework is lacking in nesosilicates). Consequently, the dissolution of nesosilicates is a comparatively simple process. For instance, forsterite dissolution takes place through exchange of Mg^{2+} ions for protons. Breaking of the Mg–O bonds completely liberates the SiO_4 tetrahedra and there is no need to break any Si–O bond for dissolving this mineral. It is not surprising, therefore, that the

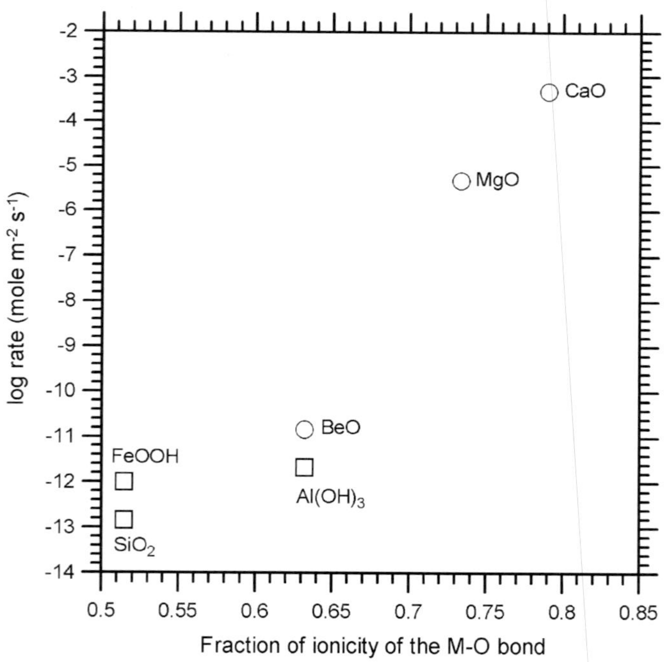

Figure 6.5. Dissolution rates for simple oxides and hyroxides of divalent metals (circles) and trivalent metals (squares) at 25°C and pH = 2 (data compiled by Oelkers, 2001) against the degree of ionicity of the M–O bond.

dissolution rates of nesosilicates, at constant temperature and pH, correlate fairly well with the dissolution rates of simple oxides, as underscored by Casey (1991).

Contrary to what is observed for nesosilicates, the dissolution of other silicates requires the breaking of more than one type of metal–oxygen bonds, and it is further complicated by two facts: (i) some of these metal–oxygen bonds break faster than others, owing to a greater degree of ionicity and (ii) some metals can be removed from the silicate structure long before it is totally destroyed by dissolution. This selective metal removal brings about the formation of a leached layer, as observed for instance, in feldspars, wollastonite and basaltic glass (see Oelkers, 2001a and references therein).

Again, similar to what occurs in (hydr)oxide dissolution, silicate dissolution also takes place through metal–proton exchange reactions, although the balance between the entering protons and the metals leaving the dissolving structure is not always respected (Casey et al., 1989).

In general, the rate of any sequence of reactions is limited by the slowest step. In silicate dissolution, the rate-limiting step is the slowest breaking of one kind of cation–oxygen bond. Based on the evidence acquired so far, Oelkers (2001a) has summarized the dissolution mechanism of some silicates under acidic conditions.

Kaolinite, enstatite and wollastonite dissolution takes place in two distinct steps. The first one consists in metal–proton exchange reactions involving Al^{3+} ion in kaolinite, Mg^{2+} ion in enstatite and Ca^{2+} ion in wollastonite. In the second step, the Si–O bonds are broken, thus completing the dissolutive destruction of these minerals.

The dissolution of alkali feldspars and muscovite is more complicated as it involves three different steps, namely (i) alkali ion for proton exchange, (ii) Al^{3+} ion–proton exchange and finally (iii) breaking of the Si–O bonds.

The dissolution of anorthite is simpler than that of alkali feldspars, as it consists of a first step involving Ca^{2+} ion–proton exchange, followed by a second step in which Al^{3+} ion–proton exchange occurs. Similar to what is observed for forsterite, the breaking of the Ca–O and Al–O bonds brings about the complete destruction of the mineral lattice and there is no need to break the Si–O bonds to complete the dissolution of anorthite.

On the contrary, basaltic glass has the most complicated dissolution mechanism, in which protons exchange for alkalis, calcium, magnesium and aluminum (in this order), and finally the Si–O bonds are broken.

6.4. The rate laws of mineral dissolution/precipitation

6.4.1. Introductory remarks

The obvious aim of dissolution/precipitation laboratory experiments and other microscopic tests (see Section 6.5) is that of obtaining a rate law; i.e. a relationship linking the dissolution/precipitation rate (the dependent variable) to one or more independent variables, comprising temperature, thermodynamic affinity (which describes the distance from mineral/solution equilibrium), ionic strength and one or more activities or concentrations of either solute species or surface species.

The dependence of the dissolution/precipitation rate on temperature is usually accounted for by means of the Arrhenius equation (see Section 6.1.3) involving the so-called apparent activation energy or activation energy term.

The variation in dissolution/precipitation rate with thermodynamic affinity is generally described by means of a TST-based rate law (see Section 6.4.2), although other type of functions have been proposed, e.g. by Lasaga and co-workers.

The change in the dissolution/precipitation rate with ionic strength has been explored for a limited number of phases, and the existence of a primary salt effect (see Section 6.1.4.1) has been observed for gibbsite dissolution by Mogollón et al. (2000, see Section 6.7.1).

Most early studies have investigated the dissolution/precipitation rates of minerals as a function of pH. The rate laws obtained in these studies were frequently called empirical, but they can be suitably taken into account in the framework of TST (see Section 6.4.3).

All previous effects are incorporated in the general form of the rate law for heterogeneous dissolution–precipitation reactions occurring at the surface of a given mineral, which was proposed by Lasaga (1995):

$$r = k_0 \cdot A_S \cdot \exp\left(-\frac{E_a}{RT}\right) \cdot a_{H^+}^{n_H} \cdot \prod_i a_i^{n_i} \cdot g(I) \cdot f(A), \tag{6-56}$$

where k_0 is the rate constant (mol m^{-2} s^{-1}); A_S is the reactive surface area of the mineral in contact with the unit volume of aqueous solution (m^2 L^{-1}); E_a is the apparent activation energy of the overall reaction (kJ mol^{-1} or kcal mol^{-1}); the term $a_{H^+}^{n_H}$ describes the pH dependence of the rate of the dissolution–precipitation reactions; the term $\prod_i a_i^{n_i}$ comprises possible catalytic or inhibitory effects linked to other solutes; and the terms $g(I)$ and $f(A)$ account for the dependence of the rate on the ionic strength I of the aqueous solution and on the distance from the equilibrium condition, which is measured by thermodynamic affinity A (see below). Note that Lasaga and co-workers use the term Gibbs free energy of reaction instead of thermodynamic affinity. In spite of their completeness, rate laws such as equation (6-56) are not entirely satisfactory for the reasons given below.

Leaving out of consideration relatively soluble minerals whose dissolution/precipitation rates are limited by diffusive transport in the aqueous solution, it must be recalled that the dissolution/precipitation kinetics of sparingly soluble minerals (e.g. silicates, oxides, hydroxides, and at least under given temperature-composition ranges, carbonates as well) are controlled by chemical reactions occurring at the surface of dissolving/precipitating minerals (see above) or close to their surface, in the so-called leached layer, in the case of mineral dissolution. Complex species forming at the mineral surface therefore play the role of activated complexes (or their precursors), whose slow decomposition is considered to be the rate-limiting step of the dissolution/precipitation process, as emphasized by several studies on oxides and hydroxides, chiefly carried out by Werner Stumm and co-workers. Consequently, dissolution/precipitation rates should be ideally described as a function of the activities or concentrations of surface complexes. However, in practice, this is not the case for most minerals, although some recent studies (e.g. those carried out by Pokrovsky, Schott and co-workers on carbonate minerals, brucite and forsterite) have been addressed to fill this gap. A further effort leading to the incorporation of this kind of rate laws in the software tools for reaction path modelling is also urgently needed.

A very interesting compromise is represented by the TST-based rate laws derived through the *multi-oxide dissolution model* developed by Oelkers, Schott and co-workers. In the approach of these authors, surface reactions are treated through the introduction of some simplifying assumptions (see Section 6.4.4), leading to rate laws in which the activity ratio, $a^{Z}_{H^+}/a_{M^{Z+}}$, is the pivotal variable, where M is a generic metal and Z^+ is the charge of the metal ion (Al^{3+} in the case of most Al-silicate crystals and glasses). It must be underscored that these rate laws can be incorporated in existing computer codes for reaction path modelling (see Sections 7.2.4 and 7.5.1) and are very useful in this respect.

6.4.2. The transition state theory-based rate laws: the distance from equilibrium

The rate-limiting mechanism of silicate dissolution can be described by means of TST, assuming that the activated complex in the transition state is represented by a localized surface configuration of atoms in a disrupted silicate network (Dibble and Tiller, 1981; Aagaard and Helgeson, 1982; Helgeson et al., 1984). Consistently with TST, the rate of surface-controlled reactions of dissolution and precipitation of a given solid phase per unit surface area, r, can be described by means of the following relation:

$$r = r_+ \cdot \left[1 - \exp\left(\frac{A}{\sigma RT}\right)\right], \tag{6-57}$$

where r_+ is the forward (dissolution) rate per unit surface area, σ the ratio between the rate of decomposition of the activated complex and the overall dissolution rate (σ is also known as Temkin's average stoichiometric number) and A is the thermodynamic affinity.

Thermodynamic affinity measures the distance from the equilibrium (saturation) condition with respect to the considered solid phase for a given aqueous solution (the same applies, in general, to any fluid). It is defined as follows:

$$A = RT \cdot \ln \frac{Q}{K_{sp}}, \tag{6-58}$$

where K_{sp} is the solubility product of the solid phase, i.e. the thermodynamic equilibrium constant of the overall dissolution reaction and Q is the corresponding ionic product:

$$Q = \prod_{i=1}^{s} a_j^{v_j}, \tag{6-59}$$

where a_j is the activity of the jth aqueous species, v_j its stoichiometric coefficient in the overall dissolution reaction and s is the total number of dissolved species in this reaction.

Based on equation (6-58), the thermodynamic affinity is positive if the aqueous solution is oversaturated with respect to the solid ($Q > K_{sp}$), negative in case of undersaturation ($Q < K_{sp}$) and zero at saturation or equilibrium ($Q = K_{sp}$). Some authors adopt a different definition, with a minus sign in the right-hand term of equation (6-58).

If so, A is still equal to zero at equilibrium, but $A < 0$ at oversaturation and $A > 0$ at undersaturation.

The Q/K_{sp} ratio is called degree of saturation and is indicated here with the letter Ω. The decimal logarithm of Ω is known as index of saturation.

For $\sigma = 1$, equation (6-57) reduces to

$$r = r_+ \cdot \left(1 - \frac{Q}{K}\right) = r_+ \cdot (1 - \Omega). \tag{6-60}$$

Under the assumption that the forward rate r_+ is positive for dissolution reactions, it follows from equations (6-57) and (6-60) that r is positive for dissolution reactions and negative for precipitation reaction. Of course, r attains zero at equilibrium.

The change in the r/r_+ ratio with thermodynamic affinity is shown in Fig. 6.6. For A lower than $-8{,}000$ J mol^{-1}, the r/r_+ ratio does not deviate significantly from unity, indicating that the dissolution rate is virtually independent of thermodynamic affinity in the dissolution region, at least in the far-from-equilibrium conditions. This is called *dissolution plateau*, if r_+ is independent of solute composition. Approaching further the equilibrium condition from below, i.e. for $-8{,}000 < A < 0$ J mol^{-1}, the r/r_+ ratio decreases significantly, and at equilibrium, the condition $r = r_+ - r_- = 0$ holds. In the precipitation region, the r/r_+ ratio continues to decrease markedly with increasing A, indicating that the precipitation rates are strongly influenced by thermodynamic affinity. As noted by Oelkers (1996), the plot of Fig. 6.6 also suggests that dissolution and precipitation rates measured in the laboratory under far-from-equilibrium conditions may deviate significantly from the close-to-equilibrium rates, which are typical of most natural systems.

6.4.3. The transition state theory-based, pH-dependent rate law

As already recalled, most of the dissolution rate experiments carried out so far have been aimed at establishing the dependence of the far-from-equilibrium dissolution kinetics of relevant mineral phases on pH. Again, although the rate laws obtained through these investigations were often termed empirical, they can be suitably considered in the framework of TST theory, as shown in the following discussion.

In the framework of TST, mineral dissolution rates (per unit surface area) can be assumed to be proportional to the mole fraction of the activated complex (or its precursor) at the mineral surface, X_P, as expressed by the equation (Wieland et al., 1988)

$$r_+ = k_+ \cdot X_P. \tag{6-61}$$

Ideally, X_P should be inferred from surface charge and surface speciation computations (e.g. Walther, 1996), but in view of the presently poor knowledge of the intrinsic constants of surface equilibria, it is preferrable (according to Oelkers et al., 1994; Oelkers, 1996, 2001a) to obtain X_P in a simplified way from the law of mass action, referring to either an adsorption reaction or an exchange reaction.

Dissolution reactions can be assumed to involve, as a first fast step, the adsorption of either H^+, or H_2O, or OH^-, depending on pH, on the surface of the dissolving solid. This

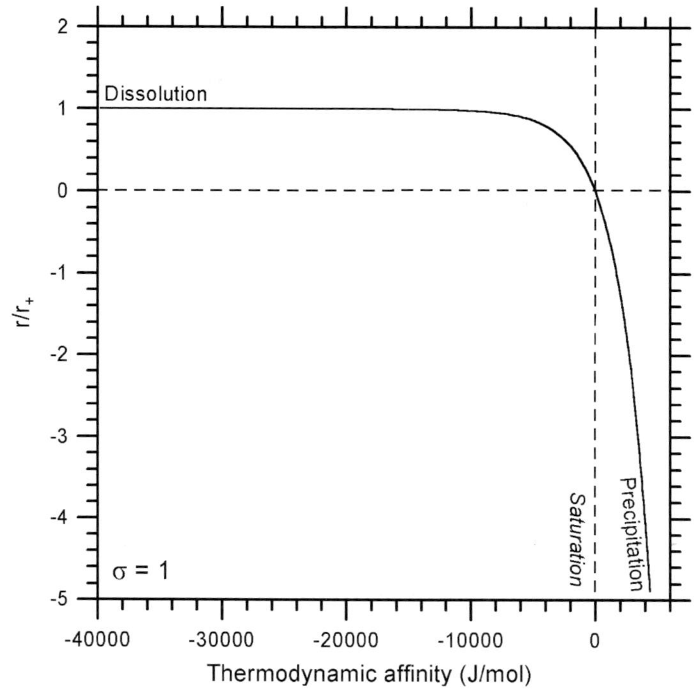

Figure 6.6. Change in the r/r_+ ratio with thermodynamic affinity. (Reprinted from Oelkers, 2001a, Copyright 2001 with permission from Elsevier, adapted.)

is expressed by the following general *adsorption reactions*:

$$n_H H^+ + \equiv M-O \rightleftharpoons \equiv M-O-H_{n_H}^{n_{H+}}, \tag{6-62}$$

$$n_W H_2O + \equiv M-O \rightleftharpoons \equiv M-O-(H-O-H)_{n_W}, \tag{6-63}$$

and

$$n_{OH} OH^- + \equiv M-O \rightleftharpoons \equiv M-O-(OH)_{n_{OH}}^{n_{OH}-}, \tag{6-64}$$

where ≡M–O is an uncomplexed generic silicate or (hydr)oxide surface, whereas the right-hand terms of reactions (6-62) to (6-64) represent the activated complexes (forming at the surface of the dissolving solid), whose decomposition is presumed to be the rate-controlling step. This second step is not indicated for simplicity.

Now we have to write the thermodynamic equilibrium constants of reactions (6-62)–(6-64). Let us do this exercise for the adsorption reaction (6-62), whose thermodynamic equilibrium constant is as follows:

$$K_H^{\ddagger} = \frac{X_{\equiv MOH} \cdot \lambda_{\equiv MOH}}{a_{H^+}^{n_H} \cdot X_{\equiv MO} \cdot \lambda_{\equiv MO}}, \tag{6-65}$$

where $\lambda_{\equiv \mathrm{MOH}}$ and $\lambda_{\equiv \mathrm{MO}}$ indicate the activity coefficient of surface species $\equiv \mathrm{M-O-H}_{n_\mathrm{H}}^{n_{\mathrm{H}^+}}$ (i.e. the activated complex or its precursor) and ≡M–O, respectively. If $\lambda_{\equiv \mathrm{MOH}}$ and $\lambda_{\equiv \mathrm{MO}}$ are assumed to be equal, equation (6-65) simplifies to

$$K_\mathrm{H}^\ddagger = \frac{X_{\equiv \mathrm{MOH}}}{a_{\mathrm{H}^+}^{n_\mathrm{H}} \cdot X_{\equiv \mathrm{MO}}}. \tag{6-66}$$

Assuming that surface sites are either in the complexed or uncomplexed forms, the mass balance

$$X_{\equiv \mathrm{MOH}} + X_{\equiv \mathrm{MO}} = 1, \tag{6-67}$$

can be written. Combination of equations (6-66) and (6-67) gives

$$X_{\equiv \mathrm{MOH}} = \frac{K_\mathrm{H}^\ddagger \cdot a_{\mathrm{H}^+}^{n_\mathrm{H}}}{1 + K_\mathrm{H}^\ddagger \cdot a_{\mathrm{H}^+}^{n_\mathrm{H}}}. \tag{6-68}$$

This result can be inserted into equation (6-61), obtaining the rate (per unit surface area) law for the acidic dissolution mechanism:

$$r_{+,\mathrm{H}} = k_{+\mathrm{H}} \cdot \frac{K_\mathrm{H}^\ddagger \cdot a_{\mathrm{H}^+}^{n_\mathrm{H}}}{1 + K_\mathrm{H}^\ddagger \cdot a_{\mathrm{H}^+}^{n_\mathrm{H}}}. \tag{6-69}$$

Following the same line of reasoning adopted for reaction (6-62), the rate laws for the neutral and basic dissolution mechanisms are easily derived. They can be finally combined in the general rate (per unit surface area) expression:

$$r_+ = k_{+\mathrm{H}} \cdot K_\mathrm{H}^\ddagger \frac{a_{\mathrm{H}^+}^{n_\mathrm{H}}}{1 + K_\mathrm{H}^\ddagger \cdot a_{\mathrm{H}^+}^{n_\mathrm{H}}} + k_{+\mathrm{W}} \cdot K_\mathrm{W}^\ddagger \cdot \frac{a_{\mathrm{H_2O}}^{n_\mathrm{W}}}{1 + K_\mathrm{W}^\ddagger \cdot a_{\mathrm{H_2O}}^{n_\mathrm{W}}} + k_{+\mathrm{OH}} \cdot K_\mathrm{OH}^\ddagger \cdot \frac{a_{\mathrm{OH}-}^{n_\mathrm{OH}}}{1 + K_\mathrm{OH}^\ddagger \cdot a_{\mathrm{OH}^-}^{n_\mathrm{OH}}}, \tag{6-70}$$

where $K_\mathrm{H}^\ddagger$, $K_\mathrm{W}^\ddagger$ and $K_\mathrm{OH}^\ddagger$ represent the thermodynamic equilibrium constants of the adsorption reactions (6-62), (6-63) and (6-64), respectively, and $k_{+\mathrm{H}}$, $k_{+\mathrm{w}}$ and $k_{+\mathrm{OH}}$ are the rate constants of the corresponding proton-, water- and hydroxyl-promoted dissolution reactions.

If the activated complexes are few, i.e. $K_\mathrm{H}^\ddagger a_{\mathrm{H}^+}^{n_\mathrm{H}} \ll 1$, $K_\mathrm{W}^\ddagger a_{\mathrm{H_2O}}^{n_\mathrm{W}} \ll 1$ and $K_\mathrm{OH}^\ddagger a_{\mathrm{HO}^-}^{n_\mathrm{OH}} \ll 1$ and assuming $a_{\mathrm{H_2O}} = 1$, equation (6-70) reduces to

$$r_+ = k_{+\mathrm{H}} \cdot K_\mathrm{H}^\ddagger \cdot a_{\mathrm{H}^+}^{n_\mathrm{H}} + k_{+\mathrm{W}} \cdot K_\mathrm{W}^\ddagger + k_{+\mathrm{OH}} \cdot K_\mathrm{OH}^\ddagger \cdot a_{\mathrm{OH}^-}^{n_\mathrm{OH}}. \tag{6-71}$$

Moreover, recalling that $a_{\mathrm{OH}^-} = K_\mathrm{W}/a_{\mathrm{H}^+}$ (where K_W is the dissociation constant of water), equation (6-71) can be rewritten as

$$r_+ = k_{+\mathrm{H}} \cdot K_\mathrm{H}^\ddagger \cdot a_{\mathrm{H}^+}^{n_\mathrm{H}} + k_{+\mathrm{W}} \cdot K_\mathrm{W}^\ddagger + k_{+\mathrm{OH}} \cdot K_\mathrm{OH}^\ddagger \cdot \left(\frac{K_\mathrm{W}}{a_{\mathrm{H}^+}} \right)^{n_\mathrm{OH}}. \tag{6-72}$$

This equation explains the typical U-shape of the log-rate of dissolution for most silicates, oxides and hydroxides as a function of pH. Several examples of this behaviour will be shown in Sections 6.6 and 6.7. At constant temperature, the slowest dissolution rate generally occurs within a few pH units from neutrality, where the mineral surface has a zero net charge. At low pH, dissolution rates increase with decreasing pH, whereas at high pH dissolution rates increase with pH.

In order to account for the temperature dependence of far-from-equilibrium dissolution rates, we may recall both the Arrhenius equation (6-25) and equation (5-40), which is a truncated form of equation (2-41) obtained assuming $\Delta_r C_P^o = 0$ (i.e. $\Delta a = \Delta b = \Delta c = 0$). Therefore, equation (6-72) can be rewritten as follows, incorporating the temperature effects on rate constants and equilibrium constants:

$$\begin{aligned} r_+ = {} & A_{a,H} \cdot \exp\left[-\frac{E_{a,H}}{R} \cdot \left(\frac{1}{T} - \frac{1}{298}\right)\right] \cdot K^{\ddagger}_{H,298} \cdot \exp\left[-\frac{\Delta H^{\ddagger o}_{H}}{R} \cdot \left(\frac{1}{T} - \frac{1}{298}\right)\right] \cdot a^{n_H}_{H^+} \\ & + A_{a,W} \cdot \exp\left[-\frac{E_{a,W}}{R} \cdot \left(\frac{1}{T} - \frac{1}{298}\right)\right] \cdot K^{\ddagger}_{W,298} \cdot \exp\left[-\frac{\Delta H^{\ddagger o}_{W}}{R} \cdot \left(\frac{1}{T} - \frac{1}{298}\right)\right] \\ & + A_{a,OH} \cdot \exp\left[-\frac{E_{a,OH}}{R} \cdot \left(\frac{1}{T} - \frac{1}{298}\right)\right] \cdot K^{\ddagger}_{OH,298} \cdot \exp\left[-\frac{\Delta H^{\ddagger o}_{OH}}{R} \cdot \left(\frac{1}{T} - \frac{1}{298}\right)\right] \\ & \times K^{n_{OH}}_{W,298} \cdot \exp\left[-\frac{n_{OH} \cdot \Delta H^o_W}{R} \cdot \left(\frac{1}{T} - \frac{1}{298}\right)\right] \frac{1}{a^{n_{OH}}_{H^+}}. \end{aligned} \tag{6-73}$$

Equation (6-73) can then be rearranged as

$$\begin{aligned} r_+ = {} & A_{a,H} \cdot K^{\ddagger}_{H,298} \cdot a^{n_H}_{H^+} \cdot \exp\left[-\frac{E_{a,H} + \Delta H^{\ddagger o}_{H}}{R} \cdot \left(\frac{1}{T} - \frac{1}{298}\right)\right] \\ & + A_{a,W} \cdot K^{\ddagger}_{W,298} \cdot \exp\left[-\frac{E_{a,W} + \Delta H^{\ddagger o}_{W}}{R} \cdot \left(\frac{1}{T} - \frac{1}{298}\right)\right] \\ & + \frac{A_{a,OH} \cdot K^{\ddagger}_{OH,298} \cdot K^{n_{OH}}_{W,298}}{a^{n_{OH}}_{H^+}} \cdot \exp\left[-\frac{E_{a,OH} + \Delta H^{\ddagger o}_{OH} + n_{OH} \cdot \Delta H^o_W}{R} \cdot \left(\frac{1}{T} - \frac{1}{298}\right)\right] \end{aligned} \tag{6-74}$$

Recently, an exhaustive compilation of dissolution rates for over 70 minerals, mostly silicates (but also including oxides, hydroxides, carbonates, sulphates, sulphides, phosphates and halides), has been performed by Palandri and Kharaka (2004), whose report is accessible at the website http://water.usgs.gov/pubs/of/2004/1068/pdf/OFR_2004_1068.pdf (see Table 6.1). They carried out piecewise linear regression of rate data, by first dividing the available data into acidic, neutral and basic regions and then fitting the obtained batches of data to partial, simplified versions of equation (6-74), namely

$$\log(r_{+,H}) = \log(A_H) - n_H \cdot \mathrm{pH} - \frac{E_H}{\ln(10)RT}, \tag{6-75}$$

$$\log(\mathrm{r}_{+,W}) = \log(A_W) - \frac{E_W}{\ln(10)RT}, \tag{6-76}$$

TABLE 6.1

Dissolution rate parameters for silicates, oxides, hydroxides, carbonates, sulphates, sulphides, phosphates and halides (from Palandri and Kharaka, 2004, except where indicated otherwise), for use with equations (6-75), (6-76) and (6-77)

Mineral	Acidic Mechanism				Neutral Mechanism			Basic (or Carbonate) Mechanism[a]			
	$\log k_{H,Tr}$[b]	E_H	$\log A_H$	n_H	$\log k_{W,Tr}$	E_W	$\log A_W$	$\log k_{OH,Tr}$[b]	E_{OH}	$\log A_{OH}$	n_{OH}
	log (mol m^{-2} s^{-1})	kJ mol^{-1}	log (mol m^{-2} s^{-1})		log (mol m^{-2} s^{-1})	kJ mol^{-1}	log (mol m^{-2} s^{-1})	log (mol m^{-2} s^{-1})	kJ mol^{-1}	log (mol m^{-2} s^{-1})	
Forsterite	−6.85	67.2	4.92	0.470	−10.64	79.0	3.20	–	–	–	–
Fayalite	−4.80	94.4	11.74	–	−12.80	94.4	3.74	–	–	–	–
Almandine	−5.20	94.4	11.34	1.000	−10.70	103.8	7.48	−13.71	37.80	−7.09	−0.350
Grossular	−5.10	85.0	9.79	1.000	−10.70	103.8	7.48	–	–	–	–
Andradite	−5.20	94.4	11.34	1.000	−10.70	103.8	7.48	–	–	–	–
Kyanite	−10.17	53.9	−0.73	1.268	−17.44	53.9	−8.00	–	–	–	–
Staurolite	−6.90	18.9	−3.59	1.000	−12.20	56.6	−2.28	−14.90	47.20	−6.63	−0.300
Epidote	−10.60	71.1	1.86	0.338	−11.99	70.7	0.40	−17.33	79.10	−3.47	−0.556
Zoisite	−7.50	66.1	4.08	0.500	−11.20	66.1	0.38	–	–	–	–
Cordierite	−3.80	113.3	16.05	1.000	−11.20	28.3	−6.24	–	–	–	–
Tourmaline	−6.50	75.5	6.73	1.000	−11.20	85.0	3.69	–	–	–	–
Enstatite	−9.02	80.0	5.00	0.600	−12.72	80.0	1.30	–	–	–	–
Bronzite	−8.30	47.2	−0.03	0.650	−11.70	66.1	−0.12	–	–	–	–
Diopside	−6.36	96.1	10.48	0.710	−11.11	40.6	−4.00	–	–	–	–
Augite	−6.82	78.0	6.84	0.700	−11.97	78.0	1.69	–	–	–	–
Wollastonite	−5.37	54.7	4.21	0.400	−8.88	54.7	0.70	–	–	–	–
Jadeite	−6.00	132.2	17.16	0.700	−9.50	94.4	7.04	–	–	–	–
Spodumene	−4.60	94.4	11.94	0.700	−9.30	66.1	2.28	–	–	–	–
Anthophyllite	−11.94	51.0	−3.01	0.440	−14.24	51.0	−5.31	–	–	–	–
Tremolite	−8.40	18.9	−5.09	0.700	−10.60	94.4	5.94	–	–	–	–
Hornblende	−11.00	75.5	2.23	0.600	−14.30	94.4	2.24	–	–	–	–
Glaucophane	−5.60	85.0	9.29	0.700	−10.10	94.4	6.44	–	–	–	–
Riebeckite	−7.70	56.6	2.22	0.700	−12.20	47.2	−3.93	–	–	–	–
Kaolinite	−11.31	65.9	0.24	0.777	−13.18	22.2	−9.29	−17.05	17.90	−13.91	−0.472
Chrysotile	–	–	–	–	−12.00	73.5	0.88	−13.58	73.50	−0.70	−0.230
Lizardite	−5.70	75.5	7.53	0.800	−12.40	56.6	−2.48	–	–	–	–

Pyrophyllite	–	–	–	–	−12.40	29.0	−7.32	–	–	–	–
Talc	–	–	–	–	−12.00	42.0	−4.64	–	–	–	–
Montmorillonite	−12.71	48.0	−4.30	0.220	−14.41	48.0	−6.00	–	–	–	–
Smectite	−10.98	23.6	−6.85	0.340	−12.78	35.0	−6.65	−16.52	58.90	−6.20	−0.400
Illite[c]	−11.72	46.0	−3.66	0.600	−15.05	14.0	−12.60	−20.71	67.00	−8.97	−0.600
Glauconite	−4.80	85.0	10.09	0.700	−9.10	85.0	5.84	–	–	–	–
Chlorite[d]	−9.79	88.0	5.63	0.490	−13.00	88.0	2.42	−16.79	88.00	−1.37	−0.430
Muscovite	−11.85	22.0	−8.00	0.370	−13.55	22.0	−9.70	−14.55	22.00	−10.70	−0.220
Paragonite	–	–	–	–	−13.00	22.0	−9.15	–	–	–	–
Biotite	−9.84	22.0	−5.99	0.525	−12.55	22.0	−8.70	–	–	–	–
Phlogopite	–	–	–	–	−12.40	29.0	−7.32	–	–	–	–
Prehnite	−10.66	80.5	3.44	0.256	−13.16	93.4	3.20	−14.86	93.40	1.50	−0.200
Albite	−9.87	65.0	1.52	0.457	−12.04	69.8	0.19	−16.98	71.00	−4.54	−0.572
Oligoclase	−9.67	65.0	1.72	0.457	−11.84	69.8	0.39	–	–	–	–
Andesine	−8.88	53.5	0.49	0.541	−11.47	57.4	−1.41	–	–	–	–
Labradorite	−7.87	42.1	−0.49	0.626	−10.91	45.2	−2.99	–	–	–	–
Bytownite	−5.85	29.3	−0.72	1.018	−9.82	31.5	−4.30	–	–	–	–
Anorthite	−3.50	16.6	−0.59	1.411	−9.12	17.8	−6.00	–	–	–	–
K-feldspar	−10.06	51.7	−1.00	0.500	−12.41	38.0	−5.75	−21.20	94.10	−4.71	−0.823
Nepheline	−2.73	62.9	8.29	1.130	−8.56	65.4	2.90	−10.76	37.80	−4.14	−0.200
Leucite	−6.00	132.2	17.16	0.700	−9.20	75.5	4.03	−10.66	56.60	−0.74	−0.200
Quartz	–	–	–	–	−13.99	87.6	1.36	−16.29	98.70	1.00	−0.500
α-Cristobalite	–	–	–	–	−12.31	65.0	−0.92	–	–	–	–
β-Cristobalite	–	–	–	–	−12.14	62.9	−1.12	–	–	–	–
Amorphous silica	–	–	–	–	−12.31	76.0	1.00	–	–	–	–
Gibbsite	−7.65	47.5	0.67	0.992	−11.50	61.2	−0.78	−16.65	80.10	−2.62	−0.784
Diaspore	–	–	–	–	−13.33	47.5	−5.01	−23.60	47.50	−15.28	−1.503
Brucite[e]	−6.38	51.0	2.55	0.200	–	–	–	–	–	–	–
Goethite	–	–	–	–	−7.94	86.5	7.21	–	–	–	–
Haematite	−9.39	66.2	2.21	1.000	−14.60	66.2	−3.00	–	–	–	–
Magnetite	−8.59	18.6	−5.33	0.279	−10.78	18.6	−7.52	–	–	–	–
Ilmenite	−8.35	37.9	−1.71	0.421	−11.16	37.9	−4.52	–	–	–	–
Uraninite	–	–	–	–	−7.98	32.0	−2.37	–	–	–	–
Calcite	−0.30	14.4	2.22	1.000	−5.81	23.5	−1.69	−3.48	35.40	2.72	1.000

(*Continued*)

TABLE 6.1 (*Continued*)

Mineral	Acidic Mechanism				Neutral Mechanism			Basic (or Carbonate) Mechanism[a]			
	$\log k_{H,Tr}$[b]	E_H	$\log A_H$	n_H	$\log k_{W,Tr}$	E_W	$\log A_W$	$\log k_{OH,Tr}$[b]	E_{OH}	$\log A_{OH}$	n_{OH}
	log (mol m^{-2} s^{-1})	kJ mol^{-1}	log (mol m^{-2} s^{-1})		log (mol m^{-2} s^{-1})	kJ mol^{-1}	log (mol m^{-2} s^{-1})	log (mol m^{-2} s^{-1})	kJ mol^{-1}	log (mol m^{-2} s^{-1})	
Dolomite-dis	−3.19	36.1	3.13	0.500	−7.53	52.2	1.61	−5.11	34.80	0.99	0.500
Dolomite-ord	−3.76	56.7	6.17	0.500	−8.60	95.3	8.10	−5.37	45.70	2.64	0.500
Magnesite	−6.38	14.4	−3.86	1.000	−9.34	23.5	−5.22	−5.22	62.80	5.78	1.000
Dawsonite	–	–	–	–	−7.00	62.8	4.00	–	–	–	–
Barite	−6.90	30.8	−1.50	0.220	−7.90	30.8	−2.50	–	–	–	–
Celestite	−5.66	23.8	−1.49	0.109	–	–	–	–	–	–	–
Gypsum	–	–	–	–	−2.79	–	–	–	–	–	–
Anhydrite	–	–	–	–	−3.19	14.3	−0.68	–	–	–	–
Anglesite	−5.58	31.3	−0.10	0.298	−6.5	31.3	−1.02	–	–	–	–
Pyrite[f,g]	−7.52	56.9	2.45	−0.500	−4.55	56.9	5.42	–	–	–	–
Pyrrhotite[f,h]	−8.04	50.8	0.86	−0.597	–	–	–	–	–	–	–
Pyrrhotite[f,i]	−6.79	63.0	4.25	−0.090	–	–	–	–	–	–	–
Amorphous As_2S_3	–	–	–	–	−9.83	8.7	−8.31	−17.39	8.70	−15.87	−1.208
Halite	–	–	–	–	−0.21	7.4	1.09	–	–	–	–
Fluorite	−7.14	73.0	5.65	1.000	−13.79	73.0	−1.00	–	–	–	–
Hydroxyapatite	−4.29	250.0	39.51	0.171	−6.00	250.0	37.80	–	–	–	–
Fluorapatite	−3.73	250.0	40.07	0.613	−8.00	250.0	35.80	–	–	–	–

[a]Carbonate mechanism for carbonate minerals and basic mechanism for all the other minerals.

[b]Log rate at pH = 0.

[c]From Köhler et al. (2003).

[d]E_H, E_W and E_{OH} from Palandri and Kharaka (2004), k_H, k_W, k_{OH}, n_H and n_{OH} from Lowson et al. (2005).

[e]From Pokrovsky and Schott (2004).

[f]The acidic mechanism of sulphide minerals depends also on Fe^{3+} activity with a reaction order of 0.500 for pyrite, 0.355 for monoclinic pyrrhotite and 0.356 for hexagonal pyrrhotite.

[g]The neutral mechanism of pyrite has a reaction order of 0.500 with respect to O_2.

[h]Monoclinic pyrrhotite.

[i]Hexagonal pyrrhotite.

and

$$\log\left(r_{+,\mathrm{OH}}\right) = \log\left(A_{\mathrm{OH}}\right) + n_{\mathrm{OH}} \cdot \mathrm{pH} - \frac{E_{\mathrm{OH}}}{\ln(10)RT}. \tag{6-77}$$

Comparison of equation (6-74) with equations (6-75), (6-76) and (6-77) shows that the pre-exponential factors A_H, A_W and A_{OH} and the activation energy terms E_H, E_W and E_{OH} in the last three equations group together several contributions, whose meaning is elucidated by equation (6-74).

Note that the pre-exponential factor in equation (6-75), A_H, is related to the rate constant $k_{H,Tr}$ through the equation

$$\log\left(A_{\mathrm{H}}\right) = \log\left(k_{\mathrm{H,Tr}}\right) + \frac{E_{\mathrm{H}}}{\ln(10) \cdot R \cdot 298.15}. \tag{6-78}$$

Similar expressions can be written for A_W and A_{OH}.

Although expressions in equations (6-75)–(6-77) are probably the most frequently used equations to describe the dissolution kinetics of solid phases, it is important to be aware of their meaning and their limitations. As shown above, these expressions are not empirical (contrary to what is frequently considered in the literature) but relatable to the adsorption reactions (6-62)–(6-64) through TST. For what concerns their limitations, they do not take into account the temperature dependence of the pre-exponential factors A_H, A_W and A_{OH}, or the possible pH dependence of the activation energy terms E_H, E_W and E_{OH}. Moreover, the bulk dissolution rates are described without considering the possible catalytic or inhibitory effects linked to other solutes, e.g. Al^{3+} ion. These represent the subject of the next section.

6.4.4. The transition state theory-based rate law involving activity ratios

Let us consider again the dissolution of a generic Al-silicate. In this case, let us assume that the rate-limiting step of the overall dissolution reaction is the breaking of the Al–O bonds, through exchange of Al^{3+} ions for protons, according to the simplified *exchange reaction* (instead of the adsorption reaction (6-62)):

$$3n\mathrm{H}^{+} + \equiv \mathrm{Al}_{n} \rightleftharpoons n\mathrm{Al}^{3+} + \equiv \mathrm{H}_{3n}, \tag{6-79}$$

where $\equiv Al_n$ represents the reactive surface site before Al–H exchange and $\equiv H_{3n}$ stands for the protonated silica-rich complex. The law of mass action for this reaction dictates

$$K_{\mathrm{Al-O}} = \left(\frac{a_{\mathrm{Al}^{3+}}}{a_{\mathrm{H}^{+}}^{3}}\right)^{n} \cdot \left(\frac{X_{\equiv \mathrm{H}_{3n}} \cdot \lambda_{\equiv \mathrm{H}_{3n}}}{X_{\equiv \mathrm{Al}_{n}} \cdot \lambda_{\equiv \mathrm{Al}_{n}}}\right), \tag{6-80}$$

which reduces to

$$K_{\mathrm{Al-O}} = \left(\frac{a_{\mathrm{Al}^{3+}}}{a_{\mathrm{H}^{+}}^{3}}\right)^{n} \cdot \frac{X_{\equiv \mathrm{H}_{3n}}}{X_{\equiv \mathrm{Al}_{n}}}, \tag{6-81}$$

under the assumption of equal activity coefficients λ for surface species. Assuming that the surface sites of the dissolving Al-silicate contain either Al^{3+} ions or protons, it follows that,

$$X_{\equiv H_{3n}} + X_{\equiv Al_n} = 1. \tag{6-82}$$

Equation (6-82) can be combined with equation (6-81) to obtain

$$X_{\equiv H_{3n}} = \frac{K_{Al-O} \cdot (a^3_{H^+}/a_{Al^{3+}})^n}{1 + K_{Al-O} \cdot (a^3_{H^+}/a_{Al^{3+}})^n}. \tag{6-83}$$

Following Oelkers (2001a), it is instructive to consider the distribution of the surface species $\equiv Al_n$ and $\equiv H_{3n}$ on the dissolving Al-silicate as a function of the parameter $K_{Al-O}\left(a^3_{H^+}/a_{Al^{3+}}\right)^n$, as shown in Fig. 6.7. For $K_{Al-O}\left(a^3_{H^+}/a_{Al^{3+}}\right)^n \ll 1$, the surface complex $\equiv Al_n$ prevails or, in other words, Al is still present in close to initial amounts on the dissolving Al-silicate surface. On the contrary, for $K_{Al-O}\left(a^3_{H^+}/a_{Al^{3+}}\right)^n \gg 1$, the protonated complex $\equiv H_{3n}$ becomes the prevailing surface species. This means that the dissolving Al-silicate surface has become depleted in Al and a leached layer has formed. Since the $a^3_{H^+}/a_{Al^{3+}}$ activity ratio increases with increasing acidity and decreasing dissolved Al

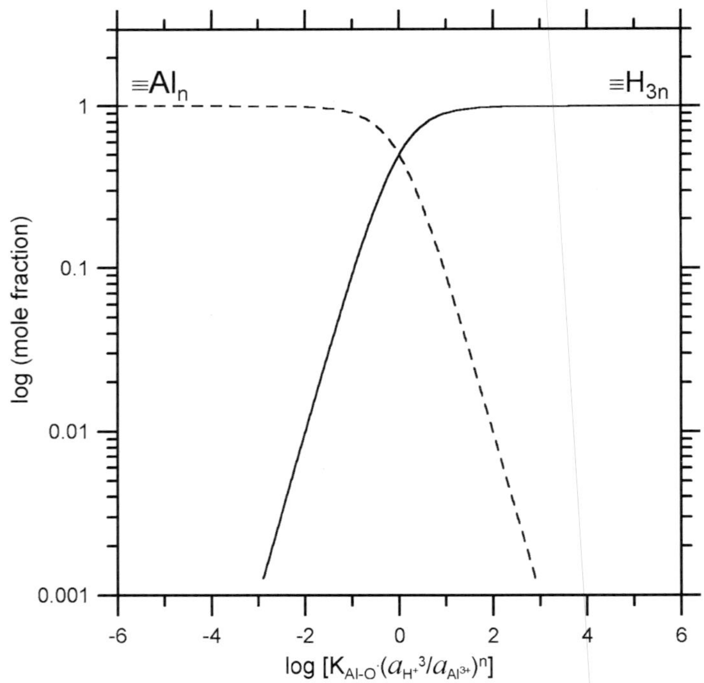

Figure 6.7. Distribution of the surface species $\equiv Al_n<$ and $\equiv H_{3n}<$ on the dissolving Al-silicate as a function of the parameter $K_{Al-O}\left(a^3_{H^+}/a_{Al^{3+}}\right)^n$. (Reprinted from Oelkers, 2001a, Copyright 2001 with permission from Elsevier, adapted.)

concentration, under these conditions the formation of the leached layer is favoured, in agreement with experimental evidence.

Combination of equations (6-83), (6-61) and (6-57) yields the following expression:

$$r = \mathrm{k}_+^{'} \cdot \frac{\left(a_{\mathrm{H}^+}^3 / a_{\mathrm{Al}^{3+}}\right)^n}{1 + K_{\mathrm{Al-O}} \cdot \left(a_{\mathrm{H}^+}^3 / a_{\mathrm{Al}^{3+}}\right)^n} \cdot \left[1 - \exp\left(\frac{A}{RT}\right)\right], \tag{6-84}$$

where $k_+^{'} = k_+ K_{\mathrm{Al-O}}$. Equation (6-84) holds for $\sigma = 1$, i.e. assuming that the stoichiometric formula of the considered Al-silicate is normalized to the formation of one rate-controlling surface complex.

For $K_{\mathrm{Al-O}} \left(a_{\mathrm{H}^+}^3 / a_{\mathrm{Al}^{3+}}\right)^n \ll 1$, i.e. when Al is still present in close to initial amounts on the surface of the dissolving Al-silicate, equation (6-84) simplifies to

$$r = k_+^{'} \cdot \left(\frac{a_{\mathrm{H}^+}^3}{a_{\mathrm{Al}^{3+}}}\right)^n \cdot \left[1 - \exp\left(\frac{A}{RT}\right)\right]. \tag{6-85}$$

On the contrary, for $K_{\mathrm{Al-O}} \left(a_{\mathrm{H}^+}^3 / a_{\mathrm{Al}^{3+}}\right)^n \gg 1$, i.e. when the protonated surface complex becomes the dominant surface species leading to the formation of a leached layer, equation (6-84) reduces to

$$r = k_+ \cdot \left[1 - \exp\left(\frac{A}{RT}\right)\right]. \tag{6-86}$$

Moreover, at far-from-equilibrium conditions, equations (6-85) and (6-86) simplify further to $r = k'_+ \left(a_{\mathrm{H}^+}^3 / a_{\mathrm{Al}^{3+}}\right)^n$ and $r = k_+$, respectively.

It is evident from previous discussion that the $a_{\mathrm{H}^+}^3 / a_{\mathrm{Al}^{3+}}$ activity ratio affects the dissolution rate of the Al-silicate if Al is present in close to initial amounts on its surface, whereas the dissolution rate becomes independent of the $a_{\mathrm{H}^+}^3 / a_{\mathrm{Al}^{3+}}$ activity ratio upon formation of a leached layer.

In the previous discussion, it was assumed that the breaking of the Al–O bonds represents the rate-limiting step of the overall dissolution reaction. In general terms, the dissolution kinetics will be governed by the breaking of a given type of M–O bond. The corresponding exchange reaction is:

$$Zn\mathrm{H}^+ + \equiv \mathrm{M}_n \rightleftharpoons \equiv \mathrm{H}_{Zn} + n\mathrm{M}^{Z+}, \tag{6-87}$$

where M is a generic metal and Z+ is the charge of the metal ion. Consequently, equations (6-84) and (6-85) can be rewritten as follows:

$$r = \mathrm{k}_+^{'} \cdot \frac{\left(a_{\mathrm{H}^+}^Z / a_{\mathrm{M}^{Z+}}\right)^n}{1 + K_{\mathrm{M-O}} \cdot \left(a_{\mathrm{H}^+}^Z / a_{\mathrm{M}^{Z+}}\right)^n} \cdot \left[1 - \exp\left(\frac{A}{\sigma RT}\right)\right] \tag{6-88}$$

TABLE 6.2

Parameters used in equations (6-88) and (6-89) for describing the dissolution rates of selected silicate minerals as a functions of water chemistry and thermodynamic affinity (from Oelkers, 1996, modified)

Mineral	M	*n*	Reference
Kyanite	Al	0.5	Oelkers and Schott (1999)
Enstatite	Mg	0.12	Oelkers and Schott (2001)
Wollastonite	Ca	0.25	Oelkers (1996)
Chlorite	Al	0.27	Lowson et al. (2005)
Muscovite	Al	0.33	Oelkers et al. (1994)
Kaolinite, acid pH	Al	1	Oelkers et al. (1994)
Kaolinite, basic pH	Al	2	Devidal et al. (1997)
K-feldspar	Al	0.33	Gautier et al. (1994)
Albite	Al	0.33	Oelkers et al. (1994)
Labradorite	Al	0.31	Carroll and Knauss (2005)
Bytownite	Al	0.33	Oelkers and Schott (1998)
Anorthite[a]	–	–	Oelkers and Schott (1995)
Analcime	Al	2	Murphy et al. (1996)
Basaltic glass	Al	0.35	Oelkers and Gislason (2001)
Brucite	Mg	0.11	See Section 6.7.2

[a]The dissolution rate of anorthite is independent of the aqueous concentration of constituents metals at far-from-equilibrium conditions.

$$r = k_+^{'} \cdot \left(\frac{a_{\mathrm{H}^+}^{\mathrm{Z}}}{a_{\mathrm{M}^{\mathrm{Z}+}}} \right)^n \cdot \left[1 - \exp\left(\frac{A}{\sigma RT} \right) \right]. \tag{6-89}$$

The identity of the metal controlling the dissolution rate of silicate minerals and the value of the stoichiometric parameter *n* cannot be predicted *a priori*, but must be established by means of suitably designed laboratory measurements. In recent years, a considerable amount of experimental work has been carried out, mainly by Oelkers, Schott and co-workers, whose main results are summarized in Table 6.2.

Some of these results are presented in Fig. 6.8, showing the experimentally measured far-from-equilibrium log-rates of dissolution of kaolinite (Devidal et al., 1997), K-feldspar (Gautier et al., 1994) and kyanite (Oelkers and Schott, 1999) against the logarithm of the $a_{\mathrm{H}^+}^3 / a_{\mathrm{Al}^{3+}}$ activity ratio. The linear dependence of the log-rate of dissolution on the $\log(a_{\mathrm{H}^+}^3 / a_{\mathrm{Al}^{3+}})$ ratio is clearly evident for all the three minerals.

In contrast, far-from-equilibrium dissolution log-rates of anorthite (Oelkers and Schott, 1995) are poorly dependent on the $\log(a_{\mathrm{H}^+}^3 / a_{\mathrm{Al}^{3+}})$ ratio (Fig. 6.9a). This is in line with the conclusions drawn by Oelkers and Schott (1995), who underscored that measured anorthite dissolution rates at constant temperature are proportional to $a_{H^+}^{1.5}$ and are independent of aqueous Al concentration. Consistently, anorthite has a dissolution mechanism different from that of alkali feldspars (see *also* Sections 6.3.2 and 6.6.4).

Similar to what observed for Al-silicates, far-from-equilibrium dissolution log-rates of enstatite (Oelkers and Schott, 2001) exhibit a linear dependence on the $\log(a_{\mathrm{H}^+}^2 / a_{\mathrm{Mg}^{2+}})$ ratio, as shown in Fig. 6.9b.

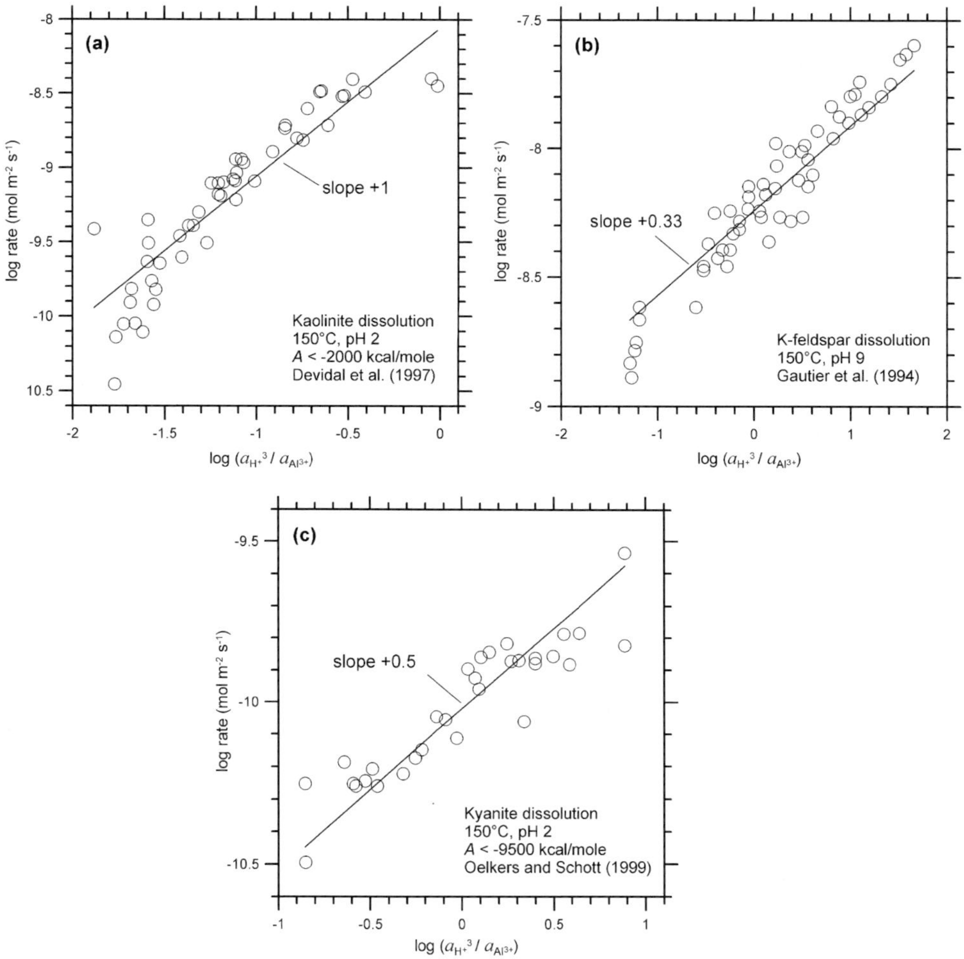

Figure 6.8. Experimentally determined far-from-equilibrium log-rates of dissolution of (a) kaolinite (data from Devidal et al., 1997), (b) K-feldspar (data from Gautier et al., 1994) and (c) kyanite (data from Oelkers and Schott, 1999) against the logarithm of the $a^3_{H^+}/a_{Al^{3+}}$ activity ratio. The linear dependence of the dissolution log-rate on the $\log(a^3_{H^+}/a_{Al^{3+}})$ ratio is clearly evident for all the three minerals.

6.5. Dissolution laboratory experiments

As anticipated above, the dissolution rates of several minerals, including silicates, oxides, hydroxides, carbonates, sulphates, sulphides, phosphates and halides, have been investigated by means of laboratory experiments under known conditions of temperature, pressure, pH, concentrations of peculiar solutes (e.g. organic acids) and composition of solids.

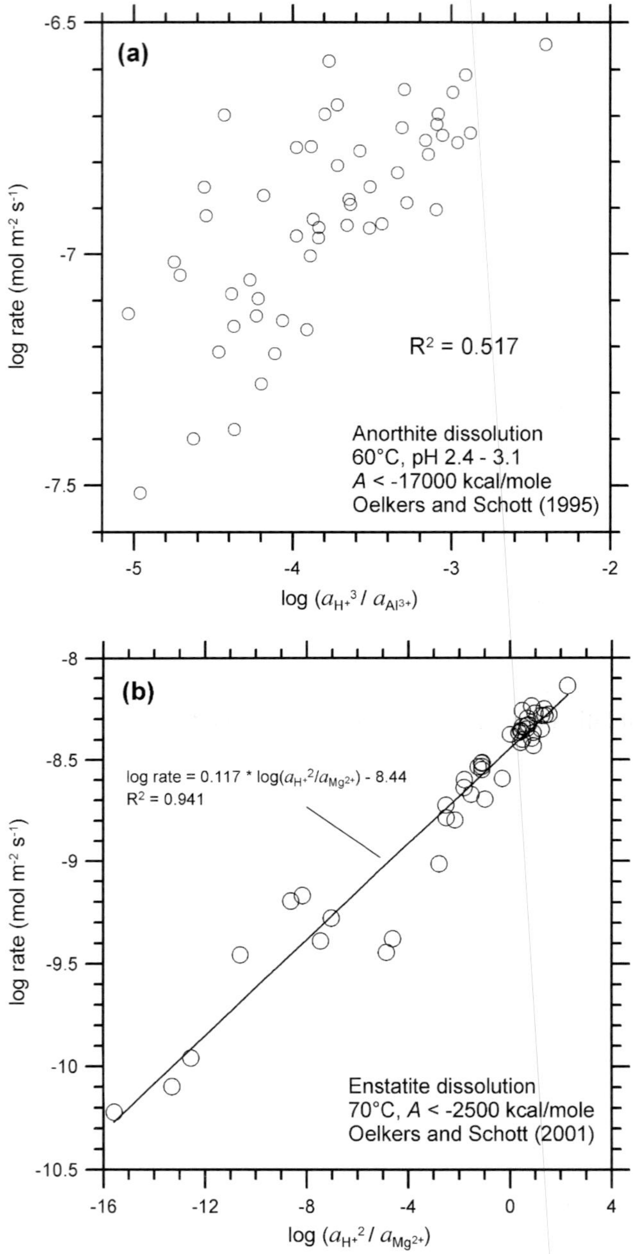

Figure 6.9. Experimentally determined far-from-equilibrium log-rates of dissolution of (a) anorthite against the logarithm of the $a^3_{H^+}/a_{Al^{3+}}$ activity ratio (data from Oelkers and Schott, 1995) and (b) enstatite against the logarithm of the $a^2_{H^+}/a_{Mg^{2+}}$ activity ratio. Again, the linear dependence of the dissolution log-rate on the log-activity ratios is evident for enstatite but not anorthite. (Reprinted from Oelkers and Schott, 2001, Copyright 2001 with permission from Elsevier, adapted.)

In contrast, growth (precipitation) rates have been the subject of limited research efforts for practical reasons, i.e. because it is much simpler to dissolve a solid than to precipitate it under controlled conditions.

Different kinds of chemical reactors were used, including the batch reactors, continuously stirred flow-through reactors, fluidized-bed reactors and plug-flow reactors.

When results of dissolution laboratory experiments are taken into account, it is important to distinguish the *release rate* [$r_{r,M}$, moles of element m^{-2} s^{-1}], which represents the rate of release of an individual metal M from the dissolving mineral, from the *dissolution rate* [r_+, moles of mineral m^{-2} s^{-1}], which is based on the stoichiometric formula of the mineral. The two rates are related through the simple relation $r_+ = r_{r,M}/v_M$, where v_M is the stoichiometric coefficient of element M in the dissolving mineral.

In recent years, macroscopic dissolution/precipitation experiments have been complemented with microscopic techniques, such as atomic force microscopy (AFM), first introduced by Binning et al. (1986), and optical interferometry (OI), which allows to observe the dynamic evolution of dissolving/precipitating mineral surfaces under water. AFM has been applied to several minerals, including quartz, feldspars, zeolites, clays, oxides and calcite. Applications of OI comprise anorthite, dolomite and calcite.

6.5.1. Experimental apparatuses

The *batch reactor* is a stirred tank containing the mineral under investigation and a known volume V of aqueous solution. The tank can be closed or open to the atmosphere. During the dissolution experiment, the molal concentration of a relevant solute product i, m_i, is monitored as a function of time, t. The dissolution rate is then computed based on the dm_i/dt ratio using the expression

$$r_+ = \frac{dm_i}{dt} \cdot \frac{V}{v_i \cdot A_S \cdot M_S}, \tag{6-90}$$

where A_S and M_S are the specific surface area (m^2 g^{-1}) and the mass (g) of the dissolving solid phase and v_i the stoichiometric coefficient of solute i in the bulk dissolution reaction, expressing the moles of solute i released by the dissolution of 1 mole of solid.

Both the mass and the specific surface area of the dissolving mineral change during the experiment and these changes may be important sources of uncertainty. Dissolution rates should be calculated with respect to the final A_S and M_S, but often only the initial values of these parameters are reported and used to normalize the dissolution rates.

If needed, analytical data are corrected for the removal of aqueous solution, which is repeatedly sampled for chemical analysis. If the concentration of solute i approaches a linear change with time during the dissolution experiment or at least a part of it, the dm_i/dt ratio is readily computed through linear regression.

Since dissolution reaction provides solutes to the aqueous phase, this may become oversaturated with respect to one or some secondary solid phases, whose precipitation can affect the experimental results. This is one of the main limitations of batch experiments. In some cases, the large variations in pH and in the concentrations of key solutes, e.g. Al, complicate the interpretation of results. Changes in pH may be avoided or at least minimized by using suitable buffers, but these may interfere with the dissolution mechanism and influence the reaction rate.

The *continuously stirred flow-through reactor* or *mixed-flow reactor* (Chou and Wollast, 1984) is a tank of known volume, which is kept under agitation by a propeller (Fig. 6.10). The aqueous solution (or in general the fluid phase) is continuously pumped through the tank at a constant flow rate, Q. The molal concentrations of some relevant solutes, produced by dissolution of the mineral phase, are monitored both in the inlet solution, m_{in}, and in the outlet solution, m_{out}. The dissolution experiment is run until m_{out} attains a constant value, or steady-state value, which is used to compute the steady-state dissolution rate through the equation

$$r_+ = \frac{Q \cdot (m_{out} - m_{in})}{v \cdot A_S \cdot M_S}. \tag{6-91}$$

If the inlet pH is significantly different from the outlet pH, the latter is considered to be representative of the system. Also in this case, variations of A_S and M_S during the dissolution

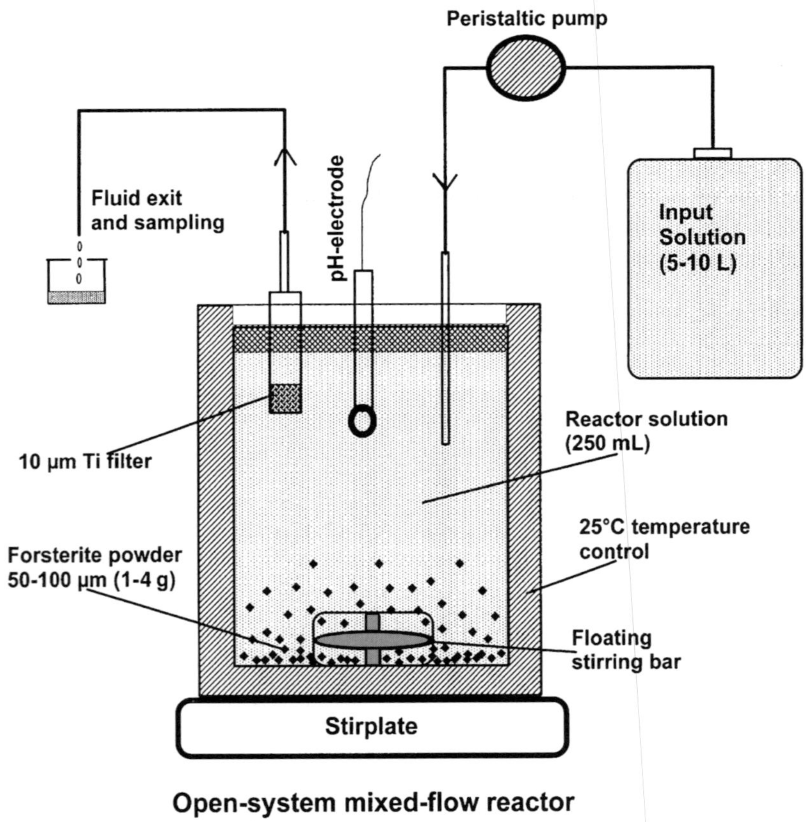

Figure 6.10. Schematic diagram of a continuously stirred flow-through reactor or mixed-flow reactor. (Reprinted from Pokrovsky and Schott, 2000b, Copyright 2000 with permission from Elsevier.)

run may determine significant uncertainties on the dissolution rates, which are often computed with respect to the initial A_S and M_S due to unavailability of the final values of these parameters.

The continuous renewal of the aqueous phase in the tank helps in maintaining the far-from-equilibrium conditions with the dissolving solid and avoiding precipitation of secondary solid phases. These are the main advantages of this kind of reactors with respect to the batch reactors.

In the *fluidized-bed reactor* there are two fluid flows: one is very fast to keep the tank under agitation and the dissolving solid particles in suspension and the other is monitored for m_{in} and m_{out}, as in the continuously stirred flow-through reactor. Again, equation (6-91) is used to obtain the dissolution rate of the solid phase.

A *plug-flow reactor* or *flow-through packed bed reactor* consists of a column containing the solid material under investigation, e.g. a sand made up of albite grains (Hajash et al., 1998). The aqueous solution is pumped through the reactor at a known flow rate and effluent fluid chemistry is monitored. In principle, for a given flow rate and inlet fluid composition and for constant mass and specific surface area of the dissolving mineral, a steady-state composition is attained in the effluent solution (Ortoleva et al., 1987; Lichtner, 1991). For these conditions, the dissolution rate is obtained by means of equation (6-91). Implicit assumptions are (i) far-from-equilibrium dissolution of the solid; (ii) ideal plug-flow without dispersion; (iii) constant porosity through the whole experiment and (iv) constant dissolution rate along the axis of the column, although a concentration gradient may establish along the reactor. Again, changes in the mass and the specific surface area of the dissolving mineral represent the main source of uncertainty in the obtained dissolution rate.

6.5.2. The surface area of solid reactants

As discussed above, the specific surface area of minerals, i.e. the surface area exposed at the solid–liquid interface by the unit mass of the solid, is a very important parameter for the correct interpretation of dissolution experiments, as dissolution rates depend on it.

There are two distinct approaches for obtaining the surface area of minerals, consisting in the use of either (i) gas sorption determination (BET method) or (ii) simple geometric calculations based on the idealized geometry of mineral grains.

6.5.2.1. The BET method

The acronym BET derives from the names of Brunauer, Emmett and Teller, who proposed this popular method for determining the specific surface area of fine-grained solid reactants (Brunauer et al., 1938). The BET method comprises, (i) the desorption of all gases from the solid surface through heating at 373 K in a vacuum and (ii) the measurement of gas sorption onto the dried solid grains at constant temperature (77 K) and variable gas pressure. The obtained gas sorption isotherm is then used to determine the surface area of the sample, based on the assumption that only a single layer of gas is sorbed during step (ii).

The simplicity of the experimental technique, its reproducibility and its speed (especially if a specialized BET equipment is used) have contributed to its popularity. Usually nitrogen or krypton are used in the BET method, as their diameters (~4–5 Å) are close to that of water (~3 Å). Thus, the BET surface area is a reasonable evaluation of the wetted surface area. However, N_2 and Kr molecules may not penetrate pores where smaller molecules or ions can easily enter. Again, the BET surface area of the starting material (before dissolution) is commonly used to normalize the result of kinetic dissolution experiments, although the surface area may vary significantly during these experiments.

6.5.2.2. The geometric approach

Suitable geometric shapes can be assumed as simple analogues of mineral grains and, consequently, surfaces of mineral grains can be obtained by simple geometric computations (Oelkers, 2002).

For instance, the surface area A_s and the volume V_s of a sphere of radius r_s are calculated from

$$A_s = 4\pi r_s^2 \tag{6-92}$$

and

$$V_s = \frac{4}{3}\pi r_s^3 . \tag{6-93}$$

The surface area A_c and the volume V_c of a cylinder of radius r_c and height h_c are computed from

$$A_c = 2\pi r_c^2 + 2\pi r_c \cdot h_c \tag{6-94}$$

and

$$V_c = \pi h \cdot r_c^2 . \tag{6-95}$$

Recalling the definition of density ($\rho = M/V$ where M is mass), we can now compute the specific surface area a_s for the sphere of radius r_s:

$$a_s = \frac{A_s}{M_s} = \frac{3}{r_s \cdot \rho_s} \tag{6-96}$$

and the specific surface area a_c for the cylinder of radius r_c and height h_c

$$a_c = \frac{A_c}{M_c} = \frac{2}{\rho_c} \cdot \left(\frac{1}{h_c} + \frac{1}{r_c} \right). \tag{6-97}$$

Assuming a density of 2.5 g cm^{-3}, the specific surface area for a sphere of radius 0.1 cm (a typical sand) is 12 cm^2 g^{-1}, whereas for a clay-sized sphere with r_s = 1 μm = 0.0001 cm, the specific surface area is 12,000 cm^2 g^{-1}, i.e. one thousand times larger than that of the sand-sized sphere.

For cylindrical grains, the specific surface area is strongly dependent on the h_c/r_c ratio, as indicated by equation (6-97). Let us consider a sand-sized cylinder with the same volume (4.19×10^{-3} cm^3) of the sand-sized sphere of 0.1 cm radius. The specific surface areas are as follows (Fig. 6.11): (i) 34 cm^2 g^{-1} for $h_c/r_c = 100$, i.e. for a long cylinder representing the analogue of a fibrous mineral; (ii) 14.5 cm^2 g^{-1} for $h_c/r_c = 1$, a value close to that computed for the spherical shape (12 cm^2 g^{-1}, see above); (iii) 158 cm^2 g^{-1} for h_c/r_c = 0.01, i.e. for a flat cylinder representative of the tabular habit typical of most phyllosilicates.

For a mixture of several grains of different size, and/or shape, and/or density, the specific surface area is equal to the weighted sum of each type of grain:

$$a_t = \sum_i x_{w,i} \cdot a_i, \tag{6-98}$$

where $x_{w,i}$ and a_i are the weight fraction and the specific surface area of the ith type of grain. Based on previous considerations, small minerals and grains of heterogeneous habit

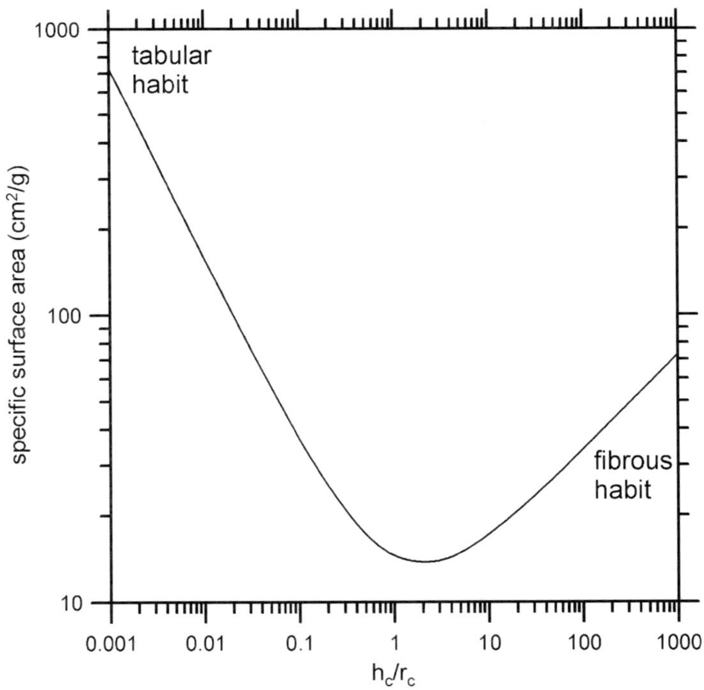

Figure 6.11. Change in the specific surface area of a cylinder (with the same volume, 4.19×10^{-3} cm^3, of a sphere of 0.1 cm radius) as a function of the height-to-radius ratio (h_c/r_c). (Reprinted from Oelkers, 2002, Copyright 2002 DipTeRis, Università di Genova, adapted.)

(either fibrous or tabular) are expected to affect strongly the specific surface area of the mixture.

Usually, dissolution experiments are carried out on size fractions obtained through sieving. For a given size fraction, therefore, only the maximum and minimum particle size, d_{max} and d_{min}, respectively, are known. Assuming that particle distribution is homogeneous, the so-called effective particle diameter d_{eff} is computed from (Tester et al., 1994)

$$d_{eff} = \frac{d_{max} - d_{min}}{\ln\left(d_{max}/d_{min}\right)}. \tag{6-99}$$

If grains have spherical shape, the effective particle diameter can be used in conjuction with equation (6-96) to calculate the "geometric" specific surface area.

6.5.2.3. The surface roughness

The application of the simple geometric approach to minerals is complicated by the usual deviation of natural grains from the ideal shapes. This deviation is taken into account through a parameter known as surface roughness, which is defined as the ratio between the true surface area and the corresponding geometric surface area of a hypothetical smooth surface encompassing the actual surface (Helgeson et al., 1984).

It is usually assumed that the true surface area is given by the gas sorption technique or BET method (see Section 6.5.2.1), whereas the geometric surface area is obtained based on size and geometry, which are derived from granulometric determinations and microscopic observations, respectively (see Section 6.5.2.2). Consequently, the surface roughness can be affected by poor knowledge of the shape and size distribution of mineral grains (Hodson et al., 1997). A further complication is brought about by internal porosity, which is the surface portion internal to the mineral grains or "out of sight" (Hochella and Banfield, 1995).

In spite of these complications, the surface roughness was found to be independent of particle size and equal to (i) 2.2 for crushed quartz grains and 6.2 for rounded natural quartz grains (Parks, 1990); (ii) 6.6 for albite and olivine (Brantley and Mellott, 1999); (iii) 7 for unweathered oxides, carbonates and silicates (White and Peterson, 1990).

Both the surface roughness and surface area of mineral grains increase dramatically with weathering, up to values of 130–2,600 for silicate minerals from glacial deposits (Anbeek, 1992) and 100–1,000 for soil grains (White and Peterson, 1990).

6.5.2.4. The reactive (effective) surface area

Both mineral grains from soils and those subjected to laboratory experiments indicate that the dissolution process affects the surface area to different extents, i.e. distinct portions of mineral surfaces dissolve with different kinetics. The formation of etch pits is a result of this differential dissolutive process. Based on this evidence, the mineral surface area actively involved in dissolution reaction has been termed effective surface area (e.g. Helgeson et al., 1984) or reactive surface area (e.g. Hochella and Banfield, 1995).

It was suggested that the reactive surface area is constituted by high-energy surface sites, such as dislocations or defects, but this suggestion was not confirmed by experimental studies on rutile (Casey et al., 1988), calcite (Schott et al., 1989) and sanidine (Murphy, 1989). This makes ambiguous and of doubtful utility the reactive surface area concept (Oelkers, 2002).

6.6. Dissolution and precipitation rates of silicates and silica minerals

6.6.1. Dissolution rates of nesosilicates and sorosilicates

From the point of view of dissolution kinetics, the nesosilicates group is extremely heterogeneous, since it includes both olivine and zircon, which are the least resistant and the most resistant silicate materials, respectively, to weathering (Velbel, 1999). Other nesosilicates have intermediate behaviour. Indeed, by comparing their relative abundance in weathered soil horizons and in the underlying bedrock, the most common nesosilicate minerals can be arranged in the following order of decreasing weathering rate (Velbel, 1999 and references therein):

olivine > garnet > staurolite > kyanite > andalusite > sillimanite > zircon.

The lattice of nesosilicate minerals is constituted by isolated SiO_4 tetrahedra. In other words, in their structure there is no corner-sharing of silica tetrahedra. The lack of covalent silicate framework makes these minerals ideal for investigating the similarity between ligand exchange reactions and silicate dissolution, as underscored by Casey and Ludwig (1995) and references therein. Indeed, the dissolution rates of endmember nesosilicates, under fixed pH and temperature conditions, are strongly correlated on a log–log scale, with the first-order rates of water exchange around the corresponding hexa-hydrated cations (Fig. 6.12). In particular, the dissolution rates of the nesosilicates of alkaline-earth metals depend on the ionic radius of the cation, with Ca^{2+} that is released more quickly than Mg^{2+} and Be^{2+} (Casey and Westrich, 1992). Similarly, the dissolution rates of the nesosilicates of first-row transition metals depend on the number of electrons in the 3d orbital, with Mn^{2+} that is removed more quickly than Fe^{2+}, Co^{2+} and Ni^{2+}.

Moreover, the dissolution rates of some olivine-group nesosilicates (i.e. forsterite, Ca-olivine, monticellite, tephroite, Co-olivine and Co-Mn-olivine) have similar pH dependencies, in spite of their different reactivities, as shown in Fig. 6.13 (Casey and Ludwig, 1995).

Present knowledge of the dissolution kinetics of nesosilicates and sorosilicates is highly variable. On the one hand, forsterite has been the subject of many laboratory experimental studies (see Section 6.6.1.1). On the other hand, few investigations were devoted to establish the dissolution rates of other minerals, such as fayalite, garnets, staurolite, kyanite and epidote.

6.6.1.1. Forsterite

Many laboratory investigations were carried out to understand the dissolution kinetics of forsteritic olivine (Luce et al., 1972; Sanemasa et al., 1972; Grandstaff, 1980, 1986; Blum and Lasaga, 1986; van Herk et al., 1989; Sverdrup, 1990; Wogelius and Walther, 1991, 1992; Westrich et al., 1993; Jonckbloedt, 1998; Xiao et al., 1999; Awad et al.,

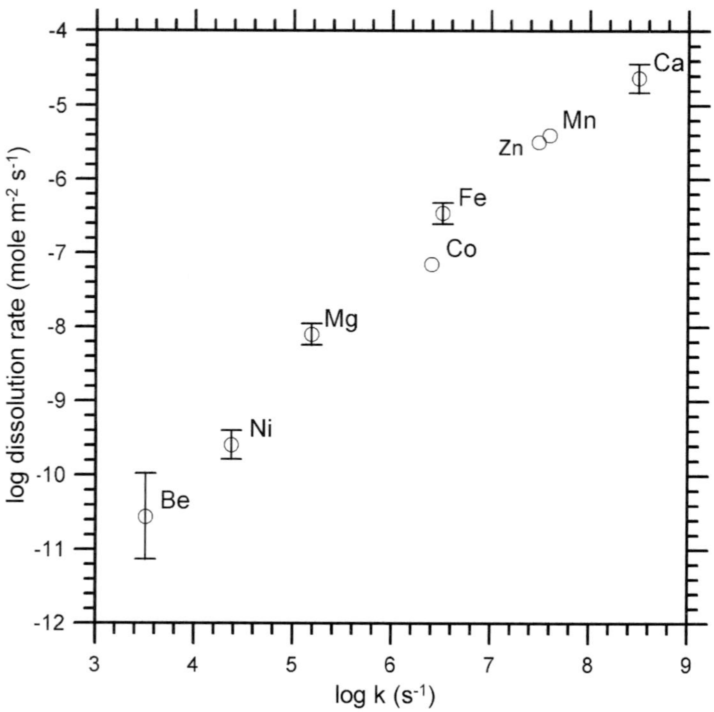

Figure 6.12. The dissolution rates of endmember nesosilicates, at pH 2 and 25°C, vs. the first-order rate of water exchange around the corresponding hexa-hydrated cations. (Reprinted from Westrich et al., 1993, Copyright 1993 American Journal of Science, adapted.)

2000; Chen and Brantley, 2000; Pokrovsky and Schott, 2000a,b; Rosso and Rimstidt, 2000; Oelkers, 2001b; Golubev et al., 2005).

Most of these studies have recognized that the dissolution of forsterite at 25°C and pH lower than 7 is congruent and the dissolution rate increases with decreasing pH, although different values of the apparent reaction order with respect to a_{H^+}, n_H, were established by different authors. For instance, n_H was found to be close to 1 by Grandstaff (1980, 1986), Murphy and Helgeson (1987) and Sverdrup (1990), whereas values of n_H close to 0.5 were recognized by Luce et al. (1972), Blum and Lasaga (1988), Wogelius and Walther (1991), Westrich et al. (1993), Pokrovsky and Schott (2000b) and Rosso and Rimstidt (2000). It must be noted that the investigation of Rosso and Rimstidt (2000) differs from other studies for the large number of produced rate measurements, 772, of which 284 were discarded to eliminate analytical errors. Based on the remaining 488 data, Rosso and Rimstidt (2000) were able to constrain the value of n_H at 0.50 ± 0.004 for the investigated ranges of pH (1.8–3.8) and temperature (25–45°C).

Conflicting data were obtained for the dissolution kinetics in basic solutions. No pH dependence was observed by Pokrovsky and Schott (2000b), whereas Blum and Lasaga (1986) and Wogelius and Walther (1991) found that the dissolution rate of forsterite

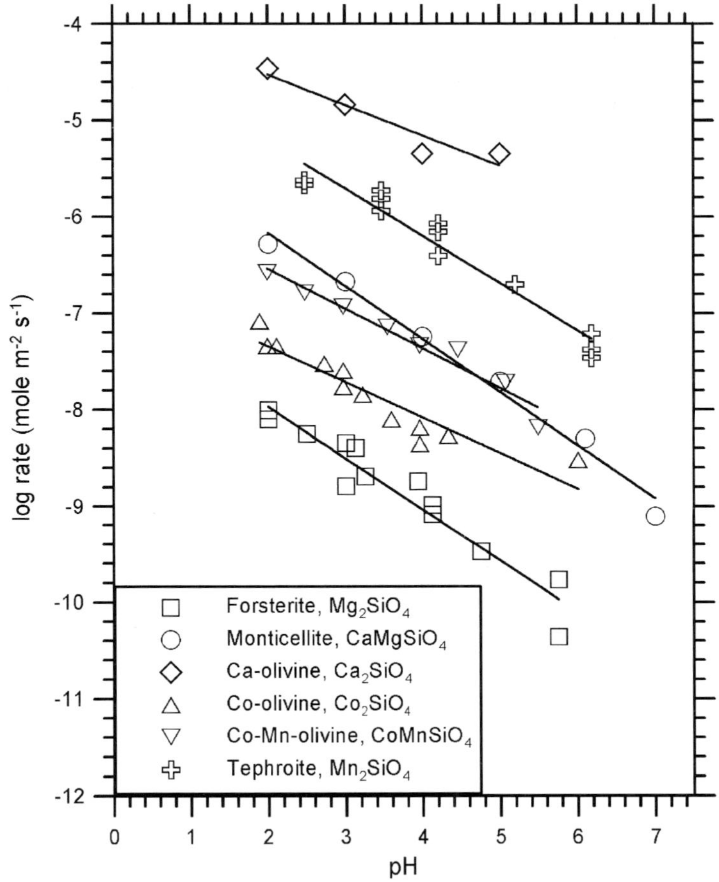

Figure 6.13. pH dependence of the logarithms of the dissolution rates of six olivine-group minerals (i.e. forsterite, monticellite, Ca-olivine, Co-olivine, Co-Mn-olivine and tephroite. In spite of their different reactivity, the slopes of log-rate vs. pH lines vary from −0.5 to −0.6. (Reprinted from Westrich et al., 1993, Copyright 1993 American Journal of Science, adapted.)

increases with pH in the basic region with a slope close to 1/3. To explain this discrepancy, Pokrovsky and Schott (2000b) noted that their dissolution experiments attained steady state, whereas those of Blum and Lasaga (1986) and Wogelius and Walther (1991) did not.

Pokrovsky and Schott (2000b) showed that the initial steps of forsterite dissolution are non-stoichiometric with preferential release of either Mg below pH 8 or Si above pH 10. Nevertheless, stoichiometric release of both Mg and Si is achieved after either few hours in acidic solutions or few days in high-pH solutions. They observed that, in the pH range 3–5, the increase in the ionic strength of the aqueous solution (from 0.001 to 0.01 M) brings about a 2-fold increase in the dissolution rate of forsterite.

As shown by Pokrovsky and Schott (2000b), the kinetics of forsterite dissolution is affected by aqueous silica concentration and is independent of aqueous Mg concentration at pH 8.9 and 11.2 and 25°C. In contrast, at pH 2 and 25°C, forsterite dissolution rates are independent of both dissolved Mg and SiO_2 concentrations, as documented by Oelkers (2001b). These findings at pH 2 were interpreted by Oelkers (2001b) as an evidence that breaking of Mg–O bonds simultaneously liberates Mg and Si from the forsterite lattice, which comprises isolated SiO_4 tetrahedra linked together by MgO_6 octahedra. As already recalled in Section 6.3.2, the SiO_4 tetrahedra are totally liberated through breaking of the Mg–O bonds and no Si–O bond has to be broken for completing forsterite dissolution.

Through an exhaustive study of the forsterite surface in contact with aqueous solutions, Pokrovsky and Schott (2000a) documented that the Mg/Si ratio at the forsterite–solution interface is always different from the stoichiometric ratio of the bulk mineral.

At pH < 9 a Si-rich, Mg-depleted altered surface layer forms, not exceeding 10–20 Å in thickness. The exchange reaction of two H^+ ions for one Mg^{2+} ion, followed by the condensation of two Si surface sites, was considered to be responsible for the formation of this surface layer. Proton penetration into this leached layer and its adsorption on the bridging oxygen of this Si-dimer generate the activated complex whose decomposition controls forsterite dissolution, according to Pokrovsky and Schott (2000b).

In contrast, at pH > 10, a Mg-rich, Si-depleted layer forms on the forsterite surface (Pokrovsky and Schott, 2000a) and forsterite dissolution is governed by the hydrolysis of Mg surface groups (Pokrovsky and Schott, 2000b).

The occurrence of these two parallel reactions at Si-rich surface sites and hydrated Mg surface groups is adequately described by the following rate equation (rate in mol m^{-2} s^{-1}):

$$r_+ = 2.38\times10^{-7}\cdot\left[\equiv Si_2O-H^+\right]+1.62\times10^{-6}\cdot\left[\equiv MgOH_2^+\right] \tag{6-100}$$

where concentrations of surface species (indicated by brackets) are in mol m^{-2}. The concentration of $\equiv Si_2O - H^+$ is computed by means of the following equation:

$$\left[\equiv Si_2O-H^+\right]=2.2\times10^{-5}\cdot\frac{K^*_{ads}\cdot a^{0.5}_{H^+}}{\left[1+K^*_{ads}\cdot a^{0.5}_{H^+}+a^2_{Mg^{2+}}/\left(K^*_{ex}\cdot a^4_{H^+}\right)\right]} \tag{6-101}$$

where K^*_{ex} and K^*_{ads} are the apparent stability constants of the H^+/Mg^{2+} exchange reaction and of the H^+ adsorption reaction, respectively. Their values are $\geq 10^{22}$ and ≤ 2.0, respectively. These two *K*s are valid only at constant I of 0.1 M. Equation (6-101) is easily derived introducing these two stability constants into the mass balance on the surface sites initially occupied by Mg, whose total concentration is assumed to be 2.2×10^{-5} mol m^{-2}. In contrast, calculation of the concentration of $\equiv MgOH_2^+$ requires a suitable surface complexation model (SCM), e.g. the constant capacitance model, and use of the MINTEQA2 code (Allison et al., 1991) or an equivalent code (see Pokrovsky and Schott, 2000b for further details).

Hydrolysis of Mg surface groups can explain the CO_3^{2-} inhibition of forsterite dissolution via generation of the $\equiv MgCO_3^-$ surface complex. Carbonate ion inhibition was detected by Pokrovsky and Schott (2000b) in basic solution for $a_{CO_3^{2-}} > 10^{-4}$ M. However, no CO_3^{2-} inhibition was observed under the same experimental conditions by Golubev et al. (2005), who attributed the different results of Pokrovsky and Schott (2000b) to experimental uncertainties.

Internally consistent activation energy data for forsterite dissolution (internally consistent means that these data were estimated without the use of external data) were produced by Sanemasa et al. (1972) at pH 1, 58.6 kJ mol^{-1} in $HClO_4$ solution and 66.9 kJ mol^{-1} in H_2SO_4 solution; Grandstaff (1986) at pH 3.25, 38.1 kJ mol^{-1}; Wogelius and Walther at pH values of 1.8–6, 79.5 ± 10.5 kJ mol^{-1}; Rosso and Rimstidt (2000) for the pH range 1.8–3.8, 42.6 ± 0.8 kJ mol^{-1}; and Oelkers (2001) at pH 2, 63.8 ± 17 kJ mol^{-1}. Based on literature data and their results at pH values of 2–5 and a constant temperature of 65°C, Chen and Brantley (2000) obtained a much higher activation energy of 126 ± 17 kJ mol^{-1}.

Selected measurements of the dissolution rate of forsterite (from Wogelius and Walther, 1992; Westrich et al., 1993; Chen and Brantley, 2000; Pokrovsky and Schott, 2000b; Rosso and Rimstidt, 2000; Oelkers, 2001b; Golubev et al., 2005) are plotted against pH in Fig. 6.14. Most of these data refer to olivines containing over 90 mol% of forsterite. The dissolution rates at 25 and 65°C obtained by summing the contributions of the acidic and neutral mechanisms reported by Palandri and Kharaka (2004, see Table 6.1) are also represented in Fig. 6.14. Inspection of Fig. 6.14 shows that the acidic dissolution mechanism prevails at pH less than 7.5–8, depending on temperature. No basic dissolution mechanism is recognizable, also due to the large scatter of the high-pH data. Figure 6.14 also shows that the dissolution rate of forsterite is significantly affected by temperature; following Palandri and Kharaka (2004), apparent activation energies were assumed to be 67 kJ mol^{-1} for the acidic dissolution mechanism and 79 kJ mol^{-1} for the neutral dissolution mechanism (Table 6.1).

Finally, it is worth recalling that forsterite does not dissolve with the same rate along the three crystallographic axes (Awad et al., 2000). This fact introduces further complications in forsterite dissolution.

6.6.1.2. Kyanite

Phase relations and reaction rates among the Al_2SiO_5 polymorphs andalusite, kyanite and sillimanite are important for the geothermometry and geobarometry of metamorphic rocks. Consequently, these subjects were extensively investigated at high temperatures and pressures. Much less is known on the rate of dissolution of these minerals at low temperatures and pressures.

Oelkers and Schott (1999) measured the dissolution rates for kyanite at pH from 1.6 to 2.2 and temperatures from 108 to 194°C. They found that dissolution was stoichiometric and that rates were linearly dependent on $a_{Al^{3+}}^{-0.5}$ and $a_{H^+}^{1.5}$. They attributed this behaviour to the destruction of a neutral, Al-deficient activated complex at the mineral surface and fitted the obtained data to the following TST, Arrhenius-type equation:

$$r_+ = A_a \cdot \exp\left(-\frac{E_a}{RT}\right) \cdot \left(\frac{a_{H^+}^3}{a_{Al^{3+}}}\right)^{0.5}, \qquad (6\text{-}102)$$

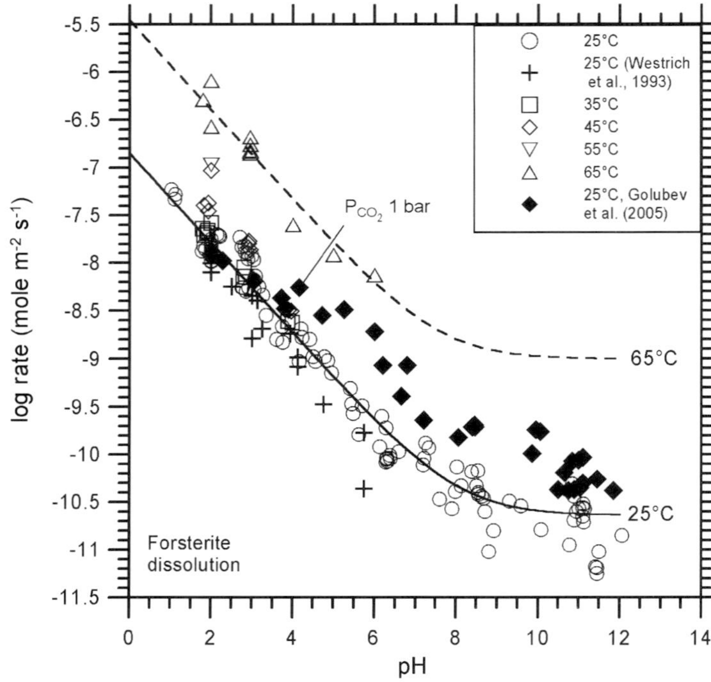

Figure 6.14. Plot of the logarithm of the dissolution rate of forsterite vs. pH, showing selected experimental data (from Wogelius and Walther, 1992; Westrich et al., 1993; Chen and Brantley, 2000; Pokrovsky and Schott, 2000b; Rosso and Rimstidt, 2000; Oelkers, 2001b; Golubev et al., 2005), as well as the dissolution rate obtained by summing the contributions of the acidic and neutral mechanisms reported by Palandri and Kharaka (2004, see Table 6.1).

where the pre-exponential factor A_a is equal to 0.2 mol m^{-2} s^{-1} and the apparent activation energy E_a is equal to 75 kJ mol^{-1}.

The experimental data by Oelkers and Schott (1999) are shown in the log-rate vs. pH plot of Fig. 6.15 together with the dissolution rates at 100, 150 and 200°C obtained by summing the contributions of the acidic and neutral mechanisms reported by Palandri and Kharaka (2004, see Table 6.1). Of course, only the rate parameters of the acidic mechanism are constrained with sufficient precision, whereas those of the neutral mechanism are highly speculative.

6.6.1.3. Epidote

The dissolution kinetics of epidote was investigated under high-temperature (250–350°C) and high-pressure (0.5–12 kbar) conditions by several authors including Blanchard (1994) and Brandon et al. (1996). Experimental studies at ambient temperature and atmospheric pressure were performed by Nickel (1973), Sverdrup (1990), Rose (1991) and Kalinowski et al. (1998).

There is a quite satisfactory agreement (±0.5 log-unit) between the low-temperature dissolution rates of Kalinowski et al. (1998) and those of Nickel (1973) and Rose

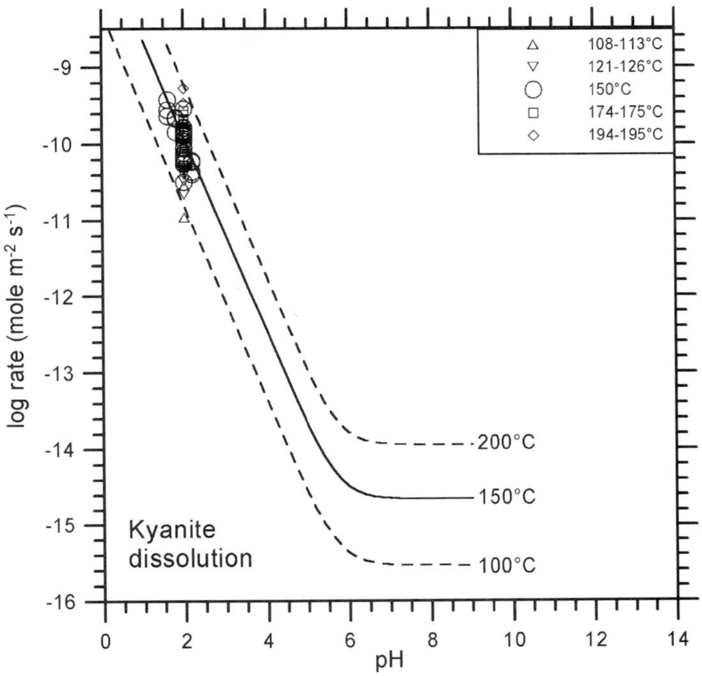

Figure 6.15. Log-rate vs. pH plot for kyanite dissolution (experimental data by Oelkers and Schott, 1999), also showing the dissolution rate obtained by summing the contributions of the acidic and neutral mechanisms reported by Palandri and Kharaka (2004, see Table 6.1).

(1991). On the contrary, the data of Sverdrup (1990) are comparable with the other ones at circumneutral pH but are significantly higher at low pH values, indicating an apparent H^+ reaction order close to 0.8 (Fig. 6.16). Use of base cations (cumulative concentration of Al^{3+} + Fe^{3+} + Ca^{2+}) for computing the dissolution rate of epidote by Sverdrup (1990) might explain these discrepancies. Indeed, Kalinowski et al. (1998), investigating the dissolution kinetics of epidote of composition $Ca_{2.07}Al_2(Al_{0.211}Fe_{0.70})Si_3O_{12}OH$, found a preferential release of Ca and Si with respect to Al and Fe. They suggested the formation of a gibbsite-goethite-like residual layer on the epidote surface.

Available experimental rates are plotted against pH in Fig. 6.16, also showing the dissolution rates at 25 and 90°C obtained by summing the contributions of the acidic, neutral and basic mechanisms reported by Palandri and Kharaka (2004, see Table 6.1). The apparent H^+ reaction orders are $n_H = 0.34$ for the acidic mechanism and $n_{OH} = -0.56$ for the basic mechanism (Table 6.1). Inspection of this plot shows that the acidic dissolution mechanism prevails below pH ~4, whereas the basic mechanism dominates above pH 9–9.5. The dissolution rate of epidote is significantly influenced by temperature, in line with apparent activation energies of 71 kJ mol^{-1} for the acidic and neutral dissolution mechanisms and 79 kJ mol^{-1} for the basic mechanism (Table 6.1).

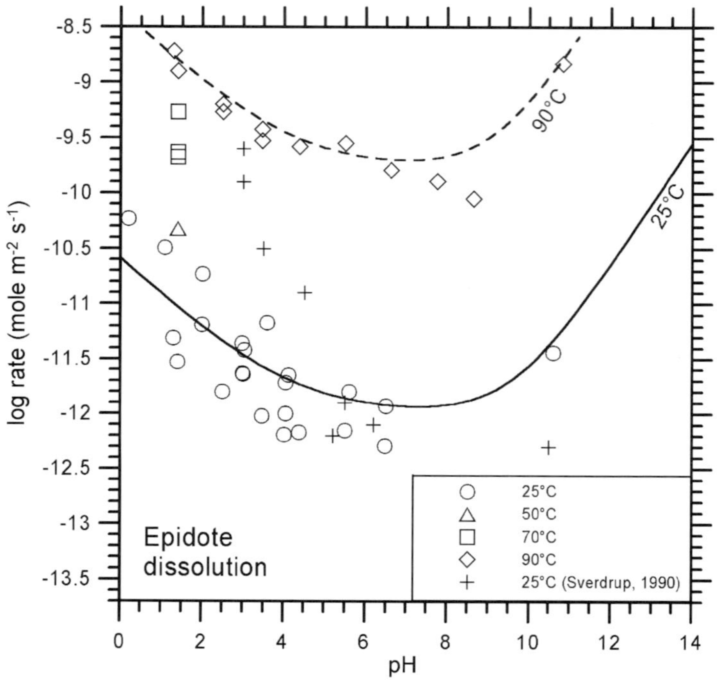

Figure 6.16. Plot of the logarithm of the dissolution rate of epidote vs. pH (experimental rates from Nickel, 1973; Sverdrup, 1990; Rose, 1991; Kalinowski et al., 1998), also showing the dissolution rate obtained by summing the contributions of the acidic, neutral and basic mechanisms reported by Palandri and Kharaka (2004, see Table 6.1).

6.6.2. Dissolution rates of chain silicates

The chain silicates considered in this section comprise the following: (i) pyroxenes, whose structure is constituted by continuous chains of composition $(SiO_3)_n$, which are obtained through linkage of SiO_4 tetrahedra by sharing two of the four corners; (ii) wollastonite, whose infinite-chain structure is made up of pairs of SiO_4 tetrahedra linked apex-to-apex alternating with a single SiO_4 tetrahedron with one edge parallel to the chain direction; (iii) amphiboles, whose structure is formed by $(Si,Al)O_4$ tetrahedra joined to constitute chains that have double the width of those in pyroxenes and have the composition $(Si_4O_{11})_n$.

Available determinations of the dissolution rates of orthopyroxenes are shown in Fig. 6.17. Data of enstatite are from Luce et al. (1972), Schott et al. (1981), Siegel and Pfannkuch (1984), Ferruzzi (1994) and Oelkers and Schott (2001). Dissolution of bronzite was measured by Grandstaff (1977) and Schott and Berner (1983).

Measured dissolution rates of the clinopyroxenes diopside and augite are depicted in Fig. 6.18. The kinetics of diopside dissolution was investigated by Schott et al. (1981), Eggleston et al. (1989), Knauss et al. (1993), Chen and Brantley (1998) and Golubev et al. (2005). The data of augite were measured by Siegel and Pfannkuch (1984), Schott and Berner (1985) and Sverdrup (1990).

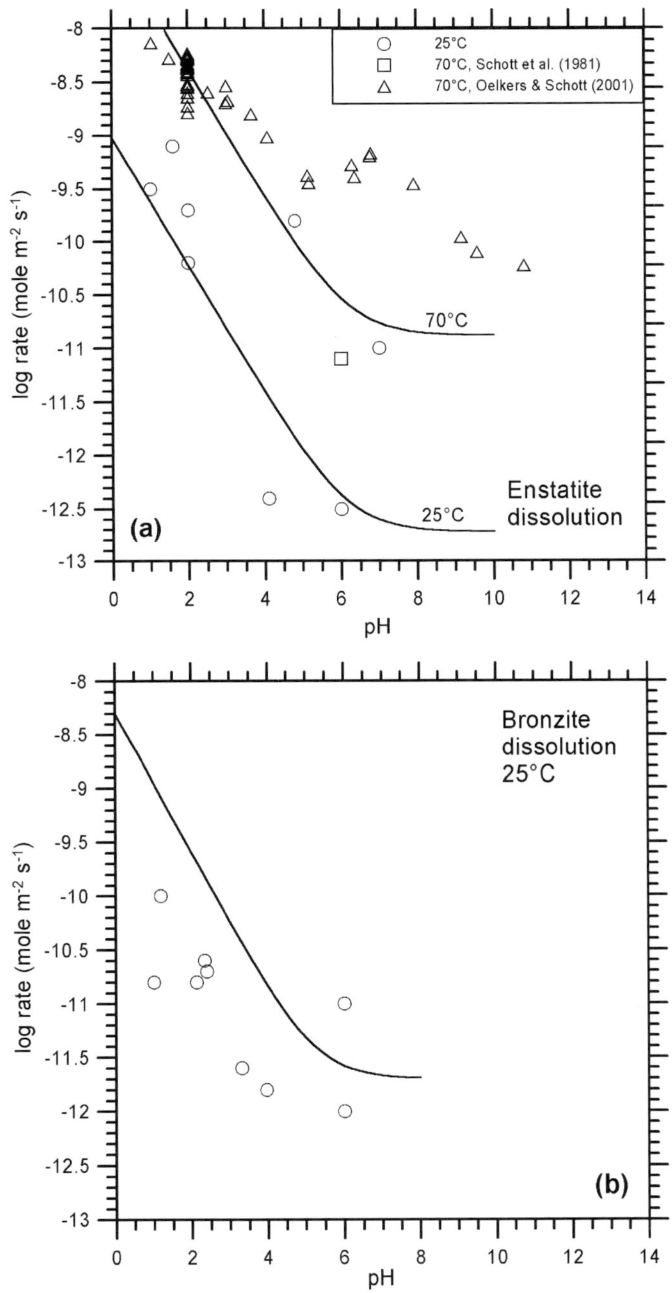

Figure 6.17. Log-rate vs. pH plot for (a) enstatite dissolution (data from Luce et al., 1972; Schott et al., 1981; Siegel and Pfannkuch, 1984; Ferruzzi, 1994; Oelkers and Schott, 2001) and (b) bronzite dissolution (data from Grandstaff, 1977; Schott and Berner, 1983), also showing the dissolution rates obtained by summing the contributions of the acidic and neutral mechanisms reported by Palandri and Kharaka (2004, see Table 6.1).

Figure 6.19 shows the rates of jadeite and spodumene dissolution (from Sverdrup, 1990) and the experimental determinations of the kinetics of wollastonite dissolution, which were performed by Rimstidt and Dove (1986), Sverdrup (1990), Ferruzzi (1994), Xie (1994), Xie and Walther (1994), Weissbart and Rimstidt (2000) and Golubev et al. (2005).

Published measurements of the dissolution rates of amphiboles are presented in Fig. 6.20. Dissolution of anthophyllite was investigated by Mast and Drever (1987) and Chen and Brantley (1998), whereas the data of hornblende dissolution are from Nickel (1973), Cygan et al. (1989), Sverdrup (1990), Zhang et al. (1993), Swoboda-Colberg and Drever (1993), Frogner and Schweda (1998) and Golubev et al. (2005).

Dissolution rate laws obtained by summing the contributions of the acidic and neutral mechanisms reported by Palandri and Kharaka (2004) are also shown in Figs. 6.17–6.20. Rate parameters are given in Table 6.1 for all the above-mentioned phases as well as for tremolite, glaucophane and riebeckite. It must be underscored that dissolution rates independent of pH were measured for wollastonite in the acid field by some authors (see below).

Although chain silicates are important contributors of Mg, Ca and Fe to natural waters, relatively few investigations on the weathering rates of these minerals have been carried out so far. Data are particularly lacking for redox-sensitive, common-chain silicates, such as augite and hornblende. In addition, the data presented by different researchers are frequently in conflict because of several reasons, including run duration, mineral composition, aging of powders, mineral surface area, thermodynamic affinity, effects of varying solution composition and non-stoichiometric dissolution, as discussed in detail by Brantley and Chen (1995). For instance, the dissolution rate law suggested by Palandri and Kharaka (2004) for hornblende at 25°C, mainly based on the data by Sverdrup (1990), is about four orders of magnitude higher than that constrained by the data of Frogner and Schweda (1998), as shown in Fig. 6.20b.

In spite of these uncertainties, the dissolution rates of Fe-free amphiboles, such as anthophyllite and tremolite, are lower than those of corresponding pyroxenes, i.e. enstatite and diopside, under the same pH and temperature conditions, in agreement with the qualitative weathering series proposed by Goldich in 1938 (Brantley and Chen, 1995).

In addition, the rate of Si release decreases with increasing connectedness, which expresses the number of bridging oxygen per SiO_4 tetrahedral unit (Liebau, 1985). This relationship agrees with the order of metal–ligand exchange predicted by Casey et al. (1993) and, therefore, can be used to predict unknown dissolution rates for endmember chain silicates, on the basis of the known rate values for nesosilicates, and to evaluate the reliability of experimental data (Brantley and Chen, 1995).

The reaction order with respect to H^+ ion, n_H, varies mostly between 0.2 and 0.8. However, wollastonite shows an intriguing behaviour (Fig. 6.19). Ferruzzi (1994) observed a slight increase in the dissolution rate with pH in the range 2–9.6 ($n_H = -0.15$) and reported non-stoichiometric dissolution in runs of 300–400 minutes duration at pH 2 and 4. Wollastonite dissolution at pH 2–6 was investigated again by Weissbart and Rimstidt (2000), who found release rates of both Si and Ca independent of pH and a quicker release of Ca with respect to Si. Based on these observations they suggested the occurrence of

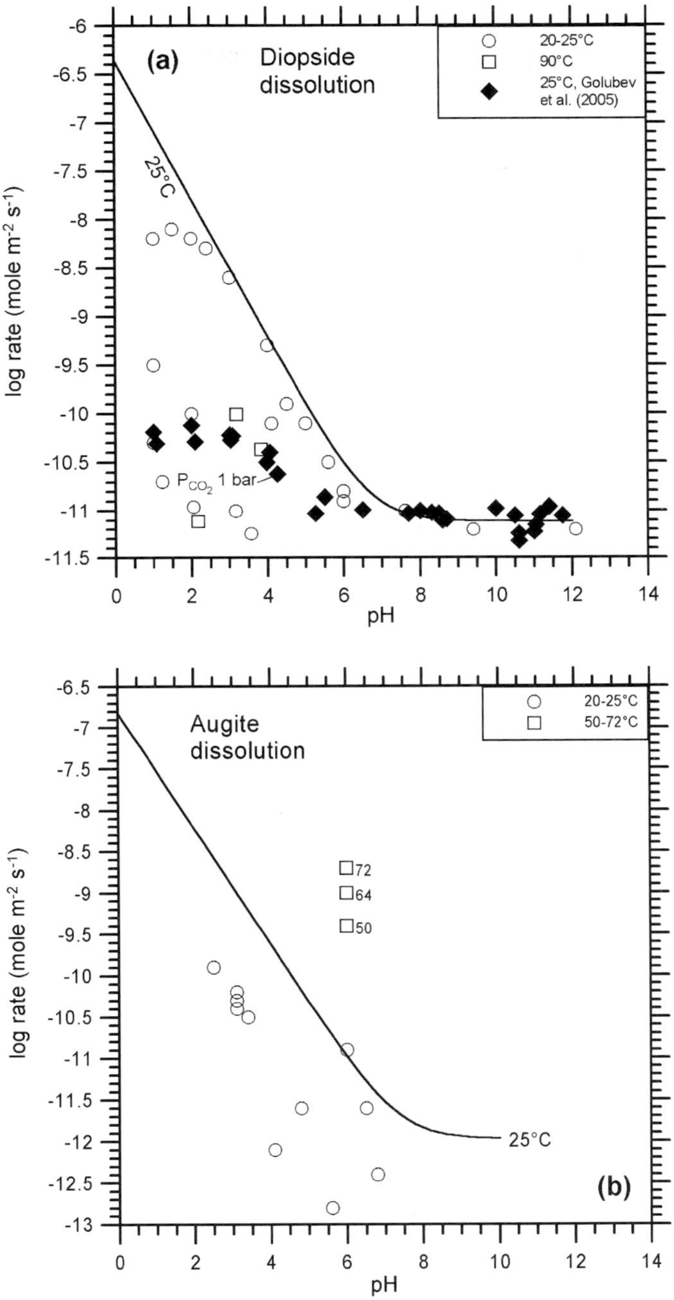

Figure 6.18. Log-rate vs. pH plot for (a) diopside dissolution (data from Schott et al., 1981; Eggleston et al., 1989; Knauss et al., 1993; Chen and Brantley, 1998) and (b) augite dissolution (data from Siegel and Pfannkuch, 1984; Schott and Berner, 1985; Sverdrup, 1990), also showing the dissolution rates obtained by summing the contributions of the acidic and neutral mechanisms reported by Palandri and Kharaka (2004, see Table 6.1).

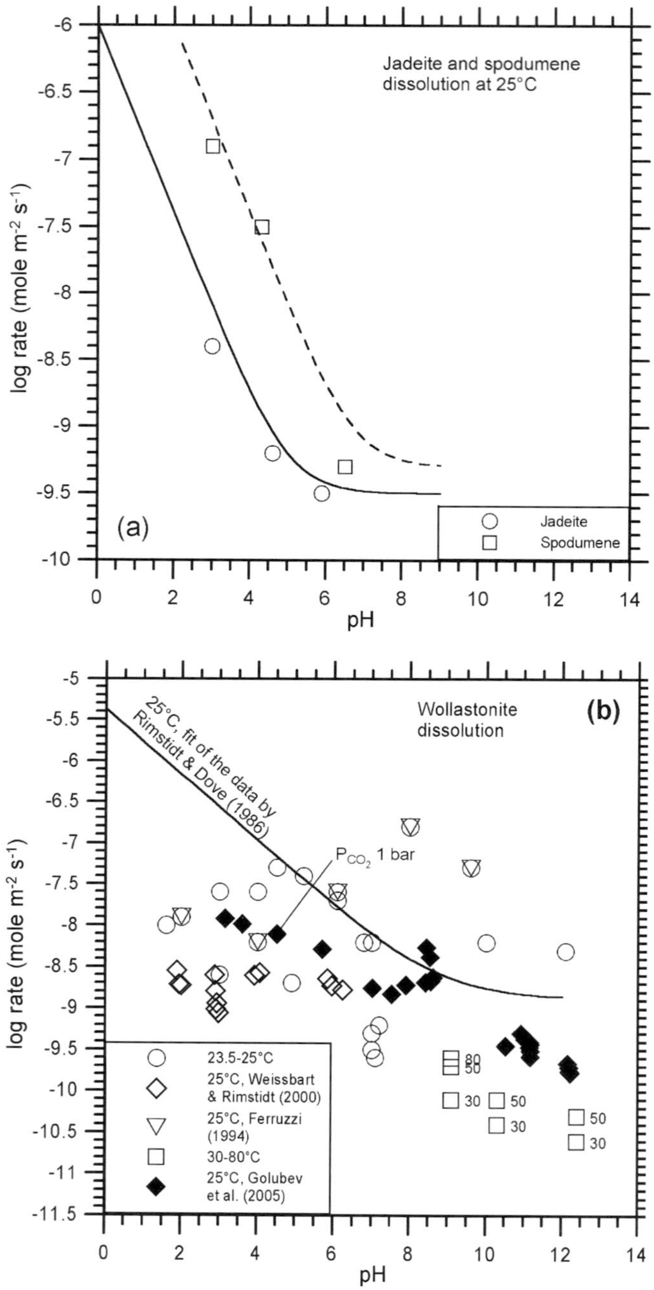

Figure 6.19. Log-rate vs. pH plot for (a) jadeite and spodumene dissolution (data from Sverdrup, 1990) and (b) wollastonite dissolution (data from Rimstidt and Dove, 1986; Sverdrup, 1990; Ferruzzi, 1994; Xie, 1994; Xie and Walther, 1994; Weissbart and Rimstidt, 2000; Golubev et al., 2005), also showing the dissolution rates obtained by summing the contributions of the acidic and neutral mechanisms reported by Palandri and Kharaka (2004, see Table 6.1).

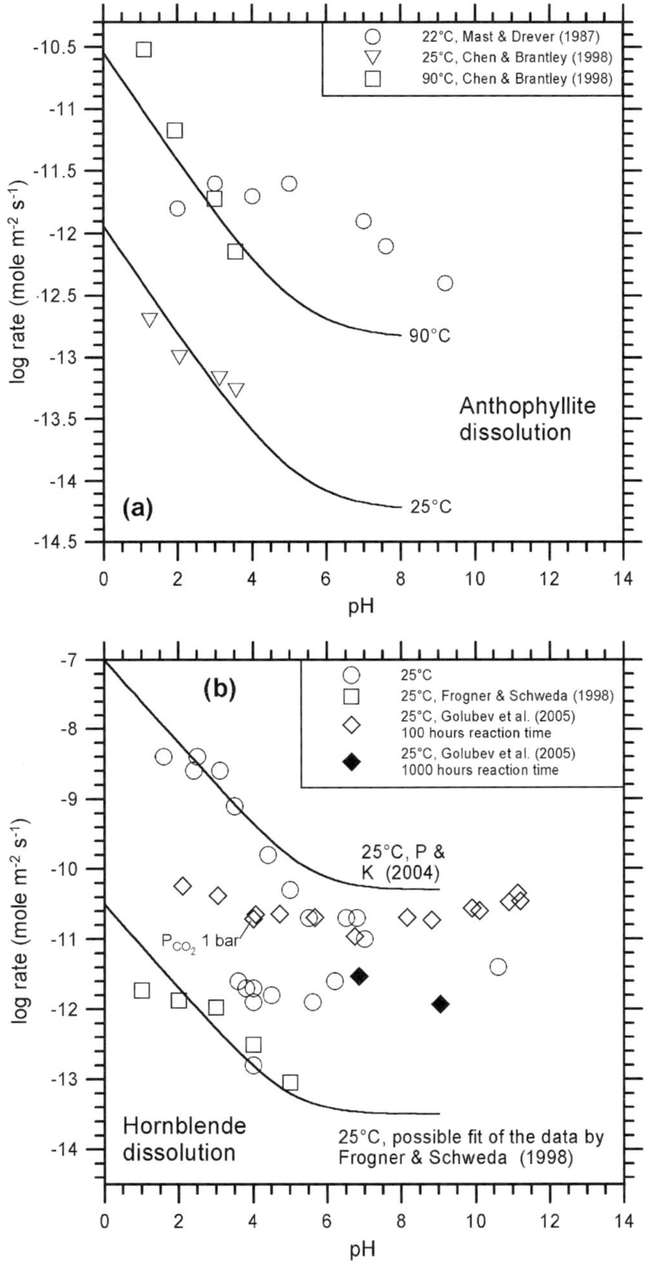

Figure 6.20. Log-rate vs. pH plot for (a) anthophyllite dissolution (data from Mast and Drever, 1987; Chen and Brantley, 1998) and (b) hornblende dissolution (data from Nickel, 1973; Cygan et al., 1989; Sverdrup, 1990; Swoboda-Colberg and Drever, 1993; Zhang et al., 1993; Frogner and Schweda, 1998; Golubev et al., 2005), also showing the dissolution rates obtained by summing the contributions of the acidic and neutral mechanisms reported by Palandri and Kharaka (2004, see Table 6.1) and a possible fit of the data by Frogner and Schweda (1998).

(a) exchange of H^+ for Ca^{2+} ion, which diffuses into the aqueous solution leaving behind a Si-rich leached layer; (b) silica polymerization in the leached layer, which evolves into a sort of less reactive vitreous silica (Casey et al., 1993); (c) dissolution of this leached layer through release of SiO_2 monomers and polymers to the aqueous phase. Note that the pH independence of the rate of silica release is not surprising, since the hydrolysis of silica does not involve H^+ ions, at least in the pH range 2–6 (see Section 6.6.5).

Preferential removal of cations from the M2 site over the M1 site in diopside and from the M4 site over the M1, M2 and M3 sites in tremolite has also been suggested by Schott et al. (1981). The extent of cation leaching increases with decreasing pH. Non-stoichiometric dissolution complicates and even prevents the evaluation of the dissolution rate.

The pH dependence of the dissolution rates of pyroxenes and amphiboles seems to decrease above about pH 7, whereas the dissolution rates in the alkaline field are poorly known. Also, the effects of solution chemistry, and in particular those of CO_2, need to be investigated.

The apparent activation energies given in Table 6.1 for dissolution of chain silicates vary between 47 and 132 kJ mol^{-1}, with the exception of tremolite, which has a value of 19 kJ mol^{-1} only. Apart from tremolite, the other values are in the typical range of most silicates, 50–120 kJ mol^{-1}.

6.6.2.1. Enstatite

Among chain silicates, whose dissolution rates are in general poorly known, enstatite is an exception. In fact, as already recalled in Section 6.4.4, its steady-state dissolution rate at far-from-equilibrium conditions was determined by Oelkers and Schott (2001) at pH from 1 to 11 and temperatures from 28 to 168°C. Stoichiometric dissolution was observed and rates were found to be linearly dependent on $a_{Mg^{2+}}^{-0.125}$ and $a_{H^+}^{0.25}$ (see Fig. 6.9b). These experimental kinetic data are described by the following TST, Arrhenius-type equation:

$$r_+ = A_a \cdot \exp\left(-\frac{E_a}{RT}\right) \cdot \left(\frac{a_{H^+}^2}{a_{Mg^{2+}}}\right)^{0.125}, \qquad (6\text{-}103)$$

where the pre-exponential factor A_a is equal to 2.4 mol m^{-2} s^{-1} and the apparent activation energy E_a is equal to 48.5 kJ mol^{-1}.

6.6.3. Dissolution rates of sheet silicates

6.6.3.1. Kaolinite

The dissolution rate of kaolinite under controlled temperature and pH conditions has been the subject of many investigations (e.g. Carroll-Webb and Walther, 1988; Carroll and Walther, 1990; Wieland and Stumm, 1992; Ganor et al., 1995; Devidal et al., 1997; Bauer and Berger, 1998; Huertas et al., 1999). Some selected results are shown in the log rate–pH plot of Fig. 6.21. Most rate data are consistent with the typical pH dependence of silicates. This behaviour is described by the dissolution rate laws at 25 and 80°C (also shown in Fig. 6.21), which were obtained by summing the contributions of the acidic, neutral and basic mechanisms reported by Palandri and Kharaka (2004). Corresponding rate parameters are listed in Table 6.1.

Kaolinite dissolution kinetics was investigated at pH 1–12 by Carroll-Webb and Walther (1988) and Carroll and Walther (1990) at temperatures of 25, 60 and 80°C, by means of batch experiments. Between pH 5 and 10, malonic acid or 2,4-pentanodione was added to keep Al in solution. Between pH 2 and 9, initial non-stoichiometric dissolution was followed by close to stoichiometric dissolution. The lowest rates were found between pH 7 and 10.

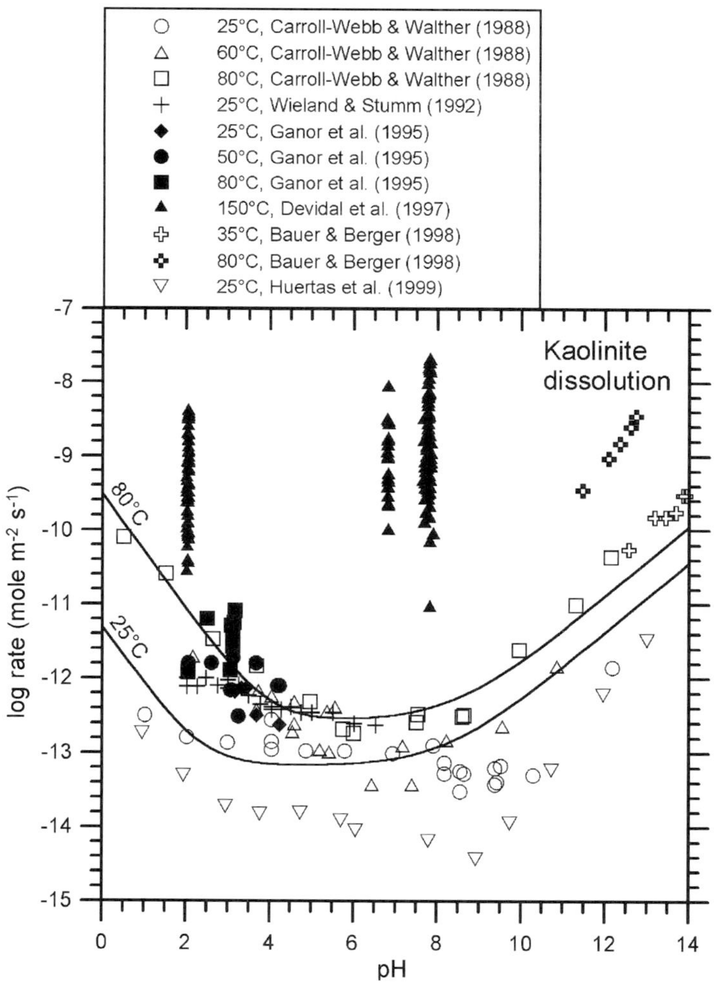

Figure 6.21. Plot of the logarithm of the dissolution rate of kaolinite vs. pH showing selected experimental data (see legend) at different temperatures and the dissolution rate obtained by summing the contributions of the acidic, neutral and basic mechanisms reported by Palandri and Kharaka (2004, see Table 6.1).

Wieland and Stumm (1992) measured kaolinite dissolution rates between pH 2 and 6.5 at 25°C, again through batch experiments, and found values 0.5–1 order of magnitude higher than those of Carroll-Webb and Walther (1988). Quite surprisingly, non-stoichiometric dissolution due to preferential release of Si was observed. However, dissolution was found to be stoichiometric in the presence of organic ligands (e.g. oxalate and salicylate) with enhanced rates.

A flow-through experimental setup was used by Ganor et al. (1995) to measure kaolinite dissolution rates at 25, 50 and 80°C, under far-from-equilibrium conditions. They observed stoichiometric dissolution.

Devidal et al. (1997) measured steady-state dissolution and precipitation rates of kaolinite at 150°C, 40 bar, and at pH 2, 6.8 and 7.8 as a function of thermodynamic affinity and solution composition by using a mixed-flow reactor. Dissolution experiments were carried out in aqueous solutions undersaturated with respect to all possible solid phases. Kaolinite dissolution was found to be stoichiometric after attainment of steady-state conditions. However, surface composition was found to be depleted in Al, through X-ray photoelectron spectroscopy (XPS) analysis, after dissolution at pH 2 and 7.8, suggesting the occurrence of initial preferential release of Al over Si. According to Devidal et al. (1997), this process may be related to the formation of a Si-rich, rate-controlling activated complex. This conclusion is also supported by the inverse relation between dissolution rates and dissolved Al concentration both at low and high pH and under different values of thermodynamic affinity. It must be underscored that the dissolution and precipitation rates determined by Devidal et al. (1997) exhibit a strongly non-linear dependence on thermodynamic affinity.

Kaolinite dissolution kinetics was investigated in 0.1–4 M KOH solutions at 35 and 80°C by Bauer and Berger (1998). Experiments were carried out in batch reactors. Dissolution rate was found to be dependent on the $Al(OH)_4^-/OH^-$ activity ratio.

Huertas et al. (1999) studied kaolinite dissolution at 25°C and at pH from 1 to 13 using batch reactors. Below pH 4 and above pH 11, stoichiometric kaolinite dissolution was observed after about 600 h, whereas kaolinite was found to dissolve incongruently between pH 5 and 10, due to precipitation of an Al-hydroxide mineral. The amount of Si released into solution was therefore used to compute the dissolution rate data reported in Fig. 6.21. In separate runs, very small amounts of CDTA (*trans*-1,2-diamino-cyclohexane-*N*,*N*,*N′*,*N′*-tetraacetic acid), an Al-complexing agent, were added to prevent precipitation of secondary minerals. Stoichiometric dissolution was found to occur. Owing to both the small concentrations of CDTA and its large size, which prevents formation of surface complexes, the dissolution rate was not enhanced significantly.

Further complications on the measurements of kaolinite dissolution rate are possibly due to stirring-induced spalling or abrasion of kaolinite grains, as suggested by Metz and Ganor (2001) and Cama et al. (2002).

Inspection of Fig. 6.21 shows that the rate data by Devidal et al. (1997) exhibit a large spread, which cannot be explained by simple pH dependence, even recognizing distinct acidic, neutral and basic dissolution mechanisms. This study deserves an especially detailed examination. According to Devidal et al. (1997), in acidic solution, dissolution rate data can be interpreted assuming the occurrence of a two-step process. The first step is the reversible surface exchange of H^+ for Al^{3+}, leading to the production of an Al-deficient, Si-rich activated complex (see Section 6.4.4). The second rate-limiting step is the irreversible hydrolysis of Si–O–Si groups.

At first, Devidal et al. (1997) tried to fit the experimental data to a modified form of equation (6-84), with the thermodynamic affinity term equal to $[1-A^{0.5}]$, $n = 1$, $K_{\text{Al–O}} = 3.83$, and $k_+ = 4.2 \times 10^{-9}$ mol m^{-2} s^{-1}. However, this equation does not describe all the experimental data satisfactorily, possibly because the assumption of equal activity coefficients λ for the two considered surface species is not always justified and, consequently, equation (6-80) cannot be simplified to equation (6-81).

Therefore, Devidal et al. (1997) approximated the surface interactions by means of a regular solution model, which is also known as parabolic Maclaurin model. The activity coefficients for the protonated surface sites ($\equiv H_{3n}$) and for those occupied by Al^{3+} ions ($\equiv Al_n$) are given by (Saxena, 1973)

$$\lambda_{\equiv H_{3n}} = \exp\left[\frac{\omega \cdot (1 - X_{\equiv H_{3n}})^2}{RT}\right], \tag{6-104}$$

and

$$\lambda_{\equiv Al_n} = \exp\left[\frac{\omega \cdot X^2_{\equiv H_{3n}}}{RT}\right], \tag{6-105}$$

respectively, where ω is the single interaction coefficient (there are no site-mixing parameters in this symmetrical model). Insertion of equations (6-104) and (6-105) into equation (6-80) leads to the following rate equation for the dissolution/precipitation of kaolinite in acidic solutions:

$$r = k_+ \cdot K_{\text{Al–O}} \cdot \left(\frac{a^3_{H^+}}{a_{Al^{3+}}}\right) \cdot (1 - X_{\equiv H_{3n}}) \cdot \exp\left[\frac{\omega}{RT} \cdot (2X_{\equiv H_{3n}} - 1)\right] \cdot (1 - A^{0.5}), \tag{6-106}$$

where ω takes the value of 3.81 kJ mol^{-1} and $K_{\text{Al–O}}$ is assumed to be given by equation (6-80) for $n = 1$.

In alkaline solutions, two sequential reactions are considered to take place, namely (1) the reversible surface exchange of OH^- for $Al(OH)_4^-$ ions, involving the breaking of Al–O–Si groups and the production of siloxane surface groups and (2) the reversible destruction of these siloxane groups (whose activity coefficient is indicated as $\lambda_{\text{Si–O–Si}}$), which releases silica into the aqueous solution and produces a Si-rich surface activated complex (whose mole fraction is $X^{\ddagger}$).

Again, Devidal et al. (1997) introduced a regular solution model to describe surface interactions and proposed the following expression for kaolinite dissolution/precipitation at pH 6.8 and 7.8:

$$r = k_+ \cdot (K_1 \cdot K_2^{\ddagger})^2 \cdot \left(\frac{a_{\mathrm{OH^-}}^2}{a_{\mathrm{Al(OH)_4^-}} \cdot a_{\mathrm{H_4SiO_4}}} \right) \cdot \left[1 - X^{\ddagger} - \frac{a_{\mathrm{H_4SiO_4}} \cdot X^{\ddagger} \cdot \exp\left[(\omega/RT) \cdot (1 - X^{\ddagger})^2 \right]}{(K_2^{\ddagger})^2 \cdot \lambda_{\mathrm{Si-O-Si}}} \right]^2$$
$$\cdot \exp\left[-\frac{\omega}{RT} \cdot (1 - 2X^{\ddagger} - X^{\ddagger 2}) \right] \cdot (1 - A) \qquad (6\text{-}107)$$

where K_1 and $K_2^{\ddagger}$ are the thermodynamic equilibrium constants of the two considered reactions. Devidal et al. (1997) give values of k_+, $K_1K_2^{\ddagger}$ and $(K_2^{\ddagger})^2\ \lambda_{\mathrm{Si-O-Si}}$ for different pH and ionic strength, whereas ω takes the same value of the acidic dissolution mechanism.

Contrasting data were reported for the apparent activation energy of kaolinite dissolution by different authors. For instance, the activation energy was considered to be strongly dependent on pH by Carroll and Walther (1990), whereas Ganor et al. (1995) proposed a constant value of 29.3 ± 4.6 kJ mol^{-1}, based on all the available pH and temperature data.

6.6.3.2. Serpentine minerals

The dissolution kinetics of serpentine minerals (chrysotile, antigorite and lizardite, average stoichiometric formula $Mg_3Si_2O_5(OH)_4$) has been the subject of a relatively limited number of experimental investigations (Luce et al., 1972; Lin and Clemency, 1981a; Bales and Morgan, 1985; Hume and Rimstidt, 1992). Available rate data have been critically reviewed by Cipolli et al. (2004), on which this section is largely based.

Luce et al. (1972) carried out batch dissolution experiments on lizardite at 25°C and different pH values. At all pHs, the Mg/Si molal ratio in the aqueous solution was significantly different from the stoichiometric value of 1.5, indicating incongruent dissolution. Parabolic kinetics was generally observed (probably due to un-proper sample preparation), except in strongly acidic solutions (pH ~1.65), where linear kinetics was found. In spite of these limitations, Cipolli et al. (2004) used the experimental data of Luce et al. (1972) to derive the dissolution rate of lizardite as a function of pH. Results are shown in Fig. 6.22, where the logarithm of the ratio $\Delta Mg/3\Delta t$ (which approximates the dissolution rate of serpentine) is plotted against the average pH measured in the corresponding time interval, Δt. Although data are highly scattered, the logarithm of the dissolution rate of lizardite decreases linearly with increasing pH with a slope close to −0.3.

Lin and Clemency (1981a) measured the dissolution rate of antigorite at 25°C at a constant P_{CO_2} of 1 bar. During the dissolution experiments, the Mg/Si molal ratio in the aqueous solution was always higher than 1.5, the expected value for congruent dissolution. According to Lin and Clemency (1981a) the reasons for incongruent dissolution might be either preferential removal of Mg^{2+} ion from the octaedral layer with respect to that of Si^{4+} ion from the tetraedral layer or the precipitation of a secondary, Si-bearing phase, e.g. amorphous silica. This second explanation is supported by equilibrium calculations, carried out by Cipolli et al. (2004) using EQ3NR, which show that amorphous silica saturation was

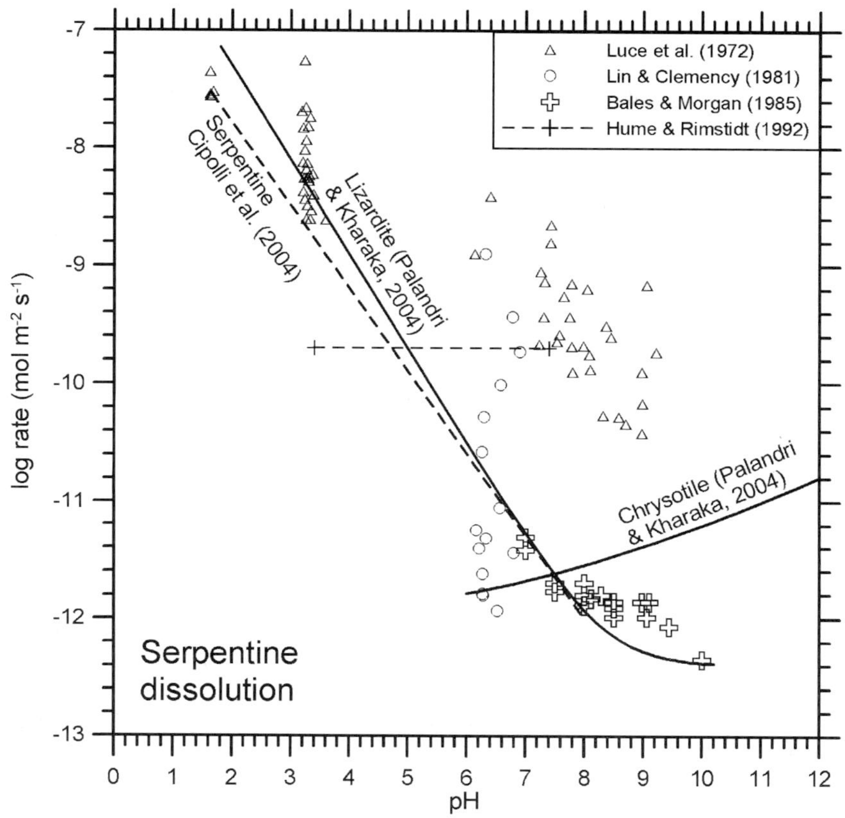

Figure 6.22. Plot of the logarithm of the dissolution rate of serpentine minerals vs. pH showing the available experimental data at 25°C (see legend) and the dissolution rate laws proposed by Palandri and Kharaka (2004) and Cipolli et al. (2004).

attained in the experiments in ~3 h and maintained afterwards. As the analytical data referring to 1, 3, 7 and 15 h are those closest to the condition of congruent dissolution, they were used to compute the ratio $\Delta Mg/3\Delta t$ (which approximates the dissolution rate of serpentine), whose logarithm is plotted against pH in Fig. 6.22.

Bales and Morgan (1985) determined the kinetics of the dissolution of chrysotile, through 5-day-long batch tests, at constant temperature, 25°C and pH in the range 7–10. The measured Mg/Si molal ratio in the aqueous solution was close to 2, i.e. significantly greater than the values of 1.4–1.5 expected for stoichiometric dissolution of the chrysotile used in the experiments. According to Bales and Morgan (1985), the release of Mg takes place at constant rate, after the first day, and is proportional to $a_{H+}^{0.24}$. The dissolution rates of chrysotile, evaluated on the basis of the Mg release, are also shown in Fig. 6.22.

Cipolli et al. (2004), assuming that the more reliable data on the dissolution kinetics of serpentine—irrespective of the polymorphic form—are those of Luce et al. (1972)

at pH 1.63–1.67 (as they are the only ones of this dataset showing a linear kinetics) and those of Bales and Morgan (1985), proposed the following relation:

$$\log r_{+} \,(\mathrm{mol\,m^{-2}\,s^{-1}}) = -0.70 \times \mathrm{pH} - 6.38, \tag{6-108}$$

for the acidic dissolution mechanism of serpentine at 25°C (Fig. 6.22). This equation is relatively similar to that suggested by Palandri and Kharaka (2004) for the acidic dissolution mechanism of lizardite. Rate laws for the neutral and basic dissolution mechanisms of chrysotile were proposed by Palandri and Kharaka (2004) on the basis of the silica release during the dissolution experiments of Bales and Morgan (1985) (see Table 6.1).

Hume and Rimstidt (1992) studied the dissolution rate of chrysotile in synthetic lung fluids at 37°C and obtained the average value of 6×10^{-10} mol m^{-2} s^{-1}, independent of pH, in the final pH range of 3.4–7.4. Considering an apparent activation energy of 70 ± 10 kJ mol^{-1} for serpentine dissolution (Thomassin et al., 1977), the rate determined by Hume and Rimstidt (1992) at 37°C corresponds to a value of 2×10^{-10} mol m^{-2} s^{-1} at 25°C, as indicated in Fig. 6.22.

6.6.3.3. Smectites

This complex group of minerals has been the subject of comparatively limited studies aimed at determining their dissolution kinetics (e.g. Hayasi and Yamada, 1990; Furrer et al., 1993; Zysset and Schindler, 1996; Bauer and Berger, 1998; Cama et al., 2000; Huertas et al., 2001).

Hayashi and Yamada (1990) measured the dissolution kinetics for Wyoming montmorillonite at 80°C in Na_2CO_3 aqueous solution.

Zysset and Schindler (1996) determined the H^+-promoted dissolution rate of SWy-1 Crook County Wyoming montmorillonite, by means of batch experiments, at 25°C in pH range 1–5 and at KCl concentrations of 0.03, 0.1 and 1 M. They also measured the concentration of adsorbed protons and its pH dependence by acid–base titrations and batch equilibration experiments, as well as the inhibition effect by added Al. This investigation represents a sort of follow-up of the early work by Furrer et al. (1993).

The structural chemical formula of SWy-1 montmorillonite was computed assuming that (1) all cation exchange sites are occupied by K and (2) Fe is entirely in the trivalent state. Results can be expressed as follows: $K_{0.636}[Si_{7.95}Al_{0.05}][Al_{3.018}Fe^{(III)}{}_{0.41}Mg_{0.566}]O_{20}(OH)_4$, on the basis of 20 O atoms and 4 OH groups.

Dissolution was mostly congruent, and deviations from congruency (especially in 0.03 M KCl) were attributed to partial readsorption of dissolved Al by ion exchange or surface complexation. Zysset and Schindler (1996) report dissolution rates in units of mol g^{-1} h^{-1}, which can be converted to units of mol m^{-2} s^{-1}, for comparison with other data, using the specific surface area (21.4 m^2 g^{-1}) obtained through the BET method using N_2 adsorption. The logarithms of dissolution rates were found to be linearly dependent on pH, with a reaction order n_H of 0.24–0.35, but dissolution kinetics resulted to be significantly faster at 1 M KCl than at lower KCl concentration (Fig. 6.23). The authors suggest that the dissolution process chiefly occurs at crystal edge faces.

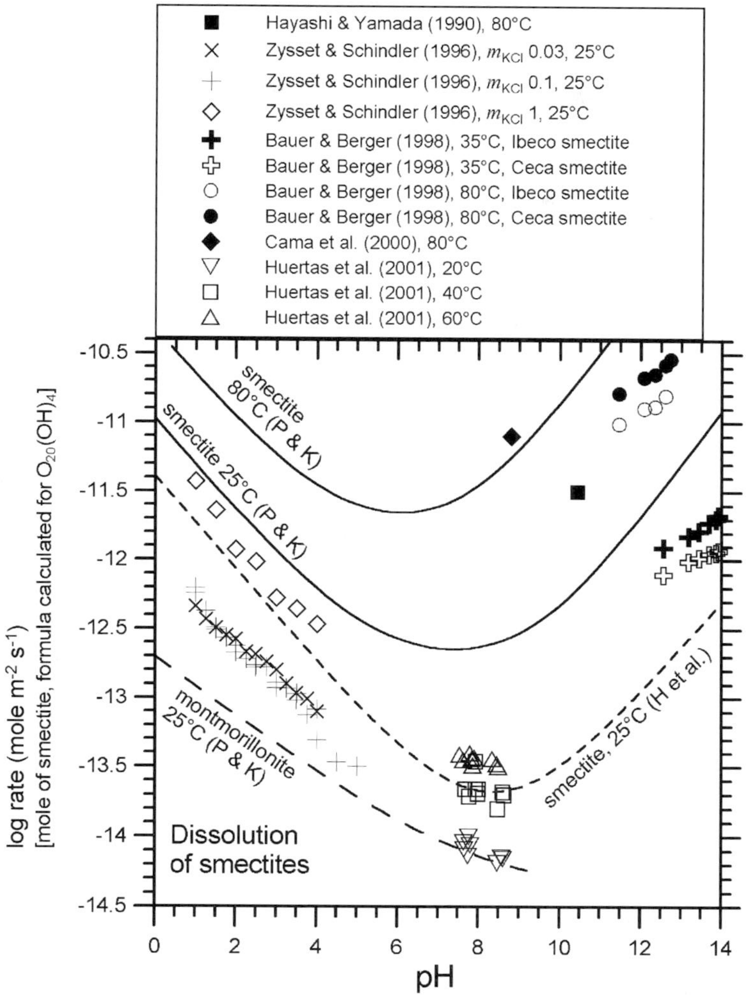

Figure 6.23. Plot of the logarithm of the dissolution rate of smectite minerals vs. pH showing the available experimental data (see legend) and the pH dependence of the theoretical dissolution rates of (i) smectite at 25°C (from Huertas et al., 2001), (ii) smectite at 25 and 80°C (from Palandri and Kharaka, 2004) and (iii) montmorillonite at 25°C (from Palandri and Kharaka, 2004).

Bauer and Berger (1998) carried out dissolution experiments on two smectites (Ceca and Ibeco), in 0.1–4 M KOH solutions, at 35 and 80°C, by means of batch reactors. The Ceca smectite (Wyoming-Na) is a nearly pure Na-montmorillonite of structural chemical formula $(Na_{0.4475}K_{0.012}Ca_{0.0984})[Si_{7.98}Al_{0.02}][Al_{3.105}Fe^{(III)}_{0.3128}Fe^{(II)}_{0.0443}Mg_{0.5109}]O_{20}(OH)_4$, whereas the Ibeco clay is a Ca-smectite with ~35% of a beidellitic component. Its structural chemical formula is $(K_{0.04}Ca_{0.5})[Si_{7.65}Al_{0.35}][Al_{2.8}Fe^{(III)}_{0.47}Fe^{(II)}_{0.06}Mg_{0.7}]O_{20}(OH)_4$. Dissolution rates are given in units of moles of Si g^{-1} h^{-1}. They have been divided by the stoichiometric coefficient of Si in the smectite formula to obtain the moles of dissolved smectite $m^{-2}\,s^{-1}$. Silicon and Al release was in the stoichiometric ratio of the dissolving smectites in all the experimental runs. Far-from-equilibrium dissolution is indicated by the absence of chemical affinity effects. Log-rates and pH resulted to be linearly correlated, with a reaction order n_{OH} of –0.15 ± 0.06, both at 35 and 80°C. The apparent activation energy of smectite dissolution was found to be 52 ± 4 kJ mol^{-1} and independent of pH. The dissolution of the tetrahedral layer was considered to be the rate-limiting step.

Cama et al. (2000) investigated the dissolution kinetics of a smectite from the Cabo de Gata volcanic rocks (Serrata de Nijar, Almería, Spain) as a function of thermodynamic affinity (which is called Gibbs free energy of reaction by the authors) at 80°C and pH 8.8, by means of stirred flow-through experiments. They also carried out batch experiments to try to obtain the equilibrium constant of the dissolution reaction of smectite at 80°C.

The untreated mineral, based on XRD analysis, is a mixed-layer illite (10–15%)and montmorillonite (90–85%), with minor amounts of cristobalite, quartz and feldspars. Its structural chemical formula, after treatment with borax at 80°C for 2 months, was calculated assuming that the sample is made up only of smectite, obtaining $(Na_{0.51}K_{0.19}Ca_{0.195}Mg_{0.09})[Si_{7.77}Al_{0.23}][Al_{2.56}Fe^{(III)}_{0.42}Mg_{1.02}]O_{20}(OH)_4$.

In the flow-through dissolution experiments, at steady state, the Al/Si and Mg/Si ratios were very close to those of the mineral, indicating stoichiometric dissolution for Si, Al and Mg, which belong to both the octahedral and tetrahedral layers of smectite. In a log–log plot, the dissolution rate was found to decrease linearly with increasing Si concentration, indicating either Si inhibition (see equation in Cama et al., 2000) or dependence on thermodynamic affinity, as expressed by the relation (rate in mol $m^{-2}\,s^{-1}$)

$$r_+ = 8.1\cdot 10^{-12}\cdot\left\{1-\exp\left[-6\cdot 10^{-10}\cdot\left(\frac{A}{RT}\right)^6\right]\right\}. \tag{6-109}$$

This type of equation was also used by Nagy and Lasaga (1992) to describe the dependence of the dissolution rate of gibbsite on thermodynamic affinity (see Section 6.7.1). Equation (6-109) indicates a far-from equilibrium dissolution rate of 8.1×10^{-12} mol $m^{-2}\,s^{-1}$, which is illustrated in Fig. 6.23.

The kinetics of montmorillonite dissolution in different aqueous solutions was studied by Huertas et al. (2001), in the pH range 7.6–8.5, at temperatures of 20, 40 and 60°C. Dissolution experiments were carried out in a batch reactor, but the mineral samples were sealed into dialysis bags and the external solution was sampled and replaced at regular intervals of time. The authors used the same smectite of Cama et al. (2000), without any pre-treatment. Therefore, the structural chemical formula of the smectite after correction for other phases, $(Na_{0.20}K_{0.20}Ca_{0.18}Mg_{0.13})[Si_{7.58}Al_{0.42}][Al_{2.83}Fe^{(III)}_{0.37}Mg_{0.89}]O_{20}(OH)_4$, is somewhat different from that of Cama et al. (2000, see above). Because of the absence of pre-treatment, a sort of parabolic kinetics was obtained, but the first data were neglected and only the subsequent linear trend was considered. Since Al was not analysed and Mg data are not reported, it is not possible to establish if dissolution was stoichiometric or not. In spite of this uncertainty, rate data appear to be virtually independent of pH in the considered pH range (7.6–8.5). Log rates are linearly correlated with the inverse of the absolute temperature, indicating an apparent activation energy of 30.5 ± 1.3 kJ mol^{-1}. This value is significantly lower than that found by Bauer and Berger (1998), 52 ± 4 kJ mol^{-1}, but for the pH ranges 12.6–13.9 at 35°C and 11.5–12.7 at 80°C.

In addition to the available rate data, also shown in Fig. 6.23 are the theoretical dissolution rates of (i) smectite at 25°C (from Huertas et al., 2001), (ii) smectite at 25 and 80°C (from Palandri and Kharaka, 2004) and (iii) montmorillonite at 25°C (from Palandri and Kharaka, 2004). In view of the large chemical and mineralogical differences among smectites, the spread in the experimental and theoretical data represented in this plot is not surprising.

6.6.3.4. Illite

In contrast to kaolinite, few dissolution rate data are available for illite. Early studies were carried out by Heydemann (1966) and Feigenbaum and Shainberg (1975), whereas a recent experimental work was performed by Köhler et al. (2003).

Heydemann (1966) measured the dissolution kinetics of illite from Fithian (Illinois, USA) in small batch reactors whose solution was replaced periodically. Her data were interpreted by Nagy (1995) who, based on Si release, obtained rates in the order of 10^{-15}–10^{-14} mol m^{-2} s^{-1}, with a low-pH dependence.

Feigenbaum and Shainberg (1975) determined dissolution rates of the same illite at 5 and 25°C and pH close to 7.5 and 2.9. Based on K release, dissolution rates in the order of 10^{-12} to 10^{-10} mol m^{-2} s^{-1} were computed by Nagy (1995).

Köhler et al. (2003) investigated the dissolution kinetics of Illite du Puy from the Massif Central (France) in batch reactors as a function of both pH, from 1.4 to 12.4, and temperature, at 5, 20, 25 and 50°C. The structural chemical formula of the Illite du Puy is $(X^{(I)}{}_{0.12}Na_{0.01}K_{0.53}Ca_{0.01})[Si_{3.55}Al_{0.45}][Al_{1.27}Fe^{(III)}{}_{0.36}Mg_{0.44}]O_{10}(OH)_2$, where $X^{(I)}$ represents the exchange site, which was saturated with Na during sample pre-treatment. Dissolution was non-stoichiometric at $4 < pH < 11$, likely due to precipitation of a secondary solid phase. In contrast, after an initial non-stoichiometric step, stoichiometric dissolution was attained below pH 4 and above pH 11.

According to Köhler et al. (2003), the measured dissolution rates are described by the following Arrhenius-type expression (rate in mol m^{-2} s^{-1}, R in kJ mol^{-1} K^{-1}, see *also* Table 6.1):

$$r_+ = 2.2\cdot 10^{-4}\cdot a_{H^+}^{0.6}\cdot \exp\left[-\frac{46}{RT}\right] + 2.5\cdot 10^{-13}\cdot \exp\left[-\frac{14}{RT}\right] + 0.27\cdot a_{OH^-}^{0.6}\cdot \exp\left[-\frac{67}{RT}\right]. \tag{6-110}$$

Dissolution rates calculated by means of equation (6-110) and experimental measurements by Köhler et al. (2003) and Heydemann (1966) are shown in Fig. 6.24. A substantial agreement between the two experimental datasets is observed.

6.6.3.5. Chlorites

A relatively limited number of studies have investigated the dissolution kinetics of chlorites, taking into account samples of different chemistry.

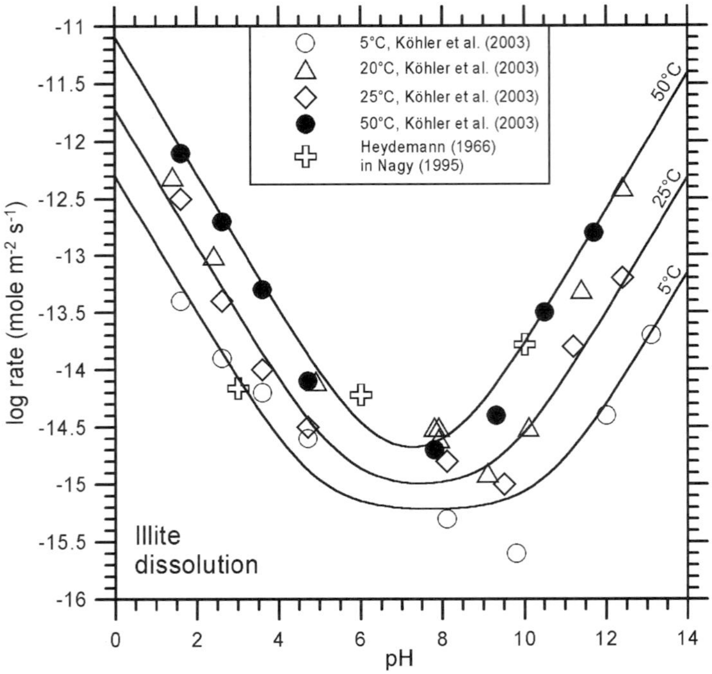

Figure 6.24. Plot of the logarithm of the dissolution rate of illites vs. pH showing the experimental measurements by Köhler et al. (2003) and Heydemann (1966), and the dissolution rates computed by means of equation (6-110).

Ross (1967) studied the dissolution rate for a chlorite ($Mg_{8.64}$ $Al_{2.10}$ $Fe^{(III)}_{0.44}$ $Fe^{(II)}_{0.54}$) [$Si_{5.96}Al_{2.04}O_{20}$]$(OH)_{16}$. Kodama and Schnitzer (1973) reported the dissolution rate for the Mg-rich leuchtenbergite ($Mg_{9.88}$ $Al_{1.42}$ $Fe^{(II)}_{1.08}$) [$Si_{5.76}Al_{2.24}O_{20}$]$(OH)_{16}$ and the Mn-rich thuringite ($Mg_{1.40}$ $Al_{1.68}$ $Fe^{(III)}_{0.30}$ $Mn_{1.04}$) [$Si_{5.08}Al_{2.92}O_{20}$]$(OH)_{16}$. Sverdrup investigated the dissolution kinetics of the Mg-rich chlorite ($Mg_{8.4}$ $Al_{2.0}$ $Fe_{1.2}$) [$Si_{5.4}Al_{2.6}O_{20}$] $(OH)_{16}$. May et al. (1995) studied the Mg-clinochlore ($Mg_{9.8}$ $Al_{1.8}$ $Fe^{(II)}_{0.2}$ $Fe^{(III)}_{0.2}$)[$Si_{6.0}$ $Al_{2.0}$ O_{20}]$(OH)_{16}$. The Mg-chlorite of composition ($Mg_{9.8}$ $Al_{1.4}$ $Fe^{(II)}_{0.2}$ $Fe^{(III)}_{0.2}$)[$Si_{7.0}$ $Al_{1.0}$ O_{20}]$(OH)_{16}$ was the subject of the research by Malmström et al. (1996). Rochelle et al. (1996) investigated the behaviour of the Fe-chlorite ($Mg_{5.36}$ $Al_{2.62}$ $Fe_{3.94}$)[$Si_{5.56}$ $Al_{2.46}$ O_{20}]$(OH)_{16}$. Salmon and Malmström (2001) did not specify the composition of the investigated sample. Hamer et al. (2003) reported on the dissolution rate of the Fe(II)-rich ripidolite ($Mg_{5.60}$ $Al_{2.52}$ $Fe^{(II)}_{3.90}$)[$Si_{5.50}$ $Al_{2.50}$ O_{20}]$(OH)_{16}$. Gustafsson and Puigdomenech (2003) investigated the dissolution kinetics of the Fe(II)-poor chlorite ($Mg_{9.80}$ $Al_{1.68}Fe^{(II)}_{0.72}$)[$Si_{6.50}$ $Al_{1.40}$ $Fe^{(III)}_{0.12}$ O_{20}]$(OH)_{16}$.

Brandt et al. (2003) examined two chlorites with intermediate Fe contents, called CCa-2 and MaIn, whose structural chemical formulae are ($Mg_{5.54}$ $Al_{2.48}$ $Fe^{(II)}_{3.02}$ $Fe^{(III)}_{0.94}$)[$Si_{5.33}$ $Al_{2.66}$ O_{20}]$(OH)_{16}$ and ($Mg_{6.33}$ $Al_{2.48}$ $Fe^{(II)}_{2.27}$ $Fe^{(III)}_{0.93}$)[$Si_{5.33}$ $Al_{2.66}$ O_{20}] $(OH)_{16}$, respectively. Brandt et al. (2003) performed dissolution experiments in mixed-flow reactors under far-from-equilibrium conditions. They observed non-stoichiometric dissolution, for pH 2–4, with Si released at a lower rate with respect to other cations, according to the sequence Al > Mg > Fe > Si. This suggests that the dissolution of the brucite-like layer is ~2–2.5 times faster than the talc-like sheets.

Lowson et al. (2005) investigated the dissolution rate of the Fe-rich chlorite ($Mg_{5.52}$ $Al_{1.94}Fe^{(II)}{}_{3.80}$ $Fe^{(III)}{}_{0.14}$) [$Si_{4.96}$ $Al_{3.04}$ O_{20}]$(OH)_{16}$ under far-from-equilibrium, quasi-steady-state conditions as a function of pH (3–10.5) at 25°C by means of a single pass flow-through cell. Based on their results and existing data, Lowson et al. (2005) derived the dissolution rate parameters reported in Table 6.1. Besides, the the dissolution rate of chlorite was found to be dependent on the $a^3_{H^+}/a_{Al^{3+}}$ ratio, as expressed by the following equation (rate in mol m^{-2} s^{-1}):

$$r_+ = 10^{-10.46 \pm 0.3} \cdot \left(\frac{a^3_{H^+}}{a_{Al^{3+}}} \right)^{0.27 \pm 0.04}.$$

Available experimental rates, mostly based on Si release, are shown in Fig. 6.25, together with the rates computed by means of the pH-dependent rate law, whose parameters are given in Table 6.1. These are partly taken from Lowson et al. (2005) and partly from Palandri and Kharaka (2004), who report an apparent activation energy of 88 kJ mol^{-1}.

6.6.3.6. Muscovite

The dissolution rates of muscovite at low temperature have been investigated through laboratory dissolution experiments by Nickel (1973), Lin and Clemency (1981b), Stumm et al. (1987), Knauss and Wolery (1989) and Kalinowski and Schweda (1996).

Nickel (1973) measured the dissolution kinetics of muscovite at 25°C and pH of 0.2, 3.6, 5.6, and 10.6 in stirred batch reactors, by daily replacing the aqueous solution. Normalizing the obtained experimental results to the initial BET surface area, Nagy (1995) computed dissolution rates of 6×10^{-13}, 3.4×10^{-13}, 2.4×10^{-13} and 7.5×10^{-13} mol m^{-2} s^{-1} in order of increasing pH. The experimental results of Nickel (1973) were also interpreted by Kalinowski and Schweda (1996), who obtained rates of 2.3×10^{-12}, 6.4×10^{-13}, 5.0×10^{-13} and 1.4×10^{-12} mol m^{-2} s^{-1}, respectively, based on Si release. Different dissolution rates were computed by the same authors based on the release of Al and K. These differences were attributed to either incongruent dissolution or effects of impurities in the mineral sample.

Lin and Clemency (1981b) investigated the dissolution kinetics of two muscovite samples (ruby mica and green mica) at 25°C under a P_{CO_2} of 1 atm in a stirred batch reactor. After 500 h, dissolved Si increased linearly with time, indicating dissolution rates of 4.0×10^{-14} mol m^{-2} s^{-1} for the ruby mica at pH 4.95 and 4.4×10^{-14} mol m^{-2} s^{-1} for the green mica at pH 5.02, according to Kalinowski and Schweda (1996).

Stumm et al. (1987) give dissolution rates of muscovite at 25°C and pH 3 and 5, which are intermediate between those of Nickel (1973) and those of Lin and Clemency (1981b).

The far-from-equilibrium dissolution rate of muscovite was determined by Knauss and Wolery (1989) at 70°C in the pH interval 1.4–11.8 by means of a flow-through reactor.

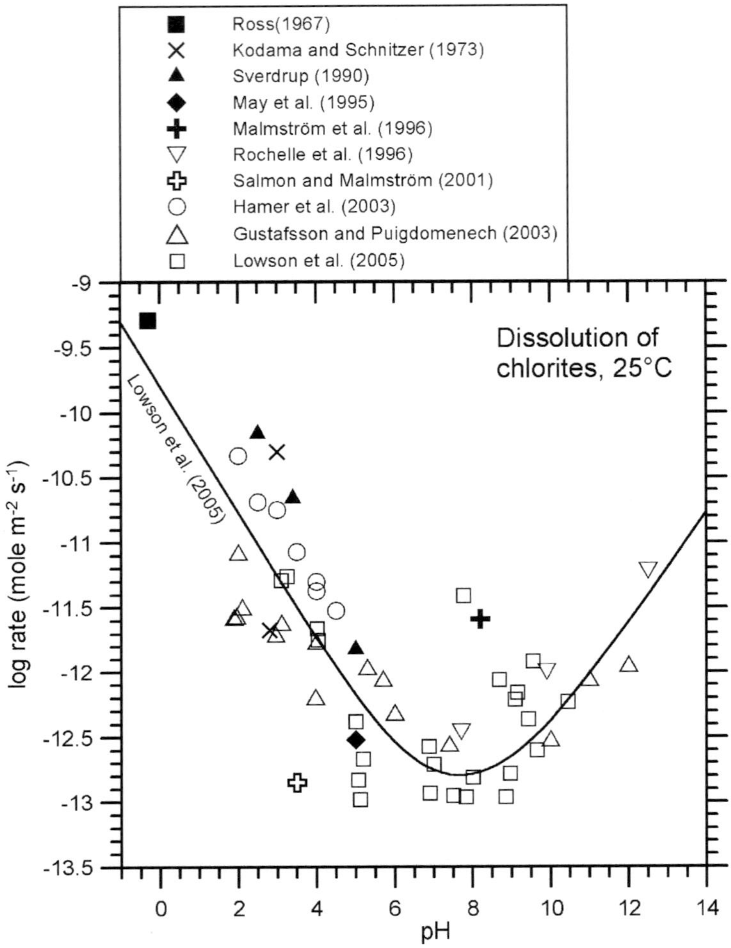

Figure 6.25. Plot of the logarithm of the dissolution rate of chlorites vs. pH showing the available experimental data (see legend), together with the rates computed by means of the pH-dependent rate law suggested by Lowson et al. (2005; see Table 6.1). (Reprinted from Lowson et al., 2005, Copyright 2005 with permission from Elsevier, adapted.)

Comparison of muscovite dissolution rates based on the release of Si, Al and K to the solution suggests that steady-state dissolution was congruent or nearly so under most pH values. The reaction order with respect to H^+ is $n_H = +0.37$ below pH ~5, whereas $n_{OH} = -0.22$ above pH ~5. These two values were adopted by Palandri and Kharaka (2004).

Kalinowski and Schweda (1996) determined the dissolution kinetics of muscovite at pH 1–4 and temperature of 22 ± 2°C. The structural chemical formula of the muscovite sample is $(Na_{0.15}K_{1.82}Ca_{0.003})[Al_{3.39}Fe^{(III)}{}_{0.40}Fe^{(II)}{}_{0.15}Mn_{0.04}Mg_{0.44}][Si_{6.18}Al_{1.82}]O_{20}(OH)_{3.91}F_{0.09}$.

These authors used a flow-through reactor in which the mineral suspension was separated from the reactant flow by dialysis membranes. The mineral suspension was kept under agitation by a stirring table as well as by pumping through a closed loop. Steady-state dissolution of muscovite was nearly stoichiometric. However, preferential release of interlayer K^+ with respect to other elements was observed at pH 2 and 3.

All the experimental dissolution rates of muscovite are plotted in Fig. 6.26, together with the rate law obtained by summing the contributions of the acidic, neutral and basic mechanisms reported by Palandri and Kharaka (2004, see Table 6.1).

6.6.3.7. Biotite

Dissolution kinetics of biotite was investigated through laboratory experiments by Acker and Bricker (1992), Turpault and Trotignon (1994), Kalinowski and Schweda (1996), Malmström et al. (1996), Malmström and Banwart (1997), Taylor et al. (2000a), Murakami et al. (2004) and Samson et al. (2005). Different kinds of experimental set-up and procedures were used by these authors.

Acker and Bricker (1992) worked with fluidized-bed reactors and flow-through columns at 25°C in the pH range 3–7 under strongly oxidizing conditions (1.5% H_2O_2), mildly oxidizing conditions (atmospheric P_{O_2}) and anoxic conditions (obtained by continuous N_2 bubbling). Turpault and Trotignon (1994) studied the dissolution of single biotite crystals through batch experiments at 24°C in 0.1 N HNO_3. Kalinowski and Schweda

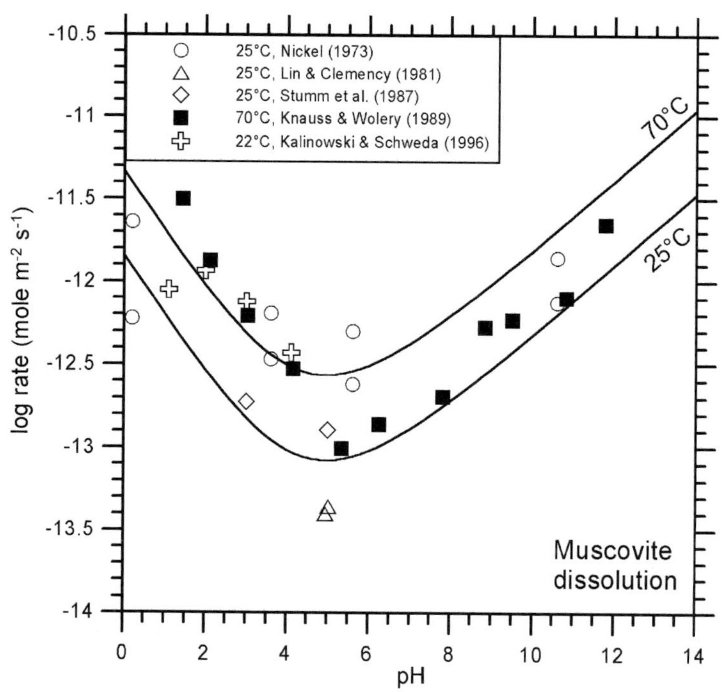

Figure 6.26. Plot of the logarithm of the dissolution rate of muscovite vs. pH showing the available experimental rates (see legend), as well as the rates computed at 25°C and 70°C by means of the rate law of Palandri and Kharaka (2004).

(1996) followed the same approach used to study the dissolution kinetics of muscovite (see Section 6.6.3.6). Malmström and co-workers carried out dissolution experiments at 25°C, in the pH range 2–10, in Na- and K-media, and at variable P_{CO_2}. They used thin-film continuous-flow reactors, in which the aqueous solution is pumped through a thin layer of the mineral powder held between two membrane filters. The study of Taylor et al. (2000a) was aimed at quantifying not only the kinetics of biotite dissolution but also the release of Sr and Sr isotopes at pH ~3 and 25°C. They utilized flow-through column reactors. Murakami et al. (2004) performed dissolution experiments in a batch reactor under anoxic conditions ($P_{O_2} < 3 \times 10^{-5}$ bar) at 1 bar P_{CO_2}, pH 4.6 and 100°C as well as under oxic conditions (atmospheric P_{O_2} and P_{CO_2}), pH 4.6 and 100°C. Samson et al. (2005) investigated transient (non-steady state) and quasi-steady-state dissolution of biotite at 22–25°C in solutions of high pH and high Na, NO_3 and Al concentrations. The term quasi-steady state was adopted to indicate that dissolution of biotite was accompanied by other reactions, including solid-state alteration of biotite and precipitation of secondary solid phases (see below). Dissolution experiments were carried out in continuously stirred flow-through reactors. Initially biotite was reacted at pH 8. Then, inlet solution of higher pH (10–14) was injected through the reactors.

In spite of the different designs and procedures, early dissolution of interlayer K was observed by all the authors who analysed the aqueous solution for this element. Fast K release, indeed, is a typical characteristic of biotite dissolution at any pH.

Another observation shared by most authors is the occurrence of non-stoichiometric dissolution. Some authors who worked under acidic conditions (e.g. Acker and Bricker, 1992; Turpault and Trotignon, 1994; Kalinowski and Schweda, 1996; Taylor et al., 2000a) report preferential release of cations from the octahedral layer (e.g. Mg) with respect to those of the tetrahedral layer (e.g. Si). The behaviour of Fe and Al is more complicated as these elements occur in both layers. However, according to Malmström and Banwart (1997), dissolution is stoichiometric for Fe, Mg, Al and Si close to pH 2, but a transition from preferential release of Fe, Mg and Al to preferential release of Si occurs at an unspecified pH in the range 4–7. In contrast, close to stoichiometric release of Si, Al and Mg, and slower release of Fe was observed by Samson et al. (2005) at high pH.

Non-stoichiometric dissolution was attributed to transformation in vermiculite by Acker and Bricker (1992) and Kalinowski and Schweda (1996), and to either alteration to vermiculite or precipitation of secondary phases by Samson et al. (2005). Interestingly, high-resolution scanning and transmission electron microscopy observations by Murakami et al. (2004) showed that secondary smectite or Fe(II)-rich vermiculite is formed at the edge of biotite crystals under anoxic conditions, whereas Fe(III)- and Al-oxyhydroxides are precipitated under oxic conditions.

Turpault and Trotignon (1994) proposed the development of a zoned residual altered layer, with an outermost zone made up of amorphous silica; this would act as a diffusion barrier controlling the release of K, Mg, Fe and Al.

According to Malmström and Banwart (1997), the non-stoichiometric dissolution of biotite is likely controlled by the irreversible mechanism of element release from the dissolving mineral (i.e. parallel dissolution of surface sites with different activation energies), rather than incongruent dissolution owing to precipitation of secondary minerals or re-adsorption, in spite of the observed formation of kaolinite and vermiculite and the possible precipitation of other minerals such as Fe(III)- and Al-oxyhydroxides. In addition, Malmström and Banwart (1997) applied the Schnoor's (1990) model to biotite dissolution, based on the following assumptions: (1) dissolution occurs in the reacting layer, i.e. in the outermost part of the solid phase, (2) multi-site dissolution takes place, i.e. the release of each element is independent from the others and (3) it is proportional to the number of sites in the reacting layer. The model suggests that stoichiometric dissolution is attained in the long term, owing to changes in the composition of the reacting layer. The thickness of the reacting layer depends on both the diffusivity of K^+ ion through the layer itself and the release rate of the sluggish element, i.e. Fe.

Turpault and Trotignon (1994), by working on single biotite crystals, emphasized that the edge surface area dissolves much faster than the basal surface area.

An interesting observation by Malmström and Banwart (1997), for the specific purposes of geological CO_2 sequestration, is that the release rate of Si is higher in CO_2-free experiments than in systems containing CO_2. In contrast, results were similar in K- and Na-media.

In the same study by Malmström and Banwart (1997), both single-pH and multi-pH experiments were carried out and the rate of release of framework ions was significant slower in the second experiments than in the first ones. These differences were attributed to depletion of metals from the reacting layer at acidic pH.

Also Samson et al. (2005), owing to their experimental procedure (see above), obtained dissolution rates generally lower than those measured in single-pH. They attributed these discrepancies to precipitation of secondary phases that passivated the biotite surface, whereas alteration to vermiculite was considered a less likely explanation.

It is important to underscore that biotite dissolution is strongly affected by the redox potential (Murakami et al., 2004), owing to the presence of Fe^{II} and Fe^{III} in its lattice (see below). Under strongly oxidizing conditions (H_2O_2), dissolution rates much faster than under anoxic conditions were measured by Acker and Bricker (1992).

Measured dissolution rates are shown in Fig. 6.27, together with the dissolution rate law obtained by summing the contributions of the acidic and neutral mechanisms reported by Palandri and Kharaka (2004, see Table 6.1), whose parameters are largely based on the results of Acker and Bricker (1992). This plot confirms that the kinetics of biotite dissolution is a rather complicated process, as discussed above.

Last but not least, it must be underscored that the differences in the dissolution rate can be ascribed, at least partly, to the differences in the chemical composition of the biotite samples. Often, this deviates significantly from that of ideal biotite $[K(Mg,Fe^{II})_3(Si_3Al)O_{10}(OH,F)_2]$. For instance, Al and Fe(III) may be present in relatively high amounts in both the octahedral and tetrahedral layers.

6.6.3.8. Phlogopite

The endmember composition of phlogopite is $KMg_3(Si_3Al)O_{10}(OH)_2$, but substitutions in the octahedral and tetrahedral layers occur, similar to biotite. Conventionally, phlogopites have Mg/Fe ratio > 2, whereas biotite has Mg/Fe ratio < 2 (Deer et al., 1992).

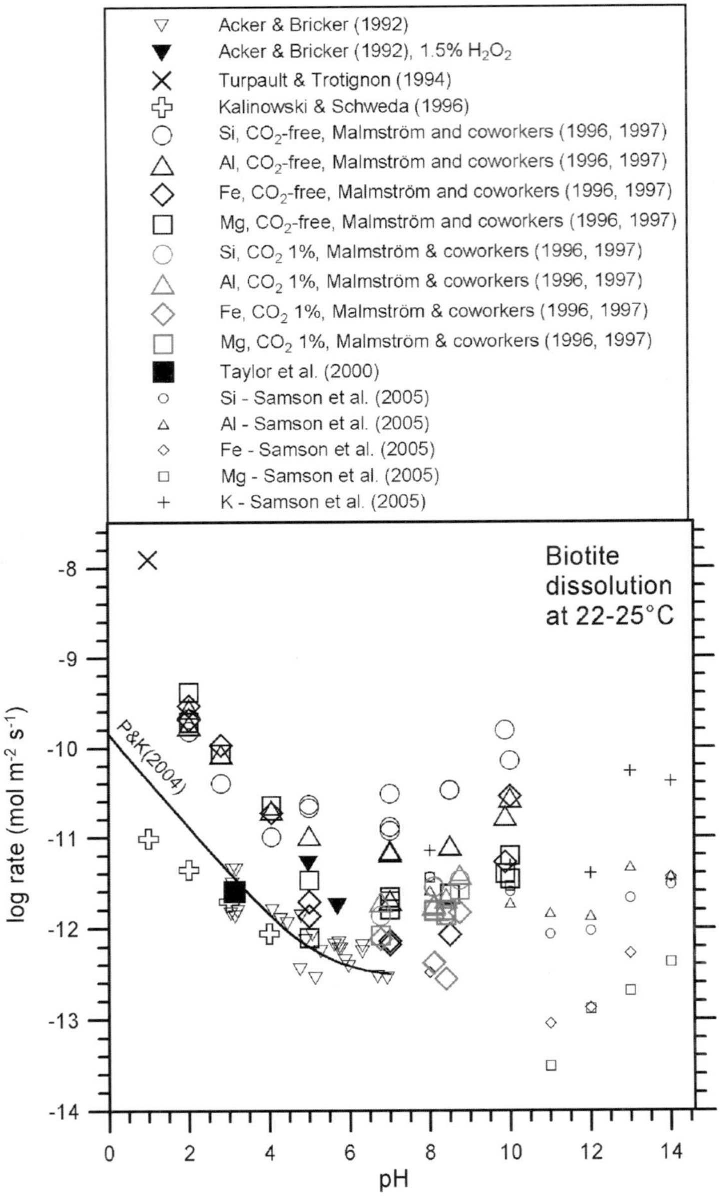

Figure 6.27. Plot of the logarithm of the dissolution rate of biotite vs. pH showing the available experimental rates (see legend), as well as the rates computed at 25°C by means of the rate law of Palandri and Kharaka (2004).

The dissolution kinetics of phlogopite was studied by Lin and Clemency (1981c), Clemency and Lin (1981), Kalinowski and Schweda (1996) and Taylor et al. (2000a).

Lin and Clemency (1981c) followed the same approach as for muscovite (see Section 6.6.3.6). Dissolution was close to stoichiometric for Mg and Si towards the end of the batch dissolution experiment. Based on Si release, a dissolution rate of 6.8×10^{-14} mol m^{-2} s^{-1} was computed at pH 5.34. These data were later revised by Nagy (1995) who proposed a dissolution rate of 3.8×10^{-13} mol m^{-2} s^{-1}.

Clemency and Lin (1981) investigated the dissolution kinetics of phlogopite in an open system, in which ionic solute products were removed through ion exchange by using a resin in H^+-form. In this way, acidic conditions were maintained, but significant pH changes occurred during the experiment. Obviously, silica was not removed and precipitated from the aqueous solution. A dissolution rate of 2.0×10^{-10} mol m^{-2} s^{-1} was estimated at pH near 3.05. Based on these data, Nagy computed a rate of 4.4×10^{-11} mol m^{-2} s^{-1} sat pH 3.3.

Kalinowski and Schweda (1996) and Taylor et al. (2000a) used the same equipments and procedures described for muscovite (see Section 6.6.3.6) and biotite (see Section 6.6.3.7), respectively. Measured dissolution rates are plotted in Fig. 6.28, together with the data mentioned above. The dissolution rate for biotite (acidic mechanism) and

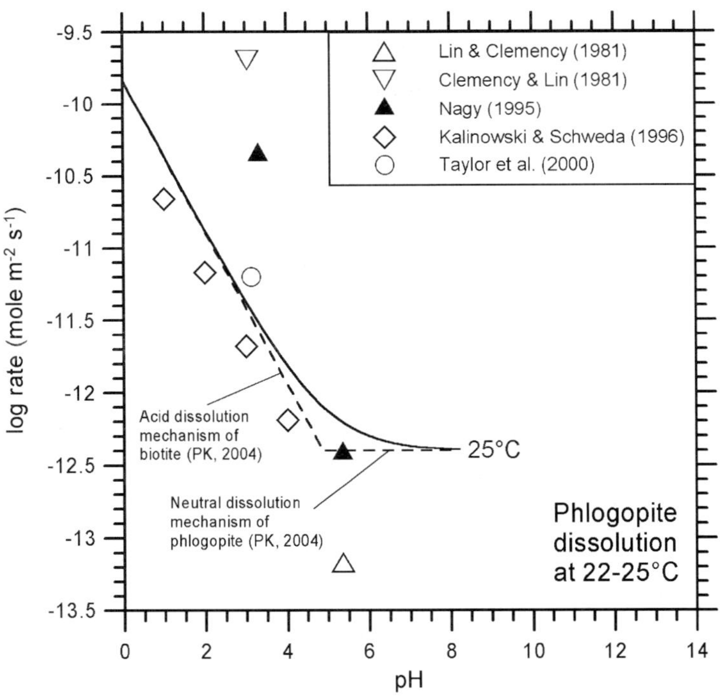

Figure 6.28. Plot of the logarithm of the dissolution rate of phlogopite vs. pH showing the measured dissolution rates (see legend). The dissolution rates for biotite (acidic mechanism) and phlogopite (neutral mechanism) computed by means of the rate laws of Palandri and Kharaka (2004) are also shown.

phlogopite (neutral mechanism) computed by means of the rate laws of Palandri and Kharaka (2004, see Table 6.1), are also shown together with the dissolution rate law obtained by summing the contributions of these two individual mechanisms.

Kuwahara and Aoki (1995) investigated phlogopite dissolution at pH 2 from 50 to 120°C, obtaining apparent activation energies of 29–42 kJ mol^{-1}.

6.6.4. Dissolution rates of feldspars

Feldspars are the most studied silicate minerals from the point of view of dissolution rates and these investigations have greatly contributed to the general understanding of the dissolution kinetics of solid phases (see Sections 6.3 and 6.4) and to ameliorate the apparatuses and techniques used in laboratory dissolution experiments (see Section 6.5).

6.6.4.1. Studies on the dissolution kinetics of feldspars

Early experiments (e.g. Evans, 1965; Lagache, 1965, 1976; Busenberg and Clemency, 1976; Holdren and Berner, 1979; Siegel and Pfannkuch, 1984) were performed in batch reactors. For example, Busenberg and Clemency (1976) investigated the dissolution rate of two potassium feldspars and six plagioclases at 1 bar P_{CO_2}, monitoring pH and the concentration of Si, Al, K, Na, Ca, Mg and Fe in the aqueous solution. As already recalled (see Section 6.5.1), in this kind of dissolution experiments, the relatively large changes in pH and solute concentrations and the precipitation of secondary solid phases complicate the interpretation of results.

To investigate the kinetics and mechanisms of albite dissolution without these problems, Chou and Wollast (1984) developed a continuous reactor based on the fluidized-bed technique. In this system, the aqueous solution is maintained undersaturated with respect to all possible secondary solid phases. Moreover, after an initial transient stage, steady state is attained, i.e. the concentrations of solutes and pH remain constant with time. Chou and Wollast (1984) observed that, at steady state, albite dissolves stoichiometrically. They also hypothesized the formation of a residual layer of a few tens of angstroms in thickness at the surface of the dissolving feldspar. This layer is enriched in Si and/or Al, and its composition depends on the pH of the solution.

At the same time, understanding of the dissolution kinetics of feldspars was improved; thanks to the theoretical work of Helgeson et al. (1984), who reviewed these early studies on feldspar dissolution kinetics with the aid of irreversible thermodynamics and TST. They underscored that far-from-equilibrium dissolution rates decrease with increasing pH for pH $\leq$ 10.6–2,300 T^{-1} (with T in K), which means for pH $\leq$ 2.9 at 25°C; rates are pH independent up to pH close to 8 at 25°C, but they increase with pH above this threshold. In the lower pH range, the rate constants for albite and adularia dissolution range from ~10^{-9} to ~10^{-3} mol m^{-2} s^{-1}, in the temperature interval 25–200°C, suggesting that apparent activation enthalpies are close to 85 kJ mol^{-1}. In the intermediate pH range, dissolution rate constants for adularia are pH independent and range from ~10^{-12} to ~ 10^{-7} mol m^{-2} s^{-1}, in the temperature interval 25–650°C, consistent with an activation enthalpy of ~40 kJ mol^{-1}.

Since the mid of the 1980s, dissolution rates of feldspars were measured in the laboratory using reactors of different type and under different conditions to investigate the effects of several factors, such as pH, temperature, concentrations of both inorganic solutes and organic ligands, ionic strength and saturation state with respect to the feldspar mineral. Some of these works, representing the sources of most rate data plotted in Figs. 6.29–6.32, are briefly recalled hereunder.

Knauss and Wolery (1986) measured the dissolution rate of albite at 25°C, in the pH range 4–10, and at 70°C, in the pH range 1.39–11.75. They used a single-pass flow-through system. Experiment duration was of 50 days. Dissolution rates were calculated on the basis of release of Si, and, in some cases, of Al and Na as well. The aqueous solution was undersaturated with respect to all possible secondary minerals. Experimental data resulted to be consistent with the TST model of Helgeson et al. (1984).

Mast and Drever (1987) investigated the influence of oxalate on the steady-state release of Si from oligoclase at different pH values, from 4 to 9, and found minor effects for oxalate concentrations of 0.5 and 1 mM.

Bevan and Savage (1989) measured the dissolution kinetics of orthoclase, both with and without 0.02 M oxalic acid, at 70 and 95°C and pH 1, 3.6 and 9. They found significant increases in dissolution rates at pH 3.6 and 9 both at 70 and 95°C, but rates decreased at pH 1.

Casey et al. (1991) carried out batch dissolution experiments in 0.01 N HCl (pH = 2) at 25°C on nine plagioclase specimens of different composition from an almost pure albite ($Ab_{0.97}Or_{0.01}An_{0.02}$) to an anorthite-rich plagioclase ($Ab_{0.08}An_{0.92}$). They found that Si-based dissolution rates increase about three orders of magnitude from $(1.1 \pm 0.6) \times 10^{-11}$ mol m^{-2} s^{-1} for the almost pure albite to $(6.4 \pm 0.4) \times 10^{-9}$ mol m^{-2} s^{-1} for the anorthite-rich plagioclase. These changes in rate with plagioclase composition are much higher than the uncertainties assigned to any single composition due to several possible factors (see discussion in Casey et al., 1991). Casey et al. (1991) proposed the following empirical relation linking the dissolution rate at pH 2 (in HCl solutions, mol m^{-2} s^{-1}) and the anorthite mole fraction in plagioclase, X_{An}:

$$\log r_{+,\mathrm{pH2}} = -10.748 - 1.3508 \cdot X_{An} + 5.0524 \cdot X_{An}^2. \tag{6-111}$$

Casey et al. (1991) also suggested a similar equation for the rates measured at pH 3 by previous authors, again in HCl solutions:

$$\log r_{+,\mathrm{pH3}} = -11.023 - 1.131 \cdot X_{An} + 3.176 \cdot X_{An}^2. \tag{6-112}$$

Amrhein and Suarez (1992) performed batch dissolution experiments at 25°C to determine the effects of several experimental parameters on the rates of anorthite (An_{93}) dissolution, comprising agitation, particle size, suspension density, wetting and drying cycles, drying temperature, sequential rinses, ionic strength, and the addition and removal of products.

Welch and Ullman (1993) investigated the steady-state rates of plagioclase dissolution in solutions containing different organic acids, including acetic acid, propionic acid and several polyfunctional acids, i.e. oxalic, citric, succinic, pyruvic and 2-ketoglutarric

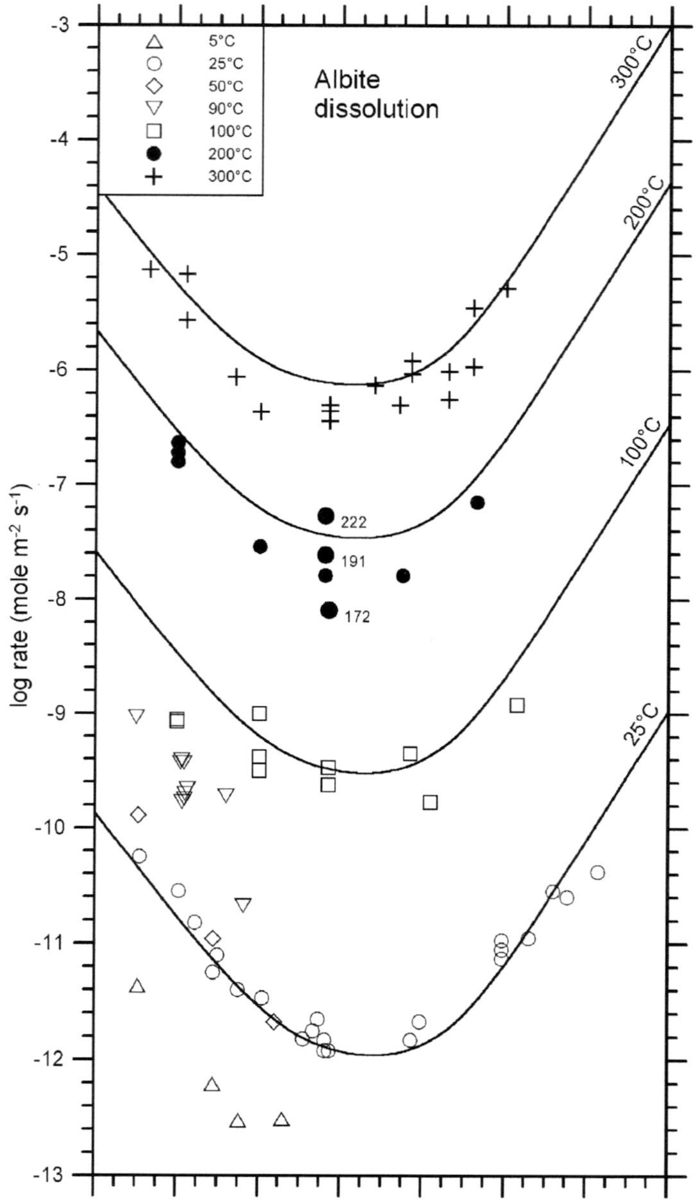

Figure 6.29. Log-rate of albite dissolution vs. pH, also showing the dissolution rate obtained by summing the contributions of the acidic, neutral and basic mechanisms reported by Palandri and Kharaka (2004, see Table 6.1). Experimental data from 100 to 300°C are from Hellmann (1984), those at 25°C are from Chou and Wollast (1985), and those at 5, 50 and 90°C are from Chen and Brantley (1997).

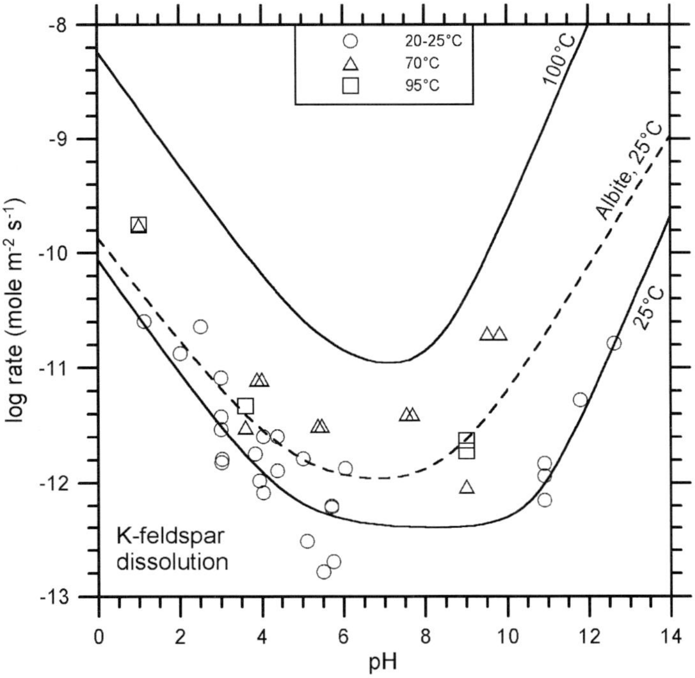

Figure 6.30. Plot of the log-rate of K-feldspar dissolution vs. pH, also showing the dissolution rate obtained by summing the contributions of the acidic, neutral and basic mechanisms reported by Palandri and Kharaka (2004, see Table 6.1) for K-feldspar and albite. Experimental data at 20–25°C are from Wollast (1967), Tan (1980), Manley and Evans (1986), Schweda (1989), Lundström and Ohman (1990), Sun (1994), Knauss and Copenhaver (1995) and Stillings and Brantley (1995); data at 70°C are from Bevan and Savage (1989), Sun (1994) and Knauss and Copenhaver (1995); data at 90°C are from Bevan and Savage (1989).

acids. Dissolution rates in solutions containing organic acids were found to be up to ten times greater than those in aqueous solutions containing inorganic acids at the same pH. The polyfunctional acids resulted to be more effective than acetate and propionate in promoting dissolution.

Oelkers et al. (1994) reported the steady-state dissolution rates of albite at 150°C and pH 9 as a function of chemical affinity (from -50 to -5 kJ mol^{-1} approximately), under different dissolved Al and Si concentrations and constant ionic strength of 0.1 mol kg^{-1} (through addition of NaCl). They fitted the experimental data to the following relationship (r in mol m^{-2} s^{-1}):

$$r = 2.9\times10^{-8}\cdot\left(\frac{a^{3}_{H^+}}{a_{Al^{3+}}}\right)^{1/3}\cdot\left[1-\exp\left(\frac{A}{3RT}\right)\right], \tag{6-113}$$

which is similar to equation (6-85), but assumes $\sigma = 3$ (see equation (6-57) and related discussion).

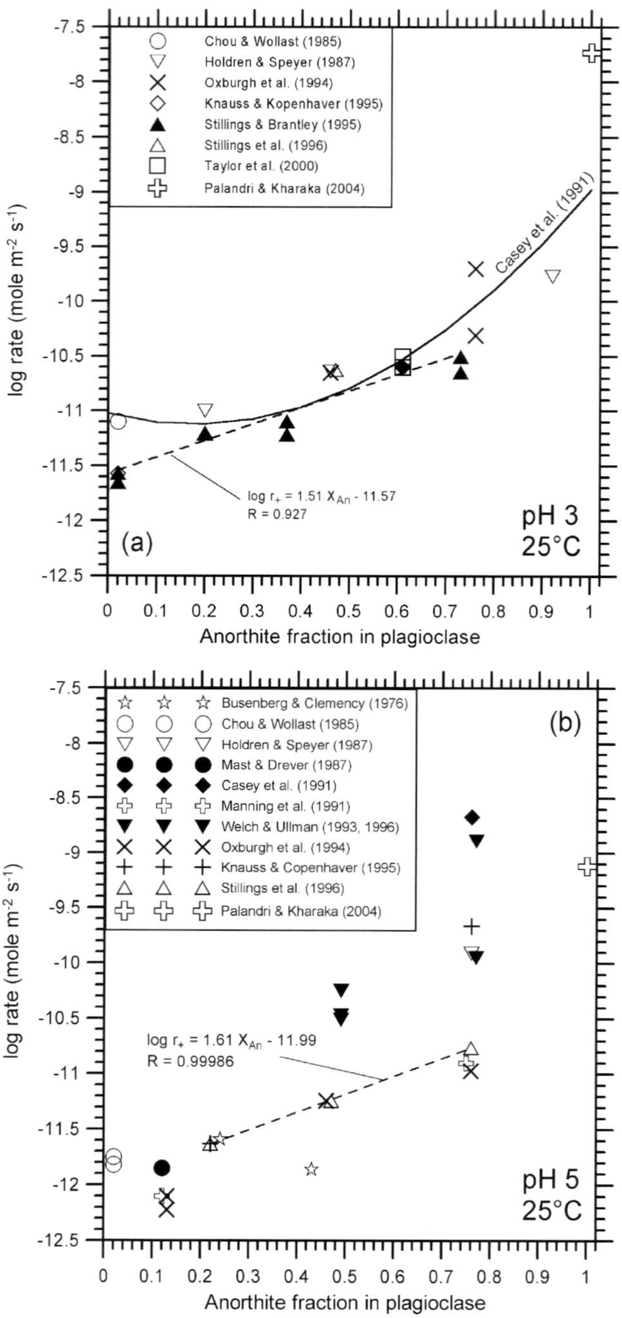

Figure 6.31. The log-rate of plagioclase dissolution at 25°C and constant pH, 3 ± 0.4 in (a) and 5 ± 0.4 in (b), is plotted against the anorthite fraction in plagioclase. All the data were measured through laboratory experiments (see legend) except those of anorthite, which were computed based on the rate equations by Palandri and Kharaka (2004).

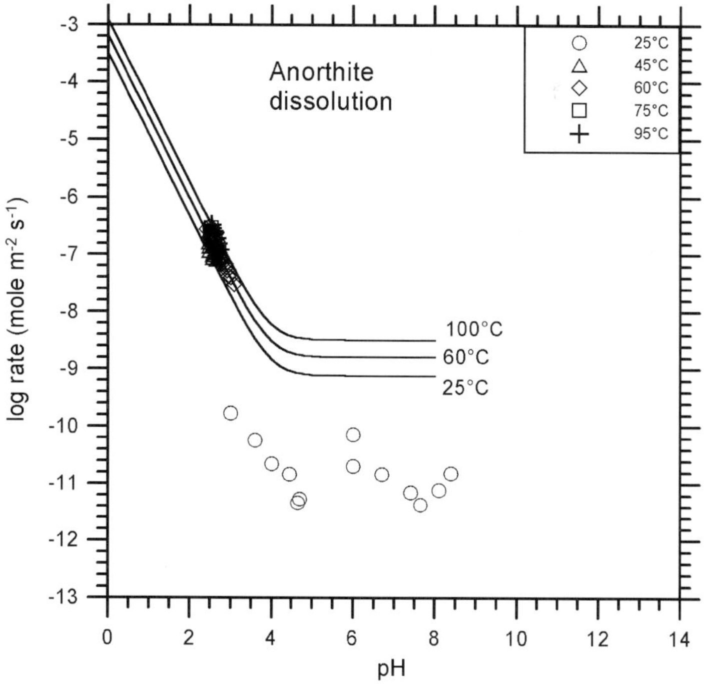

Figure 6.32. Plot of the log-rate of anorthite dissolution vs. pH, also showing the dissolution rate obtained by summing the contributions of the acidic and neutral mechanisms reported by Palandri and Kharaka (2004, see Table 6.1). Experimental data at 25°C are from Bailey (1974), Holdren and Speyer (1987), Amrhein and Suarez (1992) and Berg and Banwart (2000); those from 45 to 95°C are from Oelkers and Schott (1995).

Gautier et al. (1994) determined the steady-state dissolution rates of a K-rich feldspar ($K_{0.81}Na_{0.15}Ba_{0.03}Al_{1.05}Si_{2.96}O_8$) as a function of chemical affinity (from -90 to -5 kJ mol^{-1} approximately), dissolved Si (1×10^{-6} to 5×10^{-4} mol kg^{-1}) and dissolved Al (4×10^{-7} to 5×10^{-4} mol kg^{-1}). Total K molality was fixed at 5×10^{-3}, temperature was maintained at 150°C and pH was kept at 9.0. The experiments by Oelkers et al. (1994) and Gautier et al. (1994) were performed in titanium mixed-flow reactors. In addition, Gautier et al. (1994) carried out similar experiments in a batch reactor to prove the consistency between dissolution rates measured in open and closed systems. Gautier et al. (1994) fitted their experimental data to the following equation (r in mol m^{-2} s^{-1}):

$$r = 1.7 \times 10^{-13} \cdot \left(\frac{1}{a_{\mathrm{Al(OH)}_4^-} \cdot a_{\mathrm{H}^+}} \right)^{1/3} \cdot \left[1 - \exp\left(\frac{A}{3RT} \right) \right]. \tag{6-114}$$

Hellmann (1994) investigated the far-from-equilibrium dissolution kinetics of Amelia albite ($Ab_{100}An_0Or_0$) at 100, 200 and 300°C under different pH values, from 1.4 to 10.3. He used a single-pass tubular flow system and computed the dissolution rates based on both mass loss of the sample and dissolved Si in the output solution. Böhmite was the only precipitating secondary phase under acidic conditions. The log-rate of dissolution vs. pH curves resulted to be U-shaped, for any given temperature, with approximately equal absolute values of the slopes in the acidic- and basic-pH regions, i.e. $|n_H| \cong |n_{OH}|$; absolute values of the slopes were found to increase with temperature from 0.2 to 0.6. Data are consistent with apparent activation energies of 89, 69 and 85 kJ mol^{-1} in the acidic-, neutral- and basic-pH regions, respectively.

Oelkers and Schott (1995) measured the steady-state dissolution rates of anorthite (An_{96}) as a function of dissolved Si, Al and Ca concentrations from 45 to 95°C and in the pH range 2.4–3.2. They used a mixed-flow reactor, whose parts in contact with the high-temperature aqueous solution are made of titanium (see Section 6.4.4 for further details).

Knauss and Copenhaver (1995) investigated the effect of malonate (1, 10 and 100 mmol kg^{-1}) on the dissolution kinetics of albite and microcline in the pH range 4–10 at 70°C. Experiments were performed in flow-through reactors. Possible effects of pH buffers were taken into account by performing dissolution experiments in buffer solutions with and without malonate and comparing the obtained results. Dissolution rates were found to increase with increasing malonate concentrations, and the maximum effect was observed at pH close to 5.5, whereas the effects of malonate was insignificant at basic-pH values.

Stillings and Brantley (1995) dissolved microcline, albite, oligoclase, labradorite and bytownite in flow-through reactors at 25°C and pH 3, in both HCl–NaCl and HCl–$(CH_3)_4NCl$ aqueous media to study the effects of feldspar composition and ionic strength upon dissolution kinetics. Owing to the significant increase in BET surface area (1.5–7 times), the dissolution rates were normalized to the final surface area, contrary to the usual practice (see Section 6.5.2.1), and this resulted in rates lower than those estimated by other authors. A linear increase in the dissolution log-rates with increasing anorthite fraction was observed (see Section 6.6.4.2). Dissolution rates were found to decrease with increasing concentrations of NaCl and, to a lesser extent, of $(CH_3)_4NCl$. Stillings and Brantley (1995), as well as Brantley and Stillings (1996) and Chen and Brantley (1997) fitted feldspar dissolution rate data to a Langmuir competitive adsorption model. These authors assumed that feldspar dissolution is controlled by hydrolysis of the Al–O–Si bridging oxygen, i.e. by the reaction,

$$\equiv \mathrm{Al-O-Si} + \mathrm{H_2O} \rightleftharpoons \equiv \mathrm{SiOH} + \equiv \mathrm{AlOH}, \tag{6-115}$$

and proposed the following rate equation:

$$r = k \cdot n_s^n \cdot T_s \cdot \left(\frac{K_H^* \cdot a_{H^+}}{1 + K_H^* \cdot a_{H^+} + K_M^* \cdot a_{M^{b+}} + \sum_i K_i^* \cdot a_{C_i^+}} \right)^n, \tag{6-116}$$

where r is the steady-state, area-normalized dissolution rate of feldspar; k is the rate constant; n_s is the fraction of ≡Al–O–Si (exchange) sites at the dissolving feldspar surface; T_s is the number of surface sites per unit area; and K_H^* and K_M^* represent the adsorption constants for H^+ and M^{b+} onto the ≡Al–O–Si (exchange) site, respectively. The effect of adsorption of any other species C_i (including Al^{3+} ion) onto the ≡Al–O–Si (exchange) site is taken into account by the term $\sum_i K_i^* a_{C_i^+}$. The value of n depends on the position of the exchange sites contributing to dissolution. It varies from 1, when proton-exchange sites on the surface only contribute to dissolution, to 0.5, when proton-exchange sites throughout the layer contribute. A similar model was early proposed by Murphy and Helgeson (1987). Note that this model is also analogous to that proposed by Pokrovsky and Schott (2000b) for describing the reaction occurring at the silica-rich surface sites during forsterite dissolution (see equation (6-101) and related discussion).

Stillings et al. (1996) studied the effect of oxalic acid on six different feldspars, namely microcline ($Ab_{22}An_0Or_{78}$), albite ($Ab_{97}An_2Or_1$), oligoclase ($Ab_{73}An_{22}Or_5$), andesine ($Ab_{51}An_{47}Or_2$), labradorite ($Ab_{50}An_{43}Or_7$) and bytownite ($Ab_{24}An_{76}Or_0$). Dissolution rates were determined in continuously stirred flow-through reactors or fluidized-bed reactors, in the pH range 3–7, either with oxalic acid (1–8 mmol kg^{-1}) or without it. The ligand-promoted rate was defined as follows:

$$r_L = r_{tot} - r_H, \tag{6-117}$$

(where r_{tot} is the total dissolution rate in oxalic acid and r_H the proton-promoted rate) and was estimated to be related to the activities of bioxalate and oxalate ions through the relation (r_L in mol m^{-2} s^{-1}):

$$r_L = 10^{8.8} \cdot \left(a_{HC_2O_4^-} + a_{C_2O_4^{2-}}\right)^{0.75}. \tag{6-118}$$

Welch and Ullman (1996) investigated the dissolution kinetics of four plagioclases (an albite, two labradorites and a bytownite) in both inorganic aqueous media and solutions containing oxalic acid. Dissolution experiments were carried out in a flow-through reactor at pH values from 3 to 10. Steady-state dissolution rates in acid inorganic media were fitted to the rate equation (r_H in mol m^{-2} s^{-1}; compare with equation (6-75)):

$$r_H = k_H \cdot a_{H^+}^{n_H} \tag{6-119}$$

where k_H and n_H depend on the Al fraction, X_{Al} = [Al/(Al+Si)], as expressed by the empirical relations

$$\log k_H = -11.24 + 25.98 \cdot X_{Al}^2 \tag{6-120}$$

and

$$\log n_H = -0.052 + 4.23 \cdot X_{Al}^2. \tag{6-121}$$

Steady-state dissolution rates in oxalic acid were found to be up to one order of magnitude higher than in inorganic media and to exhibit a similar dependence on pH and

X_{Al}, apart from bytownite. According to the authors, the strong dependence of both proton-promoted and organic-promoted rates on X_{Al} suggests that proton and organic ligands attack the mineral surface at the same Al sites.

Alekseyev et al. (1997) investigated the dissolution rates of sanidine in 0.1 mol kg^{-1} $NaHCO_3$ and of albite in 0.1 mol kg^{-1} $KHCO_3$ at 300°C, 88 bar and pH close to 9. Experiments were performed in Pt-sealed capsules. Initial dissolution of both sanidine and albite was congruent, but it became incongruent with production of analcime and sanidine, respectively, as secondary mineral phases.

Blake and Walter (1999) carried out separate batch experiments to study the effects of oxalate and citrate (0.5, 3, 10 and 20 mM) on the dissolution rates of albite, orthoclase and labradorite (An_{70}) at 80°C, pH 6, in solutions of different ionic strength (0.02–2.2 M NaCl). Both the amounts of Si and Al released from the three feldspars and their dissolution rates were found to increase with increasing concentrations of organic acids. In organic acid-free media, labradorite dissolution decreased with increasing ionic strength. In aqueous solutions containing both oxalate and NaCl, the increase in dissolution rate is suppressed by the ionic-strength effect.

Berg and Banwart (2000) performed an experimental investigation on the dissolution kinetics of anorthite at 25°C, in $NaClO_4$–$NaHCO_3$ solutions of constant ionic strength (0.05 M), and different pH, in the range 5.5–8.5. The aqueous solutions were maintained at a constant P_{CO_2} of 0.1 and 0.01 bar. A separate set of experiments was carried out in 0.05 M $NaClO_4$ solutions kept under a N_2 atmosphere. Dissolution experiments were performed in a thin-film continuous-flow reactor. The authors computed the rate of carbonate-promoted dissolution of anorthite, $r_{+,C}$ (in mol m^{-2} s^{-1}), by subtracting the rate measured in the absence of CO_2 from the total dissolution rate. It was found that $r_{+,C}$ depends on pH and P_{CO_2}, as described by the following empirical rate law:

$$\log r_{+,C} = (-10.51 \pm 1) - 0.24 \cdot pK_2 + 0.24 \cdot \log P_{CO_2} + 0.48 \cdot \text{pH}, \tag{6-122}$$

where pK_2 = 17.64 (at 25°C in 0.05 M $NaClO_4$) refers to the following reaction:

$$CO_{2(g)} + H_2O = CO_3^{2-} + 2H^+. \tag{6-123}$$

Taylor et al. (2000b) determined the dissolution rates and Sr release of a labradorite sample ($Ab_{37}An_{61}Or_2$) as a function of the saturation state, at 25°C, by means of flow-through column experiments. The dependence of the dissolution rate of labradorite on the saturation state of the aqueous solution was described through the equation,

$$r_+ = -k_{\min}\left\{0.76 \cdot \left[1 - \exp\left(-1.3 \cdot 10^{-17} \cdot \left(\frac{|A|}{RT}\right)^{14}\right)\right] - 0.24 \cdot \left[1 - \exp\left(-0.35 \cdot \frac{|A|}{RT}\right)\right]\right\} \tag{6-124}$$

where $k_{\min}$ is the rate constant under far-from-equilibrium conditions ($10^{-10.6 \pm 0.1}$ mol m^{-2} s^{-1}) and A is the thermodynamic affinity with respect to labradorite, which is called Gibbs free energy of reaction by the authors. The same equation can be used to describe the dependence of the Sr release rate on the saturation state of the aqueous solution, where k_{min} is the far-from-equilibrium Sr release rate ($10^{-13.2\pm0.1}$ mol Sr m^{-2} s^{-1}).

Welch and Ullman (2000) investigated the temperature dependence of bytownite ($Ab_{23}An_{77}Or_1$) dissolution by means of batch experiments at temperatures of 5, 20 and 35°C and pH of 6.0–6.5. The mineral was dissolved in pure water and in 1 mM solutions of KNO_3, sodium acetate, sodium oxalate and sodium gluconate. The dissolution rate and temperature dependence in solutions containing acetate were virtually equal to those in the inorganic aqueous solutions, whereas a significant increase in the dissolution rate was observed for the solutions containing oxalate and gluconate, especially at low temperatures.

Carroll and Knauss (2005) measured the dissolution rate of labradorite ($An_{0.6}$) in acidic-aqueous solutions (pH close to 3.2), at variable temperature (30–130°C), and as a function of dissolved Al, 10^{-6} to 10^{-3} mol kg^{-1} (see also Carroll and Knauss, 2001). Dissolution experiments were carried out in a mixed-flow reactor. To investigate possible CO_2 effects, experimental runs were carried out both in CO_2-rich aqueous solutions (m_{CO_2} = 0.61 ± 0.07) and in dilute HCl solutions (6.5 × 10^{-4} mol kg^{-1}) equilibrated with atmospheric CO_2, which have a pH close to 3.2 but an m_{CO_2} of 1.2 × 10^{-5} only. However, no direct effect of dissolved CO_2 was observed on the dissolution rate of labradorite. In other words, the enhancement in the dissolution rate in CO_2-rich aqueous solutions is due to the increase in H^+ ion concentration derived from dissociation of dissolved CO_2. This effect as well as that of dissolved Al on the dissolution rate of labradorite were taken into account by fitting the experimental data to equation (6-84), without the affinity term, but adding an Arrhenius-type term, thus obtaining,

$$r_+ = k_+ \cdot 10^{\frac{-E_a}{2.303\,RT}} \cdot \frac{\left(a^3_{H^+}/a_{Al^{3+}}\right)^n \cdot K_T}{1 + K_T \cdot \left(a^3_{H^+}/a_{Al^{3+}}\right)^n}, \tag{6-125}$$

where $n = 0.31$, $k_+ = 10^{-1.69}$ mol m^{-2} s^{-1} and E_a = 42.1 kJ mol^{-1}. Note that K_T depends on temperature as expressed by equation (5-40), which is a truncated form of equation (2-41) obtained assuming $\Delta_r C^\circ_P = 0$ (i.e. $\Delta a = \Delta b = \Delta c = 0$) and setting Tr = 303.15 K, K_{Tr} = 4.49 and $\Delta_r H^\circ_{Tr}$ = 2.26 kJ mol^{-1}.

6.6.4.2. pH dependence of the dissolution kinetics of feldspars

Albite. Figure 6.29 shows that the dissolution rate of albite is strongly dependent upon pH and temperature. At all temperatures, from 5 to 300°C, the dissolution rate is at a minimum in the neutral range and increases in both the acidic- and basic-pH intervals. As already underscored, this behaviour is shared by all feldspars and most (if not all) silicates (e.g. Helgeson et al., 1984; Murphy and Helgeson, 1987). Besides, this pattern is in line with the occurrence of different dissolution mechanisms, which are promoted by (i) hydrogen ion, in the acidic-pH region, (ii) water, in the neutral-pH region and (iii) hydroxyl ion, in the basic-pH region (see Section 6.4.3). However, this interpretation is not universally accepted and some researchers (e.g. Brady and Walther, 1989; Schweda, 1990; Blum and Lasaga, 1991) recognized only the two pH-dependent mechanisms.

Figure 6.29 also shows the rate laws at 25, 100, 200 and 300°C obtained by summing the contributions of the acidic, neutral and basic mechanisms reported by Palandri and Kharaka (2004, see Table 6.1). These rate laws satisfactorily reproduce most available experimental data in the pH range 1–12.5, and in the temperature interval 5–300°C.

K-feldspar. The dissolution log-rate of K-feldspar is plotted as a function of pH in Fig. 6.30. At 25°C, below pH 6, the dissolution rate of K-feldspar differs from that of albite by 0.2–0.4 log-units only. This similarity in the dissolution kinetics of the two alkali feldspars suggests that the nature of the alkaline cation is of little importance, at least for the dissolution mechanism in the acidic-pH region. It is likely that for pH < 6, the surfaces of albite and K-feldspar experience relatively quick exchange of alkali cations for protons, forming a protonated-feldspar surface layer. This process implies that the acidic dissolution rates of the two alkali feldspars are primarily controlled by the nature of the tetrahedral framework and the breaking of the Si–O and/or Al–O bonds of this framework.

In apparent contrast with these laboratory experimental results, K-feldspar weathering in soils proceeds less quickly than albite alteration, as already recognized by Goldich (1938). These two observations can be reconciled considering that soil waters are usually much closer to saturation with K-feldspar than with albite, and owing to these distinct thermodynamic affinities, the dissolution rates of K-feldspar in soils are strongly decreased with respect to those of albite. A similar picture is found in natural waters (e.g. Gambardella et al., 2005). The moral of the story is that caution must be exerted when comparing far-from-equilibrium rates measured in laboratory experiments with field-weathering rates.

Although relatively few measurements of the dissolution rate are available for K-feldspar in the basic-pH region (Fig. 6.30), these rate data are significantly different from those of albite. This suggests that alkali cations influence the dissolution mechanism of feldspars in the basic-pH region, contrary to what is observed at low-pH values.

Plagioclase. The dissolution rates of plagioclase measured by several authors (Busenberg and Clemency, 1976; Chou and Wollast, 1985; Holdren and Speyer, 1987; Mast and Drever, 1987; Manning et al., 1991; Welch and Ullman, 1993; Oxburgh et al., 1994; Knauss and Copenhaver, 1995; Stillings and Brantley, 1995; Stillings et al., 1996; Welch and Ullman, 1996; Taylor et al., 2000b) at 25°C and pH 3 and 5 (± 0.4 pH units) were compiled and plotted in Fig. 6.31 as a function of the anorthite fraction in plagioclase, X_{An}. The dissolution rates of anorthite were computed based on the rate equations by Palandri and Kharaka (2004, see below).

At both pH values, the log-rate of plagioclase dissolution increases significantly with X_{An}, in spite of some disagreements among the data of different authors, which are particularly evident at pH 5. This dependence of the dissolution kinetics of plagioclase on its composition was already underscored by several authors (see the review by Blum and Stillings, 1995 and references therein). Interestingly, taking into account the data of a single study, e.g. that by Stillings and Brantley (1995) at pH 3 and that by Stillings et al.

(1996) at pH 5, the log-rate of dissolution appears to depend linearly on plagioclase composition, at least for $X_{An} < 0.75$ (see regression equations in Fig. 6.31). In spite of the discrepancies among the data of different authors, the dependence of the log-rate of dissolution on X_{An} appears to deviate from linearity above this compositional threshold, with the anorthite-rich plagioclases dissolving much faster than predicted, based upon the anorthite-poor plagioclases. A non-linear dependence of the dissolution rate on plagioclase composition was instead proposed by Casey et al. (1991) (see equations (6-111) and (6-112) and related discussion).

Available dissolution rates of anorthite are plotted against pH in Fig. 6.32. Again, although the data of different researchers are not mutually consistent, it seems likely that the reaction order with respect to hydrogen ion for the acidic dissolution mechanism, n_H, is equal to 1.5, as suggested by Oelkers and Schott (1995) (see *also* Table 6.1). This value is significantly different from the n_H of albite dissolution, 0.5 (see Table 6.1 and Fig. 6.29).

Although the plagioclase composition at which n_H deviates from 0.5, approaching 1.5 for $X_{An} \rightarrow 1$, cannot be clearly established, it is probably close to X_{An} values of 0.75–0.80 (see Blum and Stillings, 1995 and references therein). This variation in n_H represents a further evidence, in addition to the deviation from linearity of the log-rate/X_{An} relationship, for the existence of a critical Si/Al ratio at $An_{75\text{-}80}$, which remarkably influences the dissolution kinetics of plagioclase. In albite (Si/Al = 3), preferential removal of Al during dissolution leaves behind a framework of partially linked SiO_4 tetrahedra. Consequently, completion of albite dissolution requires breaking of Si–O bonds. In contrast, preferential removal of Al during dissolution of anorthite (Si/Al = 1) leaves behind fully isolated SiO_4 tetrahedra, and breaking of Si–O bonds is not required to destroy the anorthite lattice (as already noticed in Section 6.3.2). The existence of a critical Si/Al ratio below which breaking of Si–O bonds does not take place during plagioclase dissolution is therefore conceivable. In other words, plagioclase dissolution follows an anorthite-like mechanism below this critical Si/Al ratio.

It must be underscored that the data for oligoclase, andesine, labradorite and bytownite reported in Table 6.1 have to be taken with caution, since they may be affected by compositional complications.

6.6.4.3. Activation energies for the dissolution kinetics of feldspars

Apparent activation energies for the dissolution kinetics of feldspar have been compiled by Blum and Stillings (1995). Those of albite generally range between 44 and 89 kJ mol^{-1}, those of K-feldspar are usually comprised between 35 and 70 kJ mol^{-1}, and those of anorthite are mostly from 33 to 81 kJ mol^{-1}, although an apparent activation energy of only 18.4 kJ mol^{-1} was found by Oelkers and Schott (1995).

6.6.5. Dissolution precipitation rates of silica minerals

As partly anticipated in Section 6.1, a differential rate equation for the dissolution and precipitation of silica minerals was proposed by Rimstidt and Barnes (1980), based on the stoichiometry of the elementary reaction (6-15). The precipitation rate of silica minerals is expressed by equation (6-16), whereas their dissolution rate in pure water can be written as,

$$r_+ = \frac{A_{SiO_2(s)}}{M_{H_2O}} \cdot k_+ \cdot a_{SiO_2(s)} \cdot a^2_{H_2O}. \tag{6-126}$$

If the considered solid phase is pure and the aqueous solution is dilute enough, the activities of the silica mineral and water are both close to unity, and equation (6-126) reduces to,

$$r_+ = \frac{A_{SiO_2(s)}}{M_{H_2O}} \cdot k_+. \tag{6-127}$$

Therefore, the dissolution of silica minerals is expected to be a zero-order reaction, whereas their precipitation is first order with respect to the molality of H_4SiO_4, as already noted in Section 6.1 (see equation (6-16)). Incidentally, this first-order behaviour has been confirmed experimentally under hydrothermal conditions by Berger et al. (1994a).

Rimstidt and Barnes (1980) found that the temperature dependence of the precipitation (backward) rate constant, k_-, for all silica minerals is described by the following expression (temperature in K):

$$\log k_-\ (\text{s}^{-1}) = -0.707 - \frac{2598}{T}, \tag{6-128}$$

which holds up to ~300°C for quartz and to ~200°C for amorphous silica. Since the thermodynamic constants of the dissolution reaction of silica minerals (equations (6-15) and (5-31)) are well-known (see Section 5.2.3), Rimstidt and Barnes (1980) computed the dissolution (forward) rate constants by means of the principle of detailed balancing (see Section 6.1), obtaining the following expressions:

$$\log k_{+,\text{quartz}} = 1.174 - 2.028 \times 10^{-3}\,T - \frac{4158}{T} \tag{6-129}$$

and

$$\log k_{+,\text{am.SiO}_2} = -0.369 - 7.889 \times 10^{-4}\,T - \frac{3438}{T}, \tag{6-130}$$

for quartz and amorphous silica, respectively.

According to Rimstidt and Barnes (1980), the apparent activation energy for quartz dissolution is 67.4–76.6 kJ mol^{-1}, whereas that for amorphous silica dissolution is somewhat lower, 60.9–64.9 kJ mol^{-1}. The apparent activation-energy values for quartz are in satisfactory agreement with more recent evaluations, such as those by Tester et al. (1994), 89 ± 5 kJ mol^{-1}, and Dove (1994), 66.0–82.7 kJ mol^{-1}.

Tester et al. (1994) investigated the dissolution kinetics of quartz in pure water by means of carefully designed experiments carried out in five different apparatuses and took into account the available dissolution rate data from 10 different studies by Kitahara (1960), van Lier et al. (1960), Siebert et al. (1963), Weill and Fyfe (1964), Rimstidt and Barnes (1980), Blum et al. (1990), Brady and Walther (1990), Dove and Crerar (1990), Bennett (1991) and Berger et al. (1994a). Both BET and geometric surface areas were utilized by Tester et al. (1994) to obtain two series of dissolution rate data, whose temperature dependence is described by the following two Arrhenius-type equations (R in J mol^{-1} K^{-1}, T in K):

$$\log k_{+,\text{quartz,geom.}} = 2.4409 - \frac{90100}{\ln(10)\cdot R\cdot T} \tag{6-131}$$

and

$$\log k_{+,\text{quartz,BET}} = 1.3802 - \frac{87700}{\ln(10)\cdot R\cdot T}. \tag{6-132}$$

These two temperature functions hold from 25 to 625°C and fit the two datasets very well, as shown in Fig. 6.33.

Equations (6-129), (6-131) and (6-132) should be valid at circumneutral pH, since the pH of an aqueous solution containing dissolved species deriving from quartz dissolution only, and in equilibrium with quartz ranges from 6.84 at 25°C to 5.45 at 200°C, as indicated by simple speciation calculations carried out by means of the code EQ3NR (Wolery, 1992).

The pH dependence of the dissolution rates of silica minerals was investigated by numerous studies (see Dove, 1995). Among the others, Knauss and Wolery (1988) studied the dissolution of quartz as a function of pH, from 1.4 to 11.8, at 70°C. Rates resulted to be pH independent below pH 6 approximately, consistent with the findings by Rimstidt and Barnes (1980) and Tester et al. (1994), as discussed above. However above pH 6, the dissolution rate was found to increase with pH, with $n_{OH} = -0.5$ (see equation (6-77)).

Several studies have shown that alkali and alkali-earth cations increase the rate of quartz dissolution (see Dove, 1995). Dove and Elston (1992) described the kinetics of quartz dissolution in water and sodium chloride solutions adopting a surface reaction model and taking into account 79 experimental determinations. The obtained rate equation is valid at 25°C, pH 2–13 and NaCl molality from 0 to 0.5. These findings were extended by Dove (1994), who proposed a general equation for the dissolution kinetics of quartz in water and sodium chloride aqueous solutions, which is applicable from 25 to 300°C, from pH 2 to 12 and from Na molality 0 to 0.3. The surface structure of quartz in contact with an aqueous NaCl solution was assumed to be dominated by the surface complexes $\equiv$SiOH°, $\equiv$SiO$^-$ and $\equiv$SiONa°, which are involved in the reactions

$$\equiv \text{SiOH}^\circ \rightleftharpoons \equiv \text{SiO}^- + \text{H}^+ \tag{6-133}$$

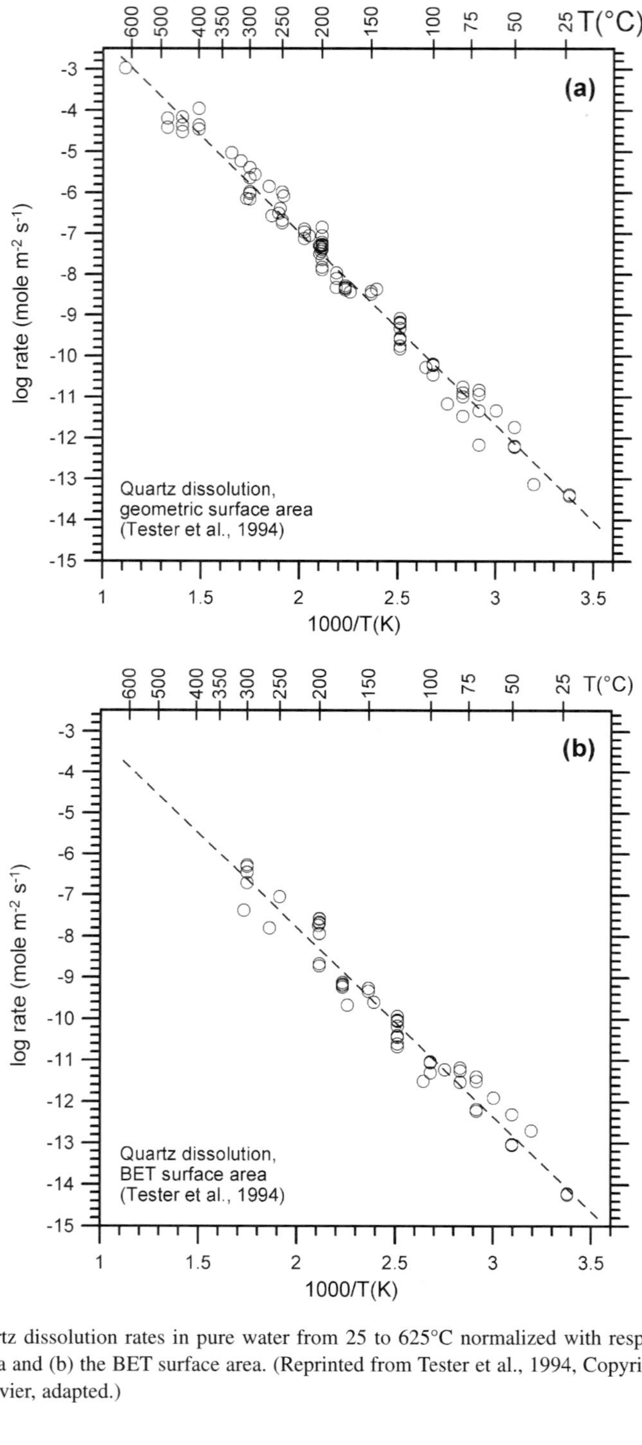

Figure 6.33. Quartz dissolution rates in pure water from 25 to 625°C normalized with respect to (a) the geometric surface area and (b) the BET surface area. (Reprinted from Tester et al., 1994, Copyright 1994 with permission from Elsevier, adapted.)

and

$$\equiv \mathrm{SiOH}^\circ + \mathrm{Na}^+ \rightleftharpoons \equiv \mathrm{SiONa}^\circ + \mathrm{H}^+, \tag{6-134}$$

whose log *K* are equal to −6.8 and −7.1 at 25°C, 1 bar. The mass balance,

$$X_{\equiv\mathrm{SiOH}^\circ} + X_{\equiv\mathrm{SiO}^-} + X_{\equiv\mathrm{SiONa}^\circ} = 1.0, \tag{6-135}$$

was also considered, where X_i denotes the mole fraction of the *i*th surface complex. The surface complexes $\equiv SiOH_2^+$ and $\equiv SiOH_2Cl^\circ$ were neglected, as they are important under strongly acidic solutions (pH < 2) only. Incidentally, the existence of the surface species $\equiv SiOH^\circ$, $\equiv SiO^-$ and $\equiv SiOH_2^+$ at the quartz–aqueous solution interface, at pH values from 0 to 10, has been recently confirmed and quantified by means of X-ray photoelectron spectroscopy (Duval et al., 2002).

Since the equilibrium constants of reactions (6-133) and (6-134) are known at 25°C, 1 bar only and are unknown at hydrothermal conditions, they were assumed to be independent of pressure and temperature by Dove (1994). However, the *in situ* pH was calculated by means of the code EQ3NR. The distribution of surface complexes was computed by means of the triple-layer model (Hayes and Leckie, 1987), assuming a constant total surface site density of 4.5 sites nm^{-2} and constant interfacial capacitances C_1 = 1.25 and C_2 = 0.20 Faraday m^{-2}. Note that some of these assumptions might be incorrect, based on the findings by Brady (1992), who carried out potentiometric titrations at 25 and 60°C and found that the surface chemistry of silica minerals is temperature-dependent.

Dove (1994) selected 271 experimental data, including available rate determinations and new rate measurements carried out in hydrothermal mixed-flow reactors. Based on the TST-equation (6-35) and separating the contributions of the different surface complexes, Dove (1994) obtained the following general rate equation (*R* in J mol^{-1} K^{-1}, *T* in K):

$$r_+ = \exp(-10.7)\times T\times\exp\left(\frac{-66000}{RT}\right)\times X_{\equiv\mathrm{SiOH}^\circ} + \exp(4.7)\times T\times\exp\left(\frac{-82700}{RT}\right)\times X^{1.1}_{\mathrm{SiO}^-_{\mathrm{TOT}}}, \tag{6-136}$$

where $X_{\equiv\mathrm{SiO}^-_{\mathrm{TOT}}} = X_{\equiv\mathrm{SiO}^-} + X_{\equiv\mathrm{SiONa}^\circ}$. It must be underscored that although the mole fraction of $\equiv SiOH^\circ$ sites is generally much greater than the mole fraction of $\equiv \mathrm{SiO}^-_{\mathrm{TOT}}$ sites (Fig. 6.34a), the contribution of the second term of equation (6-136) to the overall dissolution rate cannot be neglected, especially at high temperatures and high pH values (Fig. 6.34b,c). It is also important to underscore that the fitting of experimental data seems to be better under hydrothermal conditions than at low temperatures, at least referring to the experiments at 0 sodium molality.

Dove and Crerar (1990) investigated the dissolution kinetics of quartz at circumneutral pH in 0–0.15 molal solutions of NaCl, KCl, LiCl and $MgCl_2$ in the temperature interval of 100–300°C. They observed a marked increase in the dissolution rate, especially in NaCl and KCl solutions, which was explained adopting a Langmuir adsorption model. Further studies on this subject were carried out by Dove and Nix (1997) and Dove (1999), who proposed a competitive-adsorption rate model for the dissolution of quartz in mixed-cation

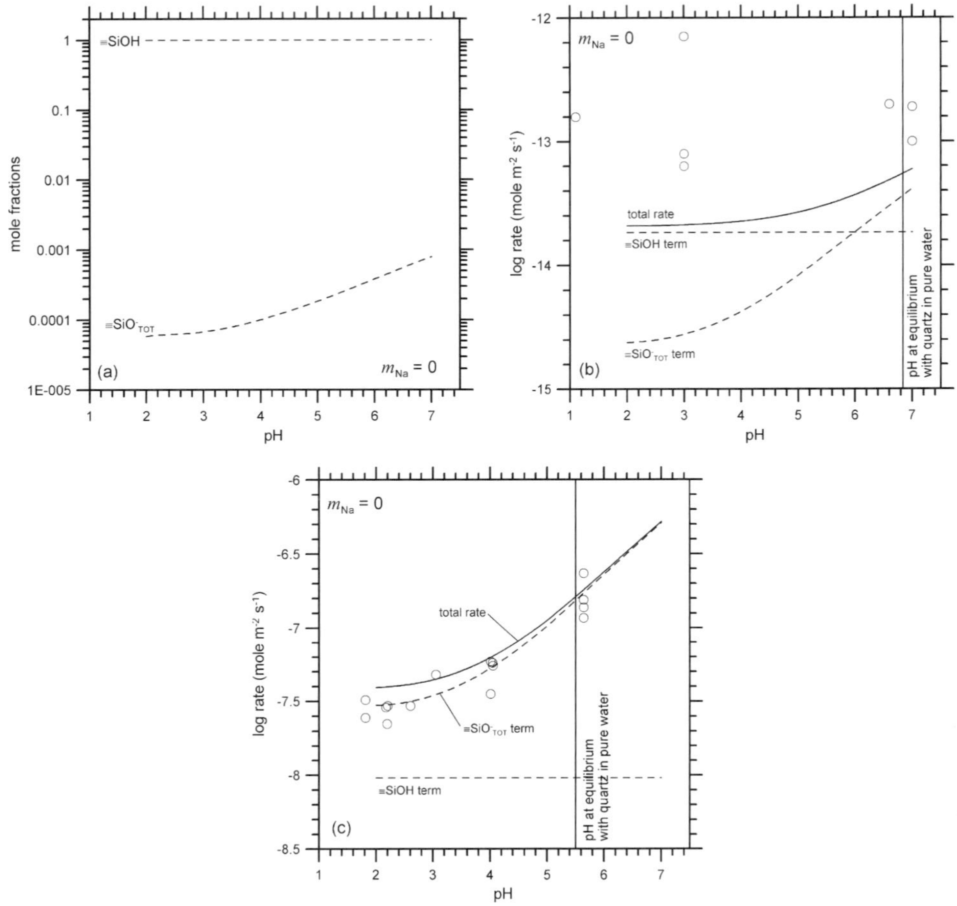

Figure 6.34. (a) Mole fractions of ≡SiOH° sites and ≡ SiO^-_{TOT} sites at 0 molal Na concentration from pH 2 to 7 (data from Dove, 1994); (b and c) Contribution of the ≡SiOH° term and of the ≡ SiO^-_{TOT} term of equation (6-136) to the overall dissolution rate as a function of pH at (b) 25°C and (c) 290°C.

solutions. These models work when (1) the dissolution rate increases strongly with the concentration of the cation, in the low-concentration region and (2) the dissolution rate becomes constant or nearly so for high concentrations of the cation, due to saturation of surface sites.

The dissolution kinetics of moganite, a recently discovered silica polymorph (see Sections 5.2.1, 5.2.2 and 5.2.3), was determined as a function of temperature from 25 to 200°C by Gislason et al. (1997). Experimental data at pH 3.5 and far-from-equilibrium conditions fit the following equation (rate constant in mol m^{-2} s^{-1}, R in J mol^{-1} K^{-1}, T in K):

$$\log k_{+,\text{moganite}} = -0.423 - \frac{70502}{\ln(10)\cdot RT}, \tag{6-137}$$

which is consistent with an apparent activation energy of 70.5 kJ mol^{-1}.

Gislason et al. (1997) also measured, at pH 3.5 and far-from-equilibrium conditions, the dissolution rate of quartz, which is described by the relation,

$$\log k_{+,\text{quartz}} = -0.0201 - \frac{80480}{\ln(10) \cdot RT}. \tag{6-138}$$

Quartz dissolution rates computed by means of this relation are comparable with those predicted by equation (6-129) at high temperatures and similar to those calculated by equation (6-132) at low temperatures. Comparison of equations (6-137) and (6-138) indicates that, at 25°C and pH 3.5, moganite dissolves 22 times faster than quartz.

6.6.6. Dissolution rates of silicate glasses

The dissolution kinetics of silicate glasses has received considerable attention, mainly for two different purposes: the modelling of radioactive wastes' confinement and the study of chemical weathering. Basaltic glasses are considered as the natural analogues for the industrial nuclear aluminoborosilicate glasses used for conditioning fission product solutions generated through reprocessing spent fuel. The chemical weathering of silicate glasses, especially those of basaltic composition, plays an important role in the geochemical cycles of several elements, owing to their high reactivity and abundance at the earth surface. A thorough review of the dissolution kinetics of silicate glasses is well beyond the purpose of this book and only a brief account on selected, recent laboratory dissolution experiments is given here.

It must be underscored that H_2O-poor glasses, prepared through melting and quenching of either natural crystalline materials or artificial oxide mixtures, were used in some of these laboratory experiments. The dissolution rates of these glasses were found to be similar to those of their crystalline counterparts (e.g. Hamilton et al., 2000, see below). These findings might seem surprising if compared to the well-known low chemical durability of natural volcanic glasses. However, the chemical durability of $Na_2O{\cdot}3SiO_2$ glasses was found to increase markedly as the H_2O content decreases from 5 to 0 wt% (Tomazawa et al., 1982). Besides, not all the natural glasses have low chemical durability. Indeed, tektite glasses have high chemical durability and, consequently, have been preserved for up to 35 million years (Glass, 1984), whereas the maximum age of natural volcanic glasses is close to 15 million years (White, 1984).

6.6.6.1. Albite, jadeite and nepheline glasses

The dissolution rates of three artificial glasses of compositions $NaAlSi_3O_8$, $NaAlSi_2O_6$ and $NaAlSiO_4$, which are those of albite, jadeite and nepheline, respectively, were studied by Hamilton et al. (2001). Note that these glasses and minerals have variable Al/Si and Na/Si molar ratios but constant Na/Al molar ratio of 1.

Batch experiments were carried out at 25°C and variable pH (1–12). Dissolution of nepheline glass was close to congruent apart from the neutral-pH region, whereas significant incongruency was observed for jadeite glass and even more for albite glass, especially at circumneutral pH. These observations suggest preferential release of Na and the generation of an Al-rich leached layer, although the precipitation of an Al-rich phase and/or sorption reactions cannot be ruled out as either alternative explanations or concurrent phenomena.

Surface analyses revealed that leached layers were ≤200–500 Å in thickness, except for jadeite glass reacted at pH 2, which shows H^+ penetration depths of ~2,000 Å, consistent with the depths of Na- and Al-depletion.

All three glasses exhibit a minimum in rate at circumneutral pH. Besides, the rates increase with increasing Al/Si ratio at all pH values (Fig. 6.35). Note that the rates reported in this figure are expressed as moles of Si m^{-2} s^{-1}; they are, therefore, release rates of Si and not dissolution rates (see Section 6.5). Figure 6.35 also shows that the slopes of the log-rate vs. pH curves, in both acidic and basic solutions, increase with increasing Al/Si ratio.

Measured rate data were compared by Hamilton et al. (2001) with those computed based on the multi-oxide dissolution model of Oelkers (2001a). Dissolution rate of nepheline glass was considered to be both dependent on and independent of the $a^3_{H^+}/a_{Al^{3+}}$ ratio. Taking into account the uncertainties in measured rates (for instance, dissolution might have not attained steady state in batch experiments) and the assumptions involved in the model of Oelkers (2001a) (see Sections 6.3.2 and 6.4.4), the differences between experimental and calculated rates are not so bad, especially for albite and jadeite.

Quite surprisingly, the dissolution rates of nepheline and jadeite glasses are slower than those of corresponding crystals measured by Tole et al. (1986) and Sverdrup (1990),

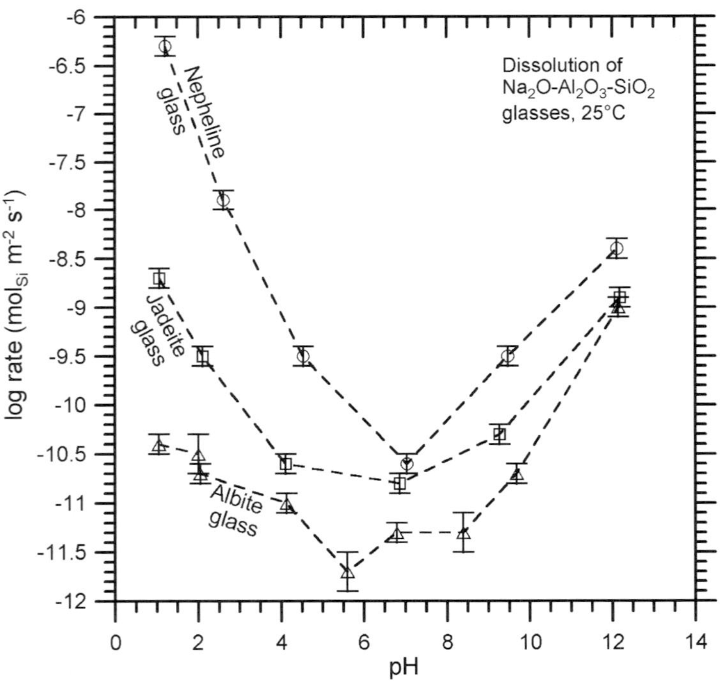

Figure 6.35. Release rate of Si from albite, jadeite and nepheline glasses vs. pH. (Reprinted from Hamilton et al., 2001, Copyright 2001 with permission from Elsevier.)

respectively. These discrepancies, however, are possibly due to compositional and structural differences as discussed by Hamilton et al. (2001).

The dissolution rates of albite glass and crystal were compared by several authors, including Shade (1981), Zellmer (1986), Zellmer and White (1986), Hellmann et al. (1990) and Hamilton et al. (2000). Shade (1981) carried out batch-dissolution experiments in deionized water at 90°C and found that the release rates of Si are similar for albite crystal and glass. Zellmer and White (1986) conducted batch experiments at 70°C and variable pH (2–10) and observed a somewhat faster Si release for albite glass compared to their crystalline counterpart. Working in flow reactors at 300°C and variable pH (3.4–11), Hellmann et al. (1990) found instead that dissolution rates are higher for albite glass than for albite crystal with comparatively large differences (0.5–1.5 log-units) at pH 5.7 and pH 11.0 and small differences (0.25–0.50 log-units) at pH 3.4.

Hamilton et al. (2000) carried out long-term dissolution experiments at 25°C in flow-through reactors at pH of 2.0, 5.6 and 8.4 (outlet values) and found that the steady-state dissolution rates of albite glass are equal to those of albite crystal within analytical error (±40%). In spite of this similarity in dissolution rates, Na and Al depletion from the glass surface is more extensive than from the crystal surface, especially in acidic solutions, probably due to the more open six-membered ring structure of glass compared to the four-membered ring structure of crystal. Sodium is released faster than Si from albite glass, whereas stoichiometric release of Al is attained at steady state for both glass and crystal.

6.6.6.2. Basaltic glass

Guy (1989) investigated the dissolution rate of basaltic glass (obtained through melting and quenching of oceanic basalt) in closed-system reactors at 50, 100 and 200°C. The surface area of the dissolving powder was estimated geometrically and no BET measurement was performed, a fact that complicates the use of these data (see discussion in Gislason and Oelkers, 2003). Dissolution rates determined at $pH < 2.5$ and $pH > 10$ were probably affected by comparatively slow transport through diffusion from the glass surface, whereas rates measured at $3 < pH > 9$ were most likely controlled by surface reactions. The dissolution rate of basaltic glass was found to increase remarkably with decreasing pH in acidic solutions, but to increase weakly with increasing pH in basic solutions. Besides, the pH of minimum dissolution rate was found to decrease with increasing temperature.

Dissolution rates of a Li-spiked synthetic basaltic glass were measured as a function of dissolved SiO_2 concentration at 150, 200 and 300°C in both batch and mixed-flow reactors by Berger et al. (1994b). Initial dissolution rate was depressed by the presence of silica in neutral aqueous solutions at all temperatures. In acidic solutions (0.1 and 0.01 M HCl), a weak dependence on silica concentration was found at 300°C, whereas at 150°C rates were found to be independent of dissolved silica but to be dependent on aqueous Al concentration. Based on these data, glass dissolution kinetics was considered to be controlled by the $\equiv SiOH^{\circ}$ surface species in neutral solutions and by charged surface species (e.g. $\equiv AlOH_2^{+}$) in acidic solutions at 150°C. Long-term dissolution rates were observed to decrease with time and eventually to approach the initial dissolution rate of amorphous silica.

Daux et al. (1997) investigated the dependence of the steady-state dissolution rate of an artificial Li-enriched basaltic glass on thermodynamic affinity (for $A > -9.8$ kJ mol^{-1}) at 90°C and pH close to 8 (7.8–8.3) by means of an open-system, mixed-flow reactor. Measured rates were found to be independent of dissolved Al and SiO_2 concentrations, but to depend on thermodynamic affinity. This parameter was defined for the hydrated basaltic glass that presumably forms during dissolution experiments, referring to the hydrolysis reaction

$$SiAl_{0.36}Fe_{0.18}(OH)_{5.62} + 0.36OH^- \\ = H_4SiO_{4(aq)} + 0.36Al(OH)_4^- + 0.18Fe(OH)_{3(aq)}, \tag{6-139}$$

whose thermodynamic equilibrium constant is 8.2×10^{-5} at 90°C. Measured rates (in mol m^{-2} s^{-1}) fit the equation

$$r = 3\times10^{-6} \cdot a_{OH^-}^{0.39}\left[1 - \exp\left(\frac{A}{RT}\right)\right], \tag{6-140}$$

which is similar in type to equation (6-77).

As anticipated in Section 6.3.2, according to the multi-oxide dissolution model of Oelkers (2001a), the dissolution of basaltic glass takes place through three sequential steps:

(1) The comparatively fast, preferential and almost complete removal of monovalent and divalent cations from the glass surface through metal/proton exchange reactions, which bring about the formation of a leached surface layer enriched in Si, Al and Fe(III) and depleted in monovalent and divalent metals, consisting in a sort of partially detached framework.

(2) The liberation of Al^{3+} ion through exchange reactions involving three H^+ ions and occurring in the glass structure near the surface.

(3) The relatively slow, rate-controlling release of the partially detached SiO_4 tetrahedra.

This dissolution mechanism is the same as that proposed by Oelkers (2001a) for crystalline silicate minerals and, consequently, the same theoretical approach discussed in Section 6.4.4 can be applied to basaltic glass also.

Oelkers and Gislason (2001) measured the steady-state, far-from-equilibrium ($A < -12$ kJ mol^{-1}) dissolution rate of a basaltic glass in mixed-flow reactors at 25°C and two different pH values, 3 and 11, as a function of dissolved SiO_2, Al and oxalic acid contents. The chemical formula normalized to one Si atom of the investigated glass coming from Stapafell mountain, Southwestern Iceland is $SiTi_{0.02}Al_{0.36}Fe(III)_{0.02}Fe(II)_{0.17}Mg_{0.28}Ca_{0.26}Na_{0.08}K_{0.008}O_{3.38}$.

Stoichiometric release of Al was observed. Measured rates were found to be independent of aqueous silica activity (in contrast with previous observations by Berger et al., 1994b; Daux et al., 1997, see above) but to decrease significantly with increasing $a_{Al^{3+}}$ at both pH values. Combining these observations with the theoretical approach outlined by Oelkers (2001a), all measured dissolution rates (in mol m^{-2} s^{-1}) at pH 3 and 11 were described by means of the following function (i.e. equation (6-85) at far-from-equilibrium condition):

$$r_+ = 10^{-7.65\pm0.13} \times \left(\frac{a_{H^+}^3}{a_{Al^{3+}}} \right)^{0.35\pm0.02} . \tag{6-141}$$

Oxalic acid was found to increase the dissolution rate of basaltic glass at pH 3 but to have negligible influence at pH 11. To explain these different effects, Oelkers and Gislason (2001) considered that oxalate forms stable aqueous complexes with Al^{3+} ion in acidic solutions, thus determining a decrease in $a_{Al^{3+}}$ and a concurrent increase in the dissolution rate of basaltic glass. In contrast, there is negligible Al-oxalate complexation in basic solutions and, consequently, little effect on the dissolution rate of basaltic glass. This is a very interesting conclusion, since it implies that the effects of other aqueous species interacting with $a_{Al^{3+}}$ can be taken into account through suitable speciation calculations.

The dependence of the far-from-equilibrium dissolution rate of basaltic glass on pH (from 2 to 11) and temperature (from 6 to 50°C and at near neutral conditions up to 150°C as well) was investigated by Gislason and Oelkers (2003) by means of experiments carried out in open-system, mixed-flow reactors. To avoid the possible influence of slow diffusional transport on measured rates (as underscored by Guy, 1989, see above), experiments at pH 2 and 3 were performed at sufficiently high stirring speeds.

The following hydrolysis reaction was assumed by Gislason and Oelkers (2003) for hydrated basaltic glass:

$$SiAl_{0.36}O_2(OH)_{1.08} + 1.08H^+ = SiO_{2(aq)} + 0.36Al^{3+} + 1.08H_2O. \tag{6-142}$$

Its thermodynamic equilibrium constant was estimated from the stoichiometrically weighted sum of the hydrolysis reactions of amorphous silica and gibbsite, obtaining log K values of 0.261, 0.027, −0.693 and −1.105 at 0, 30, 100 and 150°C, respectively. Based on these log K values, the A/RT term in equations similar in type to equation (6-57) was estimated to be <-5 for all the dissolution rates measured by Gislason and Oelkers (2003).

Selected dissolution rate data for basaltic glass are presented in Fig. 6.36, showing a pH dependence that is consistent with previous observations by Guy (1989, see above). All the dissolution rate data measured by Gislason and Oelkers (2003), as well as those from some previous studies (Guy, 1989; Berger et al., 1994b; Daux et al., 1997; Oelkers and Gislason, 2001), fit the following equation, within 0.34 log-units on average:

$$r_{+,geom} = 10^{-1.6} \times \exp\left(\frac{-E_a}{RT} \right) \times \left(\frac{a_{H^+}^3}{a_{Al^{3+}}} \right)^{1/3} , \tag{6-143}$$

where $r_{+,geom}$ is the far-from-equilibrium, steady-state dissolution rate of basaltic glass normalized to the geometric surface area and expressed in mol m^{-2} s^{-1} and E_a, the

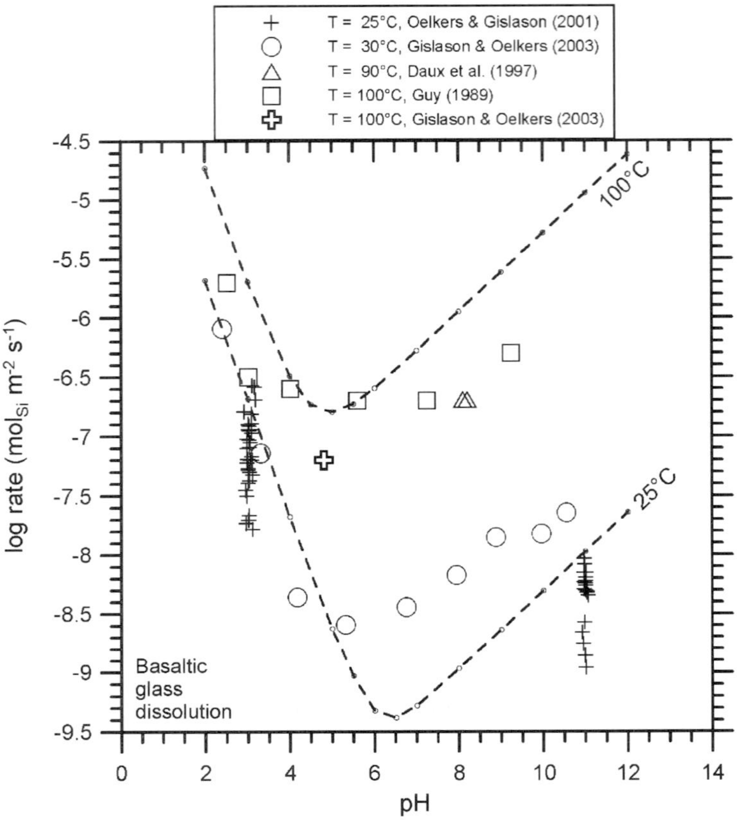

Figure 6.36. Plot of selected experimental dissolution rate data (see legend) for basaltic glass vs. pH. Curves were generated based on the $a^3_{H^+}/a_{Al^{3+}}$ activity ratios computed by means of EQ3NR for constant temperature (either 25 or 100°C), constant total Al concentration of 1 μmol kg^{-1}, and variable pH.

pH-independent apparent activation energy, is equal to 25.5 kJ mol^{-1}. Note that the values of $r_{+,geom}$ are 92 times larger than the corresponding $r_{+,BET}$, i.e. the dissolution rates normalized to the glass surface area obtained from the BET gas adsorption technique. Equation (6-143) is valid for 6 < temperature < 300°C and 1 < pH < 11. It was used to generate the curves reported in Fig. 6.36, based on the $a^3_{H^+}/a_{Al^{3+}}$ activity ratios computed by means of EQ3NR for constant temperature (either 25 or 100°C), constant total Al concentration of 1 μmol kg^{-1} and variable pH.

In principle, the following general equation can be used to describe the dissolution rates of basaltic glass as a function of thermodynamic affinity, solution chemistry and temperature (see Sections 6.4.4 and 6.1.3):

$$r = 10^{-1.6} \times \exp\left(\frac{-E_a}{RT}\right) \times \frac{\left(a^3_{H^+}/a_{Al^{3+}}\right)^{1/3}}{1 + K_{Al-O} \cdot \left(a^3_{H^+}/a_{Al^{3+}}\right)^{1/3}} \cdot \left[1 - \exp\left(\frac{A}{\sigma RT}\right)\right]. \tag{6-144}$$

In practice, the use of equation (6-144) is limited by poor knowledge of σ and K_{Al-O}, as underscored by Gislason and Oelkers (2003).

6.6.6.3. Application of the multi-oxide dissolution model to glasses of variable composition

The far-from-equilibrium, steady-state dissolution rates of seven natural glasses (one basalt, two dacites and four rhyolites) with SiO_2 contents ranging from 46.48 to 72.62 wt% were measured at 25°C and pH 4 by Wolff-Boenisch et al. (2004a), as a function of total dissolved fluoride concentration (up to 1.8×10^{-4} mol kg^{-1}) in mixed-flow reactors.

Preferential release of Al, Fe, Ca, Mg and Na was observed in the initial part of the dissolution runs. This behaviour is consistent with the multi-oxide dissolution model (Oelkers, 2001a), which comprises early exchange of these metals with H^+ ions at the glass surface and subsequent rate-limiting detachments of tetrahedrally coordinated Si atoms (see above).

Wolff-Boenisch et al. (2004a) chose to normalize measured rates to the geometric surface area, for several reasons explained by Wolff-Boenisch et al. (2004b), but these values can be easily referred to the BET surface area that is also given by the authors. The $A_{s,BET}/A_{s,geom}$ ratio (roughness factor or surface roughness, see Section 6.5.2.3) ranges from 8 to 94 for the glass samples studied by Wolff-Boenisch et al. (2004a).

At zero total fluoride concentration, the dissolution rate of the glasses is inversely correlated to their SiO_2 content, i.e. to the degree of polymerization. Dissolution rates were found to increase monotonically with increasing total fluoride concentration owing to the formation of Al^{3+}–F^- aqueous complexes, a process that decreases $a_{Al^{3+}}$. However, the trend of decreasing dissolution rate with increasing SiO_2 content of the glasses is observed at all total fluoride concentrations, although the fluoride effect is strongest on basaltic glasses.

Based on these observations, all measured rate data were described by means of the following TST-based rate equation, which is consistent with the multi-oxide dissolution model by Oelkers (2001a):

$$\log r_{+,\mathrm{geom}}\left(\mathrm{mol\cdot m^{-2}\cdot s^{-1}}\right) = \left[-0.086 \times \mathrm{SiO_2(wt\%)} - 2.25\right] - \left[0.0067 \times \mathrm{SiO_2(wt\%)} + 0.683\right] \times \log\left(\frac{a^3_{H^+}}{a_{Al^{3+}}}\right). \tag{6-145}$$

Comparison of equation (6-145) with equation (6-85) at far-from-equilibrium condition (or equation (6-141)) shows that the first term on the right-hand side of equation (6-145) is equal to n, whereas the coefficient of the $a^3_{H^+}/a_{Al^{3+}}$ log-ratio represents the rate constant parameter k'_+. Both n and k'_+ depend on the SiO_2 content of the glasses. The increase in both n and k'_+ with decreasing SiO_2 content suggests that the number of partially detached Si atoms on the glass surface increases with decreasing SiO_2 content of the glasses, which is consistent with a lower degree of polymerization. The difference between experimentally measured rates and those computed by means of equation (6-145) is 0.2 log-units on average.

Wolff-Boenisch et al. (2004b) measured the far-from-equilibrium, steady-state dissolution rates of 18 different glasses (7 basalts, 1 mugearite, 2 basaltic andesites, 3 dacites and 5 rhyolites) in mixed-flow reactors at pH 4 and 10.6 and temperatures ranging from 25 to 74°C. The value of $\log\left(a^3_{H^+}/a_{Al^{3+}}\right)$ was fixed at -6.4 ± 0.6 to suppress the effect of this parameter on the glass dissolution rate. This introduces an uncertainty in measured rates of ±0.2 log-units.

Dissolution rates normalized to the geometric surface area depend on the SiO_2 content of the glasses as described by the following linear regression equations:

$$\log r_{+,\text{geom}}\left(\text{mol}\cdot\text{m}^{-2}\cdot\text{s}^{-1}\right) = \left[-0.03\times \text{SiO}_2(\text{wt\%}) - 7.58\right], \tag{6-146}$$

for the runs at pH 4 ($r^2 = 0.75$), and

$$\log r_{+,\text{geom}}\left(\text{mol}\cdot\text{m}^{-2}\cdot\text{s}^{-1}\right) = \left[-0.02\times \text{SiO}_2(\text{wt\%}) - 7.02\right], \tag{6-147}$$

for the runs at pH 10.6 ($r^2 = 0.43$).

Dissolution rates at high temperature were utilized to evaluate apparent activation energies, which ranged from 24 to 43 kJ mol^{-1} for basaltic glasses and from 32 to 42 kJ mol^{-1} for acidic glasses, at pH 4. Somewhat higher E_a were found at pH 10.6.

6.6.7. The influence of CO_2 on the kinetics of silicate dissolution

The direct influence of dissolved CO_2 on the dissolution kinetics of silicate minerals was taken into account by Carroll and Knauss (2005) and Golubev et al. (2005).

The work by Carroll and Knauss (2005) on the dissolution kinetics of labradorite has been already mentioned above (see Section 6.6.4.1). This investigation is especially interesting as the effects of CO_2 were evaluated for an unusually high m_{CO2} of 0.6, at 30, 60 and 130°C. The corresponding P_{CO_2} values are close to 20 bar at 30°C, 36 bar at 60°C and 68 bar at 130°C. As already mentioned above, no direct CO_2 effect was found, as high CO_2 concentrations increase labradorite dissolution indirectly by enhancing the acidity of the aqueous medium.

Golubev et al. (2005) measured the dissolution rates of diopside, forsterite, wollastonite and hornblende at 25°C and reviewed previous studies (Wogelius and Walther, 1991; Knauss et al., 1993; Brady and Carroll, 1994; Brady and Gislason, 1997; Malmström and Banwart, 1997; Berg and Banwart, 2000; Pokrovsky and Schott, 2000b).

The results of all these experimental investigations are summarized in Table 6.3, whose inspection suggests that the direct effects of dissolved CO_2 (up to 1 bar P_{CO_2} apart from the study by Carroll and Knauss, 2005), HCO_3^- and CO_3^{2-} ions are generally nil to negligible, and the main controlling parameter of silicate dissolution is pH.

In addition to this pH effect, however, the effects of dissolved metals and organic and inorganic ligands can also be important. It must be underscored, again, the general lack of data under high-P_{CO_2} (>1 bar) conditions.

TABLE 6.3

Summary of the experimental data on the direct effect of CO_2 on the dissolution rates of silicate minerals (from Golubev et al., 2005, modified)

Mineral	pH Range	CO_2 Effect	Reference
Forsterite	4–11	Absent	Golubev et al. (2005)
Olivine (Fo92)	4–6	Absent	Pokrovsky and Schott (2000b)
Olivine (Fo92)	>9	Inhibition	Wogelius and Walther (1991) Pokrovsky and Schott (2000b)
Olivine (Fo92)	10.5–11	Absent	Golubev et al. (2005)
Diopside	4–11	Absent	Golubev et al. (2005)
Diopside	8–12	Absent	Knauss et al. (1993)
Augite	4	Absent	Brady and Carroll (1994)
Wollastonite	4	Absent	Golubev et al. (2005)
Wollastonite	7–8	Weak acceleration	Golubev et al. (2005)
Wollastonite	12	Absent or weak inhibition	Golubev et al. (2005)
Hornblende	4–11	Absent	Golubev et al. (2005)
Biotite	8–9	Inhibition	Malmström and Banwart (1997)
Labradorite	3.2[a]	Absent	Carroll and Knauss (2005)
Anorthite	4	Absent	Brady and Carroll (1994)
Anorthite	5.5–8.5	Acceleration	Berg and Banwart (2000)
Basalt glass	8	Absent	Brady and Gislason (1997)

[a]P_{CO_2} values of 20 bar at 30°C, 36 bar at 60°C, and 68 bar at 130°C.

6.7. Dissolution rates of oxides and hydroxides

6.7.1. Al-hydroxide and Al-oxide minerals

6.7.1.1. Gibbsite

Since the 1930s, the dissolution kinetics of gibbsite has been investigated by many authors, but acquired data are somewhat contrasting and it is difficult to obtain a consistent description of the process. Early dissolution experiments carried out by Clay and Thomas (1938) showed that the dissolution rate of hydrous alumina in 0.2 M HCl is strongly increased by moderate concentrations (0.001 M) of some anions, i.e. fluoride, phosphate, phosphite, oxalate and sulphate or the corresponding acids. Moreover, Clay and Thomas (1938) recognized that these effects are explained by the formation of a rate-determining complex at the solid–liquid interface.

Experimental studies by Scotford and Glastonbury (1971, 1972), Packter and Dhillon (1973) and Peric' et al. (1985) demonstrated that the dissolution kinetics of gibbsite has a first-order dependence on NaOH concentration (n_{OH} equal to -1), in the strongly alkaline solutions (pH $>$ 13.25) utilized in the Al-mining processes.

Packter and Dhillon (1969) carried out dissolution experiments in acidic solutions at constant molar concentration and showed that the dissolution rate of gibbsite depends on the type of acid, with faster rates in H_2SO_4, intermediate rates in HCl and slower rates in $HClO_4$. This suggests that the anion identity has a strong effect on the dissolution kinetics of gibbsite.

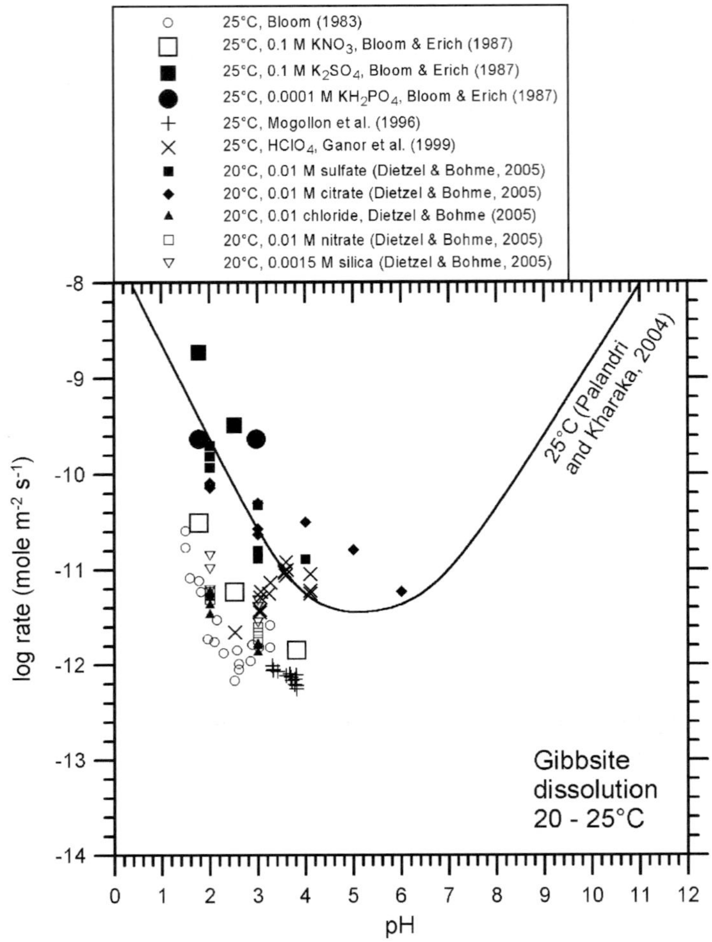

Figure 6.37. Experimental log-rates of gibbsite dissolution vs. pH in different aqueous media at 20–25°C (see legend). The dissolution rate obtained by summing the contributions of the acidic, neutral and basic mechanisms reported by Palandri and Kharaka (2004, see Table 6.1) is also shown.

Bloom (1983) measured the the dissolution rate of two synthetic gibbsite samples at 25°C in HNO_3 solutions with pH values from 1.5 to 3.2 (Fig. 6.37) and comparatively low (but variable) ionic strengths. The reaction order with respect to hydrogen ion, n_H, was found to be from 0 to 1.7 at pH < 2.1, but n_H resulted to vary from 0 to -0.7 at pH > 2.4, depending on the type of gibbsite. Bloom (1983) suggested that the rate-governing mechanism was (i) protonation of surface sites (either single-charged to double-charged or neutral to single-charged) at pH < 2.1, or alternatively (ii) water attack at pH > 2.4. However, these results could be affected by changes in NO_3 concentration, as later underscored by Bloom and Erich (1987).

In new dissolution experiments of gibbsite at 25°C, Bloom and Erich (1987) observed n_H values from 0 to 1 in the pH interval 1.7–3.9, depending on the identity of the anions present (Fig. 6.37). To explain the pH-independent rate in solutions containing phosphate, Bloom and Erich (1987) invoked phosphate adsorption as dominant reaction mechanism. In contrast, the pH-dependent rates in solutions containing nitrate and sulphate were justified, assuming that these anions are not adsorbed on gibbsite and that proton-attack was the rate-controlling mechanism.

Mogollón et al. (1996) studied the dissolution kinetics of natural gibbsitic bauxite in $HClO_4$ solutions at 25°C, by using column reactors. Output pH varied from 3.3 to 3.9 and the flow rate of the $HClO_4$ solution was also suitably chosen, to determine dissolution kinetics under different saturation-state conditions, from close-to-equilibrium to far-from-equilibrium. Steady-state dissolution rates at far-from-equilibrium (i.e. for $A < -2.9$ kJ mol^{-1}) indicate an n_H close to 0.3 (Fig. 6.37). In these experiments, ionic strength was relatively low ($\leq$0.001 mol kg^{-1}), but not constant.

Ridley et al. (1997) carried out batch-dissolution experiments at 5°C and near-constant pH (~2.0) and ionic strength (0.1 mol kg^{-1}) under extremely undersaturated conditions, showing that the dissolution kinetics in H_2SO_4–NaCl solutions is ~10 times faster than in HCl–NaCl solutions. Since the surface area of gibbsite was not measured, dissolution rates cannot be normalized to the unit surface area.

Ganor et al. (1999) performed far-from-equilibrium (i.e. for $A < -4.6$ kJ mol^{-1}) dissolution experiments of synthetic gibbsite both in a stirred flow-through reactor and in a column reactor at 25°C and outlet pH values from 2.5 to 4.1. Solutions of perchloric acid of comparatively low and variable ionic strengths were used as aqueous media. The logarithm of the dissolution rate of gibbsite was found to increase linearly with pH up to 3.5, whereas it decreased with pH above this threshold (Fig. 6.37). The latter change was attributed to proton catalysis. Hypothesizing saturation of surface sites with protons at pH close to 3.6, perchlorate inhibition was invoked to explain the decrease in rate with increasing acidity (and ClO_4^- concentration) below pH 3.5.

The influence of cation and anion identity on the dissolution kinetics of natural gibbsitic bauxite at 25°C was investigated by Mogollón et al. (2000) by means of column experiments and using input solutions of different type ($NaClO_4/HClO_4$, NaCl/HCl, KCl/HCl, $MgCl_2$/HCl, $CaCl_2$/HCl, $NaNO_3/HNO_3$ and Na_2SO_4/H_2SO_4) and variable ionic strength. They found that different cations have no influence on the far-from-equilibrium dissolution rates, whereas anions have contrasting effects. A decrease in rate was observed for monovalent anions (ClO_4^-, Cl^- and NO_3^-) as function of $m^{-0.11\pm0.001}$, but the rate was found to increase with sulphate concentration, as a function of $m_{SO_4^{2-}}^{0.4\pm0.1}$ (Fig. 6.38). Recalling the dependence of the rate on the ionic strength, as predicted by TST (see Section 6.1.4.1), these data are consistent with a change in the product of the charges of the chemical entities involved in the formation of the activated complex from negative, when monovalent anions take part in the surface reaction, to positive, when sulphate is involved in this reaction. However, the identity of this chemical entity is not clear at all.

Dietzel and Böhme (2005) measured the dissolution rates of gibbsite at 20°C under both far-from-equilibrium and close-to-equilibrium conditions, by using both batch and

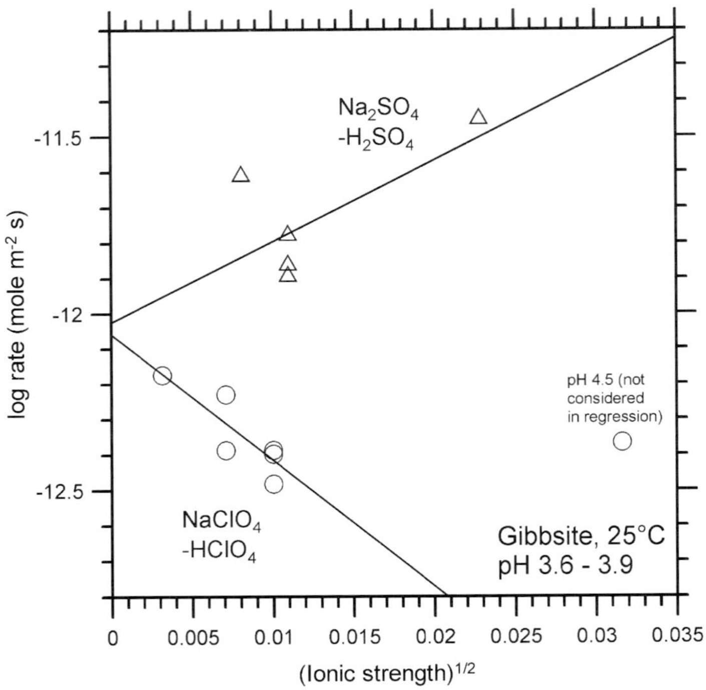

Figure 6.38. Experimental log-rates of dissolution of a gibbsitic bauxite vs. the square root of the ionic strength at 25°C and pH of 3.6–3.9 in $Na_2SO_4-H_2SO_4$ and $NaClO_4$–$HClO_4$ aqueous solutions. The two regression lines converge at zero ionic strength. (Reprinted from Mogollón et al., 2000, Copyright 2000 with permission from Elsevier, modified.)

mixed flow-through reactors. Far-from-equilibrium experiments were performed in solutions containing (i) 10 mmol L^{-1} citrate, from pH 2 to 6, every pH unit; (ii) 10 mmol L^{-1} chloride or nitrate or sulphate, at pH 2 and 3; and (iii) 1.5 mmol L^{-1} silica, at pH 2 and 3. It was found that gibbsite dissolution is faster in the presence of sulphate (as already noted by previous authors) and citrate and is slower in the presence of chloride and nitrate, whereas intermediate rates were determined in the presence of silica (Fig. 6.37). Formation of Al–ligand complexes at the gibbsite surface was invoked to explain the faster rates measured in the presence of sulphate, citrate and silica. The enhancement of proton-promoted dissolution at pH <3–3.5 was excluded based on the computed concentration of surface complexes, which confirms saturation of surface sites with protons at pH <3–3.5, consistent with the dissolution rates measured by Bloom and Erich (1987) and Ganor et al. (1999).

The dissolution kinetics of gibbsite depends strongly on thermodynamic affinity, as shown by Nagy and Lasaga (1992), Mogollón et al. (1996) and Dietzel and Böhme (2005).

Nagy and Lasaga (1992) measured dissolution and precipitation rates of gibbsite at 80°C and pH 3, in stirred flow-through reactors as a function of the saturation state. Under close-to-equilibrium conditions ($-0.8 < A < 0$ kJ mol^{-1}) dissolution rates are a linear function of A but a highly non-linear dependence was found away from these conditions, with a sharp increase over the interval $-2.1 < A < -0.8$ kJ mol^{-1} and a constant value for $A < -2.1$ kJ mol^{-1} (dissolution plateau). The dependence of the dissolution rate in the range $-4.8 < A < 0$ kJ mol^{-1} is described by the following function (rate in mol m^{-2} s^{-1}):

$$r_{+} = (4.72 \pm 0.28)10^{-10} \cdot \left\{ 1 - \exp\left[(-8.12 \pm 1.02) \cdot \left(\frac{|A|}{RT} \right)^{3.01 \pm 0.05} \right] \right\}. \tag{6-148}$$

Precipitation rates are a linear function of A for $0 \leq A \leq +1.95$ kJ mol^{-1}. The almost linear dependence of both dissolution and precipitation rates on A, under close-to-equilibrium conditions ($-0.8 \leq A \leq +0.8$ kJ mol^{-1}), corroborates the application of TST to the overall dissolution/precipitation reaction of gibbsite as well as the use of the principle of detailed balancing.

Mogollón et al. (1996) underscored that equation (6-148) describes suitably the dependence of the dissolution rate of gibbsitic bauxite on thermodynamic affinity at 25°C.

The close-to-equilibrium dissolution rates at 20°C, determined by Dietzel and Böhme (2005), are a linear function of thermodynamic affinity over the interval $-2.9 < A < -0.2$ kJ mol^{-1}, in substantial agreement with previous findings.

The apparent activation energies for gibbsite dissolution (Packter and Dhillon, 1969, 1973, 1974; Scotford and Glastonbury, 1971; Bloom, 1983; Peric' et al., 1985; Bloom and Erich, 1987; Nagy and Lasaga, 1992; Nagy, 1995) range between 45 and 68 kJ mol^{-1} in acidic solutions of different composition and are probably independent of anion identity. Somewhat higher activation energies, 73−130 kJ mol^{-1}, are reported for gibbsite dissolution in NaOH solutions.

6.7.1.2. Bayerite and diaspore

The rate of dissolution of bayerite [γ-Al(OH)$_3$] in HNO_3–HF solutions was investigated by Pulfer et al. (1984) together with the surface complexation of H^+ and F^- ions at the bayerite surface. The ionic strength of the aqueous solution was kept constant at 1 M by using KNO_3. Rates were obtained, quite unconventionally, by titrating the bayerite suspension with either HNO_3 or HF solutions and measuring the volumes of these solutions needed to maintain the concentrations of H^+ and F^- ions at the initial values (see details in Pulfer et al., 1984). The Al added to the aqueous phase by bayerite dissolution was not measured. In the absence of fluoride, the dissolution rate of bayerite is close to first order with respect to H^+ concentration (Fig. 6.39). In the presence of fluoride, dissolution kinetics is approximately first order with respect to both H^+ and F^- concentrations, for high H^+ and F^- concentrations; however, the reaction order increases with respect to both H^+ and F^- with decreasing H^+ and F^- concentrations. Figure 6.39 also shows that the dissolution rate of bayerite is enhanced by approximately three orders of magnitude by addition of fluoride. Based on their experimental results, Pulfer et al. (1984) proposed a rate law involving the intrinsic stability constants of H^+ and F^- sorption onto the bayerite surface and the electrostatic surface potential, which was computed from the surface charge assuming a Helmoltz-type double layer. Consequently, the overall model for dissolution kinetics and

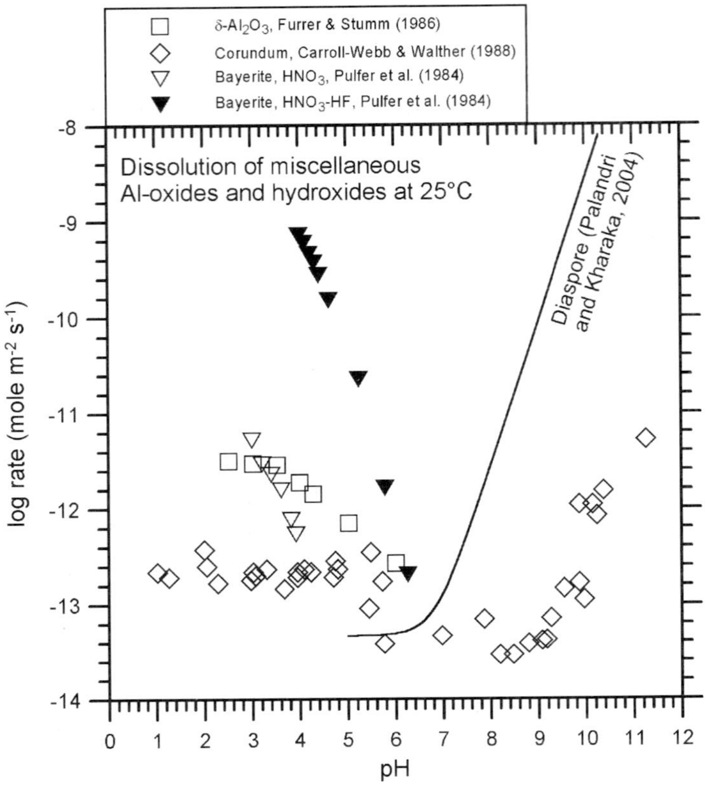

Figure 6.39. Experimental log-rates of dissolution of bayerite [γ-Al(OH)$_3$], δ-Al_2O_3 and γ-Al_2O_3 vs. pH in different aqueous solutions at 20–25°C. The dissolution rate of diaspore [α-AlO(OH)] computed at 25°C by summing the contributions of the neutral and basic mechanisms reported by Palandri and Kharaka (2004, see Table 6.1) is also shown.

sorption equilibria is necessarily based on hypotheses concerning the bayerite surface structure and the double-layer capacitance.

The dissolution rate of diaspore [α-AlO(OH)] at 25°C, obtained by summing the contributions of the basic and neutral mechanisms reported by Palandri and Kharaka (2004, see Table 6.1), is also represented in Fig. 6.39. Kinetic parameters were derived from Chang et al. (1979).

6.7.1.3. Al-oxides

The dissolution kinetics of δ-Al_2O_3 in 0.1 M $NaNO_3-HNO_3$ solutions was investigated by Furrer and Stumm (1986) from pH 2.5 to 6, whereas the dissolution rate of corundum (α-Al_2O_3) was studied by Carroll-Webb and Walther (1988) from pH 1 to 11.2. Both sets of data were obtained through batch experiments.

The log-rate of the proton-promoted dissolution of δ-Al_2O_3 is independent of pH for pH $\leq$ 3.5 (where the surface concentration of protons attains the maximum value), but it decreases with increasing pH above pH 3.5, with a reaction order close to 0.4 (Fig. 6.39). Both the pH dependence and the reaction order of the δ-Al_2O_3 dissolution are similar to those observed for gibbsite (cf. Figs. 6.37 and 6.39 and see Section 6.7.1.1). Furrer and Stumm (1986) also suggested, based on the results of acid/base surface titrations, that the reaction mechanism involves the attachment of three H^+ ions to the surface site, prior to the release of an Al species to the aqueous solution.

The log-rate of corundum dissolution is pH independent at pH < 5, it decreases with increasing pH from pH 5 to 8 (in spite of a significant scatter in the data), and increases with increasing pH from pH 9 to 11.2. The reaction order with respect to OH^- is close to +1 (n_{OH} close to -1), similar to what is observed for gibbsite dissolution in strongly alkaline solutions (see Section 6.7.1.1). Based on acid/base surface titrations, Carroll-Webb and Walther (1988) also recognized that the surface reaction mechanism involves either one proton or four hydroxyl ions.

6.7.2. Brucite and periclase

The dissolution kinetics of brucite and periclase and the precipitation kinetics of brucite have been the subject of several experimental studies (e.g. Vermilyea, 1969; Segall et al., 1978; Jones et al., 1981; Wogelius et al., 1995; Jordan and Rammensee, 1996; Jordan et al., 1999; Pokrovsky and Schott, 2004; Pokrovsky et al., 2005a).

Wogelius et al. (1995) recognized that a hydroxylated near-surface layer forms on periclase during dissolution. This hydroxylation reaction is very quick and is followed by the slower, rate-governing step of Mg-detachment from the brucite-like structure. Consequently, periclase and brucite are expected to have the same dissolution rates.

Vermilyea (1969) investigated the dissolution kinetics of magnesium oxide and hydroxide through batch experiments and documented the significant pH dependence of dissolution rates, with a reaction order with respect to the proton close to 0.5. The rate law proposed by Palandri and Kharaka (2004) for the acidic dissolution mechanism of brucite (Table 6.1 and Fig. 6.40), with an n_H of 0.5, is largely based on these data and those by Jordan and Rammensee (1996, see below). Relatively high values of n_H were also found by Segall et al. (1978) and Jordan et al. (1999), whereas small values of n_H were observed by other authors (e.g. Jones et al., 1981). In addition to these distinct values of the reaction order with respect to the H^+ ion, the dissolution rates determined by different investigators deviate in orders of magnitude, as shown in Fig. 6.40.

Vermilyea (1969) estimated an apparent activation energy of ~42 kJ mol^{-1} in the temperature range 25–75°C. A somewhat higher value, 60 ± 12 kJ mol^{-1}, was found by Jordan and Rammensee (1996) based on the microtopography of the brucite (001) surface during its dissolution in acidic solutions (pH 2.7) at temperatures of 21–35°C.

Very detailed studies were carried out by Pokrovsky and Schott (2004) and Pokrovsky et al. (2005a). By using a mixed-flow reactor, Pokrovsky and Schott (2004) determined the dissolution and precipitation kinetics of brucite at 25°C, as a function of pH (from 2.5 to 12), ionic strength (from 10^{-4} to 3 M), dissolved Mg concentration (from 10^{-6} to 5×10^{-4} M) and thermodynamic affinity (from $-68{,}600$ to $+3{,}700$ J mol^{-1}). To obtain the total net surface charge of brucite in the aqueous phase, Pokrovsky and Schott

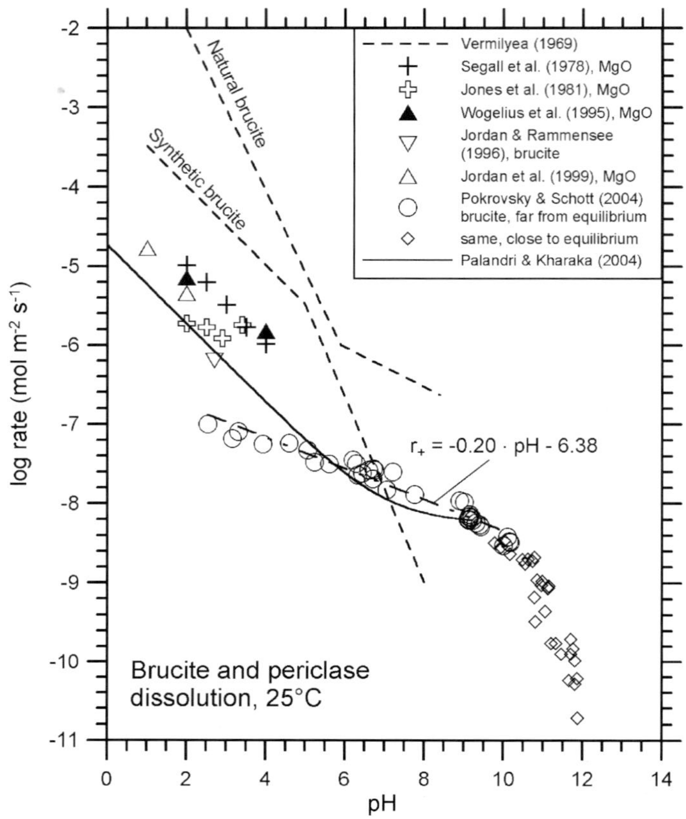

Figure 6.40. Experimental log-rates of dissolution of brucite and periclase at 25°C as a function of pH. (Reprinted from Pokrovsky and Schott, 2004, Copyright 2004 with permission from Elsevier, modified.)

(2004) performed surface titrations by using a limited-residence-time reactor (owing to the high solubility of brucite) and working under a N_2 atmosphere (to prevent carbonation of brucite). They also carried out (i) electrophoretic measurements to find the pH of the isoelectric point, which constrains the intrinsic stability constants at the brucite–solution interface and (ii) downward pH-jump experiments, from neutral and alkaline conditions to pH 2 or 3, whose results indicate the relatively fast dissolution, under acidic conditions, of the reactive surface sites produced at higher pH values. They found that the pH of zero charge and the pH of the isoelectric point are close to 11. This implies that $(pK_1 + pK_2)/2 = 11$, which is consistent with $pK_1 = 12$ and $pK_2 = 10$, where surface reactions 1 and 2 are,

$$\equiv \mathrm{MgO}^- + \mathrm{H}^+ \rightleftharpoons \equiv \mathrm{MgOH}^\circ \tag{6-149}$$

and

$$\equiv \mathrm{MgOH}^\circ + \mathrm{H}^+ \rightleftharpoons \equiv \mathrm{MgOH}_2^+, \tag{6-150}$$

respectively. Based on a two-*pK*, single-site, constant-capacitance model with a surface site density of 16×10^{-6} mol m^{-2} and an electric double-layer capacitance of 5 F m^{-2}, Pokrovsky and Schott (2004) showed that the prevailing surface species at the brucite–water interface are $\equiv MgOH_2^+$, below pH 7.5, and $\equiv MgOH^\circ$, above this threshold (see Fig. 3 of Pokrovsky and Schott, 2004). The surface species $\equiv MgO^-$ is subordinated at all pH values, but its concentration increases with pH and becomes significant above pH 12.

The log-rates of brucite dissolution measured by Pokrovsky and Schott (2004) under far-from-equilibrium conditions in the pH range 2.5–10 depend linearly on pH with a reaction order of 0.2, whereas close-to-equilibrium rates ($10 < pH < 12$) have a much steeper pH dependence (Fig. 6.40). Far-from-equilibrium log-rates are also linearly correlated with the logarithm of the concentration of $\equiv MgOH_2^+$ (Fig. 6.41):

$$\log r_+ = (1.90 \pm 0.02) \times \log\left[\equiv MgOH_2^+\right] + (1.99 \pm 0.09), \quad (6\text{-}151)$$

as expected based on the strong pH dependence of $[\equiv MgOH_2^+]$ (see Fig. 3 of Pokrovsky and Schott, 2004). In equation (6-151), r_+ is in mol m^{-2} s^{-1} and the concentration of the surface complex is in mol m^{-2}.

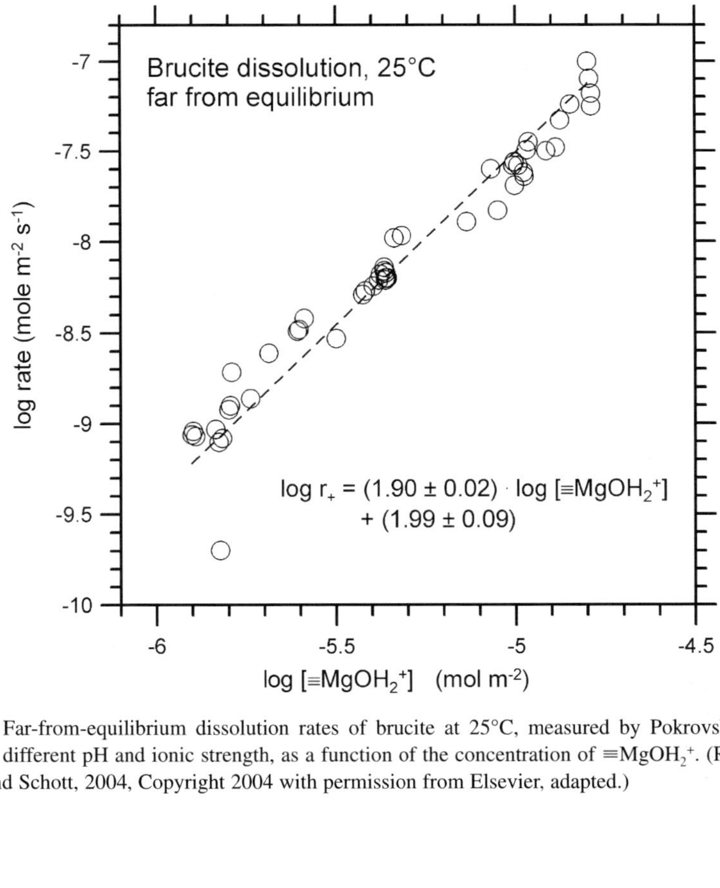

Figure 6.41. Far-from-equilibrium dissolution rates of brucite at 25°C, measured by Pokrovsky and Schott (2004) under different pH and ionic strength, as a function of the concentration of $\equiv MgOH_2^+$. (Reprinted from Pokrovsky and Schott, 2004, Copyright 2004 with permission from Elsevier, adapted.)

In contrast, the kinetics of brucite dissolution is almost independent of the ionic strength, at least in the pH range 6.5–11, in agreement with the results of surface speciation modelling.

Accepting that under close-to-equilibrium conditions, dissolution and precipitation rates are satisfactorily described by equation (6-57), and that not only dissolution rates but also precipitation rates are proportional to $[\equiv MgOH_2^+]^{1.9}$, the kinetics of both dissolution and precipitation is described by the following TST-type rate equation (Pokrovsky and Schott, 2004, modified):

$$r = k^+_{Mg} \cdot [\equiv MgOH_2^+]^{1.9} \cdot \left[1 - \exp\left(\frac{A}{0.53 \cdot RT}\right)\right] \tag{6-152}$$

where, again r is in mol m^{-2} s^{-1}, the concentration of the surface complex is in mol m^{-2}, and $k^+_{Mg} = 10^{1.99} = 97.7$ s^{-1}. Note that the Temkin's average stoichiometric number is $\sigma = 1/n_{[\equiv MgOH_2^+]} = 0.53$. Indeed, a relatively good correspondence between experimental and theoretical rates is observed in Fig. 6.42. Equation (6-152) can be used up to a thermodynamic affinity of 2,260 J mol^{-1} (corresponding to a Q/K_{sp} ratio of 2.5), but it cannot be applied under conditions of high oversaturation, where nucleation prevails and the hypothesis that the same activated complex controls both dissolution and precipitation is not acceptable anymore. Besides, equation (6-152) cannot be applied above pH 12, where the concentration of the $\equiv MgO^-$ sites cannot be neglected (Fig. 6.42).

Interestingly, the log-rates of brucite dissolution determined by Pokrovsky and Schott (2004) under far-from-equilibrium conditions in the pH range 2.5–10 are also linearly correlated with the logarithm of the $a^2_{H+}/a_{Mg^{2+}}$ activity ratio (Fig. 6.43) and fit the linear regression equation ($R^2 = 0.932$)

$$\log r_+ = 0.106 \times \log\left(a^2_{H^+}/a_{Mg^{2+}}\right) - 6.705, \tag{6-153}$$

similar to what is observed for enstatite (Oelkers and Schott, 2001) and *mutatis mutanda* for Al-silicates (see Section 6.4.1). Therefore, the experimental, far-from-equilibrium, kinetic data of brucite can also be interpreted in the light of the TST-based rate law involving the $a^2_{H+}/a_{Mg^{2+}}$ activity ratio. The following TST, Arrhenius-type equation

$$r_+ = A_a \cdot \exp\left(-\frac{E_a}{RT}\right) \cdot \left(\frac{a^2_{H^+}}{a_{Mg^{2+}}}\right)^{0.106} \tag{6-154}$$

is obtained, where the pre-exponential factor A_a is equal to 170 mol m^{-2} s^{-1} and the apparent activation energy E_a is equal to 51 kJ mol^{-1}. This E_a corresponds to the average between the values suggested by Vermilyea (1969) and Jordan and Rammensee (1996) (see above).

Pokrovsky et al. (2005a) measured the dissolution rate of brucite at 25°C in a mixed-flow reactor at different pH values (from 5 to 11), ionic strengths (from 0.01 to 0.03 M) and

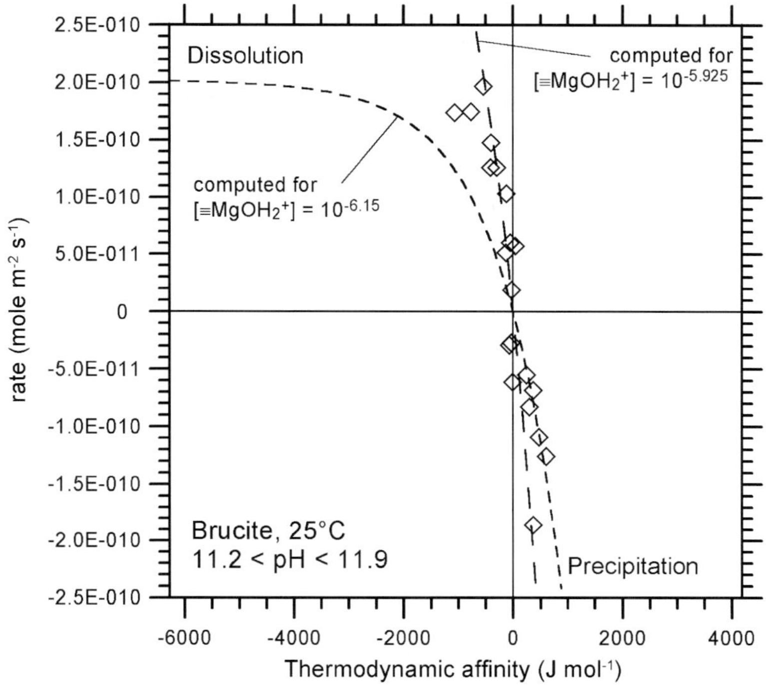

Figure 6.42. Dissolution rates of brucite at 25°C, measured by Pokrovsky and Schott (2004) at pH of 11.2–11.9, as a function of thermodynamic affinity. Lines are computed by means of equation (6-152). (Reprinted from Pokrovsky and Schott, 2004, Copyright 2004 with permission from Elsevier, adapted.)

concentrations of 5 inorganic ligands, 15 organic ligands, and 8 divalent dissolved metals, for investigating possible catalysing or inhibiting effects. In addition, ligand sorption experiments were performed for five selected ligands (fluoride, acetate, oxalate, citrate and ethylene diamine tetracetate (EDTA)) and Langmurian adsorption isotherms were obtained.

It was found that the ligands acting as catalysts (i.e. that enhance the dissolution rate of brucite) are in the following order of effectiveness: EDTA $\geq$ $H_2PO_4^-$ $>$ catechol $\geq$ HCO_3^- $>$ ascorbate $>$ citrate $>$ oxalate $>$ acetate $\cong$ lactate. In contrast, the brucite dissolution rate is decreased by H_3BO_3 at pH 7.3 and PO_4^{3-}, CO_3^{2-}, F^-, oxine, salicylate, lactate, acetate, 4-hydroxybenzoate, borate and SO_4^{2-} at pH $\geq$ 10. However, relatively high concentrations of organic ligands (0.01–0.1 M) are needed to increase or decrease the kinetics of brucite dissolution. This process was found to be unaffected by xylose ($\leq$ 0.1 M), glycine ($\leq$0.05 M), formiate ($\leq$0.3 M) and fulvic and humic acids ($\leq$40 mg L^{-1} of total dissolved organic carbon).

Fluoride has a very peculiar behaviour, in that it experiences not only surface adsorption but also irreversible incorporation in the brucite lattice, through exchange with hydroxyl ion, as documented by sorption experiments in batch and flow-through reactors.

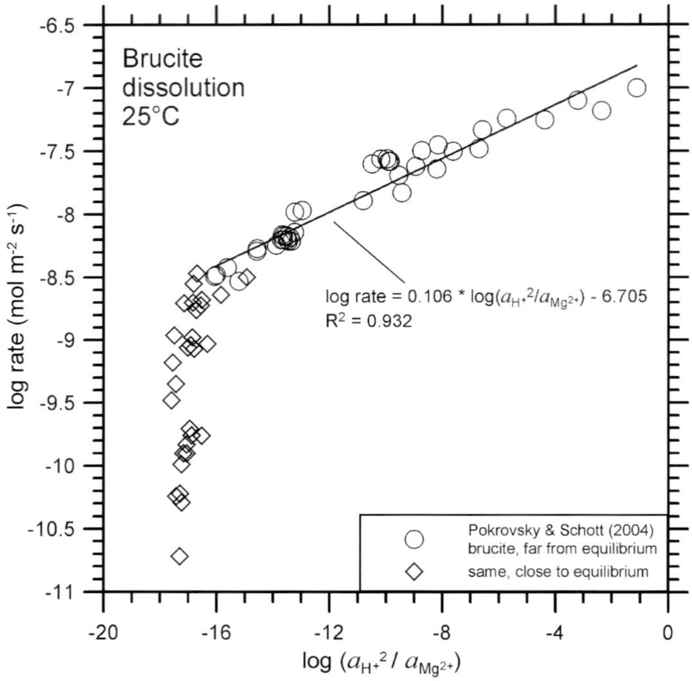

Figure 6.43. Dissolution rates of brucite at 25°C, measured by Pokrovsky and Schott (2004) under different pH, ionic strength and saturation states against the logarithm of the $a^2_{H^+}/a_{Mg^{2+}}$ activity ratio. The linear dependence of the dissolution log-rate on the $\log\left(a^2_{H^+}/a_{Mg^{2+}}\right)$ ratio is clearly evident for the far-from-equilibrium data.

The dissolution rates of brucite in 0.01 M solutions of Ca, Sr, Ba, Mn, Fe, Co, Ni and Pb at pH 4.9 resulted to be proportional, on a log–log scale, to the water exchange rates in the first hydration sphere of the corresponding hexa-hydrated cations, similar to what is observed for nesosilicates (see Section 6.6.1). In particular, Pb and Ca have a catalysing effect, whereas the other metals have an inhibiting effect, which is very strong for Ni and Co.

Pokrovsky et al. (2005a) modelled the effects of the considered ligands on the kinetics of brucite dissolution based on the surface coordination approach (Stumm, 1992). In other words, they assumed that the overall dissolution rate is the sum of the rates of all the individual dissolution reactions, which are promoted by the distinct reactants competing for the available surface sites.

6.7.3. Fe(III) (hydr)oxides

The mechanisms and rates of the dissolution of hydroxides of trivalent iron (e.g. goethite [α-FeOOH], lepidocrocite [γ-FeOOH] and ferrihydrite) and oxides of trivalent iron or partly containing it (e.g. haematite and magnetite) have been the subject of several studies. Here, all these minerals are called Fe(III) (hydr)oxides. In particular, a very important contribution has been provided by Stumm and co-workers to the understanding of not only this topic but also the important role of surface complexation in controlling the dissolution kinetics of simple oxides and multiple oxides, i.e. silicate and Al-silicate minerals.

As summarized by Stumm (1997), the dissolution of Fe(III) (hydr)oxides comprises different mechanisms, namely (i) proton-promoted, (ii) ligand-promoted and (iii) reductant-promoted pathways. Either surface protonation or the formation of a surface complex represents the first step of the entire process. This is followed by breaking of the $\equiv FeO^-$ or $\equiv FeOH^\circ$ bonds, leading to the detachment of either (i) a Fe(III)-aquo complex, or (ii) a trivalent Fe–ligand complex, or (iii) a divalent Fe–ligand complex. Since the formation of the surface complex is a fast reaction (compared to the subsequent steps), attainment of equilibrium can be assumed for the first step of the dissolution reaction. In any case, the initial surface structure of the Fe(III) (hydr)oxide is regenerated and, consequently, dissolution takes place at steady state or close to this condition. Experimental evidence suggests that the dissolution rate is proportional to the concentration of the surface complex, which is assumed to be the precursor of the activated complex.

Several studies have shown that the dissolution rate of Fe(III) (hydr)oxides is enhanced significantly by ligands (e.g. oxalate) forming bidentate surface complexes with an Fe atom on the (hydr)oxide surface. A further increase in the dissolution rate is caused by reductants, including Fe(II), which is produced in reductive dissolution of Fe(III) (hydr)oxides. Synergistic effects between complexing agents and reductants have also been observed. For obvious reasons of space, only a few studies are considered hereunder.

Banwart et al. (1989) report the results of an investigation of the effects of oxalate, acting as complex former, and L-ascorbic acid, acting as reductant, on the dissolution of haematite at 25°C, pH 3, and ionic strength 0.01 M, in a $NaNO_3$–(HNO_3) aqueous medium. Dissolution experiments were performed in the dark, to avoid light-induced reductive dissolution (see Sulzberger et al., 1989). The slowest dissolution mechanism was the proton-promoted dissolution (open circle with label P in Fig. 6.44). The rate was enhanced in the presence of 50 μM oxalate through ligand-promoted dissolution (open circle with label L in Fig. 6.44). The reductive dissolution, assisted by 10 μM ascorbate, proceeded faster than previous runs (open circle with label R in Fig. 6.44). The fastest dissolution was achieved by adding 100 μM ascorbate along with 50 μM oxalate, i.e. through the ligand-assisted reductive pathway (open circle with label LR in Fig. 6.44).

Suter et al. (1991) investigated the reductive dissolution of haematite assisted by ascorbate (10^{-4} M) in the pH range 2.5–6, as well as the reduction of Fe(III) on the goethite surface driven by oxalate complexes of Fe(II) (oxalate 10^{-3} M, Fe(II) 5×10^{-5} M). In agreement with previous results by Zinder et al. (1986), it was found that protonation of the haematite surface speeds up reductive dissolution, although the surface concentration of ascorbate decreases with decreasing pH (open squares in Fig. 6.44). In contrast, the dissolution rate of goethite increases with pH (open triangles in Fig. 6.44). To explain this effect, it must be underscored that an increase in pH enhances the concentration of deprotonated oxalate which, consequently, complexes more Fe(II) ions. This causes an increase in the concentration of the Fe(II)-oxalate complex on the goethite surface, which promotes the reduction of Fe(III) to Fe(II) and its detachment, thus causing a higher dissolution rate of goethite.

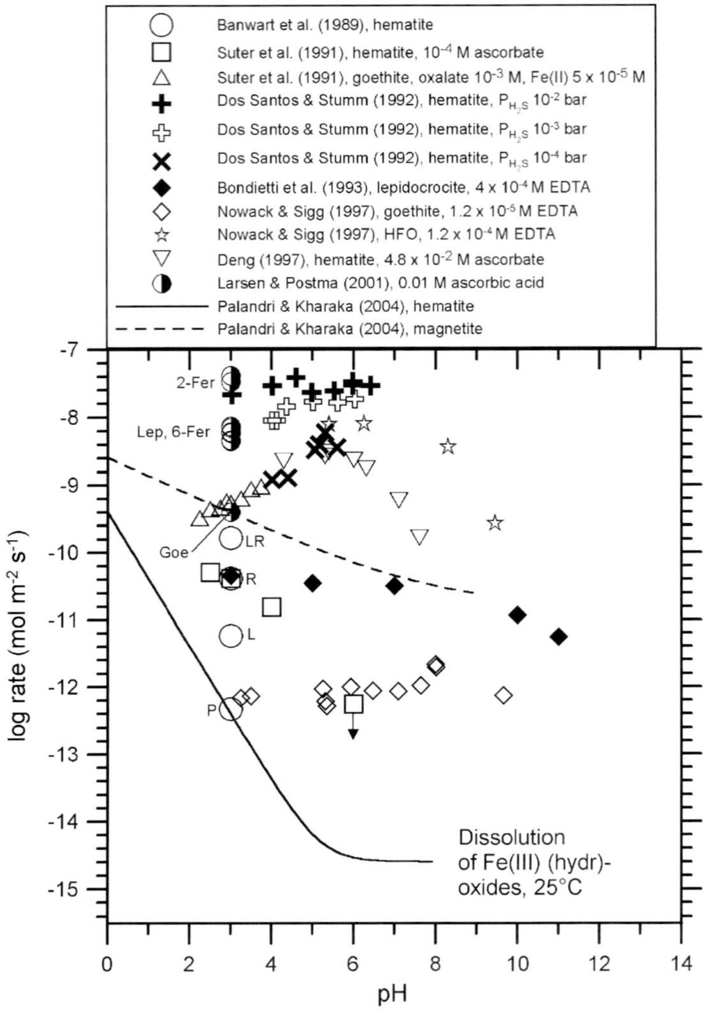

Figure 6.44. Log-rates of dissolution of Fe(III) (hydr)oxides at 25°C as a function of pH. Codes are as follows: 2-Fer = 2-line ferrihydrite, 6-Fer = 6-line ferrihydrite, Lep = lepidocrocite and Goe = goethite.

The dissolution of Fe(III) hydroxide in 4.8×10^{-2} M ascorbate was also investigated by Deng (1997) as a function of both pH (from 4.3 to 7.6) and temperature (from 20 to 60°C). He found that at 20°C, the reductive dissolution rate is pH independent from pH 4.3 to 6.0 (open inverted triangles in Fig. 6.44). In contrast, the rate decreases substantially with increasing pH in the pH interval 6.0–7.6, in agreement with what was observed by Suter et al. (1991). As shown by Deng (1997), these pH effects are consistent with the pH dependence of ascorbate adsorption onto Fe(III) hydroxide. The apparent activation energy resulted to be 27 kJ mol^{-1} at pH 6.0 and 47 kJ mol^{-1} at pH 7.5, suggesting that the overall dissolution kinetics of Fe(III) hydroxide is governed by surface reaction.

Dos Santos and Stumm (1992) studied the reductive dissolution of haematite governed by H_2S under constant P_{H_2S} of 10^{-4}, 10^{-3}, and 10^{-2} atm at 25°C and pH 3–7, below saturation with iron sulphide. Haematite dissolution proceeds very fast under these conditions (crosses in Fig. 6.44). It was shown that the dissolution rate r_+ (mol m^{-2} s^{-1}) depends on the concentrations of the surface complexes $\equiv FeS^-$ and $\equiv FeSH$, as expressed by the relation

$$r_+ = k_1 \cdot [\equiv \mathrm{FeS}^-] + k_2 \cdot [\equiv \mathrm{FeSH}], \tag{6-155}$$

where surface concentrations (indicated by brackets) are in mol m^{-2}, k_1 = 30 h^{-1} and k_2 = 400 h^{-1}.

Bruno et al. (1992) found that the dissolution rate r_+ (mol m^{-2} s^{-1}) of haematite at 25°C is increased in bicarbonate solutions, according to the expression

$$r_+ = k_1 \cdot m_{\mathrm{HCO_3^-}}^{0.23} \tag{6-156}$$

with $k_1 = 3.94 \times 10^{-11}$ s^{-1}. Alternatively, the dissolution rate can be expressed as a linear function of the concentration of the surface bicarbonate complex FeOH-HCO_3^-, with a rate constant $k_s = 3.1 \times 10^{-7}$ s^{-1}.

Bondietti et al. (1993) investigated the inhibition/catalysis of the dissolution of lepidocrocite determined by organic (benzoate and phthalate) and inorganic solutes (arsenate, phosphate, selenite and sulphate) and the inhibition of the dissolution of goethite caused by both trivalent and hexavalent chromium. The non-reductive dissolution promoted by EDTA was chosen as a reference condition. Batch experiments were carried out in the dark at constant pH (from 3 to 11) and ionic strength, 0.01 M ($NaClO_4$). The dissolution rates at various pH values in the presence of 4×10^{-4} M EDTA are plotted as closed diamonds in Fig. 6.44. It was found that benzoate, phthalate, sulphate and, at neutral pH values, arsenate, phosphate and selenite inhibit the dissolution of lepidocrocite, whereas below pH 5, arsenate, phosphate and selenite enhance the dissolution rate. Biber et al. (1994) confirmed that binuclear surface complexes, such as those produced through the adsorption of oxoanions (phosphate, arsenate, sulphate and borate) inhibit both reductive and non-reductive dissolutions of Fe(III) (hydr)oxides.

The dissolution rate of goethite and hydrous ferric oxide driven by uncomplexed EDTA (open diamonds and open stars, respectively) and metal-complexed EDTA, at 25°C, was studied by Nowack and Sigg (1997). Oxides were suspended in 0.01 M $NaNO_3$ and EDTA or metal–EDTA concentrations were 1.2×10^{-5} M for goethite and 1.0×10^{-4} M EDTA or 1.2×10^{-5} M metal–EDTA for hydrous ferric oxide. It was found that the dissolution is slower if EDTA is complexed by metals with respect to uncomplexed EDTA, in the following order: EDTA $\gg$ Ca–EDTA > Pb–EDTA > Zn–EDTA > Cu–EDTA > Co(II)–EDTA > Ni–EDTA. However, differences are small for goethite and large for hydrous ferric oxide. To explain this different behaviour of the two oxides, it must be considered that the overall dissolution process comprises the following steps: (1) sorption of metal–EDTA complexes to the mineral surface, (2) dissociation of the surface complex and (3) detachment of Fe(III)–EDTA into the aqueous solution. Consequently, Nowack and Sigg (1997) concluded that step (3) is the slowest and limits the dissolution rate of goethite, whereas step (2) constrains the dissolution kinetics of hydrous ferric oxide.

A study by Larsen and Postma (2001) was aimed at elucidating the different kinetics of reductive dissolution of 2-line ferrihydrite, 6-line ferrihydrite, lepidocrocite and goethite in O_2-free 10 mM ascorbic acid at pH 3 and 25°C. At relatively high concentrations (>5 mM), ascorbic acid saturates the surface of Fe(III) (hydr)oxides and the reductive dissolution rate becomes independent of ascorbic acid concentration (Banwart et al., 1989). These conditions, therefore, are suitable for studying the effects of crystal reactivity on dissolution kinetics. It was found that 2-line ferrihydrites and goethite are the fastest and slowest dissolving Fe(III) (hydr)oxides, respectively, whereas the reductive dissolution rates of lepidocrocite and 6-line ferrihydrite are similar and intermediate (Fig. 6.44). Larsen and Postma (2001) also showed the important variations in the rates with the remaining (undissolved) mineral fraction.

White et al. (1994) carried out short-term ($<$36 h) and long-term ($<$120 days) dissolution experiments of natural magnetite (and ilmenite) samples in anoxic solutions of pH 1–7 at temperatures of 2–65 °C. In the short-term experiments, release rates of Fe(II) from dissolving magnetite were found to be linear with time, indicating dissolution rates of 0.3 to 13×10^{-10} mol m^{-2} s^{-1} depending on pH. The dissolution rate law by Palandri and Kharaka (2004, Table 6.1) is based on these data. In long-term experiments, the release rate of Fe(II) decreased with time due to coating of magnetite grains by oxide surface layers of increasing thickness.

6.8. Dissolution and precipitation rates of carbonates

The kinetics of dissolution and precipitation of carbonate minerals represents a matter of extreme interest, which has received considerable attention during the last decades. Indeed, the kinetics of dissolution and/or precipitation of carbonate minerals plays a pivotal role in distinct environments, such as oil and gas reservoirs, geothermal aquifers, fresh-water aquifers, ocean sediments, human buildings, etc. — and, of course, it is of utmost importance for the geological storage of CO_2 as well.

6.8.1. Laboratory experimental techniques

The methodologies employed in laboratory dissolution/precipitation experiments include, in addition to those described in Section 6.5.1, the pH-stat method and two hydrodynamically well-defined systems — the rotating disk technique and the channel-flow cell method. These peculiar experimental techniques, which are different from those discussed in Section 6.5.1, are dictated by the relatively fast reaction rates of carbonate minerals compared with silicates and (hydr)oxides.

The *pH-stat method* consists of an automated pH meter that drives either a burette containing an acid solution or another burette containing a basic solution, if the pH of the solution of interest deviates from the selected value. The reaction rate is obtained from the rate of acid (or base) addition needed to maintain the pH constant. In addition, the P_{CO_2} of the solution of interest can also be kept constant by bubbling a gas of known P_{CO_2} through it, in spite of possible problems due to the relatively slow rates of the reactions of dissolution of gaseous CO_2 and of hydration of dissolved molecular CO_2 to produce true carbonic acid (see discussion in Chou et al., 1989). Dissolution rates as low as 6×10^{-9} mol m^{-2} s^{-1}, with a reproducibility better than $\pm 10\%$, can be determined by means of the pH-stat method (Morse, 1974).

The *rotating disk technique* allows the determination of dissolution rates under varying hydrodynamic conditions. It was carefully investigated by Sjöberg and Rickard (1983). At the beginning of the run, the reactive solution is charged into the reactor and the sample disk is mounted on a support that is fixed to the rotator and plunged in the aqueous solution. Critical parameters in the experimental set-up are the vessel volume, the disk diameter, and the rotation speed. The vessel volume must be large with respect to the disk diameter to avoid interference between the vessel walls and non-turbulent flow across the disk. The rotation speed must be constant and precise and wobbling of the disk must be virtually nil. The angular velocity of the disk, ω, must be adapted to maintain a Reynolds number lower than the threshold value between laminar and turbulent flow. We recall that the Reynolds number, Re, is

$$\mathrm{Re} = \frac{r^2 \cdot \omega}{\nu} \tag{6-157}$$

where r is the radius of the disk and ν the kinematic viscosity. The equations used to obtain rate data from this technique are given below (equations (6-159)–(6-164)).

In the *channel-flow cell method*, developed by Compton and co-workers, the reacting solution is flowed through a miniaturized cell under laminar flow conditions. The flow cell consists of a Perspex block closed by a cover plate, which is made up of a sample crystal. Immediately downstream of the crystal, an electrochemical sensor (either an amperometric detector or a pH electrode) determines the concentration of the H^+ ion. Transport processes are modelled by means of the backward implicit finite difference method to calculate the concentration profiles of H^+ and other solutes throughout the cell at steady state. A rate law for the dissolution/precipitation of sample mineral must be specified, as a boundary condition at the mineral–water interface, and can thus be tested.

As already recalled in Section 6.5, in recent years, new microscopic approaches, i.e. AFM and OI, have been applied to investigate the processing occurring on the surface of solid phases, including carbonate minerals, at the atomic scale. The obtained data complement the results of macroscopic dissolution/precipitation experiments.

6.8.2. Calcite

6.8.2.1. Calcite: dissolution–precipitation mechanisms

Nancollas and Reddy (1971) found the backward rate constant to be independent of the stirring rate and an activation energy of 46 ± 4 kJ mol^{-1}, both suggesting that crystal growth follows a surface-controlled mechanism, at pH of 8.5–8.8. Sjöberg (1976) observed that the apparent activation energy for calcite dissolution between pH 8.0 and 10.1, 35 kJ mol^{-1}, is consistent with a chemically controlled rather than a diffusion-controlled reaction. However, he added that transport effects may become rate-limiting at lower pH values, where calcite dissolution proceeds at a faster rate.

The relative importance of transport processes and heterogeneous reactions at the calcite surface (see Sections 6.2.1 and 6.2.2) was discussed by several authors since the 1930s, as well as by Plummer and Wigley (1976), who investigated the kinetics of calcite dissolution at ~1 bar P_{CO_2}, 25°C, and $4 < \text{pH} < 6$ in stirred aqueous solutions. They found that under these experimental conditions, the rate of the process is probably governed by surface reactions. However, they also suggested the conditions for transport-, intermediate- and surface-controlled calcite dissolution, as a function of the dimensionless $C_B/C_{S,eq}$ ratio (where C_B is the bulk fluid concentration and $C_{S,eq}$ is the theoretical equilibrium concentration at the mineral surface) and the parameter $\alpha = k_T/(k_C \times C_{S,eq})$, where k_C and k_T are the surface-(chemically)-controlled and the transport-controlled rate constants, respectively.

Rickard and Sjöberg (1983) and Sjöberg and Rickard (1984a,b) investigated the same matter and recognized the presence of three distinct rate-controlling mechanisms for calcite dissolution:

(i) A transport-controlled kinetic regime at low-pH values, where the dissolution rate shows a marked dependence on pH but is independent of the ionic strength.

(ii) A mixed kinetic regime, in neutral to alkaline solutions, where the dissolution rate is independent of pH but is dependent on ionic strength. In this region, the rate results from the interaction of a transport process and a surface chemical reaction. The latter contribution can be expressed as follows:

$$r_{+,C} = k_C \cdot \frac{A_S}{V} \cdot \left(C_{B,eq} - C_S\right)^n, \tag{6-158}$$

where $C_{B,eq}$ is the concentration in the bulk solution at mineral–solution equilibrium, C_S the actual concentration at the mineral surface, A_S the surface area of the dissolving mineral and V the volume of the aqueous solution.

(iii) A transitional regime, in which the dependence on H^+ activity varies from 0, as pH → 5.5 at 25°C, to 1, as pH → 4.0 at 25°C. However, the transitional pH depends also on the hydrodynamics of the system and increases with increasing angular velocity of the disk and concurrent decrease in the thickness of the so-called diffusion-boundary layer (see below).

In the pH-dependent, transport-controlled region, the rate of dissolution measured by means of pH-stat runs using a rotating disk apparatus was found to be proportional to the square root of the angular velocity of the disk, ω, as expected — based on the theory of Levich (1962) for a purely transport-controlled reaction. In this region, the rate of the dissolution process is given by (Nernst, 1904)

$$r_{+,T} = \frac{D \cdot A_S \cdot \left(C_S - C_B\right)}{V \cdot \delta_{DBL}} \tag{6-159}$$

where D is the Fickian diffusion coefficient and δ_{DBL} the thickness of the so-called diffusion boundary layer, i.e. the fluid layer close to the mineral surface, where mass-transport process takes place through diffusion. Setting $D/\delta_{DBL} = k_T$, and referring to unit A_S and unit V, equation (6-159) is usually rewritten as,

$$r_{+,T} = k_T \cdot (C_S - C_B). \tag{6-160}$$

Rickard and Sjöberg (1983) showed that the rate of calcite dissolution can be expressed as,

$$r_+ = \frac{k_C \cdot k_T}{k_C + k_T} \cdot (C_{B,eq} - C_B), \tag{6-161}$$

which again refers to unit A_S and unit V. If $k_T \gg k_C$, $k_+ \cong k_C$, i.e. the dissolution rate is surface controlled, whereas for $k_C \gg k_T$, $k_+ \cong k_T$, and the rate is transport controlled.

As shown by Rickard and Sjöberg (1983), the rotating disk technique can be used to discriminate and measure the contributions from transport and surface reactions to the kinetic constant. To this purpose, they considered the initial rate condition, where $C_B \rightarrow 0$, and wrote the inverse of equation (6-161), obtaining (recall that $k_T = D/\delta_{DBL}$),

$$\frac{1}{r_+} = \frac{1}{k_T \cdot C_{eq}} + \frac{1}{k_C \cdot C_{eq}} = \frac{\delta_{DBL}}{D \cdot C_{eq}} + \frac{1}{k_C \cdot C_{eq}}. \tag{6-162}$$

For the rotating disk technique, δ_{DBL} is given by (Rickard and Sjöberg, 1983),

$$\delta_{DBL} = \frac{1.61 D^{1/3} \nu^{1/6}}{\omega^{1/2}}. \tag{6-163}$$

Insertion of this relation in equation (6-162) leads to,

$$\frac{1}{r_+} = \frac{1.61 \cdot \nu^{1/6}}{D^{2/3} \cdot C_{eq} \cdot \omega^{1/2}} + \frac{1}{k_C \cdot C_{eq}}. \tag{6-164}$$

Equation (6-164) indicates that in a plot of $1/r_+$ vs. $1/\omega^{1/2}$, a linear relation is expected, with an intercept that is a function of the chemical rate constant and a slope depending on the diffusion coefficient (see, for instance, Fig. 5 in Rickard and Sjöberg, 1983). However, when the chemical rate constant is large, the second term on the right-hand side of equation (6-164) vanishes and the linear relation in the plot of $1/r_+$ vs. $1/\omega^{1/2}$ passes through the origin, preventing the determination of the chemical rate constant (see, for instance, Fig. 3 in Alkattan et al., 1998).

Rickard and Sjöberg (1983) also noticed that the chemically controlled rate constant (k_C) was different for distinct types of calcite, e.g. fine-grained precipitates vs. Iceland spar, and suggested that this fact is ascribable to differences in the number and kind of reactive surface sites or kinks on the distinct types of calcite. Strictly, this implies that it is impossible to establish a general rate equation for calcite (and for any other mineral, as well).

The complex mechanisms of calcite dissolution might seem rather odd, after having considered, in previous sections, the dissolution and precipitation reactions of silicates and (hydr)oxides, which are all surface-controlled processes under varying boundary conditions (pH, temperature, ionic strength, etc.). However, it must be underscored that calcite (and other carbonates, see below) dissolves and precipitates much faster than silicates and (hydr)oxides. Consequently, either the removal of particles from the surface of dissolving calcite or the addition of particles to the surface of precipitating calcite is so rapid that solute transport in the aqueous solution cannot keep up with these processes and represents the rate-controlling mechanism. This situation typically occurs at low-pH values, where the dissolution and precipitation reactions of calcite attain the fastest rates.

6.8.2.2. Calcite: dissolution and precipitation rates

In earlier studies of calcite dissolution and precipitation, close-to-equilibrium data were usually fitted to simple empirical functions of the degree of saturation Ω (see Section 6.4), namely

$$r_+ = k_+ \cdot (1-\Omega)^{n_+}, \tag{6-165}$$

for dissolution, and

$$r_- = k_- \cdot (\Omega-1)^{n_-}, \tag{6-166}$$

for precipitation, where n_+ and n_- are generally fractional numbers. Equations (6-165) and (6-166) can be written in logarithmic form, obtaining the linear equations

$$\log r_+ = n_+ \cdot \log(1-\Omega) + \log k_+ \tag{6-167}$$

and

$$\log r_- = n_- \cdot \log(\Omega-1) + \log k_-. \tag{6-168}$$

As shown by equations (6-167) and (6-168), log–log plots allow one to derive n_+ and n_- from the slope and log k_+ and log k_- from the intercept. For instance, dissolution experiments of deep-sea sediments from the Indian, Atlantic and Pacific Oceans in either deep-sea water or pore water yielded values of n_+ ranging between 3.0 and 5.2 (Morse and Berner, 1972, 1979; Morse, 1978).

Reddy and Nancollas (1971) studied the kinetics of calcite precipitation in simple aqueous solutions at 25°C and pH close to 8.5, and Nancollas and Reddy (1971) extended these investigations at 10 and 40°C, at pH of 8.5–8.8. Stable supersaturated solutions were inoculated with calcite seed crystals, and the concentrations of dissolved Ca^{2+} and H^+ ions were monitored as a function of time. The kinetics of calcite precipitation resulted to be described by the following rate expression:

$$r_- = k_- \cdot A_S \cdot \left(m_{Ca^{2+}} \cdot m_{CO_3^{2-}} - \frac{K_{sp}}{\gamma_{(II)}^2} \right), \tag{6-169}$$

where k_- is the apparent precipitation rate constant, A_S the calcite surface area in contact with 1 dm^3 of aqueous solution, K_{sp} the solubility product of calcite and $\gamma_{(II)}$ the individual activity coefficient of any divalent ion, computed by means of the Davies equation (4-13). Note that equation (6-169) can be rewritten in the form

$$r_- = k_- \cdot A_S \cdot \frac{K_{sp}}{\gamma_{(II)}^2} (\Omega - 1). \tag{6-170}$$

Sjöberg (1976) studied experimentally the dissolution rate of $CaCO_3$ powder at 5, 20, 40 and 50°C in a 0.7 M KCl solution (whose ionic strength is similar to that of seawater), between pH 8.0 and 10.1. He concluded that the rate of calcite dissolution was described by the equation

$$r_+ = k_+ \cdot A_S \cdot \left[\left(\frac{K_{sp}}{\gamma_{Ca^{2+}} \cdot \gamma_{CO_3^{2-}}} \right)^{1/2} - m_{Ca^{2+}}^{1/2} \cdot m_{CO_3^{2-}}^{1/2} \right], \tag{6-171}$$

where k_+ is the apparent dissolution rate constant, 9.5×10^{-2} s^{-1} m^{-2} at 20°C. Equation (6-171) can be easily converted into the form

$$r_+ = k_+ \cdot A_S \cdot \left(\frac{K_{sp}}{\gamma_{Ca^{2+}} \cdot \gamma_{CO_3^{2-}}} \right)^{1/2} \cdot \left(1 - \Omega^{1/2}\right). \tag{6-172}$$

Equations (6-170) and (6-172) look similar to equations (6-166) and (6-165), respectively. However, equations (6-170) and (6-172), without the ion activity coefficient(s), can be derived by using the principle of detailed balancing (see Section 6.1) and rest, therefore, on theoretical bases rather than on empirical grounds.

Reddy et al. (1981) applied a seeded growth method to investigate calcite precipitation from calcium bicarbonate solutions at 25°C and fixed P_{CO_2} between 0.03 and 0.3 bar. They found that the relationships between the log rate of calcite precipitation and either log Ω, or log (Ω − 1), or log ($C_B - C_{B,eq}$) (where C_B is total Ca concentration in the bulk fluid and $C_{B,eq}$ is total Ca concentration at equilibrium for each experiment) are all non-linear. This non-linearity clearly evidences the limits of equations (6-166) and (6-170) as well as equations (6-165) and (6-172). To overcome these difficulties, a different theoretical approach was needed. This was already laid down by Plummer et al. (1978).

Before presenting this important contribution, it must be recalled that, according to some authors, equation (6-166) can be used to detect the precipitation mechanism of calcite. For instance, Shiraki and Brantley (1995 and references therein) suggested that the exponent n_- of the $(\Omega - 1)^{n-}$ term is equal to 1 for adsorption-controlled growth, it becomes 2 for spiral growth at screw dislocations, and it ranges from 2 to 3 for growth by both screw and edge dislocations. In contrast, the rate becomes an exponential function of Ω in polynuclear growth, when many nuclei generate on the surface of the crystal before a whole new layer of the crystal is formed. This approach, however, was questioned by Teng et al. (2000) through both application of the classical crystal growth theory (Burton et al., 1951) and microscopic observation of surface processes. Since the only measurable variable in these rate laws is Ω, these equations cannot reflect the controls of microscopic growth parameters (step speed and hillock slope) on the growth rate or reliably point out the actual mechanism of crystal growth.

Plummer et al. (1978) studied the effects of several variables systematically, such as pH (in the range 2–7), P_{CO_2} (from 0.0003 to ~1 bar) and temperature (from 5 to 60°C), on the dissolution kinetics of calcite. For the first time, rate equations were formulated in terms of the activities of reactants and products effectively involved in the dissolution reaction of calcite and affecting its rate.

Far-from-equilibrium experiments were carried out by Plummer et al. (1978) by using the pH-stat technique. Three simultaneous, mutually independent forward reactions were identified, namely

$$CaCO_3 + H^+ \underset{k_{-1}}{\overset{k_1}{\rightleftarrows}} Ca^{2+} + HCO_3^-, \tag{6-173}$$

$$CaCO_3 + H_2CO_3^* \underset{k_{-2}}{\overset{k_2}{\rightleftarrows}} Ca^{2+} + 2HCO_3^-, \tag{6-174}$$

and

$$CaCO_3 + H_2O \underset{k_{-3}}{\overset{k_3}{\rightleftarrows}} Ca^{2+} + HCO_3^- + OH^-. \tag{6-175}$$

Forward reaction (6-173) is important in the acid range, below pH 4.5–5.5, at all P_{CO_2} values; it explains the dependence of the rate on H^+ activity (Fig. 6.45). Forward reaction (6-174) is important at high pH (above 4.5–5.5) and $P_{CO_2} > 0.1$ bar; it accounts for the linear dependence of the rate on the total concentration of dissolved CO_2, $H_2CO_3^*$, i.e. the sum of the concentrations of true carbonic acid $H_2CO_3^+$ along with true dissolved molecular $CO_{2(aq)}$. Forward reaction (6-175) is important at high pH (above 5.5) and $P_{CO_2} < 0.1$ bar. Consequently, under far-from-equilibrium conditions, the rate of calcite dissolution r_+ (in mol m^{-2} s^{-1}) is given by the sum of the three individual rates for each reaction:

$$r_+ = k_1 \cdot a_{H^+} + k_2 \cdot a_{H_2CO_3^*} + k_3 \cdot a_{H_2O}, \tag{6-176}$$

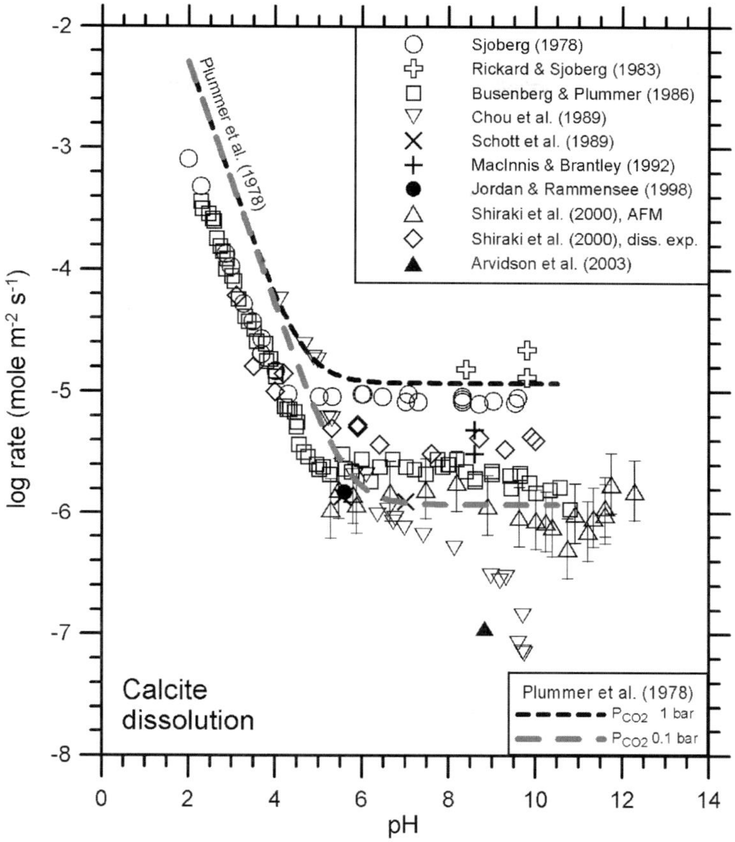

Figure 6.45. Log-rates of calcite dissolution at 25°C as a function of pH, showing the data of several macroscopic and microscopic experimental investigations and the predictions of the rate law of Plummer et al. (1978) at two different P_{CO_2} values.

where k_1, k_2 and k_3 are the rate constants, whose temperature dependence is given by (T in K),

$$\log k_1 = 1.198 - \frac{444}{T}, \tag{6-177}$$

$$\log k_2 = 3.84 - \frac{2177}{T}, \tag{6-178}$$

$$\log k_3 = -4.86 - \frac{317}{T} \qquad (T < 298\,K), \tag{6-179}$$

and

$$\log k_3 = -0.10 - \frac{1737}{T} \qquad (T > 298\,K). \tag{6-180}$$

At 25°C, the forward rate constants are $k_1 = 10^{-0.29}$, $k_2 = 10^{-3.46}$ and $k_3 = 10^{-5.93}$ mol m^{-2} s^{-1}. Note that since activities are dimensionless, rate and rate constants have the same measurement units, in this case.

Plummer et al. (1978) also studied calcite dissolution under close-to-equilibrium conditions by monitoring the pH changes in a batch reactor under constant P_{CO_2} (free-drift technique). They suggested that the backward rate is proportional to the product of Ca^{2+} and HCO_3^- activities, i.e.

$$r_- = k_{-4} \cdot a_{Ca^{2+}} \cdot a_{HCO_3^-}. \tag{6-181}$$

It can be shown that the backward rate constant k_{-4}, appearing in equation (6-181), is a function of the three forward rate constants k_1, k_2 and k_3. To this purpose, let us start to write the backward rate as follows:

$$r_- = k_{-1} \cdot a_{Ca^{2+}} \cdot a_{HCO_3^-} + k_{-2} \cdot a_{Ca^{2+}} \cdot a^2_{HCO_3^-} + k_{-3} \cdot a_{Ca^{2+}} \cdot a_{HCO_3^-} \cdot a_{OH^-}. \tag{6-182}$$

Note that equation (6-182) sums the backward contributions of the three reactions (6-173), (6-174) and (6-175). Then the principle of detailed balancing (see Section 6.1) is applied separately to each of these three reactions, assuming that they are elementary reactions and obtaining

$$k_{-1} = k_1 \cdot \frac{K_{HCO_3^-}}{K_{sp}}, \tag{6-183}$$

$$k_{-2} = k_2 \cdot \frac{K_{HCO_3^-}}{K_{H_2CO_3^*} \cdot K_{sp}}, \tag{6-184}$$

and

$$k_{-3} = k_3 \cdot \frac{K_{HCO_3^-}}{K_W \cdot K_{sp}}, \tag{6-185}$$

where $K_{H_2CO_3^*}$, $K_{HCO_3^-}$, K_W and K_{sp} are the thermodynamic equilibrium constants of the following reactions:

$$H_2CO_3^* \rightleftharpoons H^+ + HCO_3^-, \tag{6-186}$$

$$HCO_3^- \rightleftharpoons H^+ + CO_3^{2-} \tag{6-187}$$

$$H_2O \rightleftharpoons H^+ + OH^-, \tag{6-188}$$

and

$$CaCO_{3(s)} \rightleftharpoons Ca^{2+} + CO_3^{2-}, \tag{6-189}$$

respectively. Combination of equations (6-182)–(6-185) and rearrangement of the obtained relation lead to

$$r_- = \frac{K_{HCO_3^-}}{K_{sp}}\left[k_1 + \frac{1}{a_{H^+}}\cdot\left(k_2 \cdot a_{H_2CO_3^*} + k_3 \cdot a_{H_2O}\right)\right]\cdot a_{Ca^{2+}} \cdot a_{HCO_3^-}. \tag{6-190}$$

Comparison of equations (6-181) and (6-190) shows that k_{-4} is not a simple rate constant, but it is given by

$$k_{-4} = \frac{K_{HCO_3^-}}{K_{sp}}\left[k_1 + \frac{1}{a_{H^+}}\cdot\left(k_2 \cdot a_{H_2CO_3^*} + k_3 \cdot a_{H_2O}\right)\right]. \tag{6-191}$$

This is the equation sought for, which links the backward rate constant k_{-4} to the three forward rate constants k_1, k_2 and k_3.

The precipitation rates measured by Reddy et al. (1981) agree within a factor of 2 with the rates predicted by the model of Plummer et al. (1978). The validity of the backward rate equation (6-181) was later confirmed by Busenberg and Plummer (1986), who determined the rate of growth of single calcite (and aragonite) crystals in oversaturated Ca–HCO_3 aqueous solutions at different P_{CO_2} values, in the range of Ω 1.5–100. However, the conclusions of Busenberg and Plummer (1986) were questioned by Lebrón and Suárez (1998), since they could be biased by the adopted experimental method, which does not allow the assessment of possible heterogeneous nucleation contribution (see below).

Besides, equation (6-181) was rejected by Inskeep and Bloom (1985), who investigated the rate of growth of calcite crystals by means of a reproducible seeded-growth technique with a pH-stat, at 25°C, pH values from 8.25 to 8.70, P_{CO_2} from 0.0006 to 0.01 bar, and ionic strength from 0.015 to 0.10 mol kg^{-1}. Several precipitation rate equations were tested by Inskeep and Bloom (1985), including the model by Plummer and co-workers that did not fit all the experimental rates. The Nancollas and Reddy model, which is based on the reaction

$$Ca^{2+} + CO_3^{2-} \underset{k_5}{\overset{k_{-5}}{\rightleftharpoons}} CaCO_3, \tag{6-192}$$

and involves the concentrations of Ca^{2+} and CO_3^{2-} ions in the bulk aqueous phase, was found to describe all the experimental data, upon introduction of the necessary changes to account for varying ionic strength. The precipitation rate constant was estimated to be 118.2 ± 13.9 $dm^6\ mol^{-1}\ m^{-2}\ s^{-1}$.

The kinetics and mechanisms of dissolution of calcite (and other carbonate minerals) were later revisited by Chou et al. (1989), who carried out dissolution experiments in a continuous flow-through fluidized-bed reactor (see Section 6.5.1) at 25°C. In the dissolution of carbonate minerals, this experimental approach has the advantage of minimizing the influence of transport phenomena, thanks to the high turbulence inside the reactor itself. However, Chou et al. (1989) noticed that it was impossible to eliminate the influence of transport control below pH 4. The disadvantage of this experimental set-up is the impossibility of constraining the activity of H_2CO_3*, in contrast with the pH-stat technique, in which the P_{CO_2} can be kept constant (see Section 6.8.1).

The dissolution rates measured by Chou et al. (1989) show a general pH dependence very similar to the results and predictions of Plummer et al. (1978) and Busenberg and Plummer (1986), in that dissolution rates are proportional to a_{H^+} at low pH values and become nearly constant in the pH range 6–8 (Fig. 6.45).
However, the net rate decreases at higher pH values, owing to the importance of the backward reaction. At 25°C, the values of forward rate constants, obtained through best fit of experimental results, are $k_1 = 10^{-0.05}$, $k_2 = 10^{-3.30}$ and $k_3 = 10^{-6.19}$ mol $m^{-2}\ s^{-1}$. These values are in satisfactory agreement (0.16–0.26 log-units) with those computed by means of equations (6-177)–(6-180).

Chou et al. (1989) also computed the backward rate r_- by difference between the forward rate r_+ given by equation (6-176) and the net, experimentally measured rate r_{net}, based on the equation $r_{net} = r_+ - r_-$. They found that the backward rate is proportional to the activity product of Ca^{2+} and CO_3^{2-}:

$$r_- = k_{-5} \cdot a_{Ca^{2+}} \cdot a_{CO_3^{2-}}, \tag{6-193}$$

(in agreement with Inskeep and Bloom, 1985) and obtained a backward rate constant k_{-5} of $10^{+2.28}$ mol $m^{-2}\ s^{-1}$, which is very close to the values given by Inskeep and Bloom (1985), suitably corrected for the activity coefficients, $10^{+2.17}$–$10^{+2.28}$ mol $m^{-2}\ s^{-1}$.

Compton et al. (1989) investigated the kinetics of calcite dissolution in acidic aqueous solutions (pH < 4) by means of the channel-flow cell method (see Section 6.8.1) and showed that the process was governed by a reaction at the calcite–water interface exhibiting first-order kinetics with respect to the surface concentration of H^+ ion.

The same approach was used by Brown et al. (1993) to study the dissolution/precipitation kinetics of calcite in 0.5 M KCl at pH 8.1–8.8 and 25°C. They proposed that the rate of calcite dissolution is given by,

$$r_+ = k_+ \cdot \left(1 - \theta_{Ca^{2+}}\right) \cdot \left(1 - \theta_{CO_3^{2-}}\right), \tag{6-194}$$

whereas the backward rate is described by,

$$r_- = k_- \cdot \theta_{Ca^{2+}} \cdot \theta_{CO_3^{2-}}, \tag{6-195}$$

where $\theta_{Ca^{2+}}$ is given by the Langmuir sorption isotherm:

$$\theta_{Ca^{2+}} = \frac{K^*_{Ca^{2+}} \cdot [Ca^{2+}]}{1 + K^*_{Ca^{2+}} \cdot [Ca^{2+}]}, \tag{6-196}$$

in which $[Ca^{2+}]$ expresses the surface concentration of Ca^{2+}. An equation similar to (6-196) holds for $\theta_{CO_3^{2-}}$. Note that $K^*_{Ca^{2+}}$ and $K^*_{CO_3^{2-}}$ are conditional equilibrium constants, defined in terms of equilibrium species concentrations, and not thermodynamic equilibrium constants involving activities. Through suitable assumptions, Brown et al. (1993) derived the following expression for the net rate:

$$r = k_- \cdot K^*_{Ca^{2+}} \cdot K^*_{CO_3^{2-}} \cdot \left\{ \frac{K^*_{sp} - [Ca^{2+}] \cdot [CO_3^{2-}]}{(1 + K^*_{Ca^{2+}}[Ca^{2+}]) \cdot (1 + K^*_{CO_3^{2-}}[CO_3^{2-}])} \right\}, \tag{6-197}$$

where K^*_{sp} is a sort of conditional solubility product defined in terms of surface concentrations, i.e. $K^*_{sp} = [Ca^{2+}][CO_3^{2-}]$. Equation (6-197) was used to interpret the experimental data, obtaining $K^*_{Ca^{2+}} = 1$ m^3 mol^{-1}, $K^*_{CO_3^{2-}} = 30$ m^3 mol^{-1} and $k_- = 3.3 \times 10^{-7}$ mol m^{-2} s^{-1}. The value of the adsorption constant $K^*_{Ca^{2+}}$ agrees with that directly measured Huang et al. (1991), i.e. 1.8 m^3 mol^{-1}.

Several authors studied the dissolution and growth rates of calcite by using microscopic methods. Jordan and Rammensee (1998) applied AFM and reported a value of 1.5 ($\pm$0.3) $\times$ 10^{-6} mol m^{-2} s^{-1} for the dissolution rate of the $(10\bar{1}4)$ surface in water at 25°C, P_{CO_2} of 10$^{-3.5}$ bar and pH close to 5.6 (Fig. 6.45). Arvidson et al. (2003) used vertical scanning interferometry and found a calcite dissolution rate of 1.1 $\times$ 10^{-7} mol m^{-2} s^{-1} in a 4.4 meq L^{-1} $NaHCO_3$–Na_2CO_3 solution, at fixed P_{CO_2} of 10$^{-3.41}$ bar, pH of 8.8 and 25°C (Fig. 6.45). This rate is lower than that obtained by means of macroscopic dissolution experiments by more than one order of magnitude. According to the authors, this difference could be because of the fact that AFM determines the dissolution rate on a flat surface, without contributions from grain-edge boundaries (characterized by high step and kink density), which could increase with decreasing grain size.

The joint application of macroscopic and microscopic methods was attempted by MacInnis and Brantley (1992) and Shiraki et al. (2000).

MacInnis and Brantley (1992) measured far-from-equilibrium dissolution rates of 3.1 ($\pm$0.4) $\times$ 10^{-6} mol m^{-2} s^{-1} for undeformed calcite (~10^3 dislocations cm^{-2}) by means of a rotating disk experiments at 1,160 rpm, 25°C and pH 8.6 in 0.7 M KCl solution (Fig. 6.45). The dissolution rate increased by a factor of ~2.3 for a strained calcite (6 $\times$ 10^8 dislocations cm^{-2}), in agreement with earlier findings by Schott et al. (1989). MacInnis and Brantley (1992) also measured the dissolution rate of undeformed calcite based on changes in microtopography, by means of a differential-interference-microscopy technique, obtaining a value of 4.9 $\times$ 10^{-6} mol m^{-2} s^{-1} (Fig. 6.45).

Shiraki et al. (2000) used both AFM and Ca fluxes from dissolution experiments to investigate the far-from-equilibrium dissolution kinetics of calcite in 0.1 M NaCl at room temperature, in the pH interval 3.1–12.2 and for P_{CO_2} values of 10^{-12} to $10^{-3.5}$ bar (Fig. 6.45). An interesting molecular interpretation of calcite dissolution was suggested by Shiraki et al. (2000). Rates measured by means of dissolution experiments are systematically higher than AFM-based dissolution rates (Fig. 6.45). According to the authors, the largest uncertainties are related to the geometric surface area, A_S, which is utilized for computing the dissolution experiment rates by means of equation (6-91). The value of A_S was obtained from the geometry of the calcite crystal before the dissolution experiment, without considering the surface roughness. Therefore, the geometric surface area underestimates the true surface area, leading to an overestimation of the computed dissolution rates. Based on their experimental results, Shiraki et al. (2000) proposed the following empirical dissolution rate law (in mol m^{-2} s^{-1}):

$$r_+ = k_1 \cdot m_{H^+} + k_2 \cdot m_{HCO_3^-} + k_3 \cdot m_{OH^-} \tag{6-198}$$

where m_i are the molalities of dissolved species and $k_1 = 10^{-0.66\pm0.13}$, $k_2 = 10^{+0.13\pm0.06}$ and $k_3 = 10^{-3.99\pm0.08}$ m^{-2} s^{-1}. Shiraki et al. (2000) recognized that a rate law in terms of adsorbed concentrations would be more comprehensive, but the determinations of surface concentrations on dissolving calcite is very difficult according to these authors. In particular, in the experiments, it is difficult to unequivocally separate the effects of adsorption of H^+ ions to the calcite surface from the consumption of H^+ ions upon detachment of solutes, i.e. to discriminate adsorption from mineral dissolution. Equation (6-198) was derived assuming a constant total dissolved inorganic carbon molality, m_{TDIC} of 1.3×10^{-6}. Under this condition, CO_2 fugacity depends chiefly on pH, as indicated by the following relation:

$$f_{CO_2} = \frac{K_{CO_{2(g)}}^{-1} \cdot \gamma_{H_2CO_3^*} \cdot m_{TDIC}}{1 + \dfrac{K_{H_2CO_3^*} \cdot \gamma_{H_2CO_3^*}}{10^{-pH} \cdot \gamma_{HCO_3^-}} + \dfrac{K_{HCO_3^-} \cdot K_{H_2CO_3^*} \cdot \gamma_{H_2CO_3^*}}{10^{-2pH} \cdot \gamma_{CO_3^{2-}}}}, \tag{6-199}$$

which is easily obtained from carbonate equilibria, where $K_{H_2CO_3^*}$ and $K_{HCO_3^-}$ are the dissociation constants of carbonic acid and bicarbonate ion, respectively (reactions (6-186) and (6-187)), and $K_{CO_{2(g)}}$ is the equilibrium constant of the reaction

$$CO_{2(g)} + H_2O = H_2CO_3^*. \tag{6-200}$$

Equation (6-199) indicates that f_{CO_2} varied between $10^{-4.41}$ and $10^{-8.50}$ bar in the dissolution experiments carried out by Shiraki et al. (2000). Consequently, equation (6-198) cannot be applied outside this range of f_{CO_2} values.

6.8.2.3. Calcite: the influence of CO_2 partial pressure on dissolution and precipitation rates

The influence of P_{CO_2} on the kinetics and mechanisms of calcite precipitation at 25°C was studied by Lebrón and Suárez (1998), but they considered the P_{CO_2} interval from

0.00035 to 0.1 bar only. Lebrón and Suárez (1998) also investigated the influence of dissolved organic matter, but here we take into account results obtained in aqueous solutions free of dissolved organic matter. These authors observed that solutions close to equilibrium ($1 < \Omega \leq 2$) remain metastable for several days in the absence of calcite crystals, suggesting that the contribution of heterogeneous nucleation to calcite precipitation is negligible under these Ω conditions. At Ω of 3–4, the rate of calcite precipitation increases quickly with small increments of Ω. In contrast, for $\Omega > 5$, where both crystal growth and heterogeneous nucleation are active, the rate of calcite precipitation increases negligibly with increasing Ω.

Besides, for $\Omega = 5$, the rate of calcite precipitation increases linearly with both P_{CO_2} and $CaHCO_3^+$ activity. This suggests that (1) not only crystal growth increases linearly with P_{CO_2} (as predicted by the model of Plummer et al., 1978), but also heterogeneous nucleation depends linearly on P_{CO_2}; (2) $CaHCO_3^+$ has an active role in both crystal growth and heterogeneous nucleation.

Even though some of the works considered above were devoted to the effects of P_{CO_2} on the calcite dissolution/precipitation rates, the first rate data for $P_{CO_2} > 1$ bar were presented very recently by Pokrovsky et al. (2005b). They measured the dissolution kinetics of calcite (as well as of dolomite and magnesite, see below) at 25°C and pH 3–4 for investigating the possible effects of P_{CO_2}, from $\leq 10^{-3.5}$ to 55 bar in aqueous solutions of variable ionic strength (0.001–1 mol kg^{-1}, although rates are expected to be independent of I at these pH values, based on the studies by Sjöberg and Rickard).

Dissolution experiments were carried out both in a mixed-flow reactor and in a closed-system, high-pressure Ti-batch reactor. Calcite powders were used in the mixed-flow reactor, whereas the rotating disk technique was applied in the batch reactor runs. In the mixed-flow reactor, P_{CO_2} was not fixed but it was evaluated to be $\leq 10^{-3.5}$ bar, whereas CO_2 was delivered in the batch reactor to maintain a constant P_{CO_2}, which was controlled by a pressure manometer. The pH was determined *in situ* in both reactors. In particular, a solid contact Li–Sn alloy pH electrode coupled with an Ag/AgCl reference electrode was used in the batch reactor.

The inverse of the rate data obtained by means of the rotating disk technique was plotted against $1/\omega^{1/2}$ (see Section 6.8.2), thus obtaining k_C values of 9.0 ± 0.9 mol m^{-2} s^{-1} at 2 bar P_{CO_2} (surface pH 4.44), 15 ± 2 mol m^{-2} s^{-1} at 10 bar P_{CO_2} (surface pH 4.19) and 28 ± 3 mol m^{-2} s^{-1} at 50 bar P_{CO_2} (surface pH 4.30). Following Pokrovsky et al. (2005b), the corresponding surface-controlled rates are computed multiplying these k_C values times the activity of H^+ ion at the solid surface. Since these rates are ~10 times greater than the experimental rates, the dissolution kinetics of calcite is transport controlled, under these conditions.

Fig. 6.46 shows the P_{CO_2} dependence of the dissolution rate of calcite in 0.1 M NaCl at a constant stirring speed of 425 rpm. In the P_{CO_2} range 1–25 bar, the measured rate increases close to linearly with P_{CO_2}, while it remains nearly constant at ~1.9×10^{-4} mol m^{-2} s^{-1} for P_{CO_2} of 25–50 bar. An 8-fold increase in rate is observed from 1 to 25 bar P_{CO_2}. Measured total dissolution rates $r_{+,meas}$ can be corrected for the proton-promoted rates $r_{+,H}$ to obtain the CO_2-promoted rate r_{+,CO_2} by using the following relation:

$$r_{+,CO_2} = r_{+,meas} - r_{+,H} = r_{+,meas} - k_1 \cdot a_{H^+}, \qquad (6\text{-}201)$$

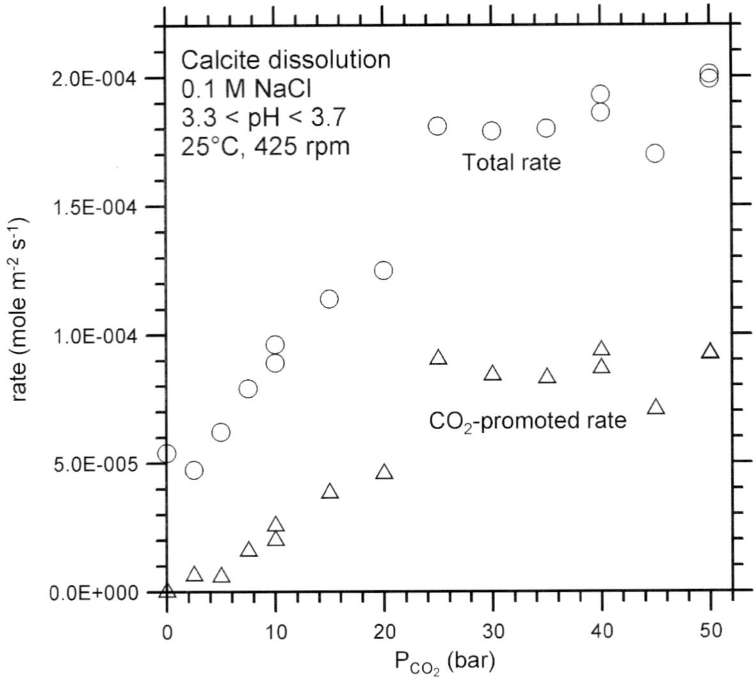

Figure 6.46. Measured overall rate of calcite dissolution and CO_2-promoted rate as a function of CO_2 partial pressure, at the conditions specified. (Reprinted from Pokrovsky et al., 2005, Copyright 2005 with permission from Elsevier, modified.)

where a_{H^+} is the activity of hydrogen ion in the bulk solution and the rate constant k_1 is equal to 0.22 mol m^{-2} s^{-1} according to Pokrovsky et al. (2005b). This value is somewhat lower than those suggested by Plummer et al. (1978), 0.51 mol m^{-2} s^{-1}, and Chou et al. (1989), 0.89 mol m^{-2} s^{-1}, possibly due to distinct hydrodynamic conditions. Not surprisingly, the CO_2-promoted rate also exhibits a linear increase with P_{CO_2}, in the P_{CO_2} range 1–25 bar, and remains nearly constant at ~8.8×10^{-5} mol m^{-2} s^{-1} for P_{CO_2} of 25–50 bar. This behaviour may be due to saturation of the surface sites of dissolving calcite.

6.8.2.4. Calcite: the influence of foreign solutes on dissolution and precipitation rates

The dissolution and precipitation rate data summarized in the previous section refer to compositionally simple aqueous solutions, representing a good starting point for investigating the kinetics of calcite dissolution and precipitation in multi-component systems, which approach the complexity of natural environments. The very important influence of different foreign solutes on the dissolution and precipitation rates of calcite was recognized by many authors (e.g. Terjesen et al., 1961), as summarized by Morse and Arvidson (2002). The following inhibitors have been identified: Mg^{2+}, Sr^{2+}, Ba^{2+}, transition metals (e.g. Sc^{3+}, Cu^{2+}, Zn^{2+}, Mn^{2+}, Ni^{2+}, Co^{2+} and Fe^{2+}), heavy metals (Pb^{2+}, Cd^{2+} and Au^{3+}), La^{3+}, Y^{3+}, Ge^{4+}, sulphate, phosphate, nitrate, silica and organic substances. The dissolution kinetics of calcite is also affected by Ca^{2+} ion, in neutral to alkaline aqueous solutions, where the dissolution reaction is controlled by mixed kinetics. The presence of aqueous Ca^{2+} decreases the rate of the transport process, whereas the chemical rate constant is independent of Ca^{2+} concentration (Sjöberg and Rickard, 1985).

The possible influence of phosphate and Mg^{2+} ion was recently revisited by Alkattan et al. (2002), who measured the dissolution rates of single calcite crystals, using the rotating disk technique, at 25°C and pH -1 to 3, as a function of the concentrations of dissolved $NaPO_3$ (up to 0.05 mol kg^{-1}) and $MgCl_2$ (up to 0.1 mol kg^{-1}).

They found that calcite dissolution kinetics in acidic solutions is unaffected by the presence of $MgCl_2$. This is somewhat unexpected, based on the earlier findings of Sjöberg (1978) in neutral and basic solutions, where the dissolution rate of calcite decreases strongly with increasing $MgCl_2$ concentration. This inhibition effect was ascribed to sorption of Mg^{2+} ion to the calcite surface by Sjöberg (1978). It is possible that Mg^{2+} sorption either does not occur or is negligible in acidic solutions, where the calcite surface has a net positive charge, although cation sorption also occurs on positively charged surfaces.

In contrast, Alkattan et al. (2002) found that the dissolution rate of calcite decreases significantly with increasing $NaPO_3$ concentration in acidic solutions, in agreement with previous findings by Sjöberg (1978) at pH 5 and 8.3 and 25°C. Again, the inhibition of calcite dissolution was attributed to sorption of $H_2PO_4^-$ and/or $H_3PO_4^\circ$ to the calcite surface. However, the concentrations of phosphate needed for inhibiting calcite dissolution are relatively high and unlikely to be present in most natural environments.

It is evident, from this discussion, that the interpretation of catalytic and inhibiting effects on the dissolution/precipitation rates of calcite requires a good knowledge of the processes occurring at the water−mineral interface. In spite of the recent experimental and theoretical efforts, the present understanding of this topic is still unsatisfactory. In addition, application of surface complexation modelling to carbonate minerals is especially challenging (e.g. Pokrovsky et al., 2005b). Discussion of this matter is postponed to Section 6.8.2.7.

6.8.2.5. Calcite: the influence of ionic strength on dissolution and precipitation rates

As recognized by Rickard and Sjöberg (1983) and Sjöberg and Rickard (1984a,b) (see Section 6.8.2.1), the dissolution rate of calcite is independent of the ionic strength in acidic solutions, where the rate is transport-controlled, whereas the dissolution rate becomes ionic-strength-dependent in neutral to alkaline solutions, under a mixed kinetic regime.

The rate data recently produced by Pokrovsky et al. (2005b) are somewhat in contrast to these earlier findings, as calcite dissolution kinetics was found to be independent of the ionic strength, both (1) between 0.001 and 1 mol kg^{-1}, at pH 4.4, 5.7 and 8, and at P_{CO_2} of $10^{-3.5}$ bar, through experiments carried out in a mixed-flow reactor and (2) between 0.01 and 0.1 mol kg^{-1}, at pH 3–4, and at P_{CO_2} of 2–4 bar, through experiments performed in a closed-system, high-pressure Ti-batch reactor.

Several studies carried out in the 1970s and 1980s documented that the kinetics of calcite precipitation in artificial seawater solutions had little or no dependence on salinity. However, this conclusion was biased by the use of equation (6-168) or similar expressions as underscored by Zuddas and Mucci (1998).

These authors performed a set of precipitation experiments in NaCl–$CaCl_2$ solutions at ionic strengths of 0.1, 0.34, 0.55 and 0.93 mol kg^{-1}, 25°C, and P_{CO_2} of 10^{-3} bar. They found that the precipitation rate of calcite increases in several orders of magnitude (up to four) when the ionic strength increases from 0.1 to 0.9 mol kg^{-1}.

Zhang and Dawe (1998) studied calcite precipitation at pH of ~6–8.6 and 25°C. They found that the rate constant of calcite precipitation increases linearly with the square root of the ionic strength from 0.1 to 2 mol kg^{-1} with a slope of +0.591. Note that the logarithm of the rate is expected to exhibit a linear dependence on the square root of the ionic strength, based on TST (see Section 6.1.4.1).

6.8.2.6. Calcite: the influence of temperature on dissolution and precipitation rates

Sjöberg and Rickard (1984b) carried out dissolution experiments under far-from-equilibrium conditions at different temperatures (1–62°C) and pH values (2.7–8.4) by means of a rotating disk reactor to investigate the empirical apparent activation energies of the different rate constants. They found that, at pH 8.4, the apparent activation energies of the surface chemical reaction rate constant k_C are 46 ± 4 and 54 ± 4 kJ mol^{-1} for Iceland spar and Carrara marble, respectively. The first value coincides with that measured by Nancollas and Reddy (1971) for crystal growth at pH of 8.5–8.8, 46 ± 4 kJ mol^{-1}, whereas a somewhat lower value, 35 kJ mol^{-1}, was determined by Sjöberg (1976) for calcite dissolution between pH 8.0 and 10.1, as already recalled above.

Sjöberg and Rickard (1984b) also found that, at pH 8.4, the apparent activation energy of the transport rate constant k_T (which depends on the rotational velocity) is 27 ± 2 kJ mol^{-1}. In the acidic range, for pH 2.7–3.7, the apparent activation energy of the transport-controlled dissolution rate is 13 ± 1 kJ mol^{-1} (Sjöberg and Rickard, 1984b), in substantial agreement with earlier findings by Plummer et al. (1978), 8.4 kJ mol^{-1}, and Sjöberg (1978), 10.5 kJ mol^{-1}. These values are consistent with the different mechanisms involved in calcite dissolution (see above).

Gutjahr et al. (1996) measured growth and dissolution rates of calcite (and aragonite) at different temperatures, from 20 to 70°C, and pH, from 7.1 to 8.9, and found activation energies of 24 ± 3 kJ mol^{-1} for dissolution (which is significantly lower than earlier results) and 42 ± 4 kJ mol^{-1} for crystal growth.

Talman et al. (1990) performed dissolution experiments of calcite in CO_2-containing aqueous solutions, at 100, 150 and 210°C along the steam–water equilibrium curve, under different calcite saturation conditions, by using a stirred batch reactor. They demonstrated that the model of Plummer et al. (1978) could be applied to calcite dissolution up to these high temperatures.

6.8.2.7. Calcite: surface complexation models

An SCM was first applied to carbonate minerals, including calcite, by Van Cappellen et al. (1993). Results of SCM were used by these authors to investigate the surface-controlled dissolution kinetics of calcite and the molecular mechanisms governing this process. Further studies of this kind were carried out by Arakaki and Mucci (1995, 1999), Pokrovsky et al. (2000) and Pokrovsky and Schott (2002). Nilsson and Sternbeck (1999) presented a model for the kinetics of calcite growth based on a surface complexation approach.

The SCM assumes that the species exposed on the cleavage planes of calcite are $\equiv Ca^+$ and $\equiv CO_3^-$, whose charges are due to partial saturation of chemical bonds with the underlying oxygens of the bulk solid. Upon hydration, the two primary sites $\equiv CaOH^\circ$ and $\equiv CO_3H^\circ$ are formed on the surface, probably in equal stoichiometric amounts. Direct, molecular-level evidence supporting the presence of these two species on fresh calcite surfaces exposed to water was provided by Stipp and Hochella (1991) through XPS investigations. Protonation /deprotonation and adsorption reactions bring about the formation of the following additional surface species: $\equiv CaOH_2^+$, $\equiv CaO^-$, $\equiv CaHCO_3^\circ$, $\equiv CaCO_3^-$, $\equiv CO_3^-$ and $\equiv CO_3Ca^+$. Their presence at the calcite (and dolomite)−aqueous solution interface is in line with the results of diffuse reflectance infrared spectroscopy (Pokrovsky et al., 2000).

Relevant surface complexation reactions and corresponding intrinsic stability constants are given in Table 6.4. These data can be inserted in the thermodynamic database of the MINTEQA2 code (Allison et al., 1991) to compute the chemical speciation in the heterogeneous system calcite–H_2O–CO_2 at 25°C, 1 bar (the temperature- and pressure-dependence of the intrinsic stability constants for surface equilibria is unknown). The original code (v. 4.02, June 2000) is available at the website http://www.epa.gov/ceampubl/mmedia/minteq/index.htm; whereas its visual basic version, VMINTEQ, containing some improvements and updated databases, is available at the website http://www.lwr.kth.se/english/OurSoftWare/vminteq/index.htm.

It is interesting to do this kind of calculation considering that, based on crystallographic data, on the $(10\bar{1}4)$ cleavage plane of calcite there are 5 sites nm^{-2} or 5×10^{18} sites m^{-2} of type $\equiv CaOH^\circ$ and the same amount of sites of type $\equiv CO_3H^\circ$ (Van Cappellen et al., 1993 and references therein). This figure is divided by the Avogadro's number ($N_A = 6.02214 \times 10^{23}$ mol^{-1}) obtaining a surface density of 8.3×10^{-6} mol m^{-2} for both types of hydration sites. Assuming that the calcite powder is composed by spherical particles of radius $r_s = 0.001$ cm, its specific surface area is computed by means of equation (6-96), where $\rho_{calcite} = 2.7$ g cm^{-3}, obtaing a value of 1,100 cm^2 g^{-1} = 0.11 m^2 g^{-1}. Considering a suspension containing 45 g of calcite per kg of water, a total concentration of 8.3×10^{-6} mol $m^{-2} \times 0.11$ m^2 $g^{-1} \times 45$ g $kg^{-1} = 4.11 \times 10^{-5}$ mol kg^{-1} is obtained for both types of hydration sites.

It must be recalled that the total concentration of the sites of type $\equiv CaOH^\circ$ is the sum of the concentrations of all the sites of this type, namely $\equiv CaOH^\circ$, $\equiv CaOH_2^+$, $\equiv CaO^-$, $\equiv CaHCO_3^\circ$ and $\equiv CaCO_3^-$, whereas the total concentration of the sites of type $\equiv CO_3H^\circ$ is the sum of the concentrations of all the surface complexes of this kind, namely $\equiv CO_3H^\circ$, $\equiv CO_3^-$ and $\equiv CO_3Ca^+$. Similar to what is usually done for computing the chemical speciation of aqueous solutions, individual concentrations of surface and aqueous species in the considered heterogeneous system are obtained through speciation calculations, including a series of mass balances (involving not only aqueous species but also surface species) and equilibrium constants, both thermodynamic constants for aqueous species and intrinsic stability constants for surface species. The electrostatic interactions acting at the solid–aqueous solution interface are taken into account in these calculations, as discussed in detail by Van Cappellen et al. (1993) and references therein. For the purpose of our exercise, the simplest electrostatic model is adopted, the so-called *constant*

TABLE 6.4

Surface complexation reactions and intrinsic stability constants at the surface–water interface for different carbonate minerals (from Van Cappellen et al., 1993; Pokrovsky et al., 1999; Pokrovsky and Schott, 1999b, 2002).

Surface Reaction	Calcite	Dolomite (Ca^{2+})	Dolomite (Mg^{2+})	Magnesite	$SrCO_3$	$BaCO_3$	$MnCO_3$	$FeCO_3$	$CoCO_3$	$NiCO_3$	$ZnCO_3$	$CdCO_3$	$PbCO_3$
$\equiv CO_3H \rightleftharpoons \equiv CO_3^- + H^+$	−5.1	−4.8 ±0.2	−4.8 ±0.2	−4.65 ±0.15	−5.0	−5.2	−5.0	−4.4	−5.0	−4.0	−5.2	−5.4	−5.0
$\equiv CO_3H + Me^{2+} \rightleftharpoons \equiv CO_3Me^+ + H^+$	−1.7	−1.8 ±0.2	−2.0 ±0.2	−2.20 ±0.15	−2.5	−2.5	−2.6	−1.6	−2.0	−2.0	−1.8	−1.8	−2.4
$\equiv MeOH \rightleftharpoons \equiv MeO^- + H^+$	−12	−12 ±2	−12 ±2	−12 ±2	−12.0	−12.0	−14.0	−10.4	−13.0	−11.7	−11.0	−12.0	−11.0
$\equiv MeOH + H^+ \rightleftharpoons \equiv MeOH_2^+$	11.85[a]	11.5 ± 0.2	10.6 ± 0.2	10.60 ± 0.15	11.9	12.1	9.9	10.2	9.2	9.5	8.8	10.2	8.3
$\equiv MeOH + CO_3^{2-} + 2H^+ \rightleftharpoons \equiv MeHCO_3 + H_2O$	23.5	24.0 ±0.5	23.5 ±0.5	22.40 ±0.40	23.0	24.5	22.65	22.75	21.8	19.5	21.15	22.15	19.65
$\equiv MeOH + CO_3^{2-} + H^+ \rightleftharpoons \equiv MeCO_3^- + H_2O$	17.1	16.6 ±0.2	15.4 ±0.2	14.40 ±0.15	16.0	16.15	14.15	14.65	14.0	14.0	13.65	13.15	14.15
EDL capacitance in 0.01 mol kg^{-1} NaCl (F m^{-2})	17	25	25	31	17	17	17	31	31	31	17	17	17
Density of $\equiv MeOH^\circ$-type surface sites (μmol m^{-2})	8.2	7.0	7.0	9.8	7.7	7.0	9.2	9.5	9.7	10.0	9.7	8.6	7.6

[a]11.5, according to Pokrovsky et al. (1999).

capacitance model, specifying a capacitance of $I^{1/2}/0.006$ F m^{-2} (Van Cappellen et al., 1993). Note that for an ionic strength $I = 0.1$ mol kg^{-1} (that we will considered in our calculation), the capacitance of the electric double layer is 52.7 F m^{-2}, which is much higher than the values reported for suspensions of (hydr)oxides in aqueous solutions, typically ~1 F m^{-2} at this ionic strength.

We have also to fix the total concentration of aqueous Ca and dissolved carbonate species. Let us assume that the latter is constrained by equilibrium with the atmosphere, whose average P_{CO_2} is $10^{-3.5}$ bar. A first run is done at high pH (e.g. 10), to decide a suitable Ca concentration, as we want to maintain the aqueous solution undersaturated with calcite. A value of 5×10^{-6} mol kg^{-1} is thus adopted. Finally, the code is run at different pH values, from 3 to 10.

The results of this exercise are shown in Fig. 6.47, whose inspection shows that the protonated species $\equiv CaOH_2^+$ and $\equiv CO_3H°$ prevail at pH values lower than 8.2 and 5.2, respectively, whereas above these thresholds the dominant surface species become $\equiv CaCO_3^-$ and $\equiv CO_3^-$, respectively. These findings are comparable with those of Van Cappellen et al. (1993) and Pokrovsky et al. (2000).

Results of SCM provide insights into the dissolution kinetics of calcite (and other carbonate minerals). For instance, the increase in the dissolution rate with decreasing pH in acidic solutions was attributed to the protonation of the $\equiv CO_3^-$ sites at the calcite–aqueous solution interface. Consequently, the proton-promoted dissolution rate can be redifined as

$$r_{+,H} = k_{CO_3H°} \cdot [\equiv CO_3H°]^m, \tag{6-202}$$

where brackets indicate the concentrations of surface species and the reaction-order coefficient m is 2 (Van Cappellen et al., 1993). Based on the SCM evidence that the $\equiv CaOH_2^+$ surface complex prevails for pH < 8.2, this surface species was assumed to govern the dissolution kinetics of calcite under neutral to mildly alkaline conditions, according to the equation (Pokrovsky and Schott, 2002)

$$r_{+,H_2O} = k_{CaOH_2^+} \cdot [\equiv CaOH_2^+]^n, \tag{6-203}$$

where the reaction order coefficient n is 1. A general rate equation for dissolution of calcite (and other divalent carbonates) can be obtained by considering the thermodynamic affinity effects also:

$$r_{+,H} = \left(k_{CO_3H°} \cdot [\equiv CO_3H°]^m + k_{CaOH_2^+} \cdot [\equiv CaOH_2^+]^n\right) \times \left[1 - \left(\frac{Q}{K}\right)^n\right]. \tag{6-204}$$

This equation can also be used to predict the precipitation rates under close-to-equilibrium conditions. The rate-governing step for crystal growth is probably the dehydration of cations sorbed at the calcite surface (see Pokrovsky and Schott, 2002, for further details).

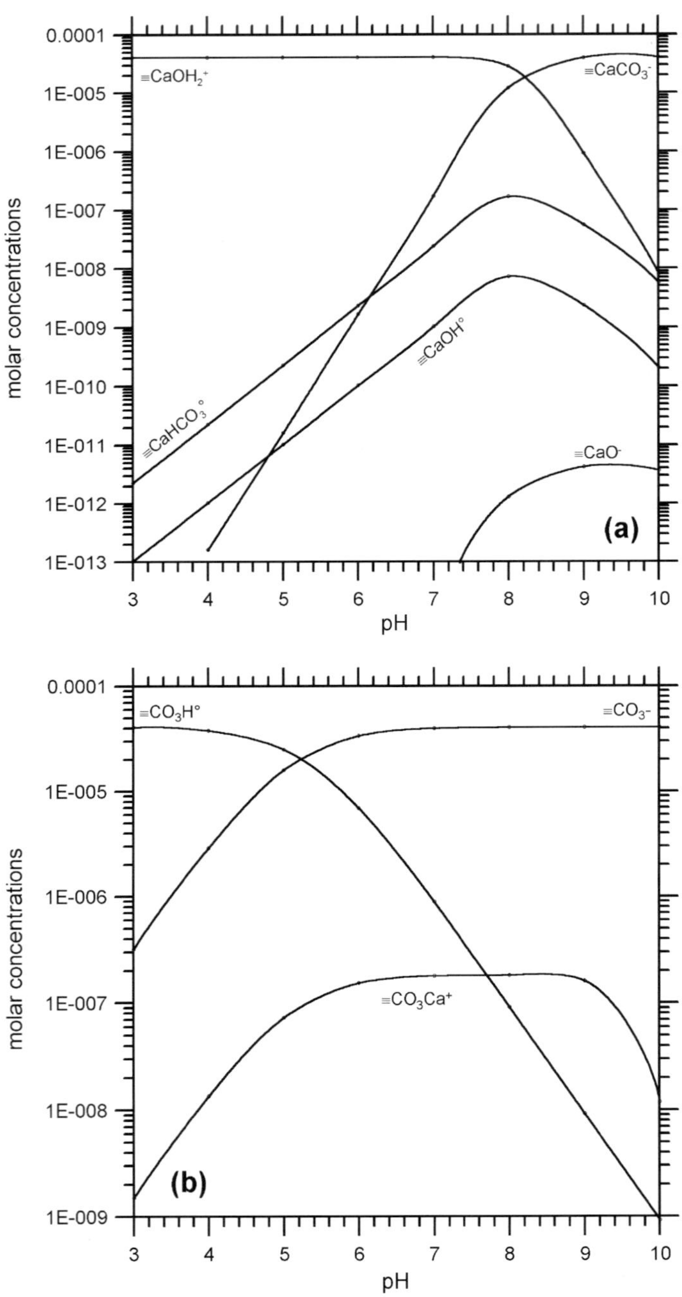

Figure 6.47. Distribution of surface species of type (a) $\equiv CaOH^\circ$ and (b) $\equiv CO_3H^\circ$ at the calcite–aqueous solution interface predicted by SCM at 25°C, $I = 0.1$ mol kg^{-1}, total dissolved Ca concentration = 5×10^{-6} mol kg^{-1}, $P_{CO_2} = 10^{-3.5}$ bar and variable pH. (Reprinted from Van Cappellen et al., 1993, Copyright 1993 with permission from Elsevier, modified.)

SCM also suggests that in the presence of significant concentrations of dissolved bicarbonate and carbonate ions, the $\equiv CaOH_2^+$ surface complex is replaced by the surface species $\equiv CaHCO_3^\circ$ and $\equiv CaCO_3^-$. Consequently, these ions may act as inhibitors of calcite dissolution under far-from-equilibrium conditions, as recognized by Lea et al. (2001) for carbonate. Also, the inhibiting effects of several adsorbed solutes (see Section 6.8.2.3) could find a theoretical support through surface complexation modelling.

Unfortunately, as noticed by Pokrovsky et al. (2005b), the changes in the dissolution rate of calcite determined by variations in P_{CO_2} (see above) cannot be analysed in the framework of SCM owing to theoretical and experimental difficulties.

6.8.3. Dolomite

6.8.3.1. Dolomite: dissolution rates

The first detailed laboratory study on dolomite dissolution kinetics was performed by Busenberg and Plummer (1982), at different pH (<1 to ~10), fixed P_{CO_2} (0 to ~1 bar), different temperature (5–65°C) and variable solution composition (addition of Ca and Mg). Selected experimental data are plotted in Fig. 6.48. Dissolution rates were determined by weight loss of single crystals of both ordered and disordered dolomite. Adopting the interpretative guidelines of Plummer et al. (1978) for calcite, Busenberg and Plummer (1982) proposed the following rate law:

$$r = k_1 \cdot a_{H^+}^{n_1} + k_2 \cdot a_{H_2CO_3^*}^{n_2} + k_3 \cdot a_{H_2O}^{n_3} - k_{-4} \cdot a_{HCO_3^-}, \tag{6-205}$$

which comprises three forward dissolution reactions, with partial order coefficients $n_1 = n_2 = n_3 = 0.5$ at 25°C. However, the partial order with respect to H^+ was found to vary with temperature above 45°C, attaining 0.6 at 55°C, 0.7 at 65°C and 0.85 at 100°C.

Similar to what was suggested for calcite, the first term of equation (6-205) is important at low pH values, either <~6 for P_{CO_2} close to 0 bar or <~1.6 for P_{CO_2} of 1 bar. The second term explains the dependence of the dissolution rate on P_{CO_2}. The third term accounts for the pH independence of the rate at pH > 8 and low P_{CO_2} values. The fourth term describes the backward (precipitation) reaction. At 25°C, the rate constants (in mol m^{-2} s^{-1}) for disordered dolomite are as follows: $k_1 = 6.5 \times 10^{-4}$, $k_2 = 7.8 \times 10^{-6}$, $k_3 = 3.0 \times 10^{-8}$ and $k_{-4} = 2.8 \times 10^{-4}$, whereas for ordered dolomite they take the values as follows: $k_1 = 1.7 \times 10^{-4}$, $k_2 = 4.3 \times 10^{-6}$, $k_3 = 2.5 \times 10^{-9}$ and again $k_{-4} = 2.8 \times 10^{-4}$ (see curves in Fig. 6.48). Apparent activation energies for disordered dolomite are 36.1 kJ mol^{-1} for the acidic mechanism, 34.8 kJ mol^{-1} for H_2CO_3*-promoted dissolution and 52.2 kJ mol^{-1} for the neutral mechanism. Apparent activation energies for ordered dolomite are 56.7 kJ mol^{-1} for the acidic mechanism, 45.7 kJ mol^{-1} for H_2CO_3*-promoted dissolution, 95.3 kJ mol^{-1} for the neutral mechanism and 44.0 kJ mol^{-1} for the backward reaction.

Note that the proton-promoted dissolution mechanism for dolomite has a k_1 value lower than calcite by three orders of magnitude. Hence, the solute transport at the surface should not be a limiting step for dolomite dissolution, as also suggested by the comparatively high values of E_a for the acidic mechanism with respect to those of calcite (8.4–13 kJ mol^{-1}, see Section 6.8.2.6).

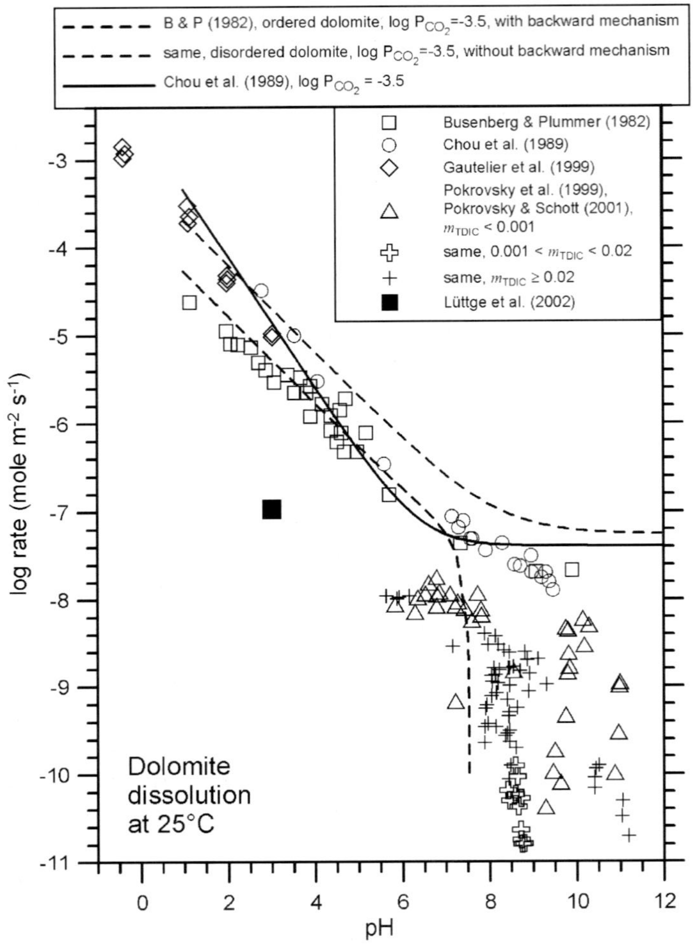

Figure 6.48. Log-rates of dolomite dissolution at 25°C as a function of pH, showing both selected experimental data and the predictions of some rate laws.

Besides, Busenberg and Plummer (1982) found the dissolution rate to be independent of the stirring rate at 45°C and pH of 3–4.4, thus confirming that dolomite dissolution is surface-controlled at least under these conditions. The possible influence of solute transport cannot be ruled out at higher pH values (see below).

Chou et al. (1989) used continuous flow-through fluidized-bed reactor (as already recalled in Section 6.8.2.2) and found dissolution rates, at 25°C, very similar to those measured by Busenberg and Plummer (1982) in the pH range 7–9, whereas the results by Chou et al. (1989) are systematically higher than those by Busenberg and Plummer (1982) below pH 6 and the difference between the two datasets attains ~1 log-unit at pH 3 (Fig. 6.48). Chou et al. (1989) fitted their data to the same rate equation suggested by Busenberg and

Plummer (1982), but they did not considered the backward term. They found partial reaction orders $n_1 = n_2 = 0.75$ and $k_1 = 2.6 \times 10^{-3}$, $k_2 = 1.0 \times 10^{-4}$ and $k_3 = 2.2 \times 10^{-8}$ mol m^{-2} s^{-1} at 25°C.

For what concerns the reaction order with respect to H^+ ion, Chou et al. (1989) demonstrated that the explanation proposed by Busenberg and Plummer (1982) is not correct. Busenberg and Plummer (1982) suggested that the $CaCO_3$ component of dolomite dissolves faster than the $MgCO_3$ component, leaving the surface of the dissolving mineral enriched in the second component (non-stoichiometric dissolution). For instance, for the proton-promoted dissolution mechanism, the two following reactions can be written:

$$CaMg(CO_3)_2 + H^+ \rightleftharpoons MgCO_3 + Ca^{2+} + HCO_3^- \tag{6-206}$$

and

$$MgCO_3 + H^+ \overset{\text{slow}}{\rightarrow} Mg^{2+} + HCO_3^-. \tag{6-207}$$

Since reaction (6-207) consumes only half the overall amount of H^+, Busenberg and Plummer (1982) concluded that this rate-determining reaction is of order 0.5 with respect to H^+ ion. However, this conclusion is wrong, as underscored by Chou et al. (1989). Let us assume that the first reaction is much faster than the second. If so, reaction (6-206) can be assumed to be close to equilibrium and its thermodynamic equilibrium constant can be written as follows:

$$K_1 = \frac{a_{Ca^{2+}} \cdot a_{HCO_3^-} \cdot a_{MgCO_3}}{a_{H^+}}, \tag{6-208}$$

under the hypothesis that dolomite is pure. Based on the second slow (and therefore rate-controlling) reaction, the dissolution rate can be expressed as

$$r_+ = k_+ \cdot a_{H+} \cdot a_{MgCO_3}. \tag{6-209}$$

Insertion of equation (6-208) into (6-209) leads to

$$r_+ = k_+ \cdot K_1 \cdot \frac{a_{H^+}^2}{a_{Ca^{2+}} \cdot a_{HCO_3^-}}. \tag{6-210}$$

Equation (6-210) shows that occurrence of a chain reaction, as postulated by Busenberg and Plummer (1982), would determine a reaction order of 2 with respect to H^+ ion, but not a fractional-order dependence. The fractional-order dependence can be explained by formation of surface complexes through sorption of H^+ ion (Chou et al., 1989). In fact, this type of reaction typically exhibits a reaction order <1 with H^+, as observed for many silicates and oxides (see above).

More recently, the dissolution rate of dolomite in acidic solutions was investigated by Orton and Unwin (1993) at room temperature and pH 3–4 through a channel-flow method, by Gautelier et al. (1999) in a wide pH range, and by Lüttge et al. (2002) who used vertical scanning interferometry.

Gautelier et al. (1999) measured the dissolution rates of polycrystalline dolomite at 25, 50 and 80°C, in the pH range –0.39 to 4.44, using an experimental set-up that combines the advantages of the mixed-flow reactor with those of the rotating disk techniques. Following the theoretical approach of Busenberg and Plummer (1982) and considering that, under acidic conditions, the contributions of the H_2CO_3*- and H_2O-promoted dissolution mechanisms are negligible, Gautelier et al. (1999) assumed the following simplified dissolution rate law:

$$r_+ = k_1 \cdot a_{H^+,S}^{n_1}. \tag{6-211}$$

Gautelier et al. (1999) also investigated the possible influence of transport on dissolution kinetics, which turned out to be negligible for pH < ~3 at 25°C, for pH < ~2 at 50°C and for pH < ~1 at 80°C (see Table 2 of Gautelier et al., 1999 and note that when dissolution kinetics is controlled by surface reactions, $a_{H^+,S} \cong a_{H^+,B}$, where subscripts S and B identify the dolomite surface and the bulk aqueous solution, respectively). The obtained values of k_1 (in mol m^{-2} s^{-1}) and n_1 are 9.82×10^{-4} and 0.63 at 25°C, 5.13×10^{-3} and 0.73 at 50°C, and 1.77×10^{-2} and 0.80 at 80°C. It must be underscored that, under very acidic conditions (i.e. $pH_S < 0$), dolomite dissolution rates tend to become independent of pH, probably owing to progressive saturation of the carbonate sites on the dolomite surface with H^+ ions. As noticed by Gautelier et al. (1999), this phenomenon might explain the low n_1 values reported by Anderson (1991), 0.2, Lund et al. (1973), 0.44 and Busenberg and Plummer (1982), 0.50, at 25°C.

Since n_1 depends on temperature, the apparent activation energy of dolomite dissolution depends on pH, decreasing from 46 to 15 kJ mol^{-1} with increasing pH_S from 0 to 5. This also indicates that the effect of solute transport becomes progressively important with increasing pH.

Lüttge et al. (2002) determined both the speed of etch-pit development on three different cleavage surfaces of dolomite and the dissolution kinetics of the nearby surfaces by using vertical scanning interferometry. Dissolution experiments were performed at 25°C and pH 3. The dissolution rate of dolomite was found to be $(1.08 \pm 0.03) \times 10^{-7}$ mol m^{-2} s^{-1}. As shown in Fig. 6.48, this value is significantly lower than those measured in macroscopic dissolution experiments, as also observed for calcite (Arvidson et al., 2003, see above).

Pokrovsky et al. (1999) and Pokrovsky and Schott (2001) performed 130 measurements of the steady-state dissolution rate of polycrystalline dolomite at 25°C as a function of pH (5–12), ionic strength (I = 0.002–0.1 mol kg^{-1}), m_{TDIC} (10^{-5}–0.1), m_{Ca} (10^{-6}–0.003), m_{Mg} (3×10^{-7}–0.005), m_{SO4} (4×10^{-4}–0.015) and molalities of several organic ligands. Although most data were obtained at far-from-equilibrium condition, 28 data refer to dissolution/precipitation rates under close-to-equilibrium conditions. These authors used a mixed-flow reactor. P_{CO_2} was computed based on pH and alkalinity, and in most experiments it was $<10^{-3}$ bar. The results of these authors are discussed in the following sections.

6.8.3.2. Dolomite: surface complexation models

Pokrovsky et al. (1999) also measured the surface charge of dolomite as a function of pH (6.5–11.5), I (0.01, 0.1 and 0.5 M NaCl) and P_{CO_2} ($10^{-3.5}$, 0.01 and 0.96 bar) by means of potentiometric titrations in a limited residence time mixed-flow reactor. They found that the pH of the point of zero charge (pH_{PZC}) varies from 6.9 to 8.0 (± 0.2), whereas the surface charge attains maximum values of ~0.02 mM m^{-2}. Although the surface charge varies with I and m_{TDIC}, the influence of I is smaller than for oxide minerals.

Pokrovsky et al. (1999) also determined the streaming potentials and electrophoretic mobility of dolomite particles in solutions of variable pH (2.5–12), I ($10^{-4}-0.1$ mol kg^{-1}), m_{Ca} and m_{Mg} ($5 \times 10^{-6}-0.1$) and m_{TDIC} ($10^{-5}-0.07$) to obtain the zeta-potential. They were able to estimate the isoelectric point of dolomite (i.e. the condition at which the electrophoretic mobility of the solid particles is equal to zero), which was found to vary from 6.0 to 8.8 and to coincide with the PZC. In particular, $pH_{ZPC} = pH_{IEP} = 8.0 \pm 0.1$ at P_{CO_2} close to $10^{-3.5}$ bar.

Based on these findings, Pokrovsky et al. (1999) developed a SCM for dolomite. Surface density was assumed to be 7×10^{-6} mol m^{-2} for each metal site of type $\equiv CaOH^\circ$ and $\equiv MgOH^\circ$ and 14×10^{-6} mol m^{-2} for the carbonate sites of type $\equiv CO_3H^\circ$, corresponding to 8 sites nm^{-2} for metal (Ca + Mg) and carbonate sites (see Section 6.8.2.7). Again, the constant capacitance model was adopted, specifying a capacitance of $I^{1/2}/0.004$ F m^{-2}, which is similar to that for calcite, in order to reproduce the high surface charge of dolomite far from the PZC and its weak dependence on I (see above). Following the approach of Van Cappellen (1993), Pokrovsky et al. (1999) selected the intrinsic stability constants of relevant surface complexation reactions (Table 6.4), which provide the best fit for all the titration data they produced.

As we have seen above for calcite, these intrinsic stability constants can be used to compute the chemical speciation at the dolomite–water interface under fixed conditions. For instance, for $I = 0.01$ mol kg^{-1}, $m_{Ca} = m_{Mg} = 10^{-3}$ and $P_{CO_2} = 10^{-3.5}$ bar, the protonated species $\equiv CO_3H^\circ$ is the prevailing carbonate site below pH 4.5, whereas the surface complex $\equiv CO_3^-$ becomes the dominating carbonate group above this pH value. Metal site speciation also changes with pH, in that $\equiv CaOH_2^+$ and $\equiv MgOH_2^+$ prevail below pH 8, but are substituted by $\equiv CaCO_3^-$ and $\equiv MgCO_3^-$ above this pH threshold. Note that the surface speciation for dolomite has a pH dependence similar to that computed for calcite (see Section 6.8.2.5). Besides, the distribution of surface complexes at varying pH is consistent with spectroscopic surface analyses of dolomite (Pokrovsky et al., 2000).

Pokrovsky et al. (1999) attributed the increase in the dolomite dissolution rate with decreasing pH below pH ~6 (Fig. 6.48) to protonation reactions occurring at the dolomite surface. Since the metal surface sites are fully protonated as $\equiv CaOH_2^+$ and $\equiv MgOH_2^+$ at pH < ~8, the enhancement in the dissolution rate must be necessarily ascribed to the protonation of the $\equiv CO_3^-$ sites, as it is the case for calcite (see Section 6.8.2.7) and

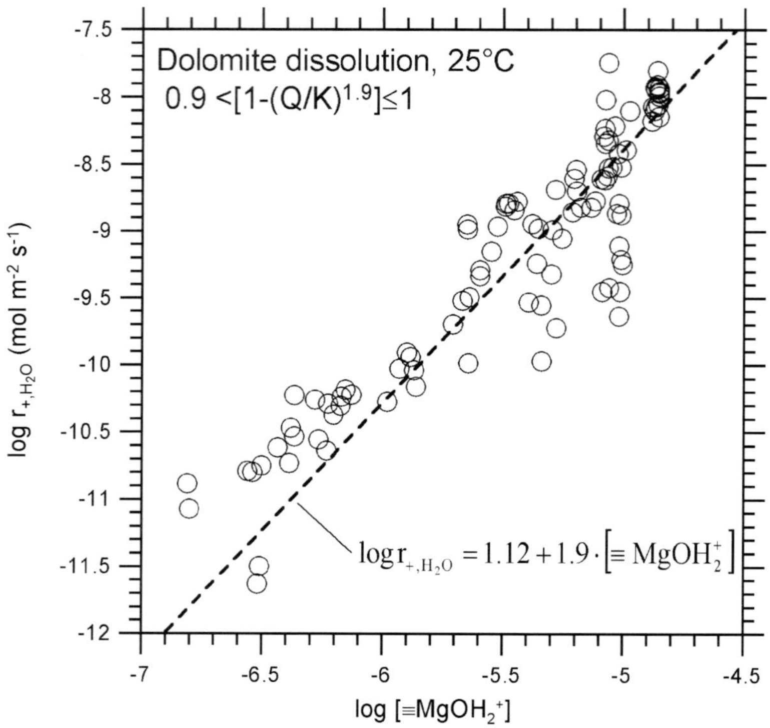

Figure 6.49. Far-from-equilibrium dissolution rates of dolomite at 25°C, measured by Pokrovsky et al. (1999) and Pokrovsky and Schott (2001), as a function of the concentration of $\equiv MgOH_2^+$. (Reprinted from both Pokrovsky et al., 1999, Copyright 1999 with permission from Elsevier and Pokrovsky and Schott, 2001, Copyright 2001 with permission from American Journal of Science, modified.)

magnesite (see Section 6.8.4). Therefore, the proton-promoted dissolution rate of dolomite can be described by means of equation (6-202), where $k_{CO_3H^\circ} = 10^{5.23}$ m^2 mol^{-1} s^{-1} and $m = 2$, based on the data by Busenberg and Plummer (1982) and Chou et al. (1989) (see Fig. 7 of Pokrovsky et al., 1999).

Pokrovsky and Schott (2001) showed that the logarithm of the far-from-equilibrium dissolution rate in the pH range 5–12 is strictly correlated with the log-concentration of the $\equiv MgOH_2^+$ surface complex, as described by the following equation (rate in mol m^{-2} s^{-1}):

$$r_{+,H_2O} = k_{MgOH_2^+} \cdot [\equiv MgOH_2^+]^n, \quad (6\text{-}212)$$

where $k_{MgOH_2^+} = 10^{1.12 \pm 0.1}$ and $n = 1.9$ (Fig. 6.49). This strong correlation indicates that the surface species $\equiv MgOH_2^+$ governs the dissolution kinetics of dolomite in neutral and alkaline aqueous solutions. Besides, the second- or close to second-order dependence of the dissolution rate on both the $\equiv CO_3H^\circ$ and $\equiv MgOH_2^+$ concentrations suggests that dolomite dissolution occurs through protonation of two adjacent surface sites.

6.8.3.3. Dolomite: the influence of foreign solutes on dissolution rates

Gautelier et al. (1999) showed that aqueous Ca and Mg have negligible to nil influence on dolomite dissolution kinetics at 50°C and pH 2. Pokrovsky and Schott (2001) found instead that dolomite dissolution in neutral and alkaline solutions is significantly inhibited by dissolved Ca, for $a_{Ca^{2+}} > 10^{-4.5}$, whereas Mg has no influence. These observations are in line with those by Busenberg and Plummer (1982) at 45°C and pH close to 4. These findings by Pokrovsky and Schott (2001) and Busenberg and Plummer (1982) are consistent with the preferential release of Ca with respect to Mg as a first step of dolomite dissolution (see Section 6.8.2.1) and imply that the surface-activated complex (or its precursor) governing dolomite dissolution is Ca-free. Indeed, the dissolution rate of dolomite in neutral and alkaline solutions depends on the concentration of the $\equiv MgOH_2^+$ surface complex but not on that of the $\equiv CaOH_2^+$ surface species (see above).

The inhibition of dolomite dissolution by CO_3^{2-} and HCO_3^- ions was documented by Pokrovsky et al. (1999) in neutral and alkaline solutions and attributed to the decrease in $\equiv MgOH_2^+$ concentration owing to the reactions

$$\equiv MgOH_2^+ + HCO_3^- \rightleftharpoons \equiv MgHCO_3^\circ + H_2O \tag{6-213}$$

and

$$\equiv MgOH_2^+ + CO_3^{2-} \rightleftharpoons \equiv MgCO_3^- + H_2O. \tag{6-214}$$

Pokrovsky and Schott (2001) considered that dolomite dissolution comprises two steps: the first one involving preferential Ca release through the following reaction:

$$\equiv CaMgCO_3^+ \rightleftharpoons \equiv MgCO_3^- + Ca^{2+} \tag{6-215}$$

whose intrinsic stability constant can be written as follows:

$$K_{Ca}^* = \frac{a_{Ca^{2+}} \cdot [\equiv MgCO_3^-]}{[\equiv CaMgCO_3^+]}, \tag{6-216}$$

neglecting the activity coefficients of surface species. The second step of dolomite dissolution consists in the hydration of surface Mg sites, i.e. the reverse of reaction (6-214), whose intrinsic stability constant can be written as follows:

$$K_{CO_3}^* = \frac{a_{CO_3^{2-}} \cdot [\equiv MgOH_2^+]}{[\equiv MgCO_3^-]}, \tag{6-217}$$

setting $a_{H_2O} = 1$ and, again, neglecting the activity coefficients of surface species. Moreover, the mass balance of metal sites requires

$$[\equiv CaMgCO_3^+] + [\equiv MgCO_3^-] + [\equiv MgOH_2^+] = 7 \times 10^{-6} \, mol \cdot m^{-2}. \tag{6-218}$$

Note that, strictly, this concentration refers to the total sites of the $\equiv$MgOH° type and should not comprise the $\left[\equiv CaMgCO_3^+\right]$ species. Anyway, combining equations (6-212), (6-216), (6-217) and (6-218) one obtains

$$r_{+,H_2O} = k_{MgOH_2^+} \cdot \left(\frac{7 \cdot 10^{-6} \cdot K_{Ca}^* \cdot K_{CO_3}^*}{K_{Ca}^* \cdot K_{CO_3}^* + K_{Ca}^* \cdot a_{CO_3^{2-}} + a_{Ca^{2+}} \cdot a_{CO_3^{2-}}} \right)^{1.9}, \tag{6-219}$$

where $K_{Ca}^* = (3.5 \pm 0.5) \times 10^{-5}$ and $K_{CO_3}^* = (4.5 \pm 0.5) \times 10^{-5}$, whereas $k_{MgOH_2^+}$ takes the value $10^{1.59}$ instead of $10^{1.12}$, as noted above. Equation (6-219) holds at constant ionic strength as electrostatic terms were not considered in its derivation. However, this equation accounts satisfactorily for the inhibiting effects of both Ca^{2+} and CO_3^{2-} ions at far-from-equilibrium conditions.

Pokrovsky and Schott (2001) also showed that the addition of sulphate and inorganic ligands to the aqueous solution brings about an increase in the dissolution rate of dolomite in the order sulphate $\approx$ formate $\approx$ tartrate $<$ acetate $<$ ascorbate $\leq$ oxalate $<$ citrate $\ll$ EDTA. These dissolution experiments were carried out at pH > 7, at which organic acids are totally dissociated.

6.8.3.4. Dolomite: transition state theory-based equations

Following Pokrovsky and Schott (2001), equations (6-212) and (6-219) can be extended, within the TST framework, by inserting the term $[1 - (Q/K_{sp})^{19}]$, which accounts for the precipitation reaction under close-to-equilibrium conditions. These two equations are thus obtained:

$$r_{+,H_2O} = k_{MgOH_2^+} \cdot [\equiv MgOH_2^+]^{1.9} \cdot \left[1 - \left(\frac{Q}{K_{sp}} \right)^{1.9} \right] \tag{6-220}$$

and

$$r_{+,H_2O} = k_{MgOH_2^+} \cdot \left(\frac{7 \cdot 10^{-6} \cdot K_{Ca}^* \cdot K_{CO_3}^*}{K_{Ca}^* \cdot K_{CO_3}^* + K_{Ca}^* \cdot a_{CO_3^{2-}} + a_{Ca^{2+}} \cdot a_{CO_3^{2-}}} \right)^{1.9} \cdot \left[1 - \left(\frac{Q}{K_{sp}} \right)^{1.9} \right]. \tag{6-221}$$

They can be used to describe the dissolution/precipitation of dolomite in neutral and alkaline solutions over a large interval of pH, $a_{CO_3^{2-}}$, and $a_{Ca^{2+}}$.

6.8.3.5. Dolomite: the influence of CO_2 partial pressure on the dissolution rate

The effects of P_{CO_2} on dolomite dissolution kinetics was investigated by Pokrovsky et al. (2005b) at 25°C and pH 3–4 on cleavage crystals of dolomite and two different polycrystalline dolomite samples (see Section 6.8.2.3 for further experimental details). Measured rates depend somewhat on stirring speed suggesting a moderate transport influence, whereas they are independent of ionic strength from 0.03–0.1 to 1 mol kg^{-1}.

For a polycrystalline dolomite, containing some organic substances, the dissolution rate was found to double with increasing P_{CO_2} from 0 to 5–10 bar, but was independent of P_{CO_2} from 10 to 50 bar. For the cleavage crystals of dolomite, the P_{CO_2} increase from 1 to 2 bar was found to bring about a 30% enhancement in the dissolution rate, which remained constant when P_{CO_2} was further enhanced up to 50 bar. Referring to equation (6-211), these results can be explained assuming $k_1 = 4.5 \times 10^{-4}$ mol m^{-2} s^{-1} and $n_1 = 0.5$ for the organic-bearing polycrystalline dolomite and $k_1 = 1.8 \times 10^{-4}$ mol m^{-2} s^{-1} and $n_1 = 0.4$ for the dolomite cleavage crystals. All in all, the influence of CO_2 partial pressure, up to 50 bar, on the dissolution rate of dolomite appears to be weaker than that of pH.

6.8.4. Magnesite

Magnesite dissolution kinetics was investigated by Chou et al. (1989) at 25°C as a function of pH, from 2 to 8 (see also Section 6.8.2.2). They found that the rate of magnesite dissolution is proportional to a_{H^+} below pH 5, whereas it becomes independent of a_{H^+} above pH 5 (Fig. 6.50). Chou et al. (1989) fitted their experimental data to equation (6-176) and proposed the following values for the rate constants: $k_1 = 2.5 \times 10^{-5}$ mol m^{-2} s^{-1}, $k_2 = 6.0 \times 10^{-6}$ mol m^{-2} s^{-1} and $k_3 = 4.5 \times 10^{-10}$ mol m^{-2} s^{-1}. Although Chou et al. (1989) were not able to observe any effect linked to the backward reaction, owing to the very slow rate of magnesite dissolution, they computed a k_{-3} value of $10^{-1.35}$ mol m^{-2} s^{-1} applying the principle of detailed balancing (see Section 6.1).

More recently, magnesite dissolution kinetics was thoroughly investigated by Pokrovsky and Schott (1999a) at 25°C as a function of pH, from 0.2 to 12, m_{TDIC}, from 10^{-5} to 0.1, and ionic strength, from 0.002 to 0.5 mol kg^{-1} by means of a mixed-flow reactor. Most experiments were carried out under far-from-equilibrium conditions, namely for $A < -5$ kJ mol^{-1}, where the influence of thermodynamic affinity on the rate is nil to negligible. Measured magnesite dissolution rates exhibit four different relations with pH and I, depending on pH (Fig. 6.50). In strongly acidic aqueous solutions (pH $\leq$ 2.5), rates are independent of both pH and I and kept constant at $(2.43 \pm 0.46) \times 10^{-8}$ mol m^{-2} s^{-1}. In the pH interval 3–5, the dissolution rate increases linearly with log a_{H^+} and increases significantly with I as well. In the pH range 5–8 rates are again independent of both pH and I and remain close to $(4.57 \pm 0.30) \times 10^{-10}$ mol m^{-2} s^{-1}. Above pH 8, rates again increase with a_{H^+} but they also decrease markedly with increasing m_{TDIC} and I. Pokrovsky and Schott (1999a) showed that the decrease in magnesite dissolution rates with pH and m_{TDIC} at pH > 8 and $m_{TDIC} > 0.01$ is not due to the backward (precipitation) reaction but is caused by the inhibitory effect of dissolved carbonate species.

The comparison of the experimental results of Chou et al. (1989) and Pokrovsky and Schott (1999a) (Fig. 6.50) shows that there is a substantial agreement between the two datasets above pH 3, whereas they diverge below this threshold, possibly because the experiments of Chou et al. (1989) were not carried out at constant ionic strength (Pokrovsky and Schott, 1999a).

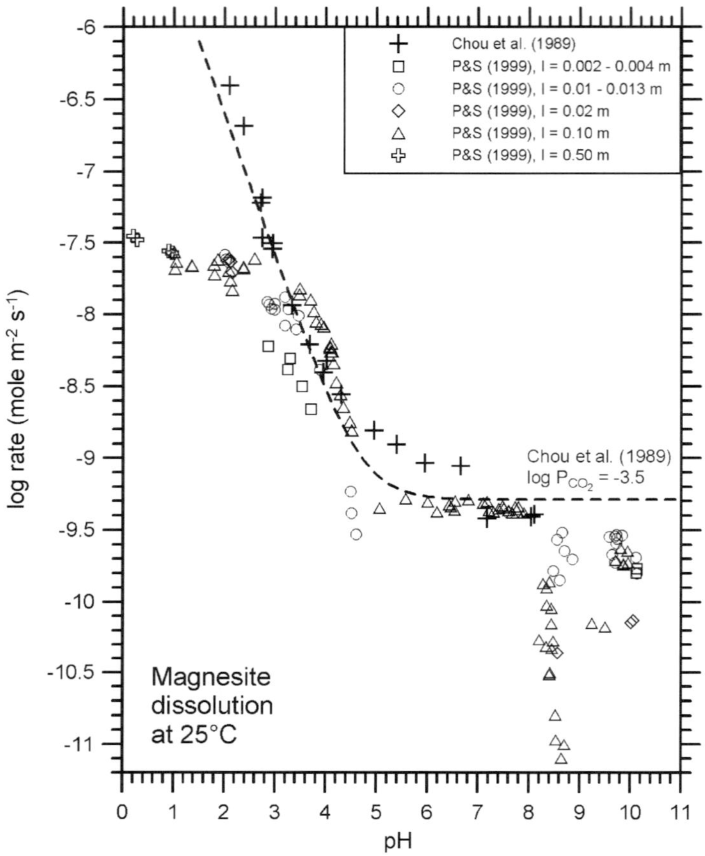

Figure 6.50. Log-rates of magnesite dissolution at 25°C as a function of pH, showing the experimental data of Chou et al. (1989) and Pokrovsky and Schott (1999a) and the prediction of the rate law proposed by Chou et al. (1989) for $P_{CO_2} = 10^{-3.5}$ bar. (Reprinted from Pokrovsky and Schott, 1999a, Copyright 1999 with permission from Elsevier, modified.)

Pokrovsky and Schott (1999a) interpreted their experimental data as a function of surface speciation at the magnesite–aqueous solution interface, based on the results achieved by Pokrovsky and Schott (1999b). Pokrovsky and Schott (1999b) measured the surface charge at the magnesite surface at 25°C as a function of pH, from 4.6 to 11, and ionic strength (0.015, 0.1 and 0.5 mol kg^{-1}), under P_{CO_2} values from $10^{-3.5}$ to 0.96 bar. Acid–base titration were carried out in a limited residence time mixed-flow reactor, as in the case of dolomite (Section 6.8.3.2). Besides, the zeta-potential of magnesite was determined at variable pH (1.6–12.4), I (0.002–0.1 mol kg^{-1}), and Mg^{2+} and HCO_3^- concentrations, by means of electrokinetic measurements, namely the streaming potential technique and the electrophoretic mobility of $MgCO_3$ particles. The isoelectric point of magnesite was found to be pH = 8.5 ± 0.2, pMg^{2+} = 3.3 ± 0.2 and alkalinity $<10^{-3}$ eq L^{-1}, and to coincide with the PZC derived from surface titration data.

Pokrovsky and Schott (1999b) developed a SCM for magnesite, postulating the presence of the two primary hydration sites $\equiv MgOH^\circ$ and $\equiv CO_3H^\circ$ and the formation of the surface species $\equiv CO_3^-$, $\equiv CO_3Mg^+$, $\equiv MgO^-$, $\equiv MgOH_2^+$, $\equiv MgHCO_3^\circ$ and $\equiv MgCO_3^-$, following the same approach outlined for calcite (Section 6.8.2.7) and dolomite (Section 6.8.3.2). Again, the pH-dependent surface charge data were used to fit the intrinsic stability constants for these surface species obtaining the values listed in Table 6.4. In these calculations, the surface site density was fixed at 19.6×10^{-6} mol m^{-2}. Again, the authors adopted the constant capacitance model and fixed the capacitance at $I^{1/2}/0.0032$ F m^{-2}, which accounts for the high surface charge of magnesite far from the PZC. At $I = 0.1$ mol kg^{-1}, $m_{TDIC} = 0.01$ and $m_{Mg} = 10^{-5}$, the SCM predicts that the main carbonate surface species are $\equiv CO_3^-$ above pH 4.5 and $\equiv CO_3H^\circ$ below this pH value. Mg-surface sites are dominated by $\equiv MgOH_2^+$ below pH ~8, which is replaced by $\equiv MgCO_3^-$, $\equiv MgOH^\circ$ and $\equiv MgO^-$, for increasing pH values.

Since metal sites are wholly protonated below pH ~8, the increase in magnesite dissolution rate with decreasing pH in the acid range was attributed to the protonation of the $\equiv CO_3^-$ sites by Pokrovsky and Schott (1999a). Accepting this explanation, the pH-independence of the magnesite dissolution rate in the strongly acidic range was ascribed to the full protonation of the carbonate sites. The proton-promoted dissolution rate (which is obtained by subtracting, from the measured rate the water-promoted dissolution rate, equal to 4.57×10^{-10} mol m^{-2} s^{-1} on average; see above) shows a good correlation with the $\equiv CO_3H^\circ$ concentration in the log–log plot of Fig. 6.51. The relation between these two variables (rate in mol m^{-2} s^{-1}, [$\equiv CO_3H^\circ$] in mol m^{-2}) is described by equation (6-202), where $k_{CO_3H^\circ} = 10^{11.2(}$ and $m = 3.97$. This value of m very close to 4 suggests that the proton-promoted dissolution of magnesite proceeds through the protonation of the four carbonate sites surrounding a hydrated metal site.

In neutral and alkaline solutions, magnesite dissolution appears to be controlled by the $\equiv MgOH_2^+$ surface complex, as indicated by the good linear correlation between the logarithm of the water-promoted dissolution rate and the logarithm of [$\equiv MgOH_2^+$] (Fig. 6.52), which is described by equation (6-212), where $k_{MgOH_2^+} = 10^{9.38}$ and $n = 3.94$ (rate in mol m^{-2} s^{-1}, [$\equiv MgOH_2^+$] in mol m^{-2}). Again, the value of n very close to 4 suggests that the water-promoted dissolution of magnesite proceeds through the hydration of the four metal sites surrounding a carbonate site.

Combining these relations linking the proton-promoted dissolution rate and the water-promoted dissolution rate to the concentrations of relevant surface species, Pokrovsky and Schott (1999a) proposed the following equation for the overall reaction rate:

$$r = \left(k_{CO_3H^\circ} \cdot [\equiv CO_3H^\circ]^{3.97} + k_{MgOH_2^+} \cdot [\equiv MgOH_2^+]^{3.94}\right) \cdot \left[1 - \left(\frac{Q}{K_{sp}}\right)^{3.94}\right], \tag{6-222}$$

which allows to describe the kinetics of magnesite dissolution/precipitation in a large range of solution composition, within the TST framework. Note that the only differences between equation (6-204) and equation (6-222), is the involvement of [$\equiv CaOH_2^+$] in the

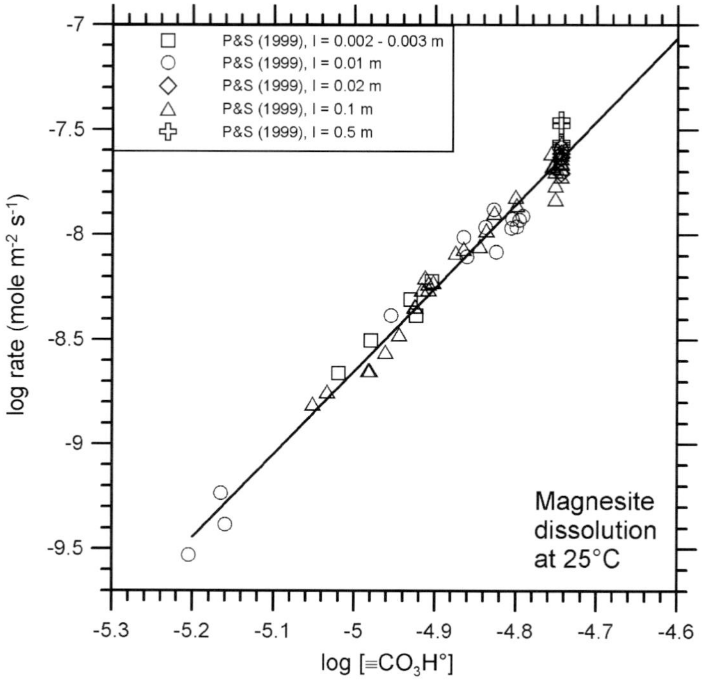

Figure 6.51. Plot of the proton-promoted dissolution rate of magnesite at 25°C and pH of 0.2–5 vs. the surface concentration of $\equiv CO_3H°$. (Reprinted from Pokrovsky and Schott, 1999a, Copyright 1999 with permission from Elsevier.)

first and of $[\equiv MgOH_2^+]$ in the second and possibly the values of the exponents, which are not defined for calcite.

The first term of equation (6-222), involving the concentration of the $\equiv CO_3H°$ surface species, refers to the proton-promoted dissolution and corresponds to the first term of equation (6-176). The second term, involving the concentration of the $\equiv MgOH_2^+$ surface species, refers to the water-promoted dissolution and corresponds to the third term of equation (6-176). The effect of the reverse (precipitation) reaction at close-to-equilibrium conditions is accounted for by the term $[1 - (Q/K_{sp})^{3.94}]$. Note that there is no provision for surface carbonation in equation (6-222), in contrast to equation (6-176), in which the CO_2-promoted dissolution is described by the second term.

The effects of variable P_{CO_2} on magnesite dissolution (as well as on calcite and dolomite, see Sections 6.8.2.3 and 6.8.3.5, respectively) was studied by Pokrovsky et al. (2005b) at 25°C, pH values of 3.05–3.9, in 0.1 mol kg^{-1} NaCl. It was found that the magnesite dissolution rate increases ~3 times for a P_{CO_2} increase from 0 to 5 bar, whereas the rate remains constant for a further increase in P_{CO_2} from 5 to 55 bar. These experimental results are well predicted by the SCM described above.

The dissolution kinetics of magnesite was investigated at the microscopic scale by Jordan et al. (2001) and Higgins et al. (2002) by means of AFM, in aqueous solutions of

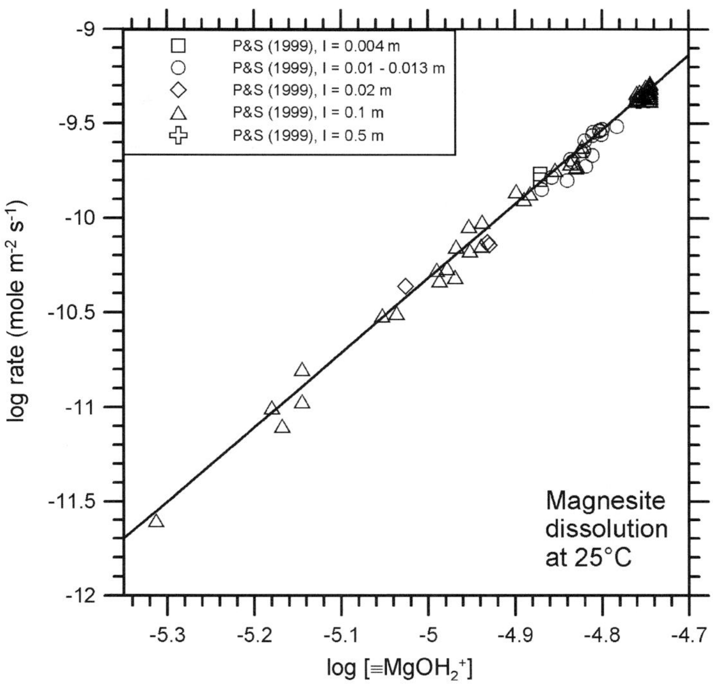

Figure 6.52. Plot of the water-promoted dissolution rate of magnesite at 25°C and pH of 5–12 vs. the surface concentration of $\equiv MgOH_2^+$. (Reprinted from Pokrovsky and Schott, 1999a, Copyright 1999 with permission from Elsevier.)

pH 2–4.2 and at temperatures between 60 and 90°C. Higgins et al. (2002) determined the rate of magnesite dissolution (surface 104) based on both the Mg concentration in the cell effluent and the AFM observations, and found different results in terms of dissolution fluxes (with those based on Mg concentration that are six to seven times larger than those derived through AFM), apparent activation energies at pH 4.2 (74 ± 22 and 41 ± 4 kJ mol^{-1}, for the AFM and chemical methods, respectively) and reaction order with respect to H^+ concentration (0.36 ± 0.13 and 0.47 ± 0.03, for the AFM and chemical methods, respectively). These differences were attributed to the different sampling length scales of the two methods (Higgins et al., 2002).

Note that the AFM-derived apparent activation energy value is similar to that estimated by Laçin et al. (2005) through dissolution experiments at 40–70°C of natural magnesite in 1.0–10 M acetic acid solutions, 78.4 kJ mol^{-1}. Both apparent activation energies for magnesite dissolution are instead lower than the value estimated by Arvidson and Mackenzie (2000) for the reverse process, i.e. magnesite precipitation, 92.9 kJ mol^{-1}, based on the apparent activation energy of calcite precipitation, 39.3 ± 3.8 kJ mol^{-1} and the difference in activation energy between magnesite and calcite, 53.6 kJ mol^{-1}. This value was suggested by Lippmann (1973) based on the energetics of cation dehydration, which represents the energetic barrier to cation incorporation into a growing carbonate lattice.

6.8.5. Other carbonate minerals

The kinetics of dissolution of other carbonate minerals was investigated by Chou et al. (1989), Greenberg and Tomson (1992) and Pokrovsky and Schott (2002).

Chou et al. (1989) measured the dissolution rates of aragonite and witherite ($BaCO_3$) at 25°C and pH 4–10, in addition to those of calcite, dolomite and magnesite (see above). The rates of aragonite and witherite dissolution resulted to be close to those of calcite and to show a similar pH dependence (Fig. 6.53). Adopting the same approach they used to interpret calcite data, Chou et al. (1989) found that the kinetic rate constants (in mol m^{-2} s^{-1}) are $k_1 = 1.2$, $k_2 = 4.0 \times 10^{-4}$, $k_3 = 1.0 \times 10^{-6}$ and $k_{-3} = 170$ for aragonite, and $k_1 = 0.76$, $k_2 = 3.0 \times 10^{-4}$, $k_3 = 7.1 \times 10^{-7}$ and $k_{-3} = 200$ for witherite.

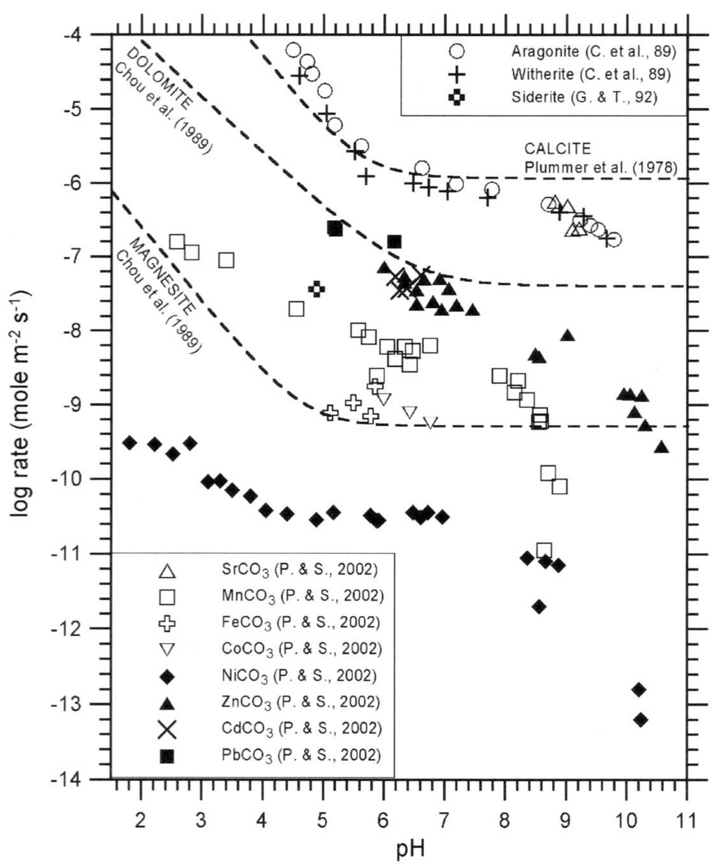

Figure 6.53. Logarithms of the dissolution rates of several divalent-metal carbonates at 25°C as a function of pH, showing the experimental data of Chou et al. (1989), Pokrovsky and Schott (2002) and Greenberg and Tomson (1992). Also shown for comparison are the dissolution rates of calcite, at $P_{CO_2} = 10^{-1}$ bar, and dolomite and magnesite at $P_{CO_2} = 10^{-3.5}$ bar.

Greenberg and Tomson (1992) determined the precipitation kinetics of siderite at temperatures from 27 to 80°C by means of a modified seeded growth kinetic method as well as its dissolution kinetics at 26 and 60°C. Precipitation was induced by lowering the P_{CO_2} of the aqueous solution, whereas dissolution was initiated by the reverse process, i.e. through CO_2 sorption. Based on the data reported by these authors, the initial dissolution rate of siderite results to be 3.67×10^{-8} mol m^{-2} s^{-1} at a pH of 4.88 and a thermodynamic affinity of -4.29 kJ mol^{-1} (pH and A were computed here by means of the software code EQ3). Apparent activation energy was found to be close to 108 kJ mol^{-1} for precipitation and 45 kJ mol^{-1} for dissolution. These high values suggest that both processes are governed by surface reactions rather than by transport in solution.

Pokrovsky and Schott (2002) measured the far-from-equilibrium, steady-state dissolution rates of several divalent-metal carbonates ($SrCO_3$, $MnCO_3$, $FeCO_3$, $CoCO_3$, $NiCO_3$, $ZnCO_3$, $CdCO_3$ and $PbCO_3$) at 25°C, ionic strength of 0.01 mol kg^{-1} and different pH values by using a mixed-flow reactor. Possible transport-control effects were observed only for $SrCO_3$ and $PbCO_3$. For multivalent elements (Fe, Co and Mn), experiments were carried out in N_2-saturated aqueous solutions. Measured dissolution rate data for all divalent-metal carbonates are plotted in Fig. 6.53 as a function of pH. The most complete dataset, i.e. that for $NiCO_3$, shows four different relations with pH, depending on pH, as observed for magnesite. In strongly acidic aqueous solutions (pH $\leq$ 3), rates are independent of pH. In the pH range 3–5, the rate increases linearly with log a_{H^+}. In the pH interval 5–8, rates are again independent of pH, whereas above pH 8, rates again increase with a_{H^+}.

Pokrovsky and Schott (2002) also performed several electrophoretic measurements and developed a SCM for divalent-metal carbonates based on these data and the correlation between the stability constants of surface reactions (see Table 6.4) and corresponding homogeneous reactions in the aqueous solution, as anticipated in Section 6.8.2.7. The SCM was then used to describe the surface-controlled dissolution rates of divalent-metal carbonates by means of equation (6-204).

6.9. Dissolution rates of sulphates, sulphides, phosphates and halides

The dissolution kinetics of other classes of minerals, i.e. sulphates, sulphides, phosphates and halides, is not reviewed in detail both owing to their limited importance for modelling the geological storage of CO_2 and to maintain this chapter to a reasonable size. However, the rate parameters of these minerals, which were compiled by Palandri and Kharaka (2004), are listed in Table 6.1.

Chapter 7

Reaction Path Modelling of Geological CO_2 Sequestration

In this chapter we will see how to predict the course of chemical reactions occurring during the geological disposal of CO_2, building on the thermodynamic and kinetic grounds that were reviewed in previous chapters.

It must be underscored that many early investigations on geological CO_2 storage took into account only trapping of gaseous or supercritical CO_2 as a separated "immiscible" phase under a low-permeability cap rock and CO_2 dissolution into the aqueous phase without considering the reactivity of the aquifer rock, although the latter effect can be very important, especially in the long term as shown below. However, recent studies devoted to geochemical modelling of CO_2 disposal in deep aquifers correctly took into account not only hydrodynamic and solubility trapping but also mineral trapping.

Let us return to suppose we want to know what happens if we inject CO_2 into a deep aquifer. Prior to CO_2 injection, it is reasonable to hypothesize that the aqueous solution stored in the deep aquifer is in (or close to) chemical equilibrium with a certain number of minerals, either the primary rock-forming minerals or, more likely, the authigenic solid phases produced through alteration of the primary minerals at *P–T–X* conditions different from those at which they formed. Although it makes a lot of difference to refer to the primary minerals rather than to the secondary minerals, here we do not want to reconstruct the evolution of the aquifer rock. Therefore, we simply hypothesize that we know somehow that the deep aqueous solution is in equilibrium with some specified minerals.

7.1. The reconstruction of the initial (before CO_2 injection) aqueous solution: speciation-saturation calculations

It was shown (e.g., Michard et al., 1981; Chiodini et al., 1991) that the chemical composition of an aqueous solution can be computed, by means of suitable speciation-saturation calculations, if we specify the temperature, the pressure, a certain number of minerals with which the aqueous solution is assumed to be in equilibrium (one for each compatible chemical component, e.g. Na^+, K^+, Ca^{2+}, Mg^{2+}, Al^{3+}, SiO_2), f_{CO_2}, which fixes HCO_3^- activity, and the total concentration of each mobile component (e.g., Cl^-). If multivalent elements, such as Fe, Mn and S, are also present in the considered system, further complications arise. A general, complete account on this subject is given by Van Zeggeren and Storey (1970) and Smith and Missen (1982). But let us proceed step by step.

7.1.1. The aqueous solution/calcite example

One of the simplest examples, which is treated in many textbooks, is that of computing the composition of the aqueous solution in equilibrium with calcite at given conditions, e.g. temperature = 25°C, pressure = 1 bar and $P_{CO_2} = 10^{-3.5}$ bar (average atmospheric value). In this case we have to take into account the thermodynamic constants of the dissociation reactions of apparent carbonic acid, bicarbonate ion and water (6-186, 6-187 and 6-188, respectively), i.e.:

$$K_{H_2CO_3^*} = \frac{a_{HCO_3^-} \cdot a_{H^+}}{a_{H_2CO_3^*}} = 10^{-6.35}, \tag{7-1}$$

$$K_{HCO_3^-} = \frac{a_{CO_3^{2-}} \cdot a_{H^+}}{a_{HCO_3^-}} = 10^{-10.33}, \tag{7-2}$$

$$K_{H_2O} = \frac{a_{OH^-} \cdot a_{H^+}}{a_{H_2O}} = 10^{-14.00}, \tag{7-3}$$

respectively, the thermodynamic constant of the dissolution reaction of gaseous CO_2 (6-200):

$$K_{CO_{2(g)}} = \frac{a_{H_2CO_3^*}}{f_{CO_2} \cdot a_{H_2O}} = 10^{-1.47}, \tag{7-4}$$

the solubility product of calcite, which is conveniently expressed as follows, referring to the sum of reaction (6-189) and the reverse of reaction (6-187):

$$K_{calcite} = \frac{K_{sp}}{K_{HCO_3^-}} = \frac{a_{HCO_3^-} a_{Ca^{2+}}}{a_{H^+} \cdot a_{calcite}} = 10^{+1.85}, \tag{7-5}$$

and the electroneutrality condition

$$2 \cdot m_{Ca^{2+}} + m_{H^+} = m_{HCO_3^-} + 2 \cdot m_{CO_3^{2-}} + m_{OH^-}. \tag{7-6}$$

Note that we have implicitly made some simplifications, in that we have neglected the formation of complex species, such as $CaHCO_3^+$ and $CaCO_3^\circ$. If we want to keep calculations to a simple level, further simplifications are needed, namely $a_{H_2O} = 1$, $a_{calcite} = 1$, $f_{CO_2} = P_{CO_2}$ and $a_i = m_i$ (i.e. $\gamma_i = 1$) for all solutes. Through suitable substitutions, the following equation is obtained:

$$\frac{2\cdot 10^{+1.85}\cdot a_{H^+}^2}{10^{-6.35}\cdot 10^{-1.47}\cdot 10^{-3.5}}+a_{H^+}-\frac{10^{-6.35}\cdot 10^{-1.47}\cdot 10^{-3.5}}{a_{H^+}}-\frac{2\cdot 10^{-10.33}\cdot 10^{-6.35}\cdot 10^{-1.47}\cdot 10^{-3.5}}{a_{H^+}^2}-\frac{10^{-14.00}}{a_{H^+}}=0. \quad (7\text{-}7)$$

This is a fourth-order polynomial equation with respect to a_{H^+} and its solution without the aid of a computer is a formidable task. By using an electronic spreadsheet we can vary a_{H^+} until the left-hand side of equation (7-7) changes sign and tends to 0. In this way a pH of 8.26 is obtained. Based on this pH value, the concentrations of other solutes are then easily computed by means of equations (7-1) to (7-5).

Alternatively, we can assume that $m_{Ca^{2+}} \gg m_{H^+}$ and $m_{HCO_3^-} \gg 2\cdot m_{CO_3^{2-}} + m_{OH^-}$, conditions that have to be verified afterwards. Under these assumptions, equation (7-7) simplifies to

$$\frac{2\cdot 10^{+1.85}\cdot a_{H^+}^2}{10^{-6.35}\cdot 10^{-1.47}\cdot 10^{-3.5}}-\frac{10^{-6.35}\cdot 10^{-1.47}\cdot 10^{-3.5}}{a_{H^+}}=0. \quad (7\text{-}8)$$

Equation (7-8) can be easily solved obtaining again pH = 8.26, the same pH value we have found by solving equation (7-7).

Although this coincidence is encouraging, we cannot think to export this approach to more complex problems. In addition, some of the assumptions introduced above may fail: for instance, at high temperature and high ionic strength $\gamma_i \neq 1$ and $a_{H_2O} \neq 1$; at high pressure $f_{CO_2} \neq P_{CO_2}$; if the considered mineral is not a pure phase its activity is different from unity.

The use of a speciation-saturation code, such as EQ3NR (Wolery, 1992) is mandatory if we want sound results, even in the simple aqueous solution/calcite example. The input file given in Table 7.1 can be used to compute the concentrations and activities of all solute species in an aqueous solution in equilibrium with calcite at temperature = 25°C, pressure = 1 bar and $f_{CO_2} = 10^{-3.5}$ bar. Without entering too much into the details of this code, we are basically saying that: (i) the activity of Ca^{2+} ion is fixed by equilibrium with calcite, (ii) the activity of HCO_3^- ion is governed by equilibrium with a hypothetical, infinite gas reservoir that fixes the f_{CO_2} of the considered system at $10^{-3.5}$ bar and (iii) pH is initially set at 8.26 ($\log a_{H^+} = -8.26$), but the code has to recalculate it by using the electroneutrality conditions. These are exactly the same conditions used above to compute the approximated pH value, which is now inserted as an input value in EQ3NR (note that it is not necessary to compute an approximated pH value to run this code; EQ3NR is able to run with any pH values or almost so). Note also that, although this exercise has no redox aspect, we set the log oxygen fugacity to −0.678 (the atmospheric value) to avoid possible computational problems.

TABLE 7.1

EQ3NR input file to compute the speciation of the aqueous solution in equilibrium with calcite at 25°C, 1 bar under a f_{CO_2} of $10^{-3.5}$ bar (required input formats are not fully respected).

```
Water in equilibrium with calcite at 25°C and atmospheric PCO2
Use SUP database

endit.
     tempc = 2.50000E+01
       rho = 1.00000E+00      tdspkg = 0.00000E+00      tdspl = 0.00000E+00
       fep = -0.678           uredox =
     tolbt = 0.00000E+00       toldl = 0.00000E+00      tolsat = 0.00000E+00
    itermx= 0
*            1      2     3     4     5     6     7     8     9    10
  iopt1-10 = 0      0     0     0     0     0     0     0     0     0
  iopg1-10 = 0      0     0     0     0     0     0     0     0     0
  iopr1-10 = 0      0     0     1     0     0     0    -1     0     0
 iopr11-20 = 0      0     0     0     0     0     0     0     0     0
  iodb1-10 = 0      0     0     0     0     0     0     0     0     0
     uebal =    H+
     nxmod =    0
data file master species = H+
   switch with species =
   jflag = 16    csp =-8.26
data file master species = Ca++
   switch with species =
   jflag = 19    csp = 0.
mineral = Calcite
data file master species = HCO3-
   switch with species =
   jflag = 21    csp = -3.5000
 mineral = CO2(g)
endit.
```

The most important parts of the EQ3NR output file are summarized in Table 7.2, which indicates that the pH sought is 8.2748, which is only 0.01 pH units higher than the approximated value computed by means of equations (7-7) and (7-8). In spite of this small difference in pH, Table 7.2 shows that (i) activity coefficients computed by EQ3NR using the extended (B-dot) Debye–Hückel equation (see Chapter 4) deviate somewhat from the value of 1, which was assumed above; these deviations are greater for the divalent ions (e.g., $\gamma_{Ca^{2+}} = 0.8445$ and $\gamma_{CO_3^{2-}} = 0.8451$) than for the monovalent ions (e.g., $\gamma_{HCO_3^-} = 0.9587$ and $\gamma_{H^+} = 0.9598$); (ii) the aqueous complexes $CaCO_3°$ and $CaHCO_3^+$ have concentrations of the same order of magnitude of CO_3^{2-} ion and, therefore, cannot be neglected.

7.1.2. The aqueous solution/multimineral paragenesis general case

A more general case is that of an aqueous solution assumed to be in equilibrium with a paragenesis comprising several solid phases. Again, in this case, the only way to

TABLE 7.2

Part of the EQ3NR output file showing the speciation of the aqueous solution saturated with calcite at 25°C, 1 bar under a f_{CO_2} of $10^{-3.5}$ bar

Species	Molality	Log molality	Log gamma	Log activity
HCO_3^-	9.5399E–04	−3.0205	−0.0183	−3.0388
Ca^{++}	4.8536E–04	−3.3139	−0.0734	−3.3874
O_2(aq)	2.6518E–04	−3.5765	0.0002	−3.5763
CO_2(aq)	1.0739E–05	−4.9691	0.0002	−4.9689
CO_3^{--}	9.5589E–06	−5.0196	−0.0731	−5.0927
$CaCO_3$(aq)	7.0307E–06	−5.1530	0.0000	−5.1530
$CaHCO_3^+$	4.3597E–06	−5.3605	−0.0189	−5.3794
OH^-	1.9873E–06	−5.7017	−0.0185	−5.7203
H^+	5.5335E–09	−8.2570	−0.0178	−8.2748

compute the speciation/saturation state of the aqueous solution is by running a suitable computer code, such as EQ3NR.

If the problem has a redox aspect, we have to assume redox equilibrium, at least if we are using version 7 of the software package EQ3/6, despite most natural systems are far from this ideal condition (e.g., Lindberg and Runnells, 1984), and we have to specify the redox potential of the system, in terms of f_{O_2}, or Eh, or pe, or through a suitable f_{O_2}-controlling reaction.

Let us assume that we want to compute the speciation/saturation state of the aqueous solution in equilibrium with analcime, calcite, kaolinite, daphnite, clinochlore, chalcedony and magnetite at 60°C, 1 bar under a f_{CO_2}, of 10^{-1} bar. Note that this paragenesis include a zeolite, analcime and some phyllosilicates, namely kaolinite, daphnite and clinochlore. The latter two minerals belong to the chlorite group and are assumed to have 14 Å basal spacing. These solid phases are typical of systems of relatively low temperature, as indicated by the relationship between hydrothermal alteration minerals and temperature in active geothermal systems and in hydrothermal ore deposits, representing their fossil analogues (e.g., Henley and Ellis, 1983).

In setting up the EQ3 input file (see Table 7.3), we state that Na^+ activity is fixed by analcime, Ca^{2+} activity by calcite, aqueous SiO_2 activity by chalcedony, Al^{3+} activity by kaolinite, Mg^{2+} activity by clinochlore [$Mg_5Al_2Si_3O_{10}(OH)_8$] and Fe^{2+} activity by daphnite [$Fe_5Al_2Si_3O_{10}(OH)_8$]. Since also magnetite [$FeO \cdot Fe_2O_3$] is part of our mineral paragenesis, the f_{O_2} of the system is uniquely fixed by daphnite–magnetite coexistence, under the redox equilibrium hypothesis. Note that the role of the daphnite–magnetite pair is explicitly declared in the input file (Table 7.3) by writing that daphnite controls Fe^{2+} activity and that magnetite constrains $O_{2(g)}$ activity (incidentally, we have also to set the switch iopt1 to −3, to follow EQ3 instructions).

Similar to the aqueous solution/calcite example: (i) the activity of HCO_3^-, ion is fixed setting f_{CO_2}, although a value of 10^{-1} bar is chosen in this case, and (ii) pH is initially set at 7.00, but the code is instructed to recompute it based on the electroneutrality conditions. In addition, total chloride concentration is assumed to be 0.03 mol kg^{-1}.

TABLE 7.3

EQ3NR input file to compute the speciation/saturation state of the aqueous solution in equilibrium with analcime, calcite, kaolinite, daphnite, clinochlore, chalcedony and magnetite at 60°C, 1 bar under a f_{CO_2} of 10^{-1} bar (required input formats are not fully respected).

```
. . . omissis
endit.
     tempc=  60.0000E+00
       rho=  1.00000E+00          tdspkg= 0.00000E+00        tdspl= 0.00000E+00
       fep=  0.00000E+00          uredox=
     tolbt=  0.00000E+00           toldl= 0.00000E+00       tolsat= 0.00000E+00
    itermx=  0
*                 1     2     3     4     5     6     7     8     9    10
  iopt1-10=      -3     0     0     0     0     0     0     0     0     0
  iopg1-10=       0     0     0     0     0     0     0     0     0     0
  iopr1-10=       0     0     0     0     0     0     0     0     0     0
 iopr11-20=       0     0     0     0     0     0     0     0     0     0
  iodb1-10=       0     0     0     0     0     0     0     0     0     0
     uebal= H+
    nxmod=  0
data file master species= Na+
    switch with species=
    jflag= 19    csp=   0.
  mineral= Analcime
data file master species= Al+++
    switch with species=
    jflag= 19    csp=   0.
  mineral= Kaolinite
data file master species= Cl-
    switch with species=
    jflag=0      csp=   3.00000E-02
data file master species= SiO2(aq)
    switch with species=
    jflag= 19    csp=   0.
  mineral= Chalcedony
data file master species= H+
    switch with species=
    jflag= 16    csp=   -7.0000
data file master species= Fe++
    switch with species=
    jflag= 19    csp=   0.
  mineral= Daphnite-14A
data file master species= O2(g)
    switch with species=
    jflag= 19    csp=   0.
  mineral= Magnetite
data file master species= Ca++
    switch with species=
    jflag= 19    csp=   0.
  mineral= Calcite
```

TABLE 7.3 (*Continued*)

```
data file master species= Mg++
   switch with species=
   jflag= 19    csp=    0.
 mineral= Clinochlore-14A
data file master species= HCO3-
   switch with species=
   jflag= 21    csp= -1.0000
      gas= CO2(g)
endit.
```

Results of the EQ3 run are listed in Table 7.4. The computed equilibrium pH is 7.68. At this pH value, $Al(OH)_4^-$, HCO_3^- and undissociated $SiO_{2(aq)}$ are by far the prevailing species of dissolved Al, carbonate and silica. Again, the $MeCO_3°$ and $MeHCO_3^+$ complex species cannot be neglected, not only for Ca (as in the previous aqueous solution/calcite example), but also for divalent Fe and Mg. In contrast, the free ions Cl^-, Na^+ and K^+ are by far the prevailing dissolved species of these chemical components.

The code also indicates that the aqueous solution is saturated (i.e. $-0.3 < \log[Q/K] < +0.3$, but these limits can be reset by the user) with respect to several minerals, comprising aragonite, Mg- and Na-beidellite, böhmite, α-cristobalite, 7 Å-cronstedtite, diaspore, huntite, paragonite, quartz, 14 Å-ripidolite, siderite and trydimite. The solid phases with which the aqueous solution is oversaturated and undersaturated are also listed.

Finally, the log f_{O_2}, resulted to be −61. Probably the $\log\left(f_{H_2}/f_{H_2O}\right)$ ratio, which is equal to −5.02 in this example, would be a better choice to describe redox conditions, for the reasons given by Giggenbach (1987).

If we also want to include S species into the geochemical model, we may add either pyrite [FeS_2] or anhydrite [$CaSO_4$] or both minerals to the input file. If we choose the pyrite constraint, total S concentration results to be 0.013 mmol kg^{-1} and the aqueous solution turns out to be strongly undersaturated with anhydrite. At equilibrium with anhydrite, total S concentration is 480 mmol kg^{-1} and the aqueous solution is strongly oversaturated with pyrite. At saturation with both pyrite and anhydrite, total S concentration results to be 472 mmol kg^{-1}. In all three cases, however, the log f_{O_2} of the system is constrained by daphnite–magnetite coexistence at −61. Looking at the thermodynamic constant of the HS^- –SO_4^{2-} equilibrium reaction, written in logarithmic form:

$$\log K_{HS^-/SO_4^{2-}} = \log a_{SO_4^-} - \log a_{HS^-} - \mathrm{pH} - 2 \cdot \log f_{O_2}, \tag{7-9}$$

it is evident that pH differences are expected between the pyrite case and the anhydrite case. Indeed, the computed pH is 7.68 in the pyrite case and 6.85 in the anhydrite case (and 6.85 again in the pyrite plus anhydrite case).

TABLE 7.4

Part of the EQ3NR output file showing the speciation of the aqueous solution in equilibrium with analcime, calcite, kaolinite, daphnite, clinochlore, chalcedony and magnetite at 60°C, 1 bar under a f_{CO_2} *of* 10^{-1} bar

. . . omissis

	pH	Eh	pe
modified NBS pH scale	7.6822	−0.2918	−4.4142E+00

. . . omissis

—- Major Aqueous Species Contributing to Mass Balances —-

Aqueous species accounting for 99% or more of Al^{+++}

Species	Factor	Molality	Per Cent
$Al(OH)_4^-$	1.00	2.3351E–07	97.40
$NaAl(OH)_4(aq)$	1.00	3.5248E–09	1.47
$Al(OH)_3(aq)$	1.00	2.6892E–09	1.12
Total		2.3973E–07	100.00

Aqueous species accounting for 99% or more of Ca^{++}

Species	Factor	Molality	Per Cent
Ca^{++}	1.00	2.7410E–05	62.86
$CaHCO_3^+$	1.00	8.5013E–06	19.50
$CaCO_3(aq)$	1.00	7.6103E–06	17.45
Total		4.3605E–05	99.81

Aqueous species accounting for 99% or more of Cl^-

Species	Factor	Molality	Per Cent
Cl^-	1.00	2.9671E–02	98.90
$NaCl(aq)$	1.00	3.2737E–04	1.09
Total		3.0000E–02	100.00

Aqueous species accounting for 99% or more of Fe^{++}

Species	Factor	Molality	Per Cent
$FeHCO_3^+$	1.00	3.0519E–06	79.01
$FeCO_3(aq)$	1.00	5.3516E–07	13.86
Fe^{++}	1.00	2.7052E–07	7.00
Total		3.8625E–06	99.87

TABLE 7.4 (*Continued*)

Aqueous species accounting for 99% or more of HCO_3^-

Species	Factor	Molality	Per Cent
HCO_3^-	1.00	5.3742E–02	92.76
$NaHCO_3(aq)$	1.00	2.1026E–03	3.63
$CO_2(aq)$	1.00	1.6092E–03	2.78
Total		5.7936E–02	99.17

Aqueous species accounting for 99% or more of Mg^{++}

Species	Factor	Molality	Per Cent
Mg^{++}	1.00	1.0965E–04	68.44
$MgHCO_3^+$	1.00	3.8034E–05	23.74
$MgCO_3(aq)$	1.00	1.1295E–05	7.05
Total		1.6022E–04	99.22

Aqueous species accounting for 99% or more of Na^+

Species	Factor	Molality	Per Cent
Na^+	1.00	8.3906E–02	97.14
$NaHCO_3(aq)$	1.00	2.1026E–03	2.43
Total		8.6379E–02	99.57

Aqueous species accounting for 99% or more of $SiO_2(aq)$

Species	Factor	Molality	Per Cent
$SiO_2(aq)$	1.00	5.8790E–04	95.45
$NaH_3SiO_4(aq)$	1.00	1.5906E–05	2.58
$H_3SiO_4^-$	1.00	1.2131E–05	1.97
Total		6.1593E–04	100.00

. . . omissis

Mineral	Log *Q*/*K*	Aff, kcal	State
Albite	0.381	0.581	ssatd
Albite high	−0.735	−1.120	
Albite low	0.381	0.581	ssatd
Amesite–14A	−2.342	−3.570	
Analcime	0.000	0.000	satd
Analcime–dehy	−5.602	−8.541	
Andalusite	−4.423	−6.742	
Anthophyllite	−5.945	−9.063	
Antigorite	6.368	9.708	ssatd
Aragonite	−0.143	−0.218	satd
Artinite	−4.258	−6.492	

TABLE 7.4 (*Continued*)

Mineral	Log *Q/K*	Aff, kcal	State
Beidellite–Ca	−0.408	−0.621	
Beidellite–H	−1.308	−1.993	
Beidellite–Mg	−0.208	−0.317	satd
Beidellite–Na	−0.162	−0.247	satd
Boehmite	−0.315	−0.480	satd
Brucite	−3.237	−4.934	
Graphite	−1.862	−2.839	
Calcite	0.000	0.000	satd
Chalcedony	0.000	0.000	satd
Chamosite–7A	−2.207	−3.364	
Chrysotile	−0.367	−0.559	
Clinochlore–14A	0.000	0.000	satd
Clinochlore–7A	−3.133	−4.776	
Clinoptilolite–Ca	−1.610	−2.455	
Clinoptilolite–Na	1.534	2.338	ssatd
Clinoptilolite–hy–Ca	−1.782	−2.717	
Clinoptilolite–hy–Na	1.545	2.355	ssatd
Coesite	−0.488	−0.744	
Corundum	−3.208	−4.891	
Cristobalite(alpha)	−0.239	−0.364	satd
Cristobalite(beta)	−0.604	−0.921	
Cronstedtite–7A	−0.062	−0.095	satd
Daphnite–14A	0.000	0.000	satd
Daphnite–7A	−3.107	−4.737	
Dawsonite	−0.366	−0.557	
Diaspore	0.036	0.056	satd
Diopside	−3.398	−5.180	
Dolomite	1.981	3.019	ssatd
Dolomite–dis	0.662	1.009	ssatd
Dolomite–ord	1.981	3.020	ssatd
Enstatite	−1.931	−2.944	
Epidote	−5.113	−7.794	
Epidote-ord	−5.115	−7.797	
Fayalite	−2.702	−4.119	
$Fe(OH)_2$	−3.777	−5.758	
$Fe(OH)_3$	−5.075	−7.736	
FeO	−3.200	−4.878	
Ferrosilite	−1.155	−1.761	
Forsterite	−5.181	−7.898	
Gaylussite	−5.918	−9.022	
Gibbsite	−0.711	−1.084	
Goethite	−0.370	−0.563	
Greenalite	−0.810	−1.235	
Halite	−4.452	−6.786	
Hedenbergite	−4.854	−7.400	
Haematite	0.380	0.579	ssatd
Hercynite	−3.760	−5.731	
Huntite	0.269	0.411	satd
Hydromagnesite	−6.282	−9.577	
Ice	−0.287	−0.437	satd

TABLE 7.4 (*Continued*)

Mineral	Log *Q*/*K*	Aff, kcal	State
Jadeite	−1.493	−2.276	
Kaolinite	0.000	0.000	satd
Kyanite	−4.232	−6.451	
Lansfordite	−2.868	−4.372	
Laumontite	−2.181	−3.325	
Lawsonite	−3.304	−5.037	
Magnesite	0.541	0.824	ssatd
Magnetite	0.000	0.000	satd
Mesolite	1.666	2.540	ssatd
Minnesotaite	0.489	0.745	ssatd
Monohydrocalcite	−0.977	−1.489	
Montmor–Ca	0.728	1.110	ssatd
Montmor–Mg	0.986	1.503	ssatd
Montmor–Na	1.033	1.575	ssatd
Mordenite	−0.865	−1.318	
Na_2CO_3	−6.512	−9.927	
Nahcolite	−2.745	−4.185	
Natrolite	−1.264	−1.927	
Nepheline	−2.821	−4.300	
Nesquehonite	−2.843	−4.335	
Nontronite–Ca	2.242	3.418	ssatd
Nontronite–H	1.343	2.048	ssatd
Nontronite–Mg	2.442	3.722	ssatd
Nontronite–Na	2.488	3.792	ssatd
Okenite	−6.026	−9.187	
Paragonite	0.064	0.097	satd
Pirssonite	−6.073	−9.259	
Prehnite	−5.542	−8.449	
Pseudowollastonite	−5.302	−8.083	
Pyrophyllite	−0.828	−1.262	
Quartz	0.243	0.370	satd
Ripidolite–14A	0.261	0.397	satd
Ripidolite–7A	−2.864	−4.366	
Saponite–Ca	2.462	3.753	ssatd
Saponite–H	1.562	2.381	ssatd
Saponite–Mg	2.657	4.051	ssatd
Saponite–Na	2.707	4.127	ssatd
Scolecite	−0.856	−1.305	
Sepiolite	−2.437	−3.715	
SiO_2(am)	−0.824	−1.256	
Siderite	0.158	0.241	satd
Sillimanite	−4.727	−7.206	
Talc	2.059	3.140	ssatd
Thermonatrite	−6.532	−9.958	
Tremolite	−3.088	−4.707	
Tridymite	0.087	0.132	satd
Wairakite	−5.593	−8.526	
Wollastonite	−5.118	−7.802	
Wustite	−3.578	−5.455	

TABLE 7.4 (*Continued*)

. . . omissis

Gas	Fugacity	Log fugacity
CH_4(g)	1.0406E–06	−5.9827
CO(g)	4.4081E–11	−10.3558
CO_2(g)	1.0000E–01	−1.0000
H_2(g)	1.5491E–06	−5.8099
H_2O(g)	1.6325E–01	−0.7872
O_2(g)	1.0378E–61	−60.9839

7.2. Reaction path modelling

7.2.1. Fundamental relationships

As discussed above, prior to CO_2 injection into the deep aquifer, it is reasonable to hypothesize that the deep aqueous solution is in chemical equilibrium with a certain number of minerals. Upon CO_2 injection, this equilibrium is altered to an extent which depends on the imposed f_{CO_2} (possible temperature changes caused by isenthalpic CO_2 compression below the Joule-Thompson inversion curve – see Section 3.9 – are neglected in this discussion). As a consequence of this induced disequilibrium, the CO_2-containing aqueous solution will start to react with the primary minerals of the hosting rock. If the hosting rock is made up of silicates, they will be dissolved and carbonates, silicates, and/or silica minerals will the produced.

Compositional changes during these processes do not depend on the initial and final states only, that is the system can follow different paths in its evolution. In multi-phase systems, different chemical reactions can take place simultaneously and the global chemical evolution can be rather complex. The evolution of the system, possibly towards the final state of chemical equilibrium, is conveniently described by means of a variable named "reaction progress" or "extent of reaction", which is usually indicated by the letter ξ and is expressed in moles (Prigogine, 1955; Denbigh, 1971; Helgeson, 1979 and references therein). If only one reaction takes place in the system under consideration, the relationship between reaction progress, change in the number of moles of a generic species i (e.g., solutes, primary phases destroyed and secondary phases produced), dn_i and stoichiometric coefficients, v_i (adimensional), is

$$dn_i = v_i d\xi \tag{7-10}$$

which gives, through integration

$$n_i = n_i^\circ + v_i \Delta\xi. \tag{7-11}$$

The derivative of equation (7-10) with respect to time is (see equation 6-2):

$$\frac{dn_i}{dt} = v_i \cdot \frac{d\xi}{dt} \tag{7-12}$$

Since the ratio $d\xi/dt$ represents the rate of the reaction, R (mol s^{-1}), equation (7-12) can be rewritten as follows:

$$\frac{dn_i}{dt} = v_i \cdot R, \tag{7-13}$$

which gives the following equation, through integration:

$$n_i = n_i^{\circ} + v_i\, R\, \Delta t, \tag{7-14}$$

under the hypothesis of constant R. If the considered reaction is the dissolution of a mineral phase, the rate of the reaction, R (mol s^{-1}), is related to the dissolution rate, r (mol m^{-2} s^{-1}), and to the reaction surface, A_s (m^2), by the simple relationship:

$$R = r\, A_s. \tag{7-15}$$

Equation (7-14) can be rewritten as

$$n_i = n_i^{\circ} + v_i\, r\, A_s\, \Delta t, \tag{7-16}$$

Equations (7-11) and (7-16) are very similar. The former describes the increase in the number of moles of component i, with respect to the initial state, based on the reaction progress, whereas the latter makes reference to the time scale. In the second case, it is necessary to define both the rate of dissolution of the solid phase and its reaction surface, as indicated by equation (7-16).

Equations (7-11) and (7-16) are the pivotal relations for reaction path modelling, a technique which has received considerable attention in geochemistry. The thermodynamic relationships describing the irreversible water–rock mass transfers were established by Helgeson (1968). PATHI, the first software code for reaction path modelling, was developed by Helgeson and coworkers, who presented several applications in excellent scientific papers (e.g., Helgeson et al., 1969, 1970). Among the other codes developed afterwards, the EQ3/6 Software Package (Wolery, 1979, 1992; Wolery and Daveler, 1992) is one of the most known and used.

7.2.2. An example of reaction path modelling: the dissolution of albite in pure water

As an example, let us take into account the reaction path in a system made up of pure high-albite [$NaAlSi_3O_8$] reacting with pure water. This is similar to the K-feldspar example

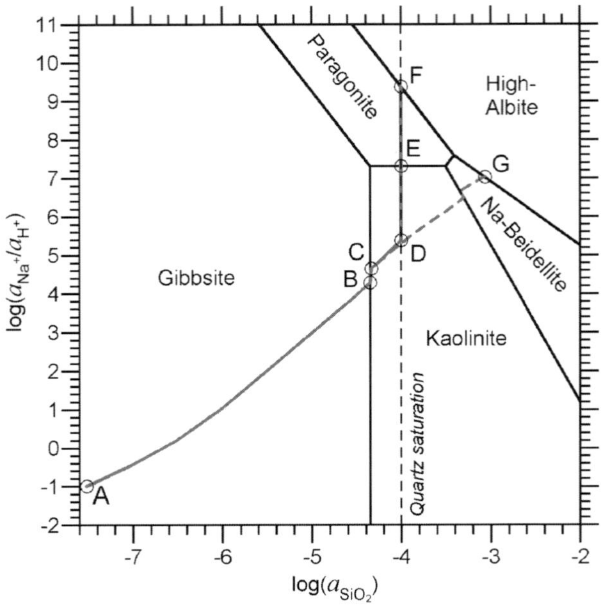

Figure 7.1. Activity plot for the system Na_2O–Al_2O_3–SiO_2–H_2O, showing two possible reaction paths for high-albite dissolution in pure water, at 25°C, 1 bar, one allowing quartz precipitation (solid grey line), the other preventing it (dashed grey line). Both reaction paths were simulated by means of EQ6.

proposed by Helgeson (1979) in one of his ground-breaking scientific papers. Initially albite and water are in disequilibrium. The activity plot for the system Na_2O–Al_2O_3–SiO_2–H_2O (Fig. 7.1) shows that the stability fields of kaolinite [$Al_2Si_2O_5(OH)_4$], Na-beidellite [$Na_{0.33}Al_{2.33}Si_{3.67}O_{10}(OH)_2$] and paragonite [$NaAl_3Si_3O_{10}(OH)_2$] are interposed between those of albite and gibbsite [$Al(OH)_3$].

The model assumes that albite dissolves congruently in pure water, at least in the initial stages, whereas albite dissolution becomes incongruent afterwards, upon attainment of saturation with respect to secondary solid phases. Speciation computations for the initial aqueous solution, containing Na, Al and Si in proportions 1:1:3 (i.e. those of albite), and in low concentrations, show that the dominant species in these initial conditions are Na^+, $Al(OH)_4^-$, $Al(OH)_3^\circ$ and $H_4SiO_4^\circ$. However, as dissolution of albite proceeds, the concentrations of other product species (e.g., $H_3SiO_4^-$, $NaH_3SiO_4^\circ$, $NaOH^\circ$, $NaAl(OH)_4^\circ$) increase too and have to be taken into account in calculations.

The ratios between the aqueous species produced by albite dissolution remain constant until the solution attains saturation with respect to solid product phases, that will precipitate during the process. This is the basic feature of the reaction path modelling. In order to recognize if the aqueous solution attains saturation with respect to a mineral phase, the solubility products of all the relevant minerals of the considered system have to be compared with the corresponding ion activity products after each increment of the reaction progress variable.

In the considered example, gibbsite is the first mineral with which the aqueous solution attains saturation, at point A of Fig. 7.1. The equilibrium between gibbsite and the aqueous solution is described by the following reaction:

$$Al(OH)_{3(s)} = Al^{3+} + 3OH^- \quad (7\text{-}18)$$

and the corresponding solubility product is

$$K_{gibbsite} = a_{Al^{3+}} \cdot a^3_{OH^-} . \quad (7\text{-}19)$$

During progressive albite dissolution, gibbsite continues to precipitate and the composition of the aqueous solution evolves along the path A–B in Fig. 7.1. During this process the concentration of dissolved silica (and Na^+) continues to increase until attainment, at point B, of saturation with respect to kaolinite, which becomes stable and precipitates. Coexistence of gibbsite and kaolinite buffers the activity of aqueous silica, as expressed by the reaction

$$2Al(OH)_{3(s)} + 2SiO_{2(aq)} \rightarrow Al_2Si_2O_5(OH)_{4(s)} + H_2O. \quad (7\text{-}20)$$

Therefore, further albite dissolution does not cause any increment in the concentration of dissolved silica, since this is consumed to convert the gibbsite, precipitated previously, in kaolinite. The Na^+ concentration continues to grow and the net result is path B–C in Fig. 7.1. At point C all the gibbsite has been consumed and the concentration of dissolved silica returns to increase with that of Na^+ following the path C–D along which kaolinite precipitates. At point D, quartz saturation is attained and this mineral starts to precipitate. Further albite dissolution moves the composition of the solution along the path D–E, along which the aqueous silica concentration is buffered by quartz saturation.

At point E the solution attains saturation with paragonite, which begins to precipitate. At point E, the aqueous solution is in equilibrium with quartz, paragonite and kaolinite. Since a maximum of four phases can coexist in a four components system (as stated above, our components are Na_2O, Al_2O_3, SiO_2, H_2O) under fixed T, P conditions (25°C, 1 bar in our case), the system has attained an invariant point. The composition of the solution cannot leave point E unless one of the three solids quartz, paragonite and kaolinite is dropped from the system. It must be underscored that albite does not have to be considered in the computation of the phases in equilibrium, since equilibrium between albite and the aqueous solution has not been attained yet and albite acts as supplier of solutes only. Therefore, the composition of the aqueous solution will remain at point E during further albite dissolution, while kaolinite reacts with the aqueous solution to form paragonite according to the reaction

$$1.5Al_2Si_2O_5(OH)_{4(s)} + Na^+ \rightarrow NaAl_3Si_3O_{10}(OH)_2 + 7.5H_2O + H^+. \quad (7\text{-}21)$$

Note that, again, minerals' coexistence buffers a parameter of the aqueous solution, this time the ratio between the activities of Na^+ and H^+.

When kaolinite has been totally consumed, further albite dissolution moves the composition of the solution along the path E–F, along which, again, the concentration of aqueous silica is fixed by quartz saturation. Finally, at point F, albite becomes stable, i.e. the condition of stable equilibrium between albite, paragonite, quartz and the aqueous solution is attained.

Note that, if quartz precipitation were prevented (for instance by kinetic constraints), the composition of the solution would evolve along path D–G. In this case, the condition of metastable equilibrium between albite, Na-beidellite and the aqueous solution would be attained.

The two reaction paths presented above and plotted in Fig. 7.1 simulate what happens if a limited amount of albite is dissolved in the aqueous solution and a constraint is then applied to prevent further dissolution and to allow attainment of equilibrium among all the chemical components. A second dissolution step is then allowed, followed by an equilibration step, and so on. After each step the system reaches a condition of metastable equilibrium for a very small progress of the irreversible reaction. The overall result is a series of metastable equilibrium states extending from the initial state of disequilibrium to the final state of stable equilibrium (point F) or metastable equilibrium (point G). The intermediate states are often named states of partial equilibrium, since the components in the aqueous solution are in mutual equilibrium but in disequilibrium with respect to albite.

7.2.3. The dissolution of albite in pure water: the numeric model

Referring to the albite example, albite dissolution accompanied by gibbsite precipitation is described by means of the following reaction, which includes the main aqueous species of interest:

$$\begin{aligned} \nu_{\mathrm{Alb}}\ \mathrm{NaAlSi_3O_{8(s)}} + \nu_{\mathrm{H_2O}}\ \mathrm{H_2O} &\rightarrow \nu_{\mathrm{Gibb}}\ \mathrm{Al(OH)_{3(s)}} + \nu_{\mathrm{Na^+}}\ \mathrm{Na^+} + \nu_{\mathrm{Al^{3+}}}\ \mathrm{Al^{3+}} \\ &+ \nu_{\mathrm{Al(OH)_3^o}}\ \mathrm{Al(OH)_3^o} + \nu_{\mathrm{Al(OH)_4^-}}\ \mathrm{Al(OH)_4^-} + \nu_{\mathrm{H_4SiO_4^o}}\ \mathrm{H_4SiO_4^o} \\ &+ \nu_{\mathrm{H_3SiO_4^-}}\ \mathrm{H_3SiO_4^-} + \nu_{\mathrm{H^+}}\ \mathrm{H^+} + \nu_{\mathrm{OH^-}}\ \mathrm{OH^-} \end{aligned} \tag{7-22}$$

Actually Al^{3+} ion is not a major dissolved species in the considered example but is retained as it is customarily used to describe the dissolution/precipitation of Al-bearing solid phases and the association/dissociation of aqueous Al-species.

Let us assume that the system contains 1,000 g of H_2O and that the reaction coefficient of albite, ν_{Alb}, is equal to 1. As already seen above, each ν_i represents the change in the number of moles of each species for an increment $d\xi$ of the reaction progress, as expressed by equation (7-10). Reactants have negative ν_i values, whereas the ν_i values of products are positive. In reaction (7-22) we could have erroneously considered as reactants some species that are actually products or vice versa. This is not important since the true situation comes out from computations. Apart from the ν_{Alb}, the other 10 ν_i are unknowns and depend on ξ. To compute these 10 unknowns a system of 10 equations is needed. Let us start to write the thermodynamic equilibrium constant for the five independent reactions between the species involved in reaction (7-22):

$$a_{OH^-} \cdot a_{H^+} = K_W \tag{7-23}$$

$$a_{Al^{3+}} \cdot a_{OH^-}^3 = K_{sp,gibbsite} \tag{7-24}$$

$$\frac{a_{Al^{3+}} \cdot a_{OH^-}^4}{a_{Al(OH)_4^-}} = K_{Al(OH)_4^-} \tag{7-25}$$

$$\frac{a_{Al^{3+}} \cdot a_{OH^-}^3}{a_{Al(OH)_3^o}} = K_{Al(OH)_3^o} \tag{7-26}$$

$$\frac{a_{H^+} \cdot a_{H_3SiO_4^-}}{a_{H_4SiO_4}} = K_{H_4SiO_4}. \tag{7-27}$$

Now we have to take the derivative of each equation with respect to ξ and rearrange the results. Let us do this exercise for equation (7-23). The derivative of this relation with respect to ξ is

$$a_{OH^-} \cdot \frac{da_{H^+}}{d\xi} + a_{H^+} \cdot \frac{da_{OH^-}}{d\xi} = 0. \tag{7-28}$$

Based on equation (7-23), we can substitute K_W/a_{H^+} for a_{OH^-} and K_W/a_{OH^-} for a_{H^+}, obtaining

$$\frac{K_W}{a_{H^+}} \cdot \frac{da_{H^+}}{d\xi} + \frac{K_W}{a_{OH^-}} \cdot \frac{da_{OH^-}}{d\xi} = 0, \tag{7-29}$$

which can be simplified as follows:

$$\frac{1}{a_{H^+}} \cdot \frac{da_{H^+}}{d\xi} + \frac{1}{a_{OH^-}} \cdot \frac{da_{OH^-}}{d\xi} = 0. \tag{7-30}$$

Considering that $m_i = a_i/\gamma_i$, equation (7-10) can be rewritten in the following form:

$$v_i = \frac{dm_i}{d\xi} = \frac{1}{\gamma_i} \cdot \frac{da_i}{d\xi}. \tag{7-31}$$

Based on this result and through suitable substitution in equation (7-30), we get

$$\frac{v_{H^+} \cdot \gamma_{H^+}}{a_{H^+}} + \frac{v_{OH^-} \cdot \gamma_{OH^-}}{a_{OH^-}} = 0 \tag{7-32}$$

or

$$\frac{v_{H^+}}{m_{H^+}} + \frac{v_{OH^-}}{m_{OH^-}} = 0 \tag{7-33}$$

Following the same line of reasoning adopted for equation (7-23), similar relationships are obtained from the expressions of the other four thermodynamic equilibrium constants, equations (7-24) to (7-27):

$$\frac{v_{Al^{3+}}}{m_{Al^{3+}}} + \frac{3 \cdot v_{OH^-}}{m_{OH^-}} = 0 \tag{7-34}$$

$$\frac{v_{Al^{3+}}}{m_{Al^{3+}}} + \frac{4 \cdot v_{OH^-}}{m_{OH^-}} - \frac{v_{Al(OH)_4^-}}{m_{Al(OH)_4^-}} = 0 \tag{7-35}$$

$$\frac{v_{Al^{3+}}}{m_{Al^{3+}}} + \frac{3 \cdot v_{OH^-}}{m_{OH^-}} - \frac{v_{Al(OH)_3^o}}{m_{Al(OH)_3^o}} = 0 \tag{7-36}$$

$$\frac{v_{H^+}}{m_{H^+}} + \frac{v_{H_3SiO_4^-}}{m_{H_3SiO_4^-}} - \frac{v_{H_4SiO_4}}{m_{H_4SiO_4}} = 0. \tag{7-37}$$

The other five needed equations are simple mass balances describing the mass exchanges of the five elements involved in reaction (7-22), with respect to albite, under closed system conditions:

$$\text{Na:} \quad v_{Na^+} = -v_{Alb} \tag{7-38}$$

$$\text{Al:} \quad v_{Al^{3+}} + v_{Al(OH)_3^o} + v_{Al(OH)_4^-} = -v_{Alb} \tag{7-39}$$

$$\text{Si:} \quad v_{H_4SiO_4} + v_{H_3SiO_4^-} = -3 \cdot v_{Alb} \tag{7-40}$$

$$\text{O:} \quad 3 \cdot v_{Al(OH)_3^o} + 4 \cdot v_{Al(OH)_4^-} + 4 \cdot v_{H_4SiO_4} + 4 \cdot v_{H_3SiO_4^-} + v_{H_2O} + v_{OH^-} = -8 \cdot v_{Alb} \tag{7-41}$$

$$\text{H:} \quad 3 \cdot v_{Al(OH)_3^o} + 4 \cdot v_{Al(OH)_4^-} + 4 \cdot v_{H_4SiO_4} + 3 \cdot v_{H_3SiO_4^-} + 2 \cdot v_{H_2O} + v_{H^+} + v_{OH^-} = 0. \tag{7-42}$$

Equations (7-33) to (7-42) make up the system of 10 equations which allows one to compute the 10 unknown v_i. The system is conveniently solved through matrix-algebra methods. The reaction progress calculations are initiated by assigning extremely small initial concentrations to the 10 aqueous species involved in reaction (7-22). The system is then solved for the 10 stoichiometric coefficients v_i. These coefficients are used to compute the new concentrations of all the species after a small reaction step $\Delta\xi$, through equation (7-11). It must be underscored that n_i in equation (7-11) stands for both the molality of aqueous species and the moles of solid phases precipitated from 1 kg of water. For each increment in the reaction progress variable, the speciation of the aqueous solution is computed to evaluate if the aqueous solution has attained saturation with respect to a new solid phase. When a new solid phase is precipitated, the system of equations is suitably modified to incorporate the new products and/or reactants. Since stoichiometric reaction coefficients change along the reaction path, small increments of ξ, e.g. 10^{-8} mol, have to be used. The best results are obtained modifying equation (7-11), that is including the first derivative of the stoichiometric coefficients with respect to the reaction progress variable:

$$\Delta m_i = v_i \cdot \Delta\xi + \frac{\mathrm{d}v_i}{\mathrm{d}\xi} \cdot \left(\frac{\Delta\xi}{2!}\right)^2 . \qquad (7\text{-}43)$$

This is simply a truncated Taylor's series in m_i. Equations describing the first derivative of the stoichiometric coefficients with respect to the reaction progress variable, $\mathrm{d}v_i/\mathrm{d}\xi$, are given by Helgeson (1979). These relationships are inserted in the system of equations and at each step of $\Delta\xi$ the system is solved for v_i and $\mathrm{d}v_i/\mathrm{d}\xi$. As already said, the reaction progress computation proceeds iteratively, using the concentrations m_i of each step to calculate the new reaction coefficients v_i. These, in turn, are used to compute the new concentrations, i.e. the m_i values pertinent to the subsequent step, and so on. The results obtained through reaction path modelling include the concentrations of all the aqueous species and the number of moles of all the minerals either produced or consumed, for each step of advancement of the reaction progress variable.

The EQ6 input file for albite dissolution is reported in Table 7.5. Results are summarized in Fig. 7.2. Letters A to F in Fig. 7.2 correspond to those in Fig. 7.1. It is clear that both the concentrations of aqueous species and the moles of product minerals vary of several orders of magnitude during the reaction path. The condition of final equilibrium is attained when the total Gibbs free energy of the system $G = \Sigma\mu_i\, n_i$ reaches the minimum value. The test concerning this condition is carried out at each step of ξ and, in the albite example, the minimization of G is achieved for $\xi = 0.01323$ mol.

7.2.4. The dissolution of albite in pure water: simulations in time frame

We have already underscored above the importance of equations (7-11) and (7-16) and their similarity. Equation (7-11) was used to simulate the dissolution of albite through a merely stoichiometric approach (i.e. in reaction progress frame) in the previous section.

TABLE 7.5

EQ6 input file for modeling the dissolution of albite in pure water at 25°C, 1 bar in stoichiometric mode (reaction progress frame). Quartz is allowed to form (required input formats are not fully respected).

```
USE the database SUP
The option switch IOPT1 is set to 0 to direct the code to compute the simulation in a reaction progress frame.
The option switch IOPT4 is set to 0 to ignore solid solutions.
The print option switch IOPR8 is set to 1 to direct the code to print a table of equilibrium gas fugacities at each
print point.
endit.
     nmodl1= 1                    nmodl2= 0
     tempc0= 2.50000E+01           jtemp= 0
        tk1= 0.00000E+00             tk2= 0.00000E+00          tk3= 0.00000E+00
     zistrt= 0.00000E+00           zimax= 1.00000E+00
      tstrt= 0.00000E+00          timemx= 0.00000E+00
            kstpmx=300             cplim= 0.00000E+00
     dzprnt= 0.00000E+00          dzprlg= 0.00000E+00        ksppmx=100
     dzplot= 0.10000E+00          dzpllg= 0.00000E+00        ksplmx= 10000
      ifile= 60
*                1     2     3     4     5     6     7     8     9    10
  iopt1-10 = 0         2     0     0     0     0     0     0     0     0
 iopt11-20 = 0         0     0     0     0     0     0     0     0     0
  iopr1-10 = 0         0     0     1     1     0    -1     1     0     0
 iopr11-20 = 0         0     0     0     0     0     0     0     0     0
  iodb1-10 = 0         0     0     0     0     0     0     0     0     0
 iodb11-20 = 0         0     0     0     0     0     0     0     0     0
      nxopt= 1
     option= all
     nxopex= 6
  exception = Gibbsite
  exception = Kaolinite
  exception = Quartz
  exception = Paragonite
  exception = Albite high
  exception = Beidellite-Na
       nffg= 0
       nrct= 1
*-----------------------------------------------------------------------------
   reactant= Albite high
      jcode= 0                    jreac = 0
       morr= 1.00000E+01          modr = 0.00000E+00
        nsk= 0                      sk = 0.00000E+00           fk= 0.00000E+00
        nrk= 1                    nrpk = 0
        rk1= 1.00000E+00           rk2 = 0.00000E+00          rk3= 0.00000E+00
*-----------------------------------------------------------------------------
     dlzidp= 0.00000E+00
      tolbt= 0.00000E+00          toldl= 0.00000E+00         tolx= 0.00000E+00
     tolsat= 0.00000E+00         tolsst= 0.00000E+00
     screw1= 0.00000E+00         screw2= 0.00000E+00       screw3= 0.00000E+00
     screw4= 0.00000E+00         screw5= 0.00000E+00       screw6= 0.00000E+00
     zklogu= 0.000               zklogl = 0.000             zkfac = 0.000
     dlzmx1= 0.00000E+00         dlzmx2= 0.00000E+00       nordlm = 0
```

TABLE 7.5 (*Continued*)

itermx= 0	ntrymx= 0	
npslmx= 0	nsslmx= 0	ioscan= 0

*

* pickup file written by EQ3NR, version 7.2b (R139)
* supported by EQLIB, version 7.2b (R168)
Pure water
endit.

tempci = 2.50000E+01		
nxmod = 0		
iopg1 = 0	iopg2 = 0	iopg3 = 0
iopg4 = 0	iopg5 = 0	iopg6 = 0
iopg7 = 0	iopg8 = 0	iopg9 = 0
iopg10 = 0		
kct = 5	ksq = 6	kmt = 6
kxt = 6	kdim = 6	kprs = 0

O	5.550896571687376E+01	
Al	9.999999999999892E–21	
H	1.110168703247900E+02	
Na	1.000000000000000E–20	
Si	9.999999999999974E–21	
Electr	−2.442121511592103E–23	
H_2O	H_2O	1.744358983526984E+00
Al^{+++}	Al^{+++}	-2.522092083305848E+01
H^+	H^+	-6.997388938332554E+00
Na^+	Na^+	-2.000000000692422E+01
SiO_2(aq)	SiO_2(aq)	-2.000048167543259E+01
O_2(g)	O_2(g)	-6.780000000000000E–01

The same process can be transposed in a time frame by using equation (7-16), which indicates that the rate of dissolution of albite, the rate of precipitation of solid product phases and the reaction surfaces of all minerals, reactant and products, have to be specified to accomplish this task.

The characterization of reaction surface areas of mineral phases is probably the highest difficulty of reaction path modelling, as recognized by many authors in water–rock interaction studies (e.g., Lichtner, 1996, 1998). In particular, the surface of contact between minerals and percolating meteoric waters in soils and rocks may deviate from the surface area under controlled mineral dissolution/precipitation experiments conducted in the laboratory. According to Appelo and Postma (1996), the estimation of mineral surface areas in field situation has hardly passed the stage of educated guessing at present. Marini et al. (2000) have modelled the chemistry of Bisagno Valley (Genoa, Italy) groundwaters, focussing on the main dissolved constituents, using an inverse approach, i.e. introducing the experimental kinetic rate constants of dissolving and precipitating mineral phases and solving for their surface areas. It turns out that computed surface areas of solid phases change substantially during progressive water–rock interaction and differ by several orders of magnitude from those obtained on the basis of modal mineralogy, grainsize and intergranular

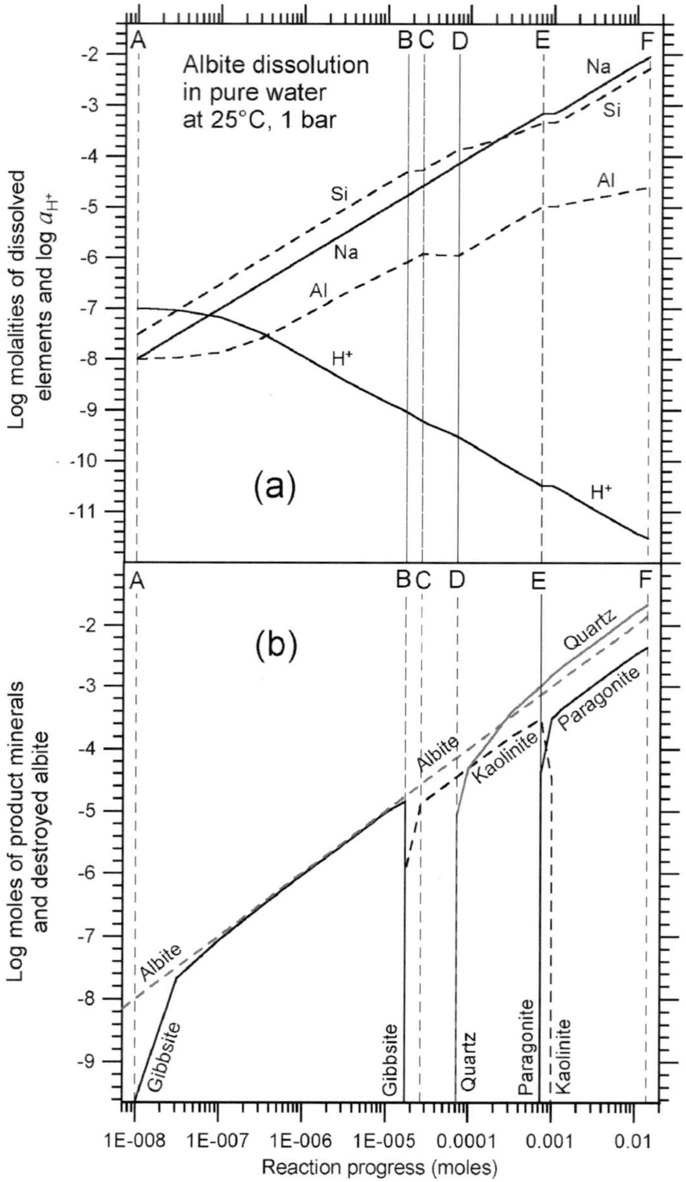

Figure 7.2. (a) Log-molalities of dissolved elements and log-activity of H^+ ion and (b) log-moles of secondary phases and of destroyed albite, as a function of the reaction progress variable, during the progressive dissolution of albite in pure water at 25°C, 1 bar, simulated by means of EQ6. Quartz is allowed to precipitate in the simulation. Letters A to F correspond to those in Figure 7.1.

porosity. Thus, even the relative surface areas do not appear to correlate with the modal abundance of minerals in the rocks.

In spite of these difficulties, it is instructive to simulate albite dissolution in kinetic mode, activating quartz precipitation. To keep these runs to a simple level, only the rate of dissolution of albite and its initial reaction surface area are specified in the EQ6 input files of Table 7.6, whereas product minerals are assumed to attain instantaneous equilibrium. This choice is dictated by the difficulty in defining the surface area for solid phases which are initially absent in the system.

In the first input file (see Table 7.6A), the dissolution of albite is described by means of a TST-based rate law, comprising the proton-, water- and hydroxyl-promoted mechanisms, whose rate constants and reaction orders are derived from Table 6.1. Note that the reaction orders in Table 6.1, +0.457 for the acid mechanism, 0.000 for the neutral mechanism and −0.572 for the basic mechanism, are all referred to proton activity, whereas in the EQ6 input file (Table 7.6) the reaction orders have to be referred to H^+, H_2O and OH^- activities, respectively, and consequently they become +0.457, +1.000 and +0.572. In addition, the rate constant for the basic mechanism given in Table 6.1 is multiplied times $K_W^{-0.572}$, to refer it to OH^- activity rather than to H^+ activity.

In the second input file (see Table 7.6B), the dissolution of albite is described by means of the rate law proposed by Oelkers et al. (1994), see equation (6-113). The purpose of this run is to show that this kind of rate laws can also be taken into account by the EQ6 code. Since equation (6-113) holds at pH 9 and 150°C, albite dissolution is simulated in a 1*N* solution of Na–HCO_3–CO_3 whose initial speciation, as computed by EQ3NR, is reported below.

Species	Molality	Log molality	Log gamma	Log activity
Na^+	9.9941E–01	−0.0003	−0.2462	−0.2464
HCO_3^-	5.9758E–01	−0.2236	−0.2313	−0.4549
CO_3^{--}	1.9873E–01	−0.7017	−0.9534	−1.6552
OH^-	4.3691E–03	−2.3596	−0.2840	−2.6436
CO_2(aq)	1.4647E–03	−2.8343	0.1157	−2.7185
NaOH(aq)	5.8880E–04	−3.2300	0.0000	−3.2300
O_2(aq)	1.4821E–04	−3.8291	0.1157	−3.7134
H^+	1.3149E–09	−8.8811	−0.1189	−9.0000

Carbonate equilibria and pH constrain CO_2 fugacity at $10^{-0.6728}$. This f_{CO_2} value is imposed on the system during the subsequent EQ6 simulation of albite dissolution, to maintain the pH constant at the desired value of 9.

As shown in Table 7.6B, the dissolution rate of albite is also described in this second case by means of a TST-based rate law, but contrary to the previous case, it includes one mechanism only, involving the activities of hydrogen ion and aluminium ion with exponents 1 and −1/3, respectively, as dictated by equation (6-113). The rate constant, expressed in mol cm^{-2} s^{-1}, is inserted in the input file as rk0.

TABLE 7.6A

EQ6 input file for modeling the dissolution of albite in pure water at 25°C, 1 bar in kinetic mode (time frame). Quartz is allowed to form (required input formats are not fully respected).

```
USE the database COM
The option switch IOPT1 is set to 1 to direct the code to compute the simulation in a TIME frame.
The option switch IOPT4 is set to 0 to ignore solid solutions.
The print option switch IOPR8 is set to 1 to direct the code to print a table of equilibrium gas fugacities at each
print point.
endit.
   nmodl1=1                    nmodl2=0
   tempc0=2.50000E+01           jtemp=0
      tk1=0.00000E+00             tk2=0.00000E+00         tk3=0.00000E+00
   zistrt=0.00000E+00          zimax=1.00000E-08
    tstrt=0.00000E+00         timemx=0.00000E+00
   kstpmx=300                  cplim=0.00000E+00
   dzprnt=0.00000E+00         dzprlg=0.00000E+00        ksppmx=100
   dzplot=0.10000E+00         dzpllg=0.00000E+00        ksplmx=10000
    ifile = 60
*          1      2      3      4      5      6      7      8      9     10
  iopt1-10=1      2      0      0      0      0      0      0      0      0
 iopt11-20=0      0      0      0      0      0      0      0      0      0
  iopr1-10=0      0      0      1      1      0     -1      1      0      0
 iopr11-20=0      0      0      0      0      0      0      0      0      0
  iodb1-10=0      0      0      0      0      0      0      0      0      0
 iodb11-20=0      0      0      0      0      0      0      0      0      0
     nxopt=1
    option=all
    nxopex=6
 exception=Gibbsite
 exception=Kaolinite
 exception=Quartz
 exception=Paragonite
 exception=Albite high
 exception=Beidellite-Na
      nffg=0
      nrct=1
*------------------------------------------------------------------------------
  reactant=Albite high
     jcode=0                   jreac=0
      morr=1.00000E+00          modr=0.00000E+00
       nsk=1                      sk=1000.00E+00         fk=0.00000E+00
       nrk=2                    nrpk=-1
     imech=3
       rk0=1.34900E-14          trk0=25.0000E+00         iact=0
      eact=0.00000E+00          hact=0.00000E+00
     ndact=1                  csigma=1.00000E+00
      udac=H+                   cdac=0.45700E+00
```

TABLE 7.6A (*Continued*)

```
rk0=9.12010E–17          trk0=25.0000E+00          iact=0
eact=0.00000E+00         hact=0.00000E+00
ndact=1                  csigma=1.00000E+00
udac=H2O                 cdac=1.00000E+00
rk0=1.06660E–13          trk0=25.0000E+00          iact=0
eact= 0.00000E+00        hact=0.00000E+00
ndact=1                  csigma=1.00000E+00
udac=OH-                 cdac=0.57200E+00
*_______________________________________________________________
dlzidp=0.00000E+00
tolbt=0.00000E+00        toldl=0.00000E+00         tolx=0.00000E+00
tolsat=0.00000E+00       tolsst=0.00000E+00
screw1=0.00000E+00       screw2=0.00000E+00        screw3=0.00000E+00
screw4=0.00000E+00       screw5=0.00000E+00        screw6=0.00000E+00
zklogu=0.000             zklogl=0.000              zkfac=0.000
dlzmx1=0.00000E+00       dlzmx2=0.00000E+00        nordlm=0
itermx=0                 ntrymx=0
npslmx=0                 nsslmx=0                  ioscan=0
*_______________________________________________________________
* pickup file written by EQ3NR, version 7.2b (R139)
* supported by EQLIB, version 7.2b (R168)
Pure water
endit.
tempci=2.50000E+01
nxmod=0
iopg1=0                  iopg2=0                   iopg3=0
iopg4=0                  iopg5=0                   iopg6=0
iopg7=0                  iopg8=0                   iopg9=0
iopg10=0
kct=5                    ksq=6                     kmt=6
kxt=6                    kdim=6                    kprs=0
O         5.550896571687376E+01
Al        9.999999999999892E–21
H         1.110168703247900E+02
Na        1.000000000000000E–20
Si        9.999999999999974E–21
electr   –2.442121511592103E–23
H2O       H2O                1.744358983526984E+00
Al+++     Al+++             –2.522092083305848E+01
H+        H+                –6.997388938332554E+00
Na+       Na+               –2.000000000692422E+01
SiO2(aq)  SiO2(aq)          –2.000048167543259E+01
O2(g)     O2(g)             –6.780000000000000E–01
```

TABLE 7.6B

EQ6 input file for modeling the dissolution of albite in a 1 N solution of Na-HCO_3–CO_3 (pH 9) at 150°C, 4.76 bar in kinetic mode (time frame). Quartz is allowed to form (required input formats are not fully respected).

```
USE the database COM
The option switch IOPT1 is set to 1 to direct the code to compute the simulation in a TIME frame.
The option switch IOPT4 is set to 0 to ignore solid solutions.
The print option switch IOPR8 is set to 1 to direct the code to print a table of equilibrium gas fugacities at each
print point.
endit.
    nmodl1= 1                          nmodl2= 0
    tempc0= 150.000E+00                 jtemp= 0
       tk1= 0.00000E+00                   tk2= 0.00000E+00        tk3= 0.00000E+00
    zistrt= 0.00000E+00                 zimax= 2.30400E-03
     tstrt= 0.00000E+00                timemx= 0.00000E+00
    kstpmx= 300                         cplim= 0.00000E+00
    dzprnt= 1.00000E-04                dzprlg= 0.00000E+00     ksppmx= 100
    dzplot= 0.10000E+00                dzpllg= 0.00000E+00     ksplmx= 10000
     ifile= 60
*            1     2     3     4     5     6     7     8     9    10
 iopt1-10= 1     2     0     0     0     0     0     0     0     0
iopt11-20= 0     0     0     0     0     0     0     0     0     0
 iopr1-10= 0     0     0     1     1     0    -1     1     0     0
iopr11-20= 0     0     0     0     0     0     0     0     0     0
 iodb1-10= 0     0     0     0     0     0     0     0     0     0
iodb11-20= 0     0     0     0     0     0     0     0     0     0
    nxopt= 1
   option= all
   nxopex= 6
exception= Gibbsite
exception= Kaolinite
exception= Quartz
exception= Paragonite
exception= Albite high
exception= Beidellite-Na
     nffg= 1
  species= CO2(g)
    moffg= 1.00000E+00                 xlkffg= -0.67280E+00
     nrct= 1
*----------------------------------------------------------------------------------
 reactant= Albite high
    jcode= 0                            jreac= 0
     morr= 1.00000E+00                   modr= 0.00000E+00
      nsk= 1                               sk= 1000.0E+00          fk= 0.00000E+00
      nrk= 2                             nrpk= -1
    imech= 1
      rk0= 2.90000E-12                   trk0= 150.000E+00       iact= 0
```

TABLE 7.6B (*Continued*)

```
      eact= 0.00000E+00                 hact= 0.00000E+00
     ndact= 2                         csigma= 1.00000E+00
      udac= H+                          cdac= 1.00000E+00
      udac= Al+++                       cdac= -0.33333E+00
*______________________________________________________________________
    dlzidp= 0.00000E+00
     tolbt= 0.00000E+00                toldl= 0.00000E+00        tolx= 0.00000E+00
    tolsat= 0.00000E+00               tolsst= 0.00000E+00
    screw1= 0.00000E+00               screw2= 0.00000E+00      screw3= 0.00000E+00
    screw4= 0.00000E+00               screw5= 0.00000E+00      screw6= 0.00000E+00
    zklogu= 0.000                     zklogl= 0.000             zkfac= 0.000
    dlzmx1= 0.00000E+00               dlzmx2= 0.00000E+00      nordlm= 0
    itermx= 0                         ntrymx= 0
    npslmx= 0                         nsslmx= 0                ioscan= 0
*______________________________________________________________________
* pickup file written by EQ3NR, version 7.2b (R139)
* supported by EQLIB, version 7.2b (R168)
Soluzione a pH 9, 150ØC
endit.
    tempci= 1.50000E+02
    nxmod= 0
     iopg1= 0                          iopg2= 0                 iopg3=  0
     iopg4= 0                          iopg5= 0                 iopg6= 0
     iopg7= 0                          iopg8= 0                 iopg9= 0
    iopg10= 0
       kct= 6                            ksq= 7                   kmt= 7
       kxt= 7                           kdim= 7                  kprs= 0
  O         5.790555475213318E+01
  Al        1.000000140335818E-20
  H         1.116194099975864E+02
  C         7.977766682638062E-01
  Na        1.000000000012290E+00
  Si        1.000000019167043E-20
  electr    1.690239705560677E-08
  H2O       H2O                     1.744358983526984E+00
  Al+++     Al+++                 - 4.121808857987895E+01
  H+        H+                    - 8.881099514946308E+00
  HCO3-     HCO3-                 - 2.236025491244946E-01
  Na+       Na+                   - 2.557881778253819E-04
  SiO2(aq)  SiO2(aq)              - 2.116808380506397E+01
  O2(g)     O2(g)                 - 6.780000000000000E-01
```

Some interesting results are summarized below:

ξ	Time days	pH	m_{Al}	m_C	m_{Na}	m_{SiO_2}	$a_{Al^{3+}}$	Affinity kcal×mol^{-1}	Rate mol×cm^{-2}×s^{-1}
0.00E+00	0.00E+00	9	*1.00E–20*	0.798	1.000	*1.00E–20*	*6.05E–42*	*−135.5676*	*4.94E–07*
1.00E–09	2.34E–11	9	1.00E–09	0.798	1.000	3.00E–09	6.05E–31	−47.6005	1.06E–10
3.16E–09	2.71E–07	9	3.16E–09	0.798	1.000	9.49E–09	1.91E–30	−43.7280	7.25E–11
1.00E–08	1.62E–06	9	1.00E–08	0.798	1.000	3.00E–08	6.05E–30	−39.8554	4.94E–11
3.16E–08	8.02E–06	9	3.16E–08	0.798	1.000	9.49E–08	1.91E–29	−35.9829	3.36E–11
1.00E–07	3.77E–05	9	1.00E–07	0.798	1.000	3.00E–07	6.05E–29	−32.1104	2.29E–11
3.16E–07	1.76E–04	9	3.16E–07	0.798	1.000	9.49E–07	1.91E–28	−28.2378	1.56E–11
1.00E–06	8.16E–04	9	1.00E–06	0.798	1.000	3.00E–06	6.05E–28	−24.3653	1.06E–11
3.16E–06	3.79E–03	9	3.16E–06	0.798	1.000	9.49E–06	1.91E–27	−20.4928	7.25E–12
1.00E–05	1.76E–02	9	1.00E–05	0.798	1.000	3.00E–05	6.05E–27	−16.6202	4.94E–12
3.16E–05	8.16E–02	8.9999	3.16E–05	0.798	1.000	9.49E–05	1.91E–26	−12.7475	3.36E–12
1.00E–04	3.79E–01	8.9999	1.00E–04	0.798	1.000	3.00E–04	6.05E–26	−8.8745	2.29E–12
3.16E–04	1.76E+00	8.9997	3.16E–04	0.797	1.000	9.49E–04	1.91E–25	−5.0006	1.56E–12
1.00E–03	8.80E+00	8.9992	1.00E–03	0.796	1.001	3.00E–03	6.10E–25	−1.1236	7.84E–13
1.10E–03	1.05E+01	8.9992	1.10E–03	0.796	1.001	3.31E–03	6.73E–25	−0.7916	6.28E–13
2.30E–03	1.34E+02	8.9990	6.89E–04	0.795	1.002	5.30E–03	4.21E–25	−0.0002	2.90E–16

The total molalities of Na and dissolved carbonates and pH remain nearly constant throughout the simulation, whereas the total molalities of dissolved Al and SiO_2 experience remarkable increases from "virtually zero" in the beginning to the final concentrations of 0.689 and 5.30 mmol kg^{-1}, respectively. This increment in total Al is accompanied by an increase in the activity of Al ion (of course the initial value of 6.05×10^{-42} is meaningless). Also the thermodynamic affinity changes substantially during the simulation attaining the equilibrium value ("virtually zero") in the last step, after 134 days.

Interestingly, the rate of albite dissolution experiences a substantial decrease (again, the initial value has no meaning) and this change is dictated by the increase in Al ion activity, apart from the decrease in rate for $A > -2$ kcal mol^{-1}. These findings are in marked contrast with the constant rate that would be predicted, at constant pH, by a TST-based rate law, comprising the proton-, water- and hydroxyl-promoted mechanisms, apart from the decrease in rate for $A > -2$ kcal mol^{-1}. These considerations underscore the importance of the approach by Oelkers, Schott and coworkers for a correct evaluation of the dissolution rate of silicate minerals. Unfortunately, this approach cannot be used extensively due to the present lack of experimental data.

The albite surface area specified in the EQ6 input files of Table 7.6, $A_{s,tot}$ = sk = 1,000, represents the area (in cm^2) in contact with (i.e. exposed to) 1,000 g of water. To understand the physical meaning of this parameter, let us consider an hypothetical system made up of N_s minerals spheres, all of the same radius r_s, and 1,000 g of water occupying the pore spaces (porosity is assumed to be 0.3), although natural systems can be different from this simple geometric analogue. The total volume of the mineral spheres, $V_{tot,s}$, can be computed based on the definition of porosity, η:

$$\eta = \frac{V_W}{V_W + V_{tot,s}}, \tag{7-44}$$

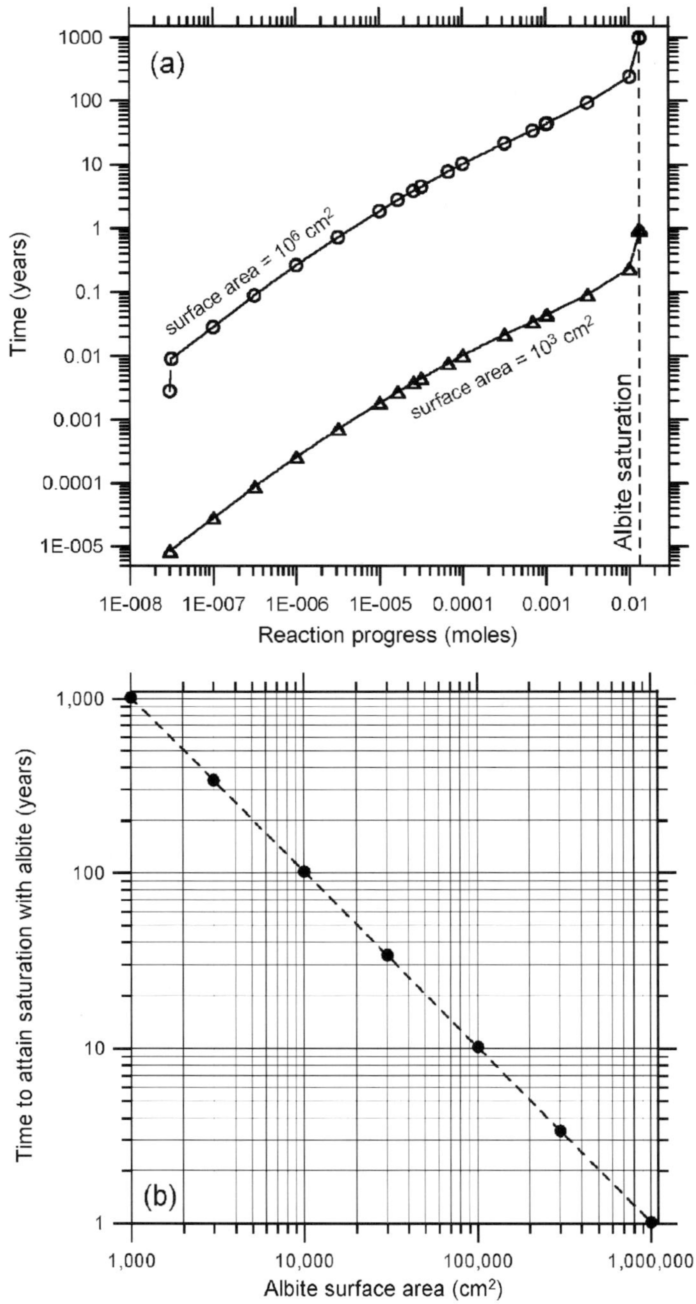

Figure 7.3. Progressive dissolution of albite in pure water at 25°C, 1 bar, simulated by means of EQ6. (a) Relations between time and reaction progress for two different values of the albite surface area and (b) Inverse correlation between the time needed to attain saturation with albite and the albite surface area.

where V_W identifies the pore volume occupied by water. For a selected radius r_s, the volume of a single mineral sphere V_s is calculated by means of equation (6-93), and the number of minerals spheres is $N_s = V_{tot,s}/V_s$. The surface of a single mineral sphere, A_s, is obtained by using equation (6-92) and the total surface area is $A_{s,tot} = A_s N_s$. These simple calculations indicate that the lowest value of the total surface area, 851 cm^2, is obtained for $N_s = 1$ (N_s cannot be less then 1!) and $r_s = 8.23$ cm. Both $A_{s,tot}$ and N_s increase as r_s decreases; for instance, a total surface area of 5,000 cm^2 pertains to a system comprising 203 spheres whose radius is 1.4 cm; $A_{s,tot}$ becomes 50,000 cm^2 for a system made up of 203,004 spheres whose r_s is 0.14 cm and attains 1,000,000 cm^2, for a system comprising 1.62×10^9 spheres whose radius is 0.007 cm.

During the EQ6 simulation, the surface area of albite can be kept constant or automatically changed by the code in proportion to its remaining mass. In the considered example, the difference between these two approaches is negligible, owing to the comparatively high mass of albite present in the system with respect to dissolved mass of albite.

In different EQ6 runs, the surface area of albite was varied from a minimum of 1,000 cm^2 up to a maximum of 1,000,000 cm^2, by changing accordingly the input file of Table 7.6a. It is important to underscore that these differences in the albite surface area do not change the results of the simulation (which are depicted in Fig. 7.2), apart from time. The reason for this is that the number of moles of any species, i.e. solutes, primary phases destroyed and secondary phases produced during the progressive dissolution of albite in pure water at 25°C, 1 bar are uniquely described by the reaction progress variable, as indicated by equation (7-11).

The relations between time and ξ are shown in Fig. 7.3a for the minimum and maximum values of the albite surface area introduced in the simulation. At the extreme right of this plot, a dashed-vertical line marks albite saturation, which is uniquely defined by $\xi = 0.01323$ mol. The different times needed to attain this condition for different values of the albite surface area are plotted in Fig. 7.3b as a function of the albite surface area itself. As expected, this plot confirms that there is a simple inverse linear relation between the logarithms of these two variables, namely, log $time_{sat}$ (years) $= -1.00 \times \log A_{s,albite}$ (cm^2) + 6.00. Therefore, uncertainties in the surface area of reacting minerals brings about corresponding uncertainties in the timing of the dissolution/precipitation process.

7.3. Reaction path modelling of geological CO_2 sequestration in ultramafic rocks

As discussed by Lackner et al. (1995, 1997), ultramafic rocks have the highest potential of CO_2 sequestration through mineral fixation, owing to their high concentrations of MgO and CaO. Ultramafic rocks are chiefly made up of four minerals, i.e. olivine, orthopyroxene, clinopyroxene and hornblende, accompanied by small amounts of biotite, garnet and spinel (Le Maitre, 2002). Ultramafic rocks are customarily divided in three broad groups, namely peridotites, pyroxenites and hornblendites, whose names evidently indicate the prevailing mineral. Peridotites contain more than 40% by volume of olivine and, based on the relative amounts of olivine (ol), orthopyroxene (opx) and clinopyroxene (cpx), they are further subdivided in dunite (ol > 90%), harzburgite (ol < 90%, cpx < 10%), wehrlite (ol < 90%, opx < 10%) and lherzolite (ol < 90%, cpx > 10%, opx > 10 %) (Le Maitre, 2002). Pyroxenites are further subdivided in clinopyroxenite, websterite, orthopyroxenite, olivine-clinopyroxenite, olivine-websterite and olivine-orthopyroxenite (see Le Maitre, 2002).

A dunite contains 49.5 wt% MgO and 0.3 wt% CaO only, a harzburgite is constituted by 45.4 wt% MgO and 0.7 wt% CaO, whereas a lherzolite is made up of 28.1 wt% MgO and 7.3 wt% CaO on average (Lackner et al., 1995). Ultramafic rocks are often affected by serpentinization and the resulting rocks, serpentinites, typically contain 40 wt% MgO and CaO close to 0 (Lackner et al., 1995).

Simple stoichiometric calculations shows that the amounts of ultramafic rocks required to bind chemically 1 kg of CO_2 through reaction with CaO and MgO are 1.8 kg of dunite, 2.0 kg of harzburgite, 2.7 kg of lherzolite and 2.3 kg of sepentinite (Lackner et al., 1995). Hence dunite has the highest potential of CO_2 sequestration among the ultramafic rocks.

Ultramafic rocks typically occur as components of ophiolite complexes, which are interpreted as slices of ancient oceanic crust and upper mantle which were tectonically uplifted and incorporated into the continental crust. Steinmann (1905) was probably the first one who recognized that ophiolites are igneous rocks emplaced on the ocean floor, an interpretation which was enriched by several details in subsequent years. Ultramafic rocks are relatively common at the earth surface and the total amount of MgO stored by these rocks far exceeds the worldwide coal reserves which are currently estimated to be ~10^{16} kg (Lackner et al., 1995, 1997).

7.3.1. Previous works

Xu et al. (2000, 2004) simulated the geological sequestration of CO_2 into different aquifers, including a dunite, by means of TOUGHREACT (Xu and Preuss, 1998). This code can be used to model reactive transport in a non-isothermal multi-phase system. In simple words, reaction path modelling is coupled with solute-transport modelling to simulate the reactive processes occurring into an aquifer. In TOUGHREACT the reaction and transport equations are solved separately. Consequently, TOUGHREACT can be used for reaction path modelling (or batch geochemical modelling to use the words of Xu and coworkers). Adopting this modelling feature, as Xu et al. (2000, 2004) did, TOUGHREACT is equivalent to EQ3/6.

The dunite case is of interest for the present discussion and is briefly presented here. Xu and coworkers considered an initial porosity of 5% for the dunite rock, which was assumed to be made up of forsterite (85.5% by volume) and fayalite (9.5%). Dissolution rates at 25°C were taken equal to 10^{-13} mol m^{-2} s^{-1} for forsterite and fayalite and precipitation rates at 25°C were fixed at 10^{-12} mol m^{-2} s^{-1} for all secondary non-carbonate minerals and 0.6×10^{-8} mole m^{-2} s^{-1} for magnesite and siderite. Apparent activation energy was considered to be 41.87 kJ mol^{-1} for magnesite and siderite and 62.76 kJ mol^{-1} for all other solid phases. Initial surface areas (in m^2 dm^{-3}) were fixed at 8.55 m^2 dm^{-3} for forsterite and 0.95 m^2 dm^{-3} for fayalite, which means to assume a total surface area of 10 m^2 dm^{-3} and to distribute it in proportion to the volume percentages of the two primary phases. Note that this total surface area value corresponds to a surface area of 2×10^6 cm^2 exposed to 1,000 g of water, owing to the relatively low porosity of the dunite rock.

In a background run, a dilute aqueous solution (with initial pH of 7 and initial Eh of −100 mV) was equilibrated with forsterite and fayalite at a temperature of 80°C. The aqueous solution thus attained stable pH and Eh values of 10.7 and −520 mV, respectively, through generation of H_2, owing to the strongly reducing conditions present in the system, and precipitation of magnetite, chrysotile and even elemental Fe. The production of these minerals would decrease porosity to 0.5% in 15,000 years.

In a distinct simulation, a CO_2 pressure of 260 bar was imposed and temperature was maintained at 80°C. This causes a remarkable decrease in pH down to 4.8 and an Eh increase up to 100 mV. Of course, under these more oxidising conditions (with respect to the background run), both H_2 generation and separation of metallic Fe do not occur anymore. The continuous supply of CO_2 to the system activates the dissolution of forsterite and fayalite and, consequently, the precipitation of magnesite and siderite and, to a lesser extent, of talc and amorphous silica as well. After ~1,000 years, ~100 kg of CO_2 are sequestered into 1 m^3 of the considered system (corresponding to approximately 2000 g kg^{-1} water) but these processes cause a decrease in porosity to 0.6%. Therefore, it seems likely that both rock alteration and geological CO_2 sequestration end owing to this reduction in porosity.

It must be underscored the prevailing reaction in the model of Xu and coworkers:

$$0.8\ \mathrm{Mg_2SiO_4} + \mathrm{CO_2} + 0.2\ \mathrm{H_2O} \rightarrow \mathrm{MgCO_3} + 0.2\ \mathrm{Mg_3Si_4O_{10}(OH)_2} \quad (7\text{-}45)$$

is characterized by a reaction volume for solids (*see* Section 5.5), ΔV_r of 20.2 cm^3. In contrast, if the dominant reaction would involve (i) a serpentine mineral instead of forsterite and (ii) chalcedony instead of talc, things might be different. This means to take into account, instead of reaction (7-45), the following reaction:

$$0.33\ \mathrm{Mg_3Si_2O_5(OH)_4} + \mathrm{CO_2} \rightarrow \mathrm{MgCO_3} + 0.67\ \mathrm{SiO_2} + 0.67\ \mathrm{H_2O}, \quad (7\text{-}46)$$

whose ΔV_r would be 7.0 cm^3 only or 11.2 cm^3, if amorphous silica were precipitated instead of chalcedony. Note that both reactions (7-45) and (7-46) involve 1 mol of CO_2 to obtain comparable results (*see* Section 5.5). Therefore, porosity being equal, a rock including a serpentine mineral as primary phase is probably able to sequester more CO_2 than a rock made up of forsteritic olivine.

Based on this expectation, the geological CO_2 sequestration in serpentinitic rocks was investigated by Cipolli et al. (2004), who modelled the irreversible mass exchanges taking place during high-pressure CO_2 injection into a deep aquifer hosted in these kind of rocks by means of the software package EQ3/6 version 7.2b.

Previous researches by Bruni et al. (2002) and Marini and Ottonello (2002) were devoted to understand the origin of the waters circulating in deep aquifers hosted in ultramafic rocks variably affected by serpentinization, which are representative of the initial (prior to CO_2 injection) aqueous solution. These waters have high pH, typically in the 11.3–11.9 range at the surface outlet, and unusual chemistry, with Ca^{2+} and OH^- as prevailing cation and anion, respectively. Other characteristics of these waters are: negative Eh (typically -460 to -520 mV), low Mg^{2+} concentrations (as low as 0.001 mg kg^{-1}), low total dissolved carbonate (generally between 0.5 and 17 mg kg^{-1} as HCO_3^-) and, consequently, very low P_{CO_2} values, from 10^{-8} to 10^{-11} bar (as already recalled in Section 5.1.4.1). Besides, Ca-OH waters are close to saturation with respect to calcite and brucite, undersaturated with forsterite and enstatite, and strongly oversaturated with respect to antigorite, chrysotile, talc and tremolite.

To understand the origin of Ca-OH waters, the dissolution of a local serpentinite was simulated by means of the EQ3/6 software package by Bruni et al. (2002) and Marini and Ottonello (2002). In this way, it was shown that Ca-OH compositions are produced through prolonged water–rock interaction in a system closed to CO_2, under strongly reducing conditions, in agreement with previous findings by Pfeifer (1977). This chemical evolution is due to C depletion in a system closed with respect to C sources. This C depletion was attributed to calcite precipitation by Bruni et al. (2002), although the possible reduction of carbonate-C to organic-C and ultimately to CH_4 might also be important. Further research is needed on this subject.

Since serpentinites are almost monomineralic rocks, stoichiometric serpentine [$Mg_3Si_2O_5(OH)_4$] was considered to be the only solid phase under dissolution by Cipolli et al. (2004). Simulation was carried out in kinetic mode, assuming constant values for both the dissolution rate of serpentine and its reactive surface area exposed to 1,000 g of water. The latter was set at 4,100 cm^2 based on geochemical evidence. Note that this value is approximately 500 times lower than that considered by Xu and coworkers in their dunite example.

7.3.2. Reaction path modelling of the geological CO_2 sequestration in a serpentinitic rock

7.3.2.1. Setting up the water–rock interaction model

Here, the geochemical model by Cipolli et al. (2004) will be extended to include other two chemical components, Al and Fe. Although serpentinites are almost monomineralic rocks, they main contain significant amounts of Fe oxides (7.9 ± 1.4 wt% as Fe_2O_3, based on the data summarized by Bruni et al., 2002) and Al_2O_3 (2.1 ± 1.0 wt%), which follow SiO_2 (40.6 ± 1.1 wt%), MgO (35.4 ± 1.4 wt%) and H_2O (12.4 ± 1.6 wt%), in order of decreasing importance. Assuming that Fe and Al are chiefly present as magnetite and kaolinite, respectively, simple mass balances shows that the molar amounts of serpentine, magnetite and kaolinite for a typical serpentinitic rock (Table 7.7) are 0.9116, 0.0282 and 0.0602, respectively. The corresponding volume fractions are 0.9317, 0.0118 and 0.0565, respectively. Accepting that the total surface area in contact with 1,000 g of water is 4,100 cm^2, as hypothesized by Cipolli et al. (2004), the initial surface areas of serpentine, magnetite and kaolinite are 3,820, 48.51 and 231.5 cm^2, respectively. Assuming that initial porosity is 0.3, geometric calculations (*see* Section 7.2.4) indicate that the initial amounts of serpentine, magnetite and kaolinite in the considered system are 20.04, 0.62 and 1.32 mol, respectively.

TABLE 7.7

Whole-rock chemical analysis of a typical serpentinitic rock from Monte Roccaprebalza (Dinelli et al., 1997) and its computed mineralogic composition

Oxide	Wt%	Element	Mol%
SiO_2	40.14	Si	10.29
TiO_2	0.08	Ti	0.02
Al_2O_3	2.20	Al	0.66
$Fe_2O_{3,tot}$	5.88	Fe	1.13
MnO	0.12	Mn	0.03
MgO	37.31	Mg	14.25
CaO	0.18	Ca	0.05
Na_2O	0.03	Na	0.01
K_2O	0.05	K	0.02
H_2O	14.00	O	49.61
		H	23.93

Minerals	Mol %	Vol %	Initial moles	Initial surfaces
Serpentine	91.16	93.17	20.04	3,820.10
Magnetite	2.82	1.18	0.62	48.51
Kaolinite	6.02	5.65	1.32	231.54

Again, following Cipolli et al. (2004), aquifer temperature was assumed to be 60°C and the aqueous solution hosted in the serpentinitic aquifer was hypothesized to be represented by the local Ca-OH spring water BR2. Heating of this aqueous solution from the emergence temperature, 20.3°C, to the aquifer temperature, 60°C ,was simulated by means of EQ6. This process does not cause any change in the total concentrations of solute species, whereas pH decreases from the outlet value, 11.73, to 10.6 and the logarithm of f_{O_2} increases from −73.6 to −67.9. Results of the heating simulation were used to prepare the EQ3NR input file of Table 7.8. Note that (i) the redox potential is assumed to be fixed by the SO_4^{2-}/HS^- redox couple and (ii) negligible molal concentrations, 10^{-20}, are hypothesized for dissolved Al and Fe, whose analytical concentrations are both below detection limits, to activate these two chemical components.

Injection of pure CO_2 at 100 bar pressure was hypothesized. To this purpose, a constant f_{CO_2} of 67.9 bar was set in EQ6 input file (*see* Table 7.9). This f_{CO_2} value is obtained considering that Γ_{CO_2} is equal to 0.679 at 60°C and 100 bar total pressure (*see* Section 3.4). The dissolution of the serpentinitic rock upon continuous CO_2 injection was simulated in kinetic mode, specifying both the dissolution rates at 60°C and the initial surface areas (see above) of the three solid reactants in the EQ6 input file.

The dissolution kinetics of reactant minerals was described by means of the TST rate law, involving the proton- and water-promoted mechanisms, whereas the hydroxyl-promoted dissolution mechanism was neglected, owing to the low pH values attained by the aqueous solution upon CO_2 injection (see below). Owing to the lack of dissolution rate parameters for the proton-promoted mechanism of chrysotile, those of lizardite were assumed to be representative of chrysotile. Dissolution rates at 60°C were computed by

TABLE 7.8

EQ3NR input file to compute the speciation/saturation state of the aqueous solution assumed to be hosted in the serpentinitic reservoir prior to CO_2 injection. This water composition was obtained through heating of spring water BR2 at 60°C (see Cipolli et al., 2004). (Required input formats are not fully respected.)

. . . omissis
endit.

tempc= 60.0000E+00
rho= 1.00000E+00 tdspkg= 0.00000E+00 tdspl= 0.00000E+00
fep= 0.00000E+00 uredox= HS-
tolbt= 0.00000E+00 toldl= 0.00000E+00 tolsat= 0.00000E+00
itermx= 0

*	1	2	3	4	5	6	7	8	9	10
iopt1–10=	1	0	0	0	0	0	0	0	0	0
iopg1–10=	0	0	0	0	0	0	0	0	0	0
iopr1–10=	0	0	0	1	0	0	1	−1	0	0
iopr11–20=	0	0	0	0	0	0	0	0	0	0
iodb1–10=	0	0	0	0	0	0	0	0	0	0

uebal=
nxmod= 18
uxmod= $S_2O_3^{--}$
0 −1
uxmod= HSO_3^-
0 −1
uxmod= SO_2(aq)
0 −1
uxmod= SO_3^{--}
0 −1
uxmod= $S_4O_6^{--}$
0 −1
uxmod= S_5^{--}
0 −1
uxmod= S_4^{--}
0 −1
uxmod= HSO_5^-
0 −1
uxmod= S_2^{--}
0 −1
uxmod= CO(aq)
0 −1
uxmod= Formaldehyde(aq)
0 −1
uxmod= Methanol(aq)
0 −1
uxmod= Acetic acid(aq)
0 −1
uxmod= Acetate
0 −1
uxmod= Formic acid(aq)
0 −1

TABLE 7.8 (*Continued*)

```
    uxmod= Formate
              0                          -1
    uxmod= Ethane(aq)
              0                          -1
    uxmod= Methane(aq)
              0                          -1
data file master species=H+
      switch with species=
      jflag=16    csp=-10.6081
data file master species=Ca++
      switch with species=
      jflag= 3    csp= 61.8
data file master species= Mg++
      switch with species=
      jflag= 3    csp= 0.001
data file master species= Na+
      switch with species=
      jflag= 3    csp= 41.1
data file master species= K+
      switch with species=
      jflag= 3    csp= 5.51
data file master species= HCO3-
      switch with species=
      jflag= 3    csp= 0.55
data file master species= SO4--
      switch with species=
      jflag= 3    csp= 0.3
data file master species= Cl-
      switch with species=
      jflag= 3    csp= 30.47
data file master species= SiO2(aq)
      switch with species=
      jflag= 3    csp= 3.1
data file master species= HS-
      switch with species=
      jflag= 3    csp= 0.22
data file master species= B(OH)3(aq)
      switch with species=
      jflag= 3    csp= 0.57
data file master species= Fe++
      switch with species=
      jflag= 0    csp= 1.00E-20
data file master species= Al+++
      switch with species=
      jflag= 0    csp= 1.00E-20
endit.
```

TABLE 7.9

EQ6 input file for modeling high-pressure ($P_{tot} = P_{CO2} = 100$ bar) CO_2 injection in a deep aquifer hosted in serpentinitic rocks at 60°C in kinetic mode (time frame). (Required input formats are not fully respected.)

```
The option switch IOPT1 is set to 1 to direct the code to compute the simulation in time frame
The print option switch IOPR8 is set to 1 to direct the code to print a table of equilibrium gas fugacities at each
print point.
endit.
   nmodl1= 1                    nmodl2= 0
   tempc0= 60.0000E+00           jtemp= 0
      tk1= 0.00000E+00             tk2= 0.00000E+00       tk3= 0.00000E+00
   zistrt= 0.00000E+00           zimax= 1.00000E-06
     tstrt= 0.00000E+00         timemx= 0.00000E+00
   kstpmx= 500                   cplim= 0.00000E+00
   dzprnt= 1.00000E+00          dzprlg= 0.00000E+00    ksppmx= 100
   dzplot= 0.10000E+00          dzpllg= 0.00000E+00    ksplmx= 10000
    ifile= 60
*              1    2    3    4    5    6    7    8    9   10
  iopt1-10=    1    0    0    0    0    0    0    0    0    0
 iopt11-20=    0    0    0    0    0    0    0    0    0    0
  iopr1-10=    0    0    0    0    1    0    0    1    0    0
 iopr11-20=    0    0    0    0    0    0    0    0    0    0
  iodb1-10=    0    0    0    0    0    0    0    0    0    0
 iodb11-20=    0    0    0    0    0    0    0    0    0    0
*________________________________________________________________
    nxopt= 1
   option= all
   nxopex= 15
exception= Gibbsite
exception= Kaolinite
exception= Chrysotile
exception= Sepiolite
exception= Hydromagnesite
exception= Nesquehonite
exception= Magnesite
exception= Calcite
exception= Dolomite
exception= Chalcedony
exception= Magnetite
exception= Siderite
exception= Pyrite
exception= Goethite
exception= Dawsonite
*________________________________________________________________
     nffg= 1
  species= CO2(g)
    moffg= 1.00000E+01          xlkffg= +1.83187E+00
*________________________________________________________________
     nrct= 3
*________________________________________________________________
 reactant= Chrysotile
    jcode= 0                     jreac= 0
```

TABLE 7.9 (*Continued*)

morr= 20.0400E+00	modr= 0.00000E+00	
nsk= 1	sk= 3,820.000E+00	fk= 0.00000E+00
nrk= 2	nrpk= 0	
imech= 2		
rk0= 4.89000E–09	trk0= 6.00000E+01	iact= 0
eact= 0.00000E+00	hact= 0.00000E+00	
ndact= 1	csigma= 1.00000E+00	
udac= H^+	cdac= 0.80000E+00	
rk0= 2.25000E–15	trk0= 6.00000E+01	iact= 0
eact= 0.00000E+00	hact= 0.00000E+00	
ndact= 1	csigma= 1.00000E+00	
udac= H_2O	cdac= 1.00000E+00	
*		
reactant= Magnetite		
jcode= 0	jreac= 0	
morr= 0.62000E+00	modr= 0.00000E+00	
nsk= 1	sk= 48.5000E+00	fk= 0.00000E+00
nrk= 2	nrpk= 0	
imech= 2		
rk0= 5.65000E–13	trk0= 5.00000E+01	iact= 0
eact= 0.00000E+00	hact= 0.00000E+00	
ndact= 1	csigma= 1.00000E+00	
udac= H^+	cdac= 0.27900E+00	
rk0= 3.65000E–15	trk0= 5.00000E+01	iact= 0
eact= 0.00000E+00	hact= 0.00000E+00	
ndact= 1	csigma= 1.00000E+00	
udac= H_2O	cdac= 1.00000E+00	
*		
reactant= Kaolinite		
jcode= 0	jreac= 0	
morr= 1.32000E+00	modr= 0.00000E+00	
nsk= 1	sk= 231.500E+00	fk= 0.00000E+00
nrk= 2	nrpk= 0	
imech= 2		
rk0= 8.00000E–15	trk0= 6.00000E+01	iact= 0
eact= 0.00000E+00	hact= 0.00000E+00	
ndact= 1	csigma= 1.00000E+00	
udac= H^+	cdac= 0.77700E+00	
rk0= 1.69000E–17	trk0= 6.00000E+01	iact= 0
eact= 0.00000E+00	hact= 0.00000E+00	
ndact= 1	csigma= 1.00000E+00	
udac= H_2O	cdac= 1.00000E+00	
*		
dlzidp= 0.00000E+00		
tolbt= 0.00000E+00	toldl= 0.00000E+00	tolx= 0.00000E+00
tolsat= 0.00000E+00	tolsst= 0.00000E+00	
screw1= 0.00000E+00	screw2= 0.00000E+00	screw3= 0.00000E+00
screw4= 0.00000E+00	screw5= 0.00000E+00	screw6= 0.00000E+00
zklogu= 0.000	zklogl= 0.000	zkfac= 0.000
dlzmx1= 0.00000E+00	dlzmx2= 0.00000E+00	nordlm= 0

TABLE 7.9 (*Continued*)

itermx= 0	ntrymx= 0	
npslmx= 0	nsslmx= 0	ioscan= 0
*		
* pickup file written by EQ3NR, version 7.2b (R139)		
* supported by EQLIB, version 7.2b (R168)		
Spring water BR2 heated at 60°C		
endit.		
tempci= 6.00000E+01		
nxmod= 18		
species= $S_2O_3^{--}$		
type= 0	option= −1	xlkmod= 0.00000E+00
species= HSO_3^-		
type= 0	option= −1	xlkmod= 0.00000E+00
species= SO_2(aq)		
type= 0	option= −1	xlkmod= 0.00000E+00
species= SO_3^{--}		
type= 0	option= −1	xlkmod= 0.00000E+00
species= $S_4O_6^{--}$		
type= 0	option= −1	xlkmod= 0.00000E+00
species= S_5^{--}		
type= 0	option= −1	xlkmod= 0.00000E+00
species= S_4^{--}		
type= 0	option= −1	xlkmod= 0.00000E+00
species= HSO_5^-		
type= 0	option= −1	xlkmod= 0.00000E+00
species= S_2^{--}		
type= 0	option= −1	xlkmod= 0.00000E+00
species= CO(aq)		
type= 0	option= −1	xlkmod= 0.00000E+00
species= Formaldehyde(aq)		
type= 0	option= −1	xlkmod= 0.00000E+00
species= Methanol(aq)		
type= 0	option= −1	xlkmod= 0.00000E+00
species= Acetic acid(aq)		
type= 0	option= −1	xlkmod= 0.00000E+00
species=Acetate		
type= 0	option= −1	xlkmod= 0.00000E+00
species= Formic acid(aq)		
type= 0	option= −1	xlkmod= 0.00000E+00
species= Formate		
type= 0	option= −1	xlkmod= 0.00000E+00
species=Ethane(aq)		
type= 0	option= −1	xlkmod= 0.00000E+00
species=Methane(aq)		
type= 0	option= −1	xlkmod= 0.00000E+00
iopg1= 0	iopg2= 0	iopg3= 0
iopg4= 0	iopg5= 0	iopg6= 0
iopg7= 0	iopg8= 0	iopg9= 0
iopg10= 0		
kct= 13	ksq= 14	kmt= 14
kxt= 14	kdim= 14	kprs= 0

TABLE 7.9 (*Continued*)

O	5.551282557733262E+01	
Al	1.000000000335209E–20	
B	9.218375990262418E–06	
Ca	1.541993186674240E–03	
Cl	8.594550329665458E–04	
Fe	1.000000000144592E–20	
H	1.110211073777142E+02	
C	9.013861125756066E–06	
K	1.409268500602602E–04	
Mg	4.114379952680602E–08	
Na	1.787751767510004E–03	
S	9.774694209498928E–06	
Si	5.159417928970046E–05	
electr	−1.150065192877400E–04	
H_2O	H_2O	1.744358983526984E+00
Al^{+++}	Al^{+++}	– 4.257641064725003E+01
$B(OH)_3(aq)$	$B(OH)_3(aq)$	– 6.725193420264618E+00
Ca^{++}	Ca^{++}	– 2.815981346143646E+00
Cl^-	Cl^-	– 3.066044127812017E+00
Fe^{++}	Fe^{++}	– 2.130818472540851E+01
H^+	H^+	– 1.057158162952603E+01
HCO_3^-	HCO_3^-	– 6.323495220189614E+00
K^+	K^+	– 3.851088497828405E+00
Mg^{++}	Mg^{++}	– 7.387095228892170E+00
Na^+	Na^+	– 2.748464060961117E+00
SO_4^{--}	SO_4^{--}	– 5.564691145737220E+00
$SiO_2(aq)$	$SiO_2(aq)$	– 5.509600627218712E+00
$O_2(g)$	$O_2(g)$	– 6.354977441383422E+01

means of equation (6-26), based on the pre-exponential factors, A_i, and apparent activation energies, E_i, given in Table 6.1. Again, note that the reaction order of the water-promoted mechanism was set to +1.000 in the EQ6 input file (Table 7.9). The EQ6 code was instructed to change the surface areas of primary minerals during the simulation in proportion to their remaining masses.

Instantaneous partial equilibrium was assumed for precipitating, secondary solid phases. Again, this choice is due to the difficulty in defining the surface area for minerals which are initially absent in the considered system. The hypothesis of instantaneous equilibrium for secondary minerals means to assume that the dissolution of reactants is the rate limiting step of the overall process. Results are presented against both the reaction progress variable, ξ, and time in the plots of Figs. 7.4 to 7.7.

7.3.2.2. Solid reactants

The amounts of reactant minerals destroyed during continuous CO_2 injection are shown as a function of time and reaction progress in Fig. 7.4. The solid phase contributing most chemical components to the considered system through its destruction is chrysotile. This reflects not only its high abundance and surface area (see Table 7.7) but

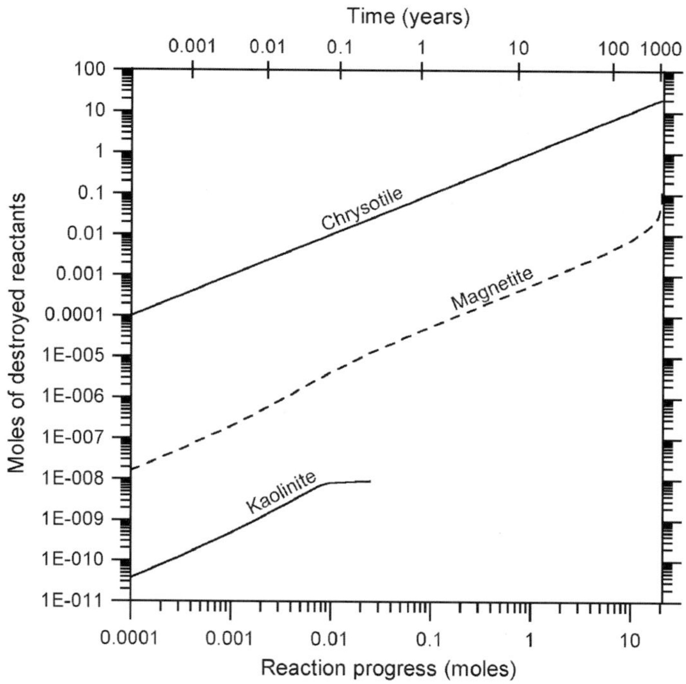

Figure 7.4. Moles of reactant minerals destroyed during high-pressure ($P_{tot} = P_{CO_2} = 100$ bar) CO_2 injection in a deep aquifer ($T = 60°C$) hosted in serpentinitic rocks as a function of time and the reaction progress variable.

also the comparatively high dissolution rate (in mol cm^{-2} s^{-1}) of serpentine minerals in acidic solutions (e.g., lizardite in Table 6.1, since the dissolution rate of chrysotile in acidic solutions was never investigated, as already recalled).

Although kaolinite is much more abundant than magnetite (see Table 7.7), the moles of magnetite destroyed are over two orders of magnitude higher than those of kaolinite, owing to the higher dissolution rate (in mol cm^{-2} s^{-1}) of magnetite with respect to kaolinite (see Table 6.1).

The aqueous solution remains strongly undersaturated with respect to chrysotile ($A < -16$ kcal mol^{-1}) and magnetite ($A < -3.4$ kcal mol^{-1}) throughout the process. For these low values of A, the dissolution rate is virtually independent of thermodynamic affinity (see Section 6.4.2) and is a function of pH only.

In contrast, the aqueous solution saturates with kaolinite for ξ of 0.025 mol. Before attainment of this condition the thermodynamic affinity with respect to kaolinite increases significantly, determining a substantial decrease in the dissolution rate. For instance, the dissolution rate of kaolinite is 1.90×10^{-17} mol cm^{-2} s^{-1} at ξ of 3.16 mmol, where A is -3.59 kcal mol^{-1} and pH is 4.49, but it reduces to 3.00×10^{-18} mol cm^{-2} s^{-1} at ξ of 10 mmol, where A is -0.12 kcal mol^{-1} and pH is 4.85. Upon attainment of equilibrium with kaolinite, this phase was dropped from the considered primary paragenesis.

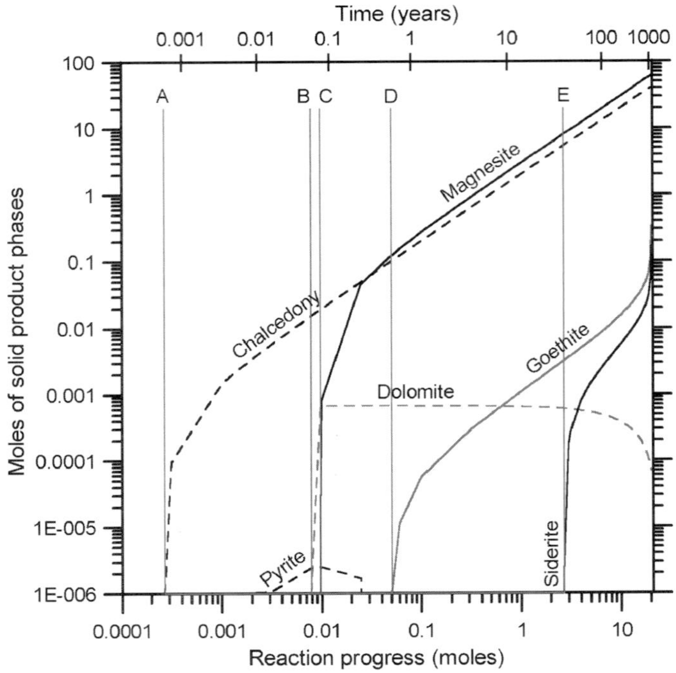

Figure 7.5. Moles of product minerals formed during high-pressure ($P_{tot} = P_{CO_2} = 100$ bar) CO_2 injection in a deep aquifer ($T = 60°C$) hosted in serpentinitic rocks as a function of time and the reaction progress variable. Note that letters A to E mark the onset of the precipitation of different secondary minerals.

Chrysotile is exhausted after 4,300 years, for ξ of 20.04 mol, which corresponds to the initial amount of this mineral in the considered serpentinitic rock. The simulation ends for ξ of 20.66 mol, upon exhaustion of magnetite.

7.3.2.3. Solid product phases

The solid phases which are considered to be produced through high-pressure ($P_{tot} = P_{CO_2} = 100$ bar) CO_2 injection in a deep aquifer hosted in serpentinitic rocks are, in order of appearance, chalcedony, pyrite, dolomite, magnesite, goethite and siderite (Fig. 7.5). However, pyrite is produced in very small amounts (< 3 μmol) and has ephemeral existence; dolomite never exceeds 0.001 mol; goethite and siderite are always ≤ 0.1 mol. As already recognized by Cipolli et al. (2004), only chalcedony and magnesite are produced in significant quantities and for $\xi > 0.05$ mol the stoichiometry of the process is adequately described by reaction (7-46). In other words, 3 mol of magnesite and 2 mol of chalcedony are produced for each mole of serpentine dissolved. The masses of magnesite and chalcedony increase almost linearly with time (not shown) and the magnesite mass is 2.32 mol kg^{-1} water (196 g) after 10 years and 10.4 mol kg^{-1} water (874 g) after 50 years, and so on, indicating the high CO_2 sequestration capacity of this process. A time of tens of years is apparently large, but it is actually small if compared with the residence times of high-pH Ca-OH waters in deep aquifers, which is estimated to be in the order of 100 to 10,000 years (Cipolli et al., 2004).

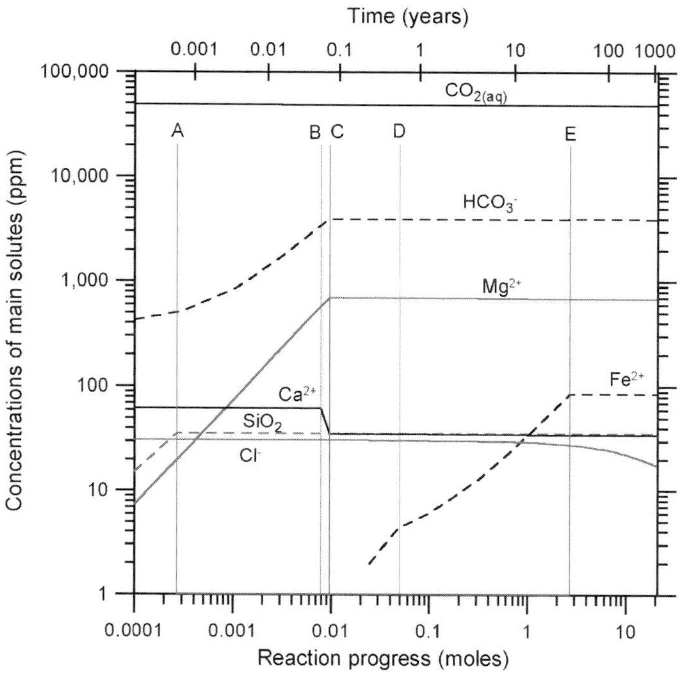

Figure 7.6. Concentrations of solutes during high-pressure ($P_{tot} = P_{CO_2} = 100$ bar) CO_2 injection in a deep aquifer ($T = 60°C$) hosted in serpentinitic rocks as a function of time and the reaction progress variable. Note that letters A to E marks the onset of the precipitation of different secondary minerals as shown in Figure 7.5.

7.3.2.4. The aqueous solution

Throughout continuous CO_2 injection, the main solute is aqueous CO_2, whose concentration is fixed at 1.1 mol kg^{-1} (48,400 ppm) by constant f_{CO_2}, under specified temperature–pressure conditions (Fig. 7.6). Aqueous CO_2 is followed by HCO_3^- and Mg^{2+}, whose concentrations, after an initial increase (for $\xi < 0.01$), stabilize at 0.062 and 0.028 mol kg^{-1}, respectively, when magnesite begins to form at point C. Note that at this point also Ca^{2+}concentration is fixed by saturation with dolomite which is attained at point B.

Also plotted in Fig. 7.6 is chloride, the most typical mobile solute, whose concentration experiences a significant decrease for high values of ξ. This effect is caused by a substantial increase in the mass of water, which is provided to the considered system by the dissolution of chrysotile, a H_2O-containing mineral. This effect is exactly the opposite of dry-up, a phenomenon due to incorporation of water in hydrous minerals (see Reed, 1997).

During the considered process, the pH of the aqueous solution changes from 3.83 to 4.85 (Fig. 7.7a). Most of this pH increase occurs between the onset of chalcedony precipitation (point A) and the beginning of magnesite precipitation (point C), after which pH is buffered by carbonate equilibria, similar to what was discussed for the aqueous solution/calcite example in Section 7.1.1.

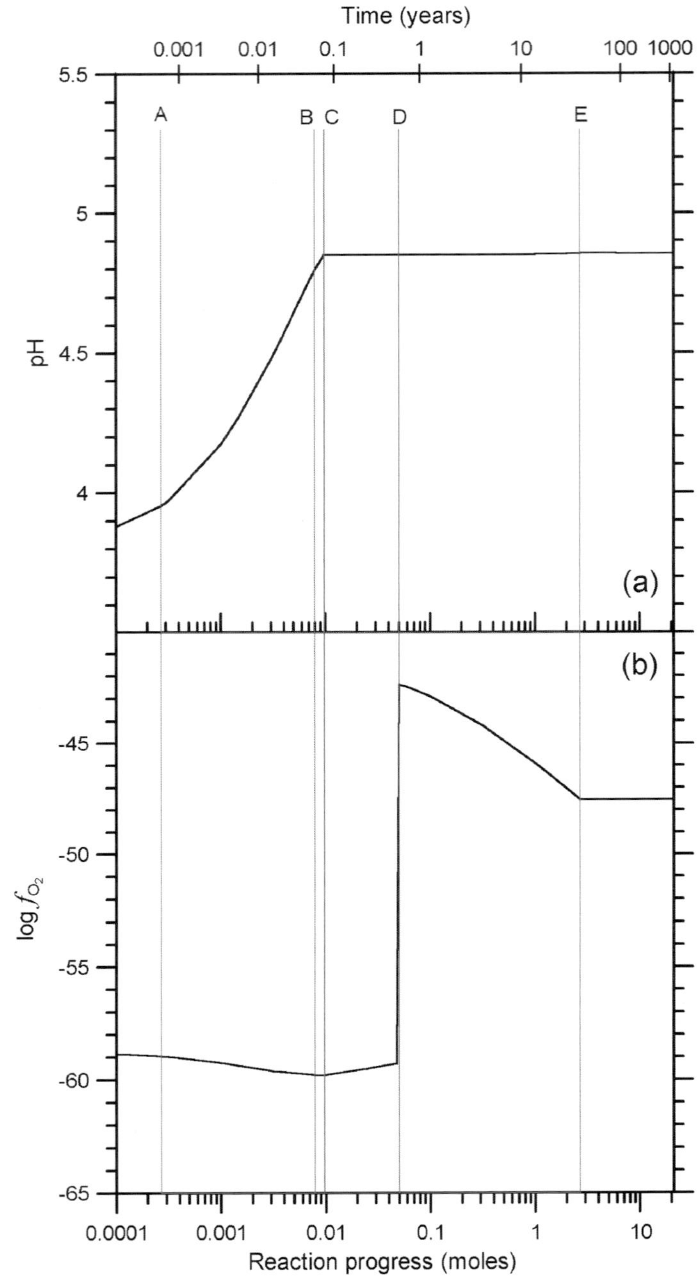

Figure 7.7. Evolution of (a) pH and (b) f_{O_2} during high-pressure ($P_{tot} = P_{CO_2} = 100$ bar) CO_2 injection in a deep aquifer ($T = 60°C$) hosted in serpentinitic rocks as a function of time and the reaction progress variable. Note that letters A to E mark the onset of the precipitation of different secondary minerals as shown in Figure 7.5.

In spite of its relatively low concentration in the aqueous phase and the small amounts of Fe-bearing secondary minerals produced during continuous CO_2 injection, equilibria between Fe product minerals and the aqueous solution govern the f_{O_2} of the system, at least after the onset of goethite precipitation (Fig. 7.7b, point D). This event brings about a sharp, remarkable increase in the logarithm of oxygen fugacity from −59.3 to −42.4. The log f_{O_2} is finally fixed at −47.53, for $\xi > 2.7$ (to the right of point E), by coexistence of goethite and siderite at constant f_{CO_2}, temperature and pressure, as expressed by the reaction:

$$FeOOH_{(s)} + CO_{2(g)} = FeCO_{3(s)} + \frac{1}{4}O_{2(g)} + \frac{1}{2}H_2O. \quad (7\text{-}47)$$

7.3.2.5. The CO_2 sequestration

The masses of CO_2 sequestered during the considered process through dissolution (and speciation) in the aqueous solution (solubility trapping) and through incorporation in solid carbonates, mainly magnesite and subordinately siderite and dolomite (mineral fixation) are plotted against both the reaction progress and time in Fig. 7.8. The total mass of sequestered CO_2, corresponding to the sum of these two terms, is also shown.

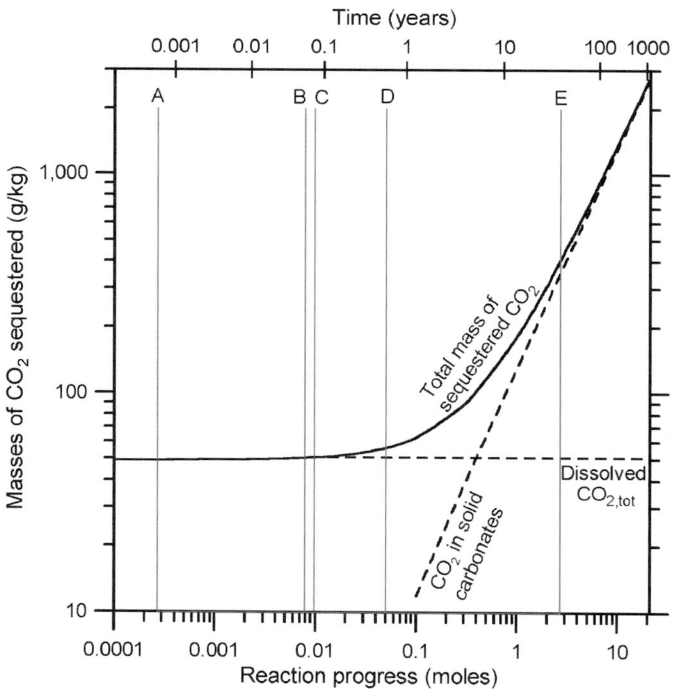

Figure 7.8. High-pressure ($P_{tot} = P_{CO_2} = 100$ bar) CO_2 injection in a deep aquifer ($T = 60°C$) hosted in serpentinitic rocks: masses of CO_2 sequestered through dissolution in the aqueous solution (solubility trapping) and incorporation in solid carbonates (mineral fixation), and total mass of sequestered CO_2 as a function of time and the reaction progress variable. Note that letters A to E mark the onset of the precipitation of different secondary minerals as shown in Figure 7.5.

Inspection of Fig. 7.8 shows that CO_2 sequestration through dissolution in the aqueous solution is instantaneous, but relatively limited, as the maximum amount of CO_2 which can be dissolved in 1 kg of water is ~50 g, at the specified conditions of 100 bar $P_{tot} = P_{CO_2}$ and 60° C. Carbon dioxide incorporation in secondary solid carbonates begins later on, namely at points B (dolomite), C (magnesite) and E (siderite). The log-mass of CO_2 incorporated in precipitating solid carbonates increases linearly with log ξ, attaining values much greater than the mass of CO_2 dissolved in the aqueous solution, but this process requires comparatively long time intervals. For instance, the contribution of mineral fixation attains the same level of solubility trapping after 5 years. However, mineral carbonation becomes more and more important afterwards, whereas solubility trapping does not change with time. Again, the sequestration capacity of the process is large and time is less than the residence times of high-pH waters in deep aquifers. In particular the value of 2,000 g kg^{-1} water is attained in 350 years. The same sequestration capacity is attained in 1,000 years in the dunite model by Xu and coworkers, who used somewhat different figures for several parameters (*see* Section 7.3.1).

7.3.2.6. Changes in the porosity of aquifer rocks

To evaluate the changes in the porosity of aquifer rocks in response to high-pressure ($P_{tot} = P_{CO2} = 100$ bar) CO_2 injection we continue to make reference to 1 kg of water. In this way, the model is kept completely independent of the effective porosity of the aquifer. For our purpose, it is useful to take into account the percentual change in the volume of solid phases, $\Delta V/V$. This parameter was already defined in equation (5-125). Here it is re-defined, more correctly, as follows:

$$\Delta V/V = \frac{\sum_P n_P \cdot v_P^o - \sum_R n_R \cdot v_R^o}{\sum_R n_R \cdot v_R^o} \cdot 100 \qquad (7\text{-}48)$$

where v_i^o and n_i represent the molar volume and the moles of the *i*th solid phase, respectively, and subscripts *P* and *R* indicate products and reactants, respectively. Indeed equation (7-48) is similar to the simplified relation (5-125), in which the stoichiometric coefficients of solid phases were used instead of their moles.

The variation of the $\Delta V/V$ parameter as a function of both reaction progress and time is shown in Fig. 7.9. During the first stages of the process (for $\xi < 0.0003$ moles) the $\Delta V/V$ is -100% as rock-forming minerals dissolve congruently, i.e. without precipitation of secondary solid phases. Formation of chalcedony (point A) brings about an initially sharp increase in $\Delta V/V$ and its subsequent stabilization close to -60%. Upon the onset of carbonate minerals precipitation, dolomite at point B and magnesite at point C, the $\Delta V/V$ parameter experiences a new initially sharp increase and gradually approaches $+19.3\%$, which is the limiting value constrained by dissolution of chrysotile accompanied by stoichiometric precipitation of magnesite and quartz (or chalcedony; note that the molar volumes of chalcedony and quartz are virtually equal; see Table 5.11). A value of $\Delta V/V$ close to 17% is attained in ~4 years (Fig. 7.9).

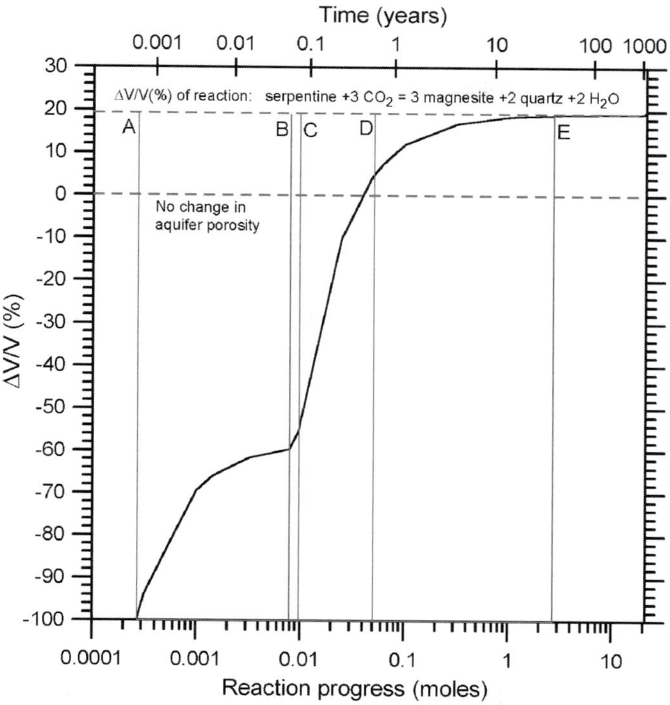

Figure 7.9. Variation of the $\Delta V/V$ parameter as a function of both reaction progress and time during high-pressure ($P_{tot} = P_{CO_2} = 100$ bar) CO_2 injection in a deep aquifer ($T = 60°C$) hosted in serpentinitic rocks. Note that letters A to E mark the onset of the precipitation of different secondary minerals as shown in Figure 7.5.

Obviously, for $\Delta V/V = 0$, the rock-forming minerals dissolved (mainly serpentine) are substituted by an equal volume of secondary solid phases and the porosity of the system does not change. Unfortunately, in the case under consideration, the $\Delta V/V$ is positive and significantly different from zero. Consequently, the occurring reactions, mainly serpentine dissolution accompanied by precipitation of magnesite and chalcedony causes a progressive reduction in the effective porosity of the aquifer. This would drop to zero if its initial value were 19%, whereas effective porosities >0 are only possible for initial values $>19\%$. If precipitation of amorphous silica takes place instead of chalcedony, the reduction in porosity could be even larger, with a maximum $\Delta V/V$ of 30.9%.

Although an initial porosity of 19% seems to be too high a value for serpentinites, it must be emphasized that (1) the natural system might behave as an open, flow-through system (rather than as a closed system), in which case secondary solid phases might deposit outside the reference volume affected by serpentine dissolution; (2) fracturing induced by injection of pressurized CO_2 might occur, with consequent increases in the effective porosity and permeability. As a matter of fact, magnesite veins with nodules of opal are common in the serpentinitic rocks of Southern Tuscany (Arisi Rota et al., 1971), a region characterized by high thermal fluxes (>100 mW m^{-2}, Baldi et al., 1994) and high

CO_2 fluxes (Marini and Chiodini, 1994), indicating that reaction (7-46) occurred naturally in these environments.

In contrast, as the magnesite and silica precipitate in the serpentinitic aquifer, these minerals will likely armor the remaining reactants, thereby further decreasing effective surface area in addition to decreasing porosity. These effects could represent serious obstacles for the implementation of this methodology of CO_2 sequestration and their importance must be evaluated by means of laboratory experiments first and field tests afterwards.

7.3.3. Reaction path modelling of the geological CO_2 sequestration in a serpentinitic aquifer: salinity effects

7.3.3.1. Setting up a water–rock interaction model involving a brine

To investigate the influence of salinity on the geological sequestration of CO_2, it would be desirable to carry out the same simulation of Section 7.3.2, but involving a brine instead of a dilute aqueous solution. As discussed in Sections 4.4 and 4.5, the best way to take into account the interactions among solutes in very concentrated electrolytes is by use of the Pitzer's equations. This capability is available in the software package EQ3/6 and we took advantage of it to carry out some calculations in Section 4.5. Unfortunately, the Pitzer's interaction parameters are not available for fundamental chemical components such as $SiO_{2(aq)}$ and Al^{3+} and are also restricted at relatively low temperatures, sometimes at 25°C only. This is the case, for instance, of the model by Harvie, Møller and Weare (1984) and the corresponding HMW database of EQ3/6, which is based on the system Na–K–Mg–Ca–H–Cl–SO_4–OH–HCO_3–CO_3–CO_2–H_2O and works at 25°C only, as already recalled in Section 4.4.2.

If we want to perform the simulation of Section 7.3.2 but taking into account a brine, we have to accept some limitations and introduce some assumptions. One possibility, that is further detailed below, is to carry out a simplified simulation at the temperature of 25°C, considering only chrysotile as reactant. Accepting to work at 25°C, the HMW database can be used, but $SiO_{2(aq)}$ has to be introduced into it. Strictly speaking, this requires the time-consuming search and processing of the experimental data (admitting that they do exist !) which are needed to derive the Pitzer's interaction parameters between $SiO_{2(aq)}$ and other chemical species, an exercise that goes well beyond the purpose of this book. This serious obstacle can be overcome through a sort of crafty trick, that is hypothesizing that the interaction parameters of $SiO_{2(aq)}$ are equal to those of dissolved CO_2, since both species are neutral. The HMW database was therefore changed accordingly. Of course, this is a very rough choice, but still it is better than nothing.

Again, we assume to inject pure CO_2 at a fugacity of 67.9 bar. Note that since at the considered temperature of 25°C, the fugacity coefficients of CO_2 are smaller than at 60°C (*see* Figure 3.5b), a total pressure ($=P_{CO_2}$) of 263 bar (where Γ_{CO_2} is 0.258) is needed to attain this f_{CO_2} value.

Two simulations were carried out, one involving a brine, initially at saturation with halite, the other with pure water. The results of the pure-water run represent the term of comparison to evaluate salinity effects. Indeed, the simulation of Section 7.3.2 cannot be used for reference due to differences not only in salinity but also in temperature, pressure, primary phases and secondary minerals (these are really too many differences). The EQ3 input file for the brine was prepared setting a pH of 7, imposing halite saturation to constrain the concentration of Na^+, and assuming $m_{Na^+,tot} = m_{Cl^-,tot}$ Negligible total concentrations (10^{-20} mol kg^{-1}) were assigned to Mg^{2+}, HCO_3^- and $SiO_{2(aq)}$ to activate these chemical components. Pure water was assumed to have initial pH 7 and initial negligible concentrations (10^{-20} mol kg^{-1}) of Mg^{2+}, HCO_3^- and $SiO_{2(aq)}$.

As anticipated, chrysotile was considered to be the only primary solid phase. Values of 4,100 cm^2 and 20.04 mol were assigned to its initial surface area exposed to 1,000 g of water and to its initial amount (see Section 7.3.2.1). Dissolution rate constants were set at 2.00×10^{-10} and 1.00×10^{-16} mol cm^{-2} s^{-1} for the proton- and water-promoted mechanisms, respectively. These are the values for lizardite and chrysotile, respectively, listed in Table 6.1. The reaction orders with respect to H^+ and H_2O are 0.8 and 1.0, respectively, as reported in Table 7.9. Chalcedony, magnesite and halite (in the brine run) were considered to be the only precipitating solid-product phases and instantaneous partial equilibrium was assumed for all secondary minerals.

7.3.3.2. Solid-product phases

The amounts of secondary solid phases produced through high-pressure ($P_{tot} = P_{CO_2} = 263$ bar) CO_2 injection in a system comprising chrysotile and an aqueous solution are displayed in Figure 7.10 for the two runs, involving either a brine, initially saturated with halite, or pure water. In this way the two sets of results can be compared and possible salinity effects can be evaluated. In the pure water run, the onset of chalcedony and magnesite precipitation is somewhat delayed with respect to the brine run, as indicated by the spacings between points A and A′ and between points B and B′. However, the amounts of chalcedony and magnesite produced are equal for $\xi > 0.003$ and $\xi > 0.1$ mol, respectively. The most striking difference between the two simulations, however, is the appearance, in the brine run, of halite whose precipitation occurs immediately upon CO_2 injection. Since the brine was initially saturated with halite under a negligible f_{CO_2}, close to 10^{-20} bar, the increase in f_{CO_2} makes the brine to become oversaturated with halite and a significant amount of this mineral, 0.24 moles, is immediately precipitated. Upon progressive chrysotile dissolution (for $\xi < 0.1$ mol approximately), no further precipitation of halite occurs and this mineral simply remains in the system. However, the dissolution of chrysotile also causes an increase in the mass of water (as already recalled in Section 7.3.2.4, see Figure 7.11a), which becomes significant for $\xi > 0.1$ mol approximately. This increase in water brings about re-dissolution of halite, which disappears from the system at $\xi = 1$ mol (point C). For higher values of the reaction progress, the two runs provide the same results in terms of precipitating solid phases.

7.3.3.3. The aqueous solution

The concentrations of relevant solutes and pH during progressive chrysotile dissolution triggered by high-pressure ($P_{tot} = P_{CO_2} = 263$ bar) CO_2 injection are shown in Figure 7.11 for both the brine run and the pure-water run, to facilitate the comparison of the two series of results and to infer salinity effects.

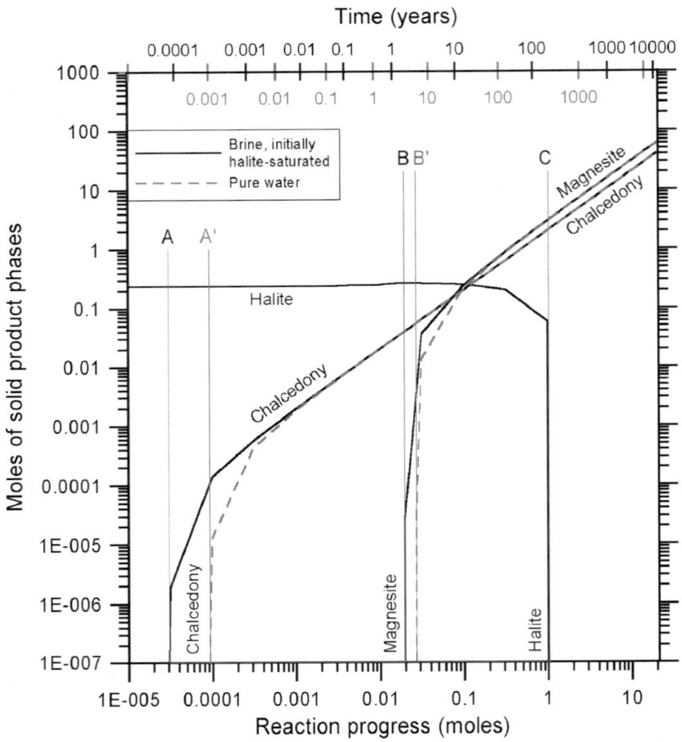

Figure 7.10. Moles of product minerals formed during high-pressure ($P_{tot} = P_{CO_2} = 263$ bar) CO_2 injection in a system ($T = 25°C$) comprising chrysotile and an aqueous solution, either a brine, initially halite-saturated (solid black lines and black time scale) or pure water (dashed grey lines and grey time scale), as a function of time and the reaction progress variable.

First, it must be underscored that after halite re-dissolution (point C), the addition of water to the system through chrysotile dissolution continues and brings about a significant decrease in the concentrations of Na^+ and Cl^-, which behave as mobile components in the final part of the brine simulation.

The main differences between the two simulations, especially before halite re-dissolution, are in pH and in the concentrations of $CO_{2(aq)}$ and $SiO_{2(aq)}$ (although the latter ones can be biased by the assumption on the interaction parameters between $SiO_{2(aq)}$ and the other dissolved species), whereas the differences in the concentrations of Mg^{2+} and HCO_3^- are comparatively small.

Dissolved CO_2 concentration in the halite-saturated brine is 32.3 g kg^{-1}, approximately 1/3 of that in pure water, 101 g kg^{-1}. The $CO_{2(aq)}$ concentration increases somewhat upon halite re-dissolution and brine dilution, i.e. after point C, and attain the maximum value of 48.8 g kg^{-1}, about ½ of the pure water value, at the end of the simulation.

The pH of the halite-saturated brine increases from the initial value of 2.27 to 4.48 at point C but remains 0.7 to 0.5 pH-units lower than the pure-water pH. The maximum

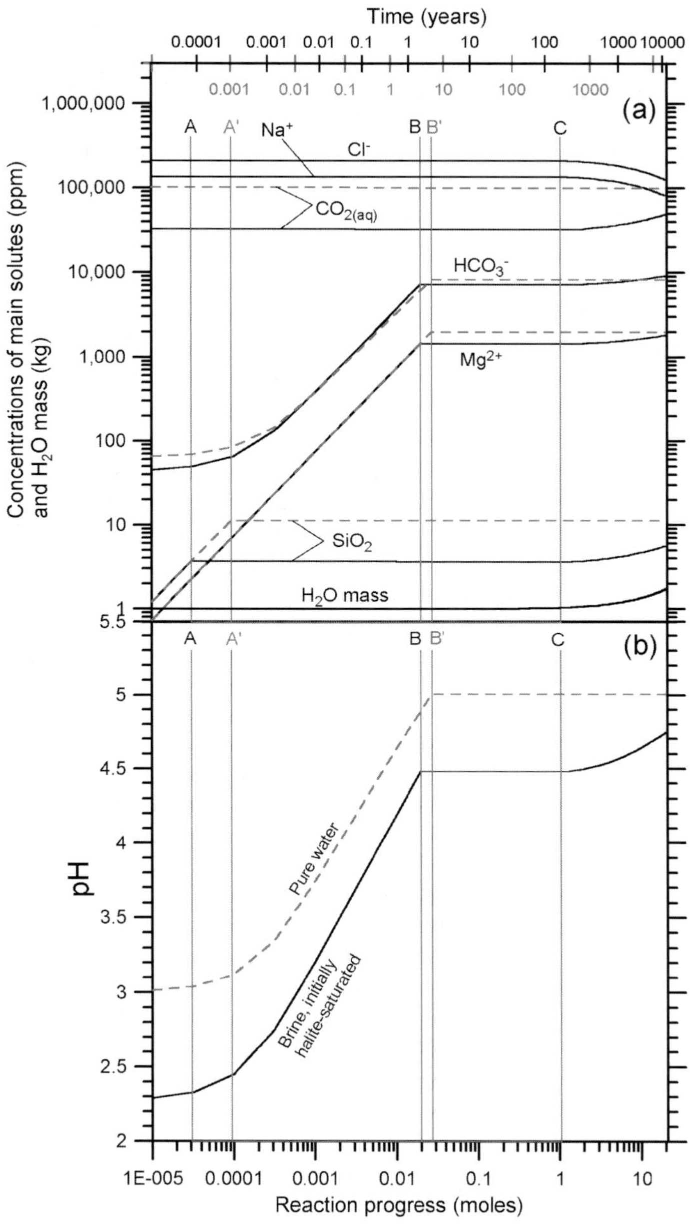

Figure 7.11. (a) concentrations of relevant solutes, mass of water, and (b) pH during high-pressure ($P_{tot} = P_{CO_2} =$ 263 bar) CO_2 injection in a system ($T = 25°C$) comprising chrysotile and an aqueous solution, either a brine, initially halite-saturated (solid black lines and black time scale) or pure water (dashed grey lines and grey time scale), as a function of time and the reaction progress variable. Letters A, A', B, B' and C correspond to those in Figure 7.10.

increase in pH takes place between the beginning of chalcedony precipitation (point A, pH = 2.33) and the onset of magnesite precipitation (point B, pH = 4.48), after which pH is fixed by carbonate equilibria, as already noticed in Sections 7.1.1 and 7.3.2.4. The difference in pH between the brine and pure water reduces after point C due to dilution of the brine.

7.3.3.4. The CO_2 sequestration

The total mass of CO_2 sequestered through high-pressure ($P_{tot} = P_{CO_2} = 263$ bar) CO_2 injection in a system made up of chrysotile and a brine, initially saturated with halite is compared with that sequestered in a reference system constituted by chrysotile and pure water in Figure 7.12. Also shown are the separated contributions of CO_2 dissolution (and speciation) in the aqueous solution (solubility trapping) and incorporation in precipitating magnesite (mineral fixation).

The plot shows that solubility trapping is strongly dependent on the salinity of the aqueous solution. Nevertheless, the spacing between the two lines outlining solubility trapping, for the brine and pure water, decreases with increasing reaction progress owing

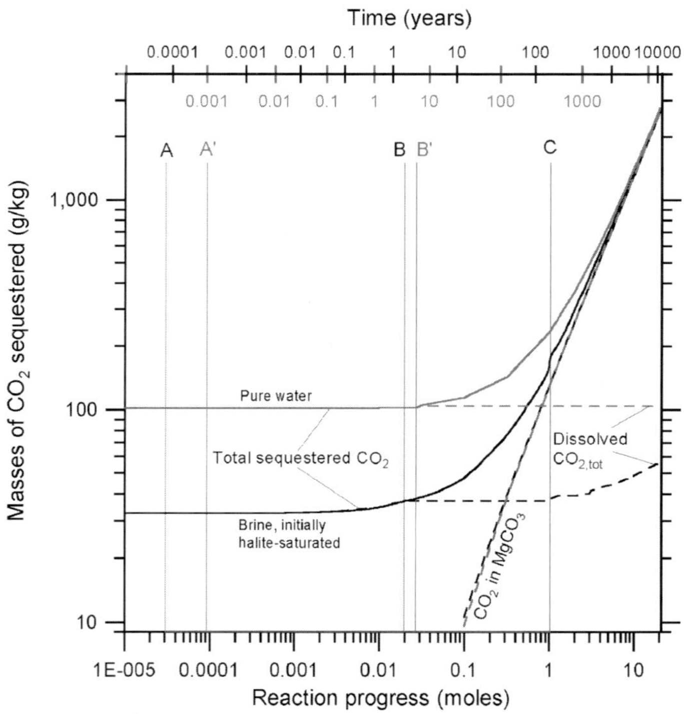

Figure 7.12. High-pressure ($P_{tot} = P_{CO_2} = 263$ bar) CO_2 injection in a system ($T = 25$°C) comprising chrysotile and an aqueous solution, either a brine, initially halite-saturated (solid black lines and black time scale) or pure water (dashed grey lines and grey time scale). The masses of CO_2 sequestered through both dissolution in the aqueous solution (solubility trapping) and incorporation in precipitating magnesite (mineral fixation) as well as the total mass of sequestered CO_2 are shown as a function of time and the reaction progress variable. Letters A, A', B, B' and C correspond to those in Figure 7.10.

to progressive dilution of the brine, as discussed in the previous section. In contrast, mineral fixation is similar for the chrysotile–brine and the chrysotile–pure water systems. Consequently, the difference between the total masses of CO_2 undergoing geological sequestration in these two distinct systems decreases with increasing reaction progress owing to the decreasing contribution of solubility trapping and the increasing contribution of mineral fixation.

Finally, it must be underscored that the time scales are different for the two considered systems. The time needed to exhaust 95% of chrysotile is 12,650 years for the chrysotile–brine system, whereas it amounts to 23,080 years for the chrysotile–pure water system. These values are much higher than the corresponding 758 years of the simulation presented in Section 7.3.2, owing to the different temperatures, 60°C in Section 7.3.2 against 25°C in this section.

7.4. Reaction path modelling of geological CO_2 sequestration in continental tholeiitic flood basalts

Continental tholeiitic flood basalts represent another type of rocks with a relatively high potential of CO_2 sequestration, owing to relatively high contents of MgO and CaO, namely 6.2 and 9.4 wt% on average, respectively (Lackner et al., 1995, 1997). The CO_2 sequestration potential is enhanced further by the significant Na_2O and FeO contents, 2.7 and 9.2 wt% on average, respectively, based on the data reported by Wilson (1989).

A typical example of this kind of rocks is represented by the Columbia River tholeiitic flood basalts in northwestern USA, which extend over an area of ~200,000 km^2 with an average thickness of ~1 km and a maximum thickness> 1.5 km (Wilson, 1989). This volume of rocks would be enough to fix over twice the amount of CO_2 produced through combustion of the global coal reserves, which are estimated to be in the order of 10^{16} kg (Lackner et al., 1995), as already recalled above.

Among the other major continental flood-basalt provinces, we may quote the Siberian Platform, with an extension > 1,500,000 km^2 and a maximum thickness of 3.5 km and the Paranà (Brazil)-Etendeka (Namibia) province, covering a total area of 1,200,000 km^2 with a maximum thickness of 1.8 km (Wilson, 1989).

Therefore it is instructive to model the sequestration of CO_2 through reaction with this type of rock.

7.4.1. Setting up the water–rock interaction model

Starting from the whole-rock chemical analysis of a continental tholeiitic flood basalt from the Columbia River Province (Table 7.10) and assuming that its main constituting mineral phases are a labradoritic plagioclase ($X_{\mathrm{Anorthite}} = 0.55$, $X_{\mathrm{Albite}} = 0.45$), clinopyroxene ($X_{\mathrm{Diopside}} = 0.75$, $X_{\mathrm{Hedenbergite}} = 0.25$), orthopyroxene ($X_{\mathrm{Enstatite}} = 0.60$, $X_{\mathrm{Ferrosilite}} = 0.40$), magnetite, apatite and ilmenite (as suggested by available petrographic data, see Wilson, 1989), simple mass balance calculations show that the molar fractions of these phases are 0.5454, 0.1968, 0.1464, 0.0253, 0.0047 and 0.0210, respectively. Apatite and ilmenite were neglected from subsequent data elaboration owing to the low content and relatively high chemical durability, respectively; rock composition was then recomputed imposing that the sum of the molar fractions of the other minerals is equal to 1. In this way, the mineralogic composition reported in Table 7.10 was obtained.

The total surface area in contact with 1,000 g of water was set again at 4,100 cm^2, as in the serpentinite case (Sections 7.3.2 and 7.3.3), for comparative purposes. It was distributed in proportion to the volume percentages of the main rock-forming minerals obtaining the initial surface areas reported in Table 7.10. Their initial amounts in the considered system (which are also given in Table 7.10) were then computed through simple geometric computations (*see* Section 7.2.4), assuming an initial porosity of 0.3.

Again, for comparative purposes, aquifer temperature was kept at 60°C and injection of pure CO_2 at a constant pressure of 100 bar (f_{CO_2} of 67.9 bar) was imposed. The initial aqueous solution was assumed to be in equilibrium with analcime, calcite, kaolinite, daphnite, clinochlore, chalcedony and magnetite at 60°C, 1 bar under a f_{CO_2} of 10^{-1} bar (*see* Section 7.1.2).

The computer simulation was performed in kinetic mode, specifying the initial surface areas of primary minerals (which are then changed by the code in proportion to their remaining masses) and describing their dissolution kinetics by means of the TST-based rate law (see Table 7.11). Again, the proton- and water-promoted mechanisms were considered but the hydroxyl-promoted dissolution mechanism was neglected, owing to the

TABLE 7.10

Whole-rock chemical analysis of a typical continental tholeiitic flood basalt from the Columbia River Province (Wilson, 1989) and its computed mineralogic composition

Oxide	Wt%	Element	Mol%
SiO_2	48.35	Si	17.19
TiO_2	1.57	Ti	0.42
Al_2O_3	15.49	Al	6.49
Fe_2O_3	3.26	Fe	3.27
FeO	8.05		
MnO	0.17	Mn	0.05
MgO	7.03	Mg	3.72
CaO	9.92	Ca	3.78
Na_2O	2.76	Na	1.90
K_2O	0.51	K	0.23
P_2O_5	0.24	P	0.07
CO_2	0.05	C	0.02
H_2O	1.52	O	59.29
		H	3.55

Minerals	Mol %	Vol %	Initial moles	Initial surfaces
Plagioclase	59.68	73.86	17.14	3,028.3
Clinopyroxene	21.54	18.32	6.19	751.3
Orthopyroxene	16.02	6.30	4.60	258.2
Magnetite	2.77	1.52	0.80	62.2

TABLE 7.11

EQ6 input file for modeling high-pressure ($P_{tot} = P_{CO2} = 100$ bar) CO_2 injection in a deep aquifer hosted in continental tholeiitic flood basalts at 60°C in kinetic mode (time frame). (Required input formats are not fully respected.)

The option switch IOPT1 is set to 1 to direct the code to compute the simulation in time frame.
The option switch IOPT4 is set to 1 to activate solid solutions.
The print option switch IOPR8 is set to 1 to direct the code to print a table of equilibrium gas fugacities at each print point.
endit.

```
     nmodl1= 1                    nmodl2= 0
     tempc0= 60.0000E+00           jtemp= 0
        tk1= 0.00000E+00             tk2= 0.00000E+00     tk3= 0.00000E+00
     zistrt= 0.00000E+00           zimax= 1.00000E-06
      tstrt= 0.00000E+00          timemx= 0.00000E+00
     kstpmx= 500                   cplim= 0.00000E+00
     dzprnt= 1.00000E+00          dzprlg= 0.00000E+00  ksppmx= 100
     dzplot= 0.10000E+00          dzpllg= 0.00000E+00  ksplmx= 10000
      ifile= 60
*             1    2    3    4    5    6    7    8    9    10
  iopt1-10=   1    0    0    1    0    0    0    0    0    0
 iopt11-20=   0    0    0    0    0    0    0    0    0    0
  iopr1-10=   0    0    0    0    1    0    0    1    0    0
 iopr11-20=   0    0    0    0    0    0    0    0    0    0
  iodb1-10=   0    0    0    0    0    0    0    0    0    0
 iodb11-20=   0    0    0    0    0    0    0    0    0    0
*____________________________________________________________
      nxopt= 1
     option= all
     nxopex= 10
  exception= Gibbsite
  exception= Kaolinite
  exception= Chalcedony
  exception= Magnetite
  exception= Goethite
  exception= Dawsonite
  exception= Calcite
  exception= Magnesite
  exception= Siderite
  exception= Dolomite
*____________________________________________________________
       nffg= 1
    species= CO2(g)
      moffg= 1.00000E+01          xlkffg= +1.83187E+00
*____________________________________________________________
       nrct= 4
*____________________________________________________________
   reactant= Plagioclase
      jcode= 1                     jreac= 0
       morr= 17.1407E+00            modr= 0.00000E+00
       Albite high     4.53500E-01
       Anorthite       5.46500E-01
endit.
```

TABLE 7.11 (*Continued*)

```
       nsk= 1                         sk= 3028.29E+00          fk= 0.00000E+00
       nrk= 2                       nrpk= -1
     imech= 2
       rk0= 8.03000E-12             trk0= 6.00000E+01        iact= 0
      eact= 0.00000E+00             hact= 0.00000E+00
     ndact= 1                     csigma= 1.00000E+00
      udac= H+                      cdac= 0.62600E+00
       rk0= 8.35000E-15             trk0= 6.00000E+01        iact= 0
      eact= 0.00000E+00             hact= 0.00000E+00
     ndact= 1                     csigma= 1.00000E+00
      udac= H2O                     cdac= 1.00000E+00
*------------------------------------------------------------------------------
  reactant= Clinopyroxene
     jcode= 1                      jreac= 0
      morr= 6.1863E+00              modr= 0.00000E+00
    Diopside          7.50000E-01
    Hedenbergite      2.50000E-01
endit.
       nsk= 1                         sk= 751.265E+00          fk= 0.00000E+00
       nrk= 2                       nrpk= -1
     imech= 2
       rk0= 4.13000E-10             trk0= 6.00000E+01        iact= 0
      eact= 0.00000E+00             hact= 0.00000E+00
     ndact= 1                     csigma= 1.00000E+00
      udac= H+                      cdac= 0.70000E+00
       rk0= 2.92000E-15             trk0= 6.00000E+01        iact= 0
      eact= 0.00000E+00             hact= 0.00000E+00
     ndact= 1                     csigma= 1.00000E+00
      udac= H2O                     cdac= 1.00000E+00
*------------------------------------------------------------------------------
  reactant= Orthopyroxene
     jcode= 1                      jreac= 0
      morr= 4.60010E+00             modr= 0.00000E+00
     Enstatite       6.00000E-01
     Ferrosilite     4.00000E-01
endit.
       nsk= 1                         sk= 258.226E+00          fk= 0.00000E+00
       nrk= 2                       nrpk= -1
     imech= 2
       rk0= 4.57000E-12             trk0= 6.00000E+01        iact= 0
      eact= 0.00000E+00             hact= 0.00000E+00
     ndact= 1                     csigma= 1.00000E+00
      udac= H+                      cdac= 0.65000E+00
       rk0= 6.02000E-15             trk0= 6.00000E+01        iact= 0
      eact= 0.00000E+00             hact= 0.00000E+00
     ndact= 1                     csigma= 1.00000E+00
      udac= H2O                     cdac= 1.00000E+00
*------------------------------------------------------------------------------
```

TABLE 7.11 (*Continued*)

```
   reactant= Magnetite
      jcode= 0                       jreac= 0
       morr= 0.79530E+00              modr= 0.00000E+00
        nsk= 1                          sk= 62.2180E+00              fk= 0.00000E+00
        nrk= 2                        nrpk= -1
      imech= 2
        rk0= 5.65000E-13              trk0= 6.00000E+01            iact= 0
       eact= 0.00000E+00              hact= 0.00000E+00
      ndact= 1                      csigma= 1.00000E+00
       udac= H+                       cdac= 0.27900E+00
        rk0= 3.65000E-15              trk0= 5.00000E+01            iact= 0
       eact= 0.00000E+00              hact= 0.00000E+00
      ndact= 1                      csigma= 1.00000E+00
       udac= H2O                      cdac= 1.00000E+00
*------------------------------------------------------------------------------
     dlzidp= 0.00000E+00
      tolbt= 0.00000E+00             toldl= 0.00000E+00            tolx= 0.00000E+00
     tolsat= 0.00000E+00            tolsst= 0.00000E+00
     screw1= 0.00000E+00            screw2= 0.00000E+00          screw3= 0.00000E+00
     screw4= 0.00000E+00            screw5= 0.00000E+00          screw6= 0.00000E+00
     zklogu= 0.000                  zklogl= 0.000                 zkfac= 0.000
     dlzmx1= 0.00000E+00            dlzmx2= 0.00000E+00          nordlm= 0
     itermx= 0                      ntrymx= 0
     npslmx= 0                      nsslmx= 0                    ioscan= 0
*------------------------------------------------------------------------------
* pickup file written by EQ3NR, version 7.2b (R139 )
* supported by EQLIB, version 7.2b (R168 )
Water in equilibrium with analcime, calcite, kaolinite,
chalcedony, daphnite, clinochlore, magnetite
at 60øC, 0.1 bar PCO2
endit.
     tempci= 6.00000E+01
      nxmod= 0
      iopg1= 0                       iopg2= 0                     iopg3= 0
      iopg4= 0                       iopg5= 0                     iopg6= 0
      iopg7= 0                       iopg8= 0                     iopg9= 0
     iopg10= 0
        kct= 10                        ksq= 11                      kmt= 11
        kxt= 11                       kdim= 11                     kprs= 0
O            5.568190070346440E+01
Al           2.397285511182285E-07
Ca           4.360523578853289E-05
Cl           3.000000000355616E-02
Fe           3.862527706607725E-06
H            1.110727982477859E+02
C            5.793621436309842E-02
Mg           1.602243822990042E-04
Na           8.637858497535496E-02
Si           6.159324420874708E-04
electr       1.323234709603221E-11
```

TABLE 7.11 (*Continued*)

H_2O	H_2O	1.744358983526984E+00
Al^{+++}	Al^{+++}	– 1.710395611381481E+01
Ca^{++}	Ca^{++}	– 4.562090256614116E+00
Cl^-	Cl^-	– 1.527663478385729E+00
Fe^{++}	Fe^{++}	– 6.567807046642986E+00
H^+	H^+	– 7.592429061239226E+00
HCO_3^-	HCO_3^-	– 1.269686271112627E+00
Mg^{++}	Mg^{++}	– 3.959985610775256E+00
Na^+	Na^+	– 1.076209250999970E+00
$SiO_2(aq)$	$SiO_2(aq)$	– 3.230700000000000E+00
$O_2(g)$	$O_2(g)$	– 6.098389925982678E+01

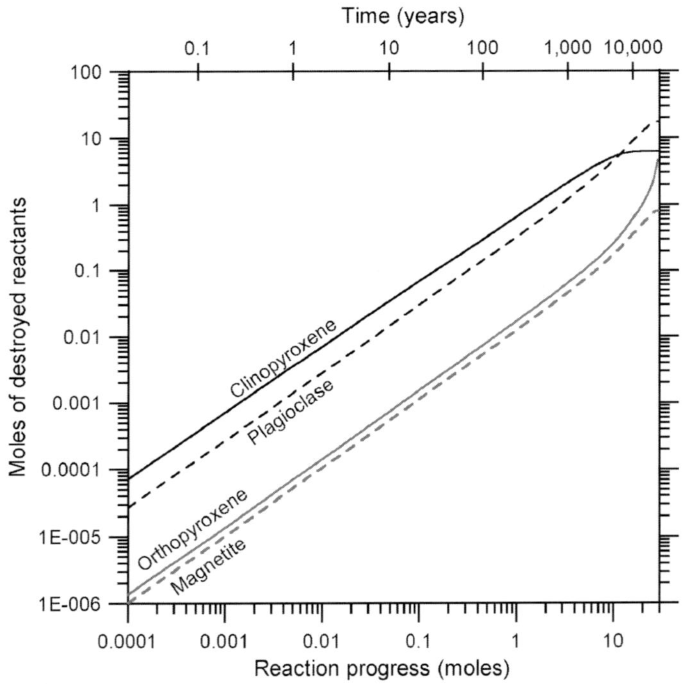

Figure 7.13. Moles of reactant minerals destroyed during high-pressure (P_{tot} = P_{CO_2} = 100 bar) CO_2 injection in a deep aquifer (T = 60°C) hosted in continental tholeiitic flood basalts as a function of time and the reaction progress variable.

low pH values constrained by CO_2 injection (see below). As in previous cases, selected secondary minerals were assumed to attain instantaneous partial equilibrium, owing to the difficulty in determining the surface area for solid phases which are not present in the considered system at the beginning of the simulation.

The diagrams of Figures 7.13 to 7.18 depict the main results of this reaction path modelling exercise against both the reaction progress variable, ξ, and time.

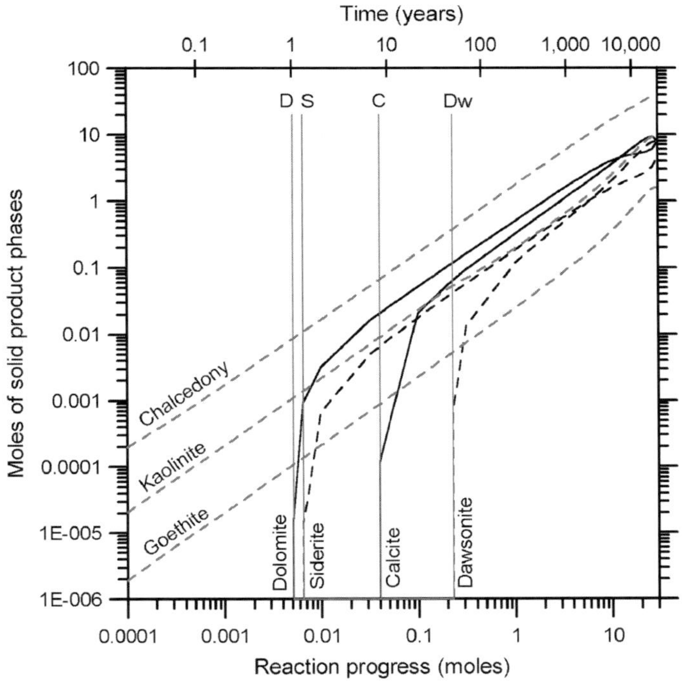

Figure 7.14. Amounts of product minerals formed during high-pressure ($P_{tot} = P_{CO_2} = 100$ bar) CO_2 injection in a deep aquifer ($T = 60°C$) hosted in continental tholeiitic flood basalts as a function of time and the reaction progress variable. Letters D, S, C and Dw mark the onset of the precipitation of secondary dolomite, siderite, calcite and dawsonite, respectively.

7.4.2. Solid reactants

Throughout CO_2 injection coupled with basalt dissolution, the aqueous solution remains strongly undersaturated with respect to plagioclase ($A \leq -9.6$ kcal mol^{-1}), clinopyroxene ($A \leq -16.3$ kcal mol^{-1}), orthopyroxene ($A \leq -8.3$ kcal mol^{-1}) and magnetite ($A \leq -3.4$ kcal mol^{-1}). Owing to these relatively low values of A, the dissolution rates of primary phases are always independent of thermodynamic affinity (*see* Section 6.4.2) and depend on pH only.

The moles of primary minerals consumed during continuous CO_2 injection and basalt dissolution are presented in Figure 7.13. The two most abundant solid phases, plagioclase and clinopyroxene, contribute most chemical components to the considered system, but their contributions are generally reversed with respect to their surface areas, owing to the overwhelming weight of the dissolution rate (in mol cm^{-2} s^{-1}), which is much higher for clinopyroxene than for the labradoritic plagioclase, in acidic solutions (see Tables 6.1 and 7.11). Only in the final part of the simulation, namely upon exhaustion of clinopyroxene, plagioclase becomes the top provider of chemical components.

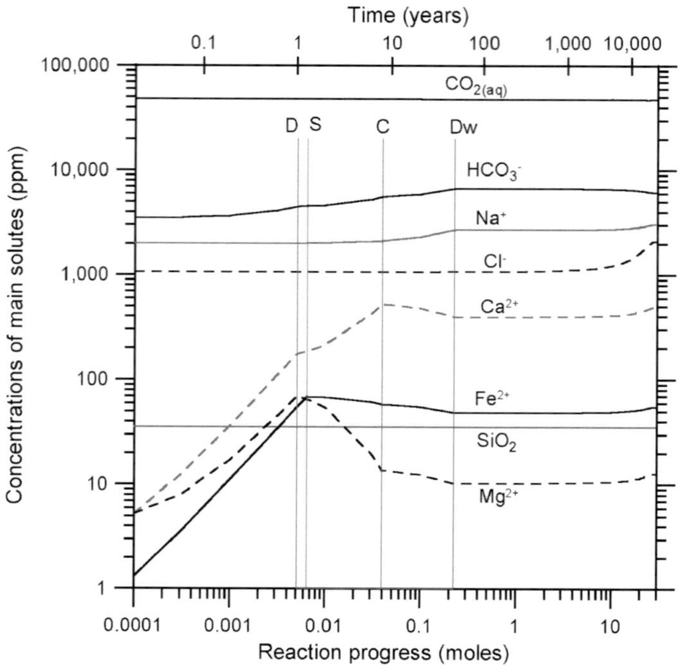

Figure 7.15. Concentrations of solutes during high-pressure (P_{tot} = P_{CO_2} = 100 bar) CO_2 injection in a deep aquifer (T = 60°C) hosted in continental tholeiitic flood basalts as a function of time and the reaction progress variable. Letters D, S, C and Dw mark the onset of the precipitation of secondary dolomite, siderite, calcite and dawsonite, respectively, as shown in Figure 7.14.

The contributions of orthopyroxene and magnetite are generally subordinate owing to their comparatively low dissolution rates (in mol cm^{-2} s^{-1}) and amounts; however, orthopyroxene becomes an important supplier of chemical components in the final part of the simulation, upon exhaustion of other solid reactants.

Clinopyroxene is totally consumed after 26,100 years, for ξ of 22.9 mol, plagioclase is exhausted after 191,300 years, for ξ of 28.4 mol, and magnetite is completely consumed after 280,600 years, for ξ of 28.6 mol. The simulation ends for ξ of 28.7 mol, upon exhaustion of orthopyroxene.

7.4.3. Solid product phases

The solid phases which are precipitated during continuous CO_2 injection in a deep aquifer hosted in tholeiitic continental flood basalts are, in order of appearance, chalcedony, kaolinite, goethite, dolomite, siderite, calcite and dawsonite [$NaAlCO_3(OH)_2$] (Figure 7.14). Both chalcedony and kaolinite form immediately upon CO_2 injection, goethite production begins after 9 h and 10 min (ξ = 5.76 μmol), whereas the precipitation of carbonate minerals starts much later: that of dolomite after 1 year (ξ = 5.12 mmol),

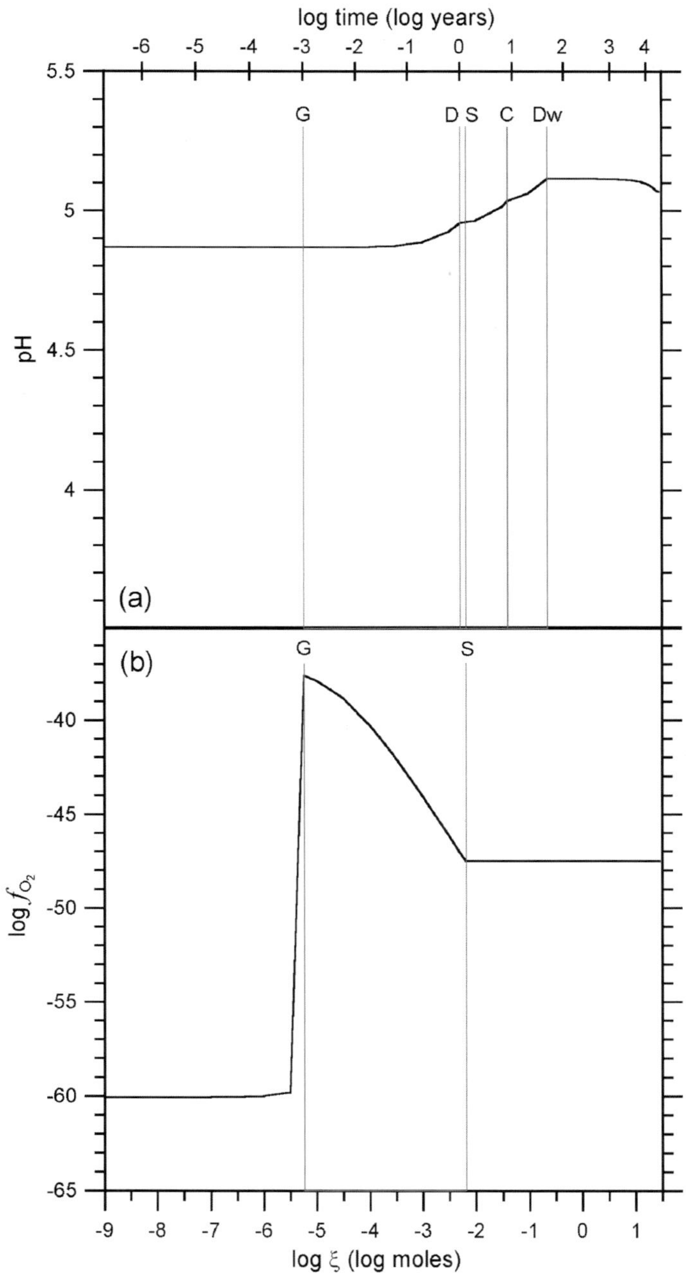

Figure 7.16. Changes in (a) pH and (b) f_{O_2} during high-pressure (P_{tot} = P_{CO_2} = 100 bar) CO_2 injection in a deep aquifer (T = 60°C) hosted in continental tholeiitic flood basalts as a function of time and the reaction progress variable. Letters G, D, S, C and Dw mark the onset of the precipitation of secondary goethite, dolomite, siderite, calcite and dawsonite, respectively.

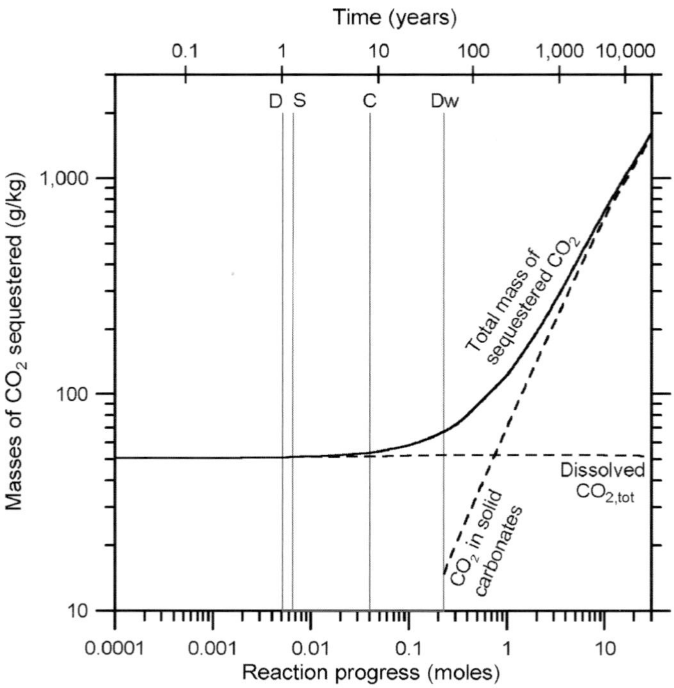

Figure 7.17. High-pressure ($P_{tot} = P_{CO_2} = 100$ bar) CO_2 injection in a deep aquifer ($T = 60°C$) hosted in continental tholeiitic flood basalts: masses of CO_2 sequestered through dissolution in the aqueous solution (solubility trapping) and incorporation in solid carbonates (mineral fixation), and total mass of sequestered CO_2 as a function of time and the reaction progress variable. Letters D, S, C and Dw mark the onset of the precipitation of secondary dolomite, siderite, calcite and dawsonite, respectively, as shown in Figure 7.14.

that of siderite after 1 year and 3 months ($\xi = 5.12$ mmol), that of calcite after 8 years and 5 months ($\xi = 39.8$ mmol) and that of dawsonite after 53 years and 5 months ($\xi = 227$ mmol).

The log-moles of each solid phase increase close to linearly with log ξ after an initial non-linear trend, which is evident for carbonate minerals only in Figure 7.14. Chalcedony is by far the most important solid product phase throughout the process, carbonate minerals and kaolinite are produced in similar amounts (after an initial settling-down period), whereas goethite is the least important secondary phase.

The masses of produced carbonate minerals amount to: (i) 91.7 g of dolomite, 31.9 g of calcite, 20.9 g of siderite and 16.6 g of dawsonite, for a total of 161.1 g, after 255 years ($\xi = 1$ mol); (ii) 736 g of dolomite, 355 g of calcite, 180 g of siderite and 288 g of dawsonite, for a total of 1,559 g, after 4,080 years ($\xi = 10$ mol); (iii) 1,015 g of dolomite, 871 g of calcite, 319 g of siderite and 960 g of dawsonite, for a total of 3,165 g, after 26,100 years ($\xi = 22.9$ mol), upon exhaustion of clinopyroxene. These figures confirm the high potential of CO_2 sequestration of these basaltic rocks, including the important role of dawsonite, in spite of the long time required for mineral fixation.

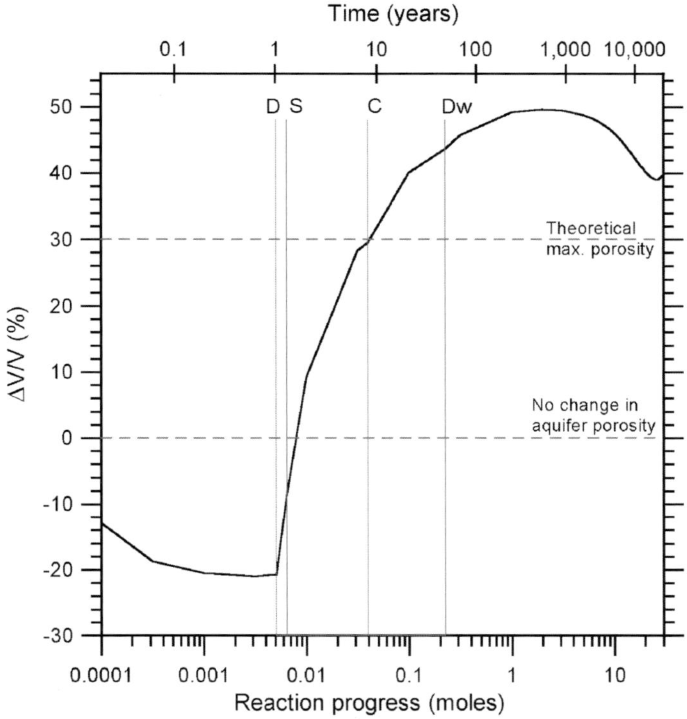

Figure 7.18. Variation of the $\Delta V/V$ parameter as a function of both reaction progress and time during high-pressure ($P_{tot} = P_{CO_2} = 100$ bar) CO_2 injection in a deep aquifer ($T = 60°C$) hosted in continental tholeiitic flood basalts. Letters D, S, C and Dw mark the onset of the precipitation of secondary dolomite, siderite, calcite and dawsonite, respectively, as shown in Figure 7.14.

7.4.4. The aqueous solution

Also throughout this computer experiment, the main solute is aqueous CO_2, with a concentration varying between 48,100 and 47,200 ppm (Figure 7.15). In other words, CO_2 concentration remains practically constant at 1.1 mol kg^{-1}, the same figure of the serpentinite experiment (*see* Section 7.3.2.4.), owing to the constancy in f_{CO_2}, at fixed temperature and pressure.

Aqueous CO_2 is followed in order of decreasing importance (at least for $\xi > 0.01$ mol) by HCO_3^-, Na^+, Cl^-, Ca^{2+}, Fe^{2+}, SiO_2 and Mg^{2+}. Dissolved SiO_2 is kept constant throughout the simulation by saturation with chalcedony, a condition which is attained immediately upon CO_2 injection. Chloride concentration remains constant at the initial value, 1,060 mg kg^{-1}, for over 22 years and then it increases. Such an increase is initially weak but becomes more and more important with time, especially after 4,000 years ($\xi > 10$ mol). This effect is caused by the progressive incorporation of water in precipitating hydrous minerals (kaolinite, dawsonite and goethite), a phenomenon known as dry-up (Reed, 1997), as already recalled above.

The concentrations of all the other solutes, i.e. HCO_3^-, Na^+, Ca^{2+}, Fe^{2+} and Mg^{2+} experience an initial increase (very pronounced for Ca^{2+}, Fe^{2+} and Mg^{2+} and rather weak for HCO_3^- and Na^+), before the attainment of saturation with respect to carbonate minerals. This is followed by a general stabilization after the onset of dawsonite precipitation (i.e. after 53 years and 5 months, for ξ of 227 mmol, point Dw in Fig. 7.15), with more or less important changes in the intermediate period. It must be underscored that for $\xi > 227$ mmol, calcite, dolomite, siderite and dawsonite are in equilibrium with the aqueous solution, a condition which fixes the activities of free Ca^{2+}, Mg^{2+}, Fe^{2+} and Na^+, respectively. Only limited changes in the total concentrations of these solutes are possible, owing to variations in salinity and consequent changes in the concentrations of complex species.

The simulation predicts that upon CO_2 injection, the pH of the aqueous solution attains immediately the value of 4.87 (Fig. 7.16a), which is higher by one unit than the initial pH value of the serpentinite case, 3.83, and comparable with its final pH value, 4.85 (*see* Section 7.3.2.4). This relatively high initial pH value of the basalt case is consistent with the high concentrations of Na^+ and HCO_3^- in the aqueous solution, which are the two prevailing solutes (see above). During progressive CO_2 injection and basalt dissolution, the pH of the aqueous solution increases up to the maximum value of 5.12 at point Dw and remains stable afterwards, since it is buffered by coexistence of carbonate minerals, apart from a minor decrease to 5.07 in the final part of the simulation.

Changes in log f_{O_2} are similar to those of the serpentinite case, with a sharp increase from −59.8 to −37.6 concurrent with the attainment of goethite saturation (Fig. 7.16b, point G), which is followed by a gradual decrease until the aqueous solution saturates with siderite (point S). From this point, coexistence of goethite and siderite at constant f_{CO_2}, temperature and pressure (reaction 7-47) fixes the log f_{O_2} at −47.53, as already mentioned in Section 7.3.2.4.

7.4.5. The CO_2 sequestration

The masses of CO_2 sequestered through both solubility trapping (i.e. as dissolved CO_2 and related carbonate aqueous species) and mineral fixation (i.e. the CO_2 stored in secondary carbonate minerals, namely dolomite, siderite, calcite and dawsonite in the considered case), as well as their sum are shown in Figure 7.17 as a function of the reaction progress and time.

As already discussed above, CO_2 sequestration through solubility trapping is instantaneous, but quite small, as it amounts to ~50 g in 1 kg of water at the specified conditions of 60°C and $P_{tot} = P_{CO_2} = 100$ bar. Incorporation of CO_2 in product carbonates begins at the onset of dolomite precipitation (point D) and continues with the contributions of siderite from point S, calcite from point C and dawsonite from point Dw. Although mineral fixation is a late process, it becomes more and more important with time and the overall sequestration potential of the process is considerable, with 1,580 g kg^{-1} of water upon exhaustion of primary minerals. However, the contribution of mineral fixation equals that of solubility trapping only after 175 years, against the 5 years of the serpentinite case (*see* Section 7.3.2.5).

7.4.6. Changes in the porosity of aquifer rocks

Also for the tholeiitic basalt case, the $\Delta V/V$ parameter was computed and plotted as a function of both the reaction progress and time (Fig. 7.18) to estimate the variations in the porosity of aquifer rocks consequent to high-pressure CO_2 injection (*see* equation (7-48) and Section 7.3.2.6). The $\Delta V/V$ parameter exhibits a sharp increase when carbonate minerals begin to precipitate, achieves the 30% value soon after point C, where the aqueous solution saturates with calcite after 8 years and 5 months (ξ = 39.8 mmol), and attains the maximum value close to 50% after 250−1,500 years. Considering that the 30% value represents a sort of upper threshold even for high-porosity rocks, continuous CO_2 injection in a deep aquifer hosted in tholeiitic continental flood basalts is expected to reduce dramatically its porosity. The long-term sustainability of this process is therefore dubious.

These volume changes are not surprising, since the considered primary phases (plagioclase, clinopyroxene, orthopyroxene and magnetite) have very low molar volumes, a situation quite similar to the dunite example of Xu et al. (2000, 2004).

All in all, the basalt example provide contrasting results, as the positive findings concerning the potential of CO_2 sequestration of Na-bearing minerals and rocks, in the form of secondary dawsonite, are in striking contrast with the negative findings regarding the huge reduction in the effective porosity of the aquifer.

7.5. Reaction path modelling of geological CO_2 sequestration in basaltic glass

In Section 7.4, the continental tholeiitic flood basalt involved in CO_2 sequestration was considered holocrystalline, that is composed entirely of crystals. This is a limiting, somewhat unlikely case for volcanic rocks, which are usually composed by varying amounts of glass, mostly depending on their cooling history and age. Glasses are typically formed by quenching of silicate melts that prevents the organization of the random liquid structure into ordered crystalline structures. Glasses are metastable and are doomed to crystallize, soon or later. They are often believed to be more reactive than their crystalline counterparts, but this is not always the case, as discussed in Section 6.6.6.

The dissolution kinetics of silicate glasses has received considerable attention (*see* Sections 6.3.2 and 6.6.6) and, in particular, that of a basaltic glass from Stapafell mountain in southwestern Iceland was investigated experimentally and described in the framework of the multi-oxide dissolution model by Oelkers, Gislason and coworkers (Section 6.6.6.2). As already recalled, the chemical composition normalized to 1 Si atom of this basaltic glass is Si $Ti_{0.02}$ $Al_{0.36}$ $Fe(III)_{0.02}$ $Fe(II)_{0.17}$ $Mg_{0.28}$ $Ca_{0.26}Na_{0.08}$ $K_{0.008}$ $O_{3.38}$. As underscored by Oelkers and Gislason (2001), the basaltic glass from Stapafell is similar in chemical composition to both the average oceanic crust [Si $Ti_{0.02}$ $Al_{0.37}$ $Fe(III)_{0.04}$ $Fe(II)_{0.13}$ $Mg_{0.23}$ $Ca_{0.27}$ $Na_{0.10}$ $K_{0.008}$ $O_{3.35}$] and the average mid ocean ridge basalt [Si $Ti_{0.02}$ $Al_{0.36}$ $Fe(III)_{0.04}$ $Fe(II)_{0.13}$ $Mg_{0.24}$ $Ca_{0.25}Na_{0.10}$ $K_{0.003}$ $O_{3.33}$], which is usually indicated by the acronym MORB. Note that since the separate contents of Fe(III) and Fe(II) are not reported for the average MORB, its Fe(III)/Fe(II) ratio was assumed to be equal to that of average oceanic crust to distribute total Fe between the ferric and ferrous reservoirs. The basaltic glass from Stapafell mountain is also quite similar to the continental tholeiitic flood basalt considered in Section 7.4, whose chemical composition normalized to 1 Si atom is Si $Ti_{0.02}$ $Al_{0.38}$ $Fe(III)_{0.05}$ $Fe(II)_{0.14}$ $Mg_{0.22}Ca_{0.22}$ $Na_{0.11}$ $K_{0.01}$ $O_{3.34}$.

In spite of the compositional similarity, the basaltic glass from Stapafell mountain is obviously different from the holocrystalline continental tholeiitic flood basalt considered in previous section from the mineralogical point of view and, consequently, a different reaction path is expected upon interaction with water and CO_2. To evaluate these differences, in this section we simulate the geological storage of CO_2 in a hypothetical aquifer hosted in the basaltic glass from Stapafell mountain.

7.5.1. Setting up the water–rock interaction model

7.5.1.1. The need for two solid reactants

As anticipated in Section 6.6.6.2, Gislason and Oelkers (2003) considered the hydrolysis reaction (6-142) for the hydrated basaltic glass of composition $SiAl_{0.36}O_2(OH)_{1.08}$, which presumably forms a thin layer at the surface of the basaltic glass during its dissolution. Incidentally, note that the chemical formula of hydrated basaltic glass can also be expressed as $SiO_2 \cdot 0.18Al_2O_3 \cdot 0.54\ H_2O$, in terms of the component oxides, or $SiO_2 \cdot 0.36Al(OH)_3$, in terms of the hypothetical component minerals. The equilibrium constants of the hydrolysis reaction of hydrated basaltic glass (reaction 6-142) was evaluated by Gislason and Oelkers (2003) based on the latter chemical formula. First, they assumed that amorphous silica and gibbsite are the two hypothetical component minerals of hydrated basaltic glass, in a molar ratio 1:0.36. Then, they computed the equilibrium constants of the hydrolysis reaction of hydrated basaltic glass by summing log $K_{\text{sp, amorphous silica}}$ plus 0.36 log $K_{\text{sp, gibbsite}}$. The obtained log K_{sp} values of hydrated basaltic glass at varying temperatures are reported in Section 6.6.6.2.

Before running EQ6, the log K_{sp} of hydrated basaltic glass were re-computed at the temperature, pressure grid required by EQ3/6, through simple interpolation, and inserted in the thermodynamic database of EQ3/6. After this necessary preparatory step, the saturation state of any aqueous solution with respect to this hydrated basaltic glass can be computed by means of EQ3/6. Moreover, this hydrated basaltic glass can be treated as a pure mineral reactant and its dissolution can be properly simulated by running EQ6 (Table 7.12).

To this purpose, the dissolution rate law proposed by Gislason and Oelkers (2003), equation (6-143), is inserted in the EQ6 input file as a TST rate law involving the activities of both H^+ and Al^{3+} ions, as we have already seen in the albite example (*see* Section 7.2.4). It must be recalled that Gislason and Oelkers (2003) report the dissolution rate of the basaltic glass normalized to its geometric surface area, $r_{+,\text{GEOM}}$. However, they also underscore that these $r_{+,\text{GEOM}}$ are 92 times larger than the dissolution rates normalized to the BET surface area, $r_{+,\text{BET}}$. Both values were inserted in two separated EQ6 input files, to evaluate the effects of these limiting dissolution rates on the EQ6 simulation.

Only Al and Si are released to the aqueous solution through dissolution of the hydrated basaltic glass involved so far in the model. In contrast, we know that dissolution

TABLE 7.12

EQ6 input file for modeling high-pressure (P_{tot} = P_{CO_2} = 100 bar) CO_2 injection in a hypothetical deep aquifer hosted in basaltic glass at 60°C in kinetic mode (time frame). The rk0 value of 2.52370 × 10^{-10} mol cm^{-2} s^{-1} is the dissolution rate constant normalized to the geometric surface of the basaltic glass at 60°C (from Oelkers and Gislason, 2003). (Required input formats are not fully respected.)

```
The option switch IOPT1 is set to 1 to direct the code to compute the simulation in time frame.
The option switch IOPT4 is set to 1 to activate solid solutions.
The print option switch IOPR8 is set to 1 to direct the code to print a table of equilibrium gas fugacities at each
print point.
endit.
    nmodl1= 1                       nmodl2= 0
    tempc0= 60.0000E+00              jtemp= 0
       tk1= 0.00000E+00                tk2= 0.00000E+00        tk3= 0.00000E+00
    zistrt= 0.00000E+00              zimax= 1.00000E-01
     tstrt= 0.00000E+00             timemx= 0.00000E+00
    kstpmx= 500                      cplim= 0.00000E+00
    dzprnt= 1.00000E-02             dzprlg= 0.00000E+00     ksppmx= 100
    dzplot= 0.10000E+00             dzpllg= 0.00000E+00     ksplmx= 10000
     ifile= 60
*              1    2    3    4    5    6    7    8    9    10
  iopt1-10=    1    0    0    1    0    0    0    0    0    0
 iopt11-20=    0    0    0    0    0    0    0    0    0    0
  iopr1-10=    0    0    0    0    1    0    0    1    0    0
 iopr11-20=    0    0    0    0    0    0    0    0    0    0
  iodb1-10=    0    0    0    0    0    0    0    0    0    0
 iodb11-20=    0    0    0    0    0    0    0    0    0    0
*___________________________________________________________________________
     nxopt= 1
    option= all
    nxopex= 10
 exception= Gibbsite
 exception= Kaolinite
 exception= Chalcedony
 exception= Magnetite
 exception= Goethite
 exception= Dawsonite
 exception= Calcite
 exception= Magnesite
 exception= Siderite
 exception= Dolomite
*___________________________________________________________________________
      nffg= 1
   species= CO2(g)
     moffg= 1.00000E+01             xlkffg= +1.83187E+00
*___________________________________________________________________________
      nrct= 2
*___________________________________________________________________________
  reactant= Hyd-basalt-glass
     jcode= 0                        jreac= 0
      morr= 56.09000E+00              modr= 0.00000E+00
       nsk= 1                           sk= 4100.00E+00        fk= 0.00000E+00
```

TABLE 7.12 (*Continued*)

```
       nrk= 2                      nrpk= -1
     imech= 1
       rk0= 2.52370E-10            trk0= 6.00000E+01          iact= 0
      eact= 0.00000E+00            hact= 0.00000E+00
     ndact= 2                    csigma= 1.00000E+00
      udac= H+                     cdac= 1.00000E+00
      udac= Al+++                  cdac= -0.33333E+00
*-----------------------------------------------------------------------------
  reactant= rock
     jcode= 2                     jreac= 0
      morr= 56.09000E+00           modr= 0.00000E+00
     vreac= 0.00000E+00
      Fe        1.9000E-01
      Mg        2.8000E-01
      Ca        2.6000E-01
      Na        8.0000E-02
      O         7.8000E-01
endit.
       nsk= 0                        sk= 0.00000E+00            fk= 0.00000E+00
       nrk= 1                      nrpk= 0
       rk1= 1.00000E+00             rk2= 0.00000E+00           rk3= 0.00000E+00
*-----------------------------------------------------------------------------
    dlzidp= 0.00000E+00
     tolbt= 0.00000E+00           toldl= 0.00000E+00          tolx= 0.00000E+00
    tolsat= 0.00000E+00          tolsst= 0.00000E+00
    screw1= 0.00000E+00          screw2= 0.00000E+00        screw3= 0.00000E+00
    screw4= 0.00000E+00          screw5= 0.00000E+00        screw6= 0.00000E+00
    zklogu= 0.000                zklogl= 0.000               zkfac= 0.000
    dlzmx1= 0.00000E+00          dlzmx2= 0.00000E+00        nordlm= 0
    itermx= 0                    ntrymx= 0
    npslmx= 0                    nsslmx= 0                  ioscan= 0
*-----------------------------------------------------------------------------
* pickup file written by EQ3NR, version 7.2b (R139 )
* supported by EQLIB, version 7.2b (R168 )
Water in equilibrium with analcime, calcite, kaolinite,
chalcedony, daphnite, clinochlore, magnetite
at 60ØC, 0.1 bar PCO2
endit.
    tempci= 6.00000E+01
    nxmod= 0
     iopg1= 0                     iopg2= 0                   iopg3= 0
     iopg4= 0                     iopg5= 0                   iopg6= 0
     iopg7= 0                     iopg8= 0                   iopg9= 0
    iopg10= 0
       kct= 10                      ksq= 11                    kmt= 11
       kxt= 11                     kdim= 11                   kprs= 0
O         5.568190070346440E+01
Al        2.397285511182285E-07
Ca        4.360523578853289E-05
Cl        3.000000000355616E-02
Fe        3.862527706607725E-06
```

TABLE 7.12 (*Continued*)

H	1.110727982477859E+02	
C	5.793621436309842E–02	
Mg	1.602243822990042E–04	
Na	8.637858497535496E–02	
Si	6.159324420874708E–04	
electr	1.323234709603221E–11	
H_2O	H_2O	1.744358983526984E+00
Al^{+++}	Al^{+++}	– 1.710395611381481E+01
Ca^{++}	Ca^{++}	– 4.562090256614116E+00
Cl^-	Cl^-	– 1.527663478385729E+00
Fe^{++}	Fe^{++}	– 6.567807046642986E+00
H^+	H^+	– 7.592429061239226E+00
HCO_3^-	HCO_3^-	– 1.269686271112627E+00
Mg^{++}	Mg^{++}	– 3.959985610775256E+00
Na^+	Na^+	– 1.076209250999970E+00
SiO_2(aq)	SiO_2(aq)	– 3.230700000000000E+00
O_2(g)	O_2(g)	– 6.098389925982678E+01

of basaltic glass adds also several other metals to the aqueous phase. To take into account the fate of all these metals, it is necessary to involve a second reactant in the EQ6 simulation. This second reactant can be treated as a so-called *special reactant* (in EQ6 meaning), that is a substance whose chemical composition is known but whose thermodynamic properties are not. Obviously, substances of this kind (e.g., any rock) cannot be inserted in the thermodynamic database of EQ3/6.

In the basaltic glass case, the special reactant has composition $Ti_{0.02}$ $Fe(III)_{0.02}$ $Fe(II)_{0.17}$ $Mg_{0.28}$ $Ca_{0.26}$ $Na_{0.08}$ $K_{0.008}$ $O_{0.82}$, since it is the Si-free, Al-free portion of the whole basaltic glass. Titanium and K are not considered in the EQ6 simulation owing to the low solubility of Ti minerals and the low content of K, which was also neglected in the tholeiitic basalt case. The number of oxygen atoms have to be decreased, therefore, to 0.78 to balance the metal charges. Note that it is very important to maintain the chemical formula of special reactants stoichiometrically balanced by adjusting the number of oxygen atoms, otherwise incorrect f_{O_2} values are computed by EQ6. Besides, it is necessary to instruct EQ6 to dissolve this special reactant at the same rate of the hydrated basaltic glass. This is accomplished by setting the rsk1 flag to 1.

7.5.1.2. The molar volume of basaltic glass

Let us assume again that the considered system comprises 1,000 g of water and has an initial porosity of 0.3. Equation (7-44) dictates that initially basaltic glass occupies a volume of 2,333 cm^3. The molar volume of the basaltic glass is needed to compute both the moles of basaltic glass initially present in the system and the changes in the $\Delta V/V$ parameter during the simulation by using equation (7-48).

The molar volume of the basaltic glass can be calculated based on the partial molar volumes and the volumetric thermal expansion coefficients of the component oxides, whereas their compressibilities can be neglected at the relatively low pressures of interest.

These volumetric parameters of the component oxides of silicate melts and glasses have been the subject of several investigations (e.g., Bottinga et al., 1982; Lange and Carmichael, 1987, 1990; Knoche et al., 1995; Lange, 1997; Toplis and Richet, 2000). Inserting the partial molar volumes $\bar{V}_i$ (cm^3 mol^{-1}) and the thermal expansions $d\bar{V}_i/dT$ (cm^3 mol^{-1} K^{-1}) given by Lange and Carmichael (1987) into the simplified relation (*T* in *K*):

$$v_{\mathrm{liq}} = \sum_i X_i \cdot \left[\bar{V}_{i,1673K} + \frac{d\bar{V}_i}{dT} \cdot (T - 1673) \right] \quad (7\text{-}49)$$

the molar volume of the basaltic glass of interest at 60°C, 1 bar results to be 21.05 cm^3 mol^{-1}, corresponding to 41.60 cm^3 mol^{-1} for the chemical formula normalized to 1 Si atom. Therefore, the moles of basaltic glass initially contained in the considered system amount to 2,333 cm^3/41.60 cm^3 mol^{-1} = 56.09 mol. Some words of caution are needed on the possible contribution of water to the molar volume of the basaltic glass. Water was not considered in previous calculations since its content in the basaltic glass of Stapafell is not reported by Oelkers and Gislason (2001) and Gislason and Oelkers (2003). Owing to its low molecular weight (18.015 g mol^{-1}) compared to other component oxides of silicate melts and glasses, relatively small amounts by weight of water correspond to much higher amounts on a molar scale. Since the partial molar volume of water in silicate glasses is relatively low (Richet and Polian, 1998 estimated a value of 12 ± 0.5 cm^3 mol^{-1} for andesite glasses), the molar volume estimated for the anhydrous basaltic glass of Stapafell is doomed to decrease, albeit to an unknown extent, upon consideration of water.

7.5.1.3. Completion of the EQ6 input file

The preparation of the input file for the EQ6 simulation is completed by setting temperature at 60°C, f_{CO_2} at 67.9 bar (which means to assume injection of pure CO_2 at a constant pressure of 100 bar) and the surface area of hydrated basalt glass in contact with 1,000 g of water at the initial value of 4,100 cm^2 (which is automatically changed by EQ6 proportionally to its remaining mass). Note that the values of these variable are the same as in previous cases. As in the tholeiitic basalt run, the initial aqueous solution was assumed to be in equilibrium with analcime, calcite, kaolinite, daphnite, clinochlore, chalcedony and magnetite at 60°C, 1 bar under a f_{CO_2} of 10^{-1} bar (*see* Section 7.1.2).

Again, instantaneous precipitation was hypothesized for the secondary solid phases, as it is difficult to constrain the initial value of the surface area for minerals which are not present in the system at the beginning of the simulation.

7.5.2. The time scales

The main results of reaction path modelling are depicted against both the reaction progress variable, ξ, and time in the diagrams of Figures 7.19 to 7.23, in which two time scales are reported. They refer to the dissolution rate of the basaltic glass normalized to either the geometric surface area, $r_{+,GEOM}$, or the BET surface area, $r_{+,BET}$. As expected on

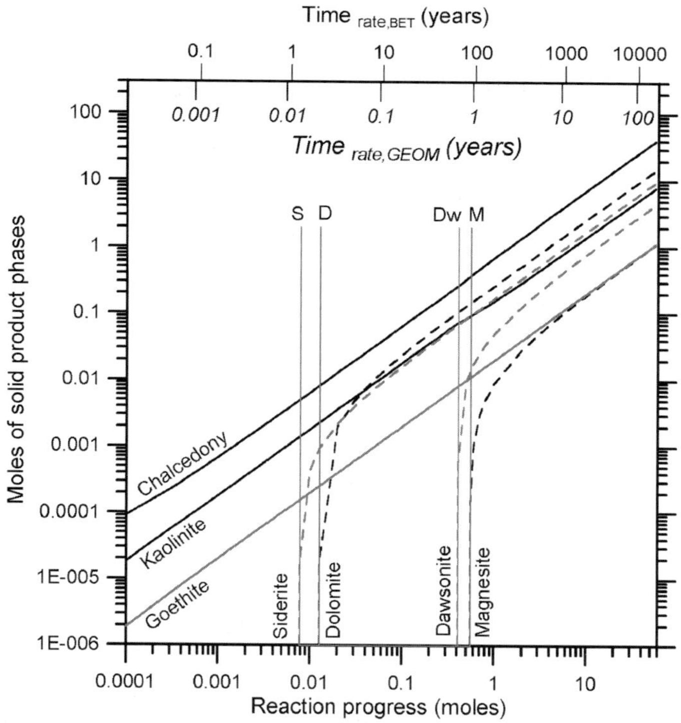

Figure 7.19. Amounts of secondary product phases generated during high-pressure ($P_{tot} = P_{CO_2} = 100$ bar) CO_2 injection in a hypothetical deep aquifer ($T = 60°C$) hosted in basaltic glass as a function of time and the reaction progress variable. Letters S, D, Dw and M mark the onset of the precipitation of secondary siderite, dolomite, dawsonite and magnesite, respectively.

the basis of equation (7-16), the $r_{+,GEOM}/r_{+,BET}$ ratio of 92 constrains the $time_{rate,GEOM}/time_{rate,BET}$ ratio at 1/92 and brings about a corresponding shift in the two time scales.

The results of the two separate simulations (concentrations of solutes, pH of the aqueous phase, oxygen fugacity, moles of secondary solid phases, etc.) are exactly the same for any given value of the reaction progress variable, ξ. However, the same values of ξ are attained at different times in the two simulations.

This example demonstrates once more that it is advisable to consider ξ as the true master variable, owing to possible uncertainties in the time scale brought about by poor knowledge of the dissolution rates and/or reactive surface areas of solid phases.

7.5.3. Solid product phases

The secondary carbonate minerals generated during high-pressure CO_2 injection in a hypothetical deep aquifer hosted in basaltic glass are, in order of appearance, siderite, dolomite, dawsonite and magnesite (Fig. 7.19). The precipitation of these solid phases

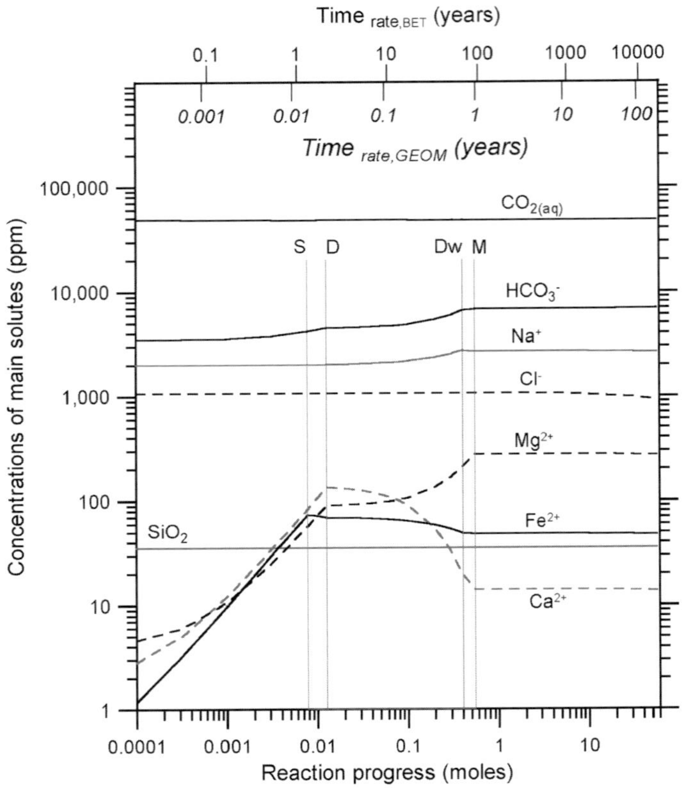

Figure 7.20. Concentrations of solutes during high-pressure ($P_{tot} = P_{CO_2} = 100$ bar) CO_2 injection in a hypothetical deep aquifer ($T = 60°C$) hosted in basaltic glass as a function of time and the reaction progress variable. Letters S, D, Dw and M mark the onset of the precipitation of secondary siderite, dolomite, dawsonite and magnesite, respectively, as shown in Figure 7.19.

begins at values of ξ indicated by the labels S, D, Dw and M, respectively. These labels are also reported in the following plots. The most important of these solid phases, in terms of mass, is dolomite followed by siderite, dawsonite and magnesite.

Comparison of Fig. 7.19 with Fig. 7.14 shows that magnesite precipitation occurs in the basaltic glass run instead of calcite production during the tholeiitic basalt example. At first sight, this difference might seem surprising, since the Ca/Mg molar ratio of the basaltic glass, 0.93, is similar to that of the tholeiitic basalt, 1.000. In order to explain this difference in secondary carbonate minerals let us consider the two dissolution processes in detail.

As discussed in Section 7.4.2, the dissolution of the holocrystalline tholeiitic basalt is dominated by destruction of clinopyroxene (10% of Ca and 7.5% of Mg on an elemental basis) and plagioclase (4.2% of Ca and no Mg), with a subordinate contribution of orthopyroxene (12% of Mg and no Ca), apart from the last steps of the simulation, upon exhaustion of clinopyroxene, when orthopyroxene dissolution becomes important. This

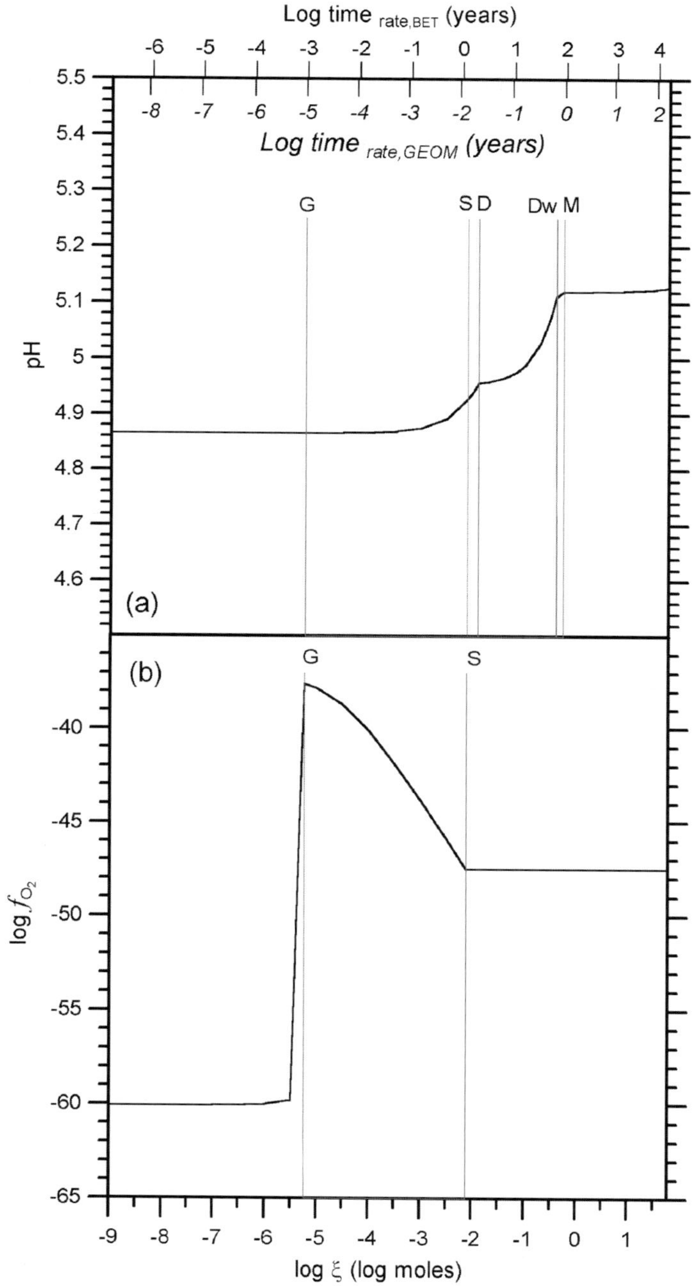

Figure 7.21. Changes in (a) pH and (b) f_{O_2} during high-pressure ($P_{tot} = P_{CO_2} = 100$ bar) CO_2 injection in a hypothetical deep aquifer (T = 60°C) hosted in basaltic glass as a function of time and the reaction progress variable. Letters G, S, D, Dw and M mark the onset of the precipitation of secondary goethite, siderite, dolomite, dawsonite and magnesite, respectively.

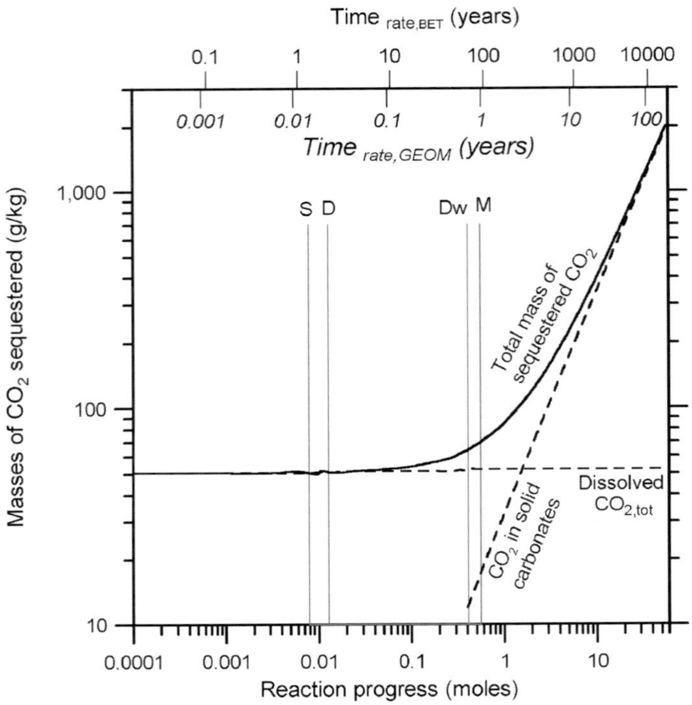

Figure 7.22. High-pressure ($P_{tot} = P_{CO_2} = 100$ bar) CO_2 injection in a hypothetical deep aquifer ($T = 60°C$) hosted in basaltic glass: masses of CO_2 sequestered through dissolution in the aqueous solution (solubility trapping) and incorporation in solid carbonates (mineral fixation), and total mass of sequestered CO_2 as a function of time and the reaction progress variable. Letters S, D, Dw and M mark the onset of the precipitation of secondary siderite, dolomite, dawsonite and magnesite, respectively, as shown in Figure 7.19.

larger availability of Ca with respect to Mg during the dissolution of tholeiitic basalt is in line with the production of calcite and dolomite in similar amounts.

In contrast, dissolution of basaltic glass contributes chemical constituents to the aqueous phase in the same stoichiometric proportions of the glass itself, i.e. with a Ca/Mg molar ratio of 0.93. This explains the overwhelming production of dolomite accompanied by subordinate magnesite.

Comparison of Fig. 7.19 with Fig. 7.14 also shows that during the dissolution of the basaltic glass, precipitating carbonate minerals are accompanied by secondary chalcedony, kaolinite and goethite in amounts comparable with those of the tholeiitic basalt case.

7.5.4. The aqueous solution

Inspection of Figs. 7.20 and 7.15 indicates that the concentrations of aqueous CO_2, HCO_3^-, Na^+, Cl^-, Fe^{2+} and SiO_2 during high-pressure CO_2 injection in a hypothetical deep aquifer hosted in basaltic glass are comparable with those of the tholeiitic basalt simulation. Also the concentration of dissolved Al^{3+} (not shown) is similar in the two runs. This

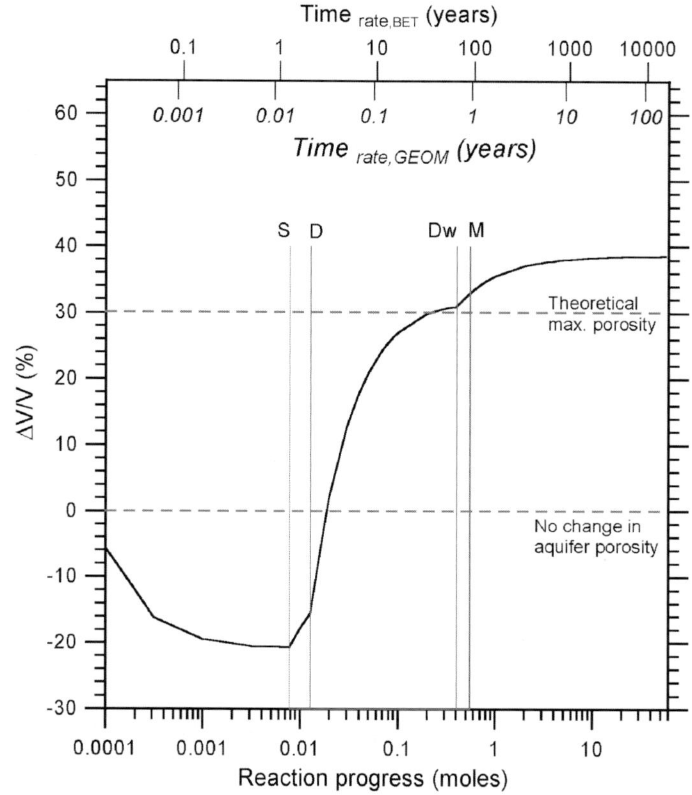

Figure 7.23. Variation of the $\Delta V/V$ parameter as a function of both reaction progress and time during high-pressure ($P_{tot} = P_{CO_2} = 100$ bar) CO_2 injection in a hypothetical deep aquifer ($T = 60°C$) hosted in basaltic glass. Letters S, D, Dw and M mark the onset of the precipitation of secondary siderite, dolomite, dawsonite and magnesite, respectively, as shown in Figure 7.19.

is due to the fact that imposed temperature and f_{CO_2} and the relevant product solid phases (chalcedony, kaolinite, siderite and dawsonite) are the same.

In contrast, the aqueous solution interacting with the basaltic glass is richer in dissolved Mg^{2+} and poorer in Ca^{2+} with respect to the aqueous phase circulating into the tholeiitic basalt. This difference in dissolved Mg^{2+} and Ca^{2+}concentrations reflects the different dissolution behaviour of the two basaltic rocks, as discussed in the previous section.

Not surprisingly, also the pH values during the basaltic glass simulation span a limited range, from 4.87 to 5.13, (Fig. 7.21a) which coincides with that of the tholeiitic basalt run, from 4.87 to 5.12 (Fig. 7.16a).

The f_{O_2} values predicted by the two simulations are also identical (see Figs. 7.21b and 7.16b), and are largely controlled by precipitation of goethite (after point G) and coexistence of goethite and siderite (after point S), as already noted above.

7.5.5. The CO_2 sequestration

Figure 7.22 depicts the amounts of CO_2 sequestered through both solubility trapping (i.e. as aqueous CO_2 and other dissolved carbonates) and mineral fixation (i.e. the CO_2 stored in dolomite, siderite, dawsonite and magnesite), as well as the total mass of sequestered CO_2 as a function of the reaction progress. The three CO_2–ξ curves are almost identical to those of the tholeiitic basalt case (Fig. 7.17) and the same considerations of Section 7.4.5 apply also to the basaltic glass example.

Considering the process with respect to time, it must be underscored that the time scale of basaltic glass carbonation, based on the results obtained for $r_{+,\mathrm{BET}}$, is comparable to that of the carbonation of holocrystalline tholeiitic basalt. In contrast, basaltic glass carbonation occurs two orders of magnitude faster than the carbonation of holocrystalline tholeiitic basalt if the results computed for $r_{+,\mathrm{GEOM}}$ are taken into account. This matter requires further investigation for its important practical implications, in that basaltic glass and other silicate glasses could be very interesting materials for the geological storage of CO_2 and mineral carbonation, if the $r_{+,\mathrm{GEOM}}$ values were representative of their dissolution rates.

7.5.6. Changes in the porosity of aquifer rocks

During high-pressure CO_2 injection in a hypothetical deep aquifer hosted in basaltic glass the $\Delta V/V$ parameter experiences a pronounced increase upon the onset of siderite and dolomite precipitation (points S and D, respectively) and a corresponding marked decrease in aquifer porosity is expected. Upon the onset of dawsonite precipitation (point Dw), the $\Delta V/V$ parameter exceeds 30%, which is a sort of upper threshold even for high-porosity rocks. Therefore, total clogging of pore spaces and voids is expected after point Dw.

Although the $\Delta V/V$ parameter remains below 40% throughout the simulation and it is somewhat lower than in the tholeiitic basalt case, the same negative consideration of Section 7.4.6 applies also in this case. To be quite explicit, the long-term sustainability of the process under consideration appears unlikely, owing to the dramatic reduction in the effective porosity of the aquifer. However, further investigations are warranted on the molar volume of silicate glasses and the role of water in these potentially interesting materials (*see* Section 7.5.1.2).

7.6. Reaction path modelling of geological CO_2 sequestration in sedimentary basins

As shown above, ultramafic and mafic rocks have very high CO_2 sequestration capacities. Nevertheless, they are not in general the best targets for geological CO_2 disposal, owing to an unfavourable geological–hydrogeological framework.

First, rock matrix has low porosity and permeability, which determine low availability of water. The permeability of lavas develops mainly along fractures, cooling joints and discontinuous scoriaceous layers. The permeability of ophiolitic complexes is largely due to fracturing induced by the tectonic events that have determined the uplift of these bodies towards the surface and their emplacement at relatively shallow levels.

Second, effective cap rocks are unlikely to be present at the top of tholeiitic basalt lava successions and ultramafic rocks variably affected by serpentinization, although relatively impervious layers may be locally present within these rock sequences. For instance, tuffs affected by argillic alteration may be associated with tholeiitic basalt lavas and clay-rich, deep-sea sediments usually occur together with ultramafic rocks in ophiolitic sequences. However, these impervious levels are randomly interposed into them, have limited lateral continuity and/or irregular shape due to the effects of tectonic events.

Favourable situations may be locally present, such as in southern Piedmont, Italy (Roberto Bencini, written communication). In this region, CO_2 injection could be performed in the conglomerate layers, chiefly made up of serpentinitic clasts, which occur at the top of the serpentinites of the Voltri Group. Clogging of pore spaces should not be a major problem in these conglomerates, due to their high porosity and permeability. Besides, the overlying sedimentary rocks of the Piedmont Tertiary Basin should act as an efficient cap rock, preventing the escape of gaseous CO_2 towards the surface. This fact must be absolutely avoided for obvious safety reasons! Further studies are needed to prove the feasibility of geological CO_2 disposal in this region.

In contrast to mafic and ultramafic rocks, sedimentary basins are very promising targets for geological CO_2 disposal. They have a favourable geological–hydrogeological framework, owing to the presence of both (i) high-porosity, high-permeability layers typically represented by sandstones and (ii) low-permeability layers of shales, potentially limiting the transfer of CO_2 towards the surface. Besides, sedimentary basins host huge amounts of waters. At sufficiently deep levels, these waters typically have high salinities and, therefore, they are of no interest as sources of drinking water. The long residence times of these waters in the deep aquifers of sedimentary basins is an other requirement for the occurrence of carbonation reactions (mineral trapping) that are very slow processes, as already discussed above.

Sedimentary basins were often drilled at depth for the exploitation of oil and natural gas. Upon completion of the extraction of these valuable resources and perhaps after a first phase in which high-pressure CO_2 is injected for enhanced oil recovery, abandoned wells can be used for geological CO_2 disposal, without the need for drilling any new borehole. This is a very attractive situation, from the economic point of view, and money often makes things move.

For instance, 1 million ton of waste CO_2 per year has been injected since September 1996 into the Utsira aquifer in the North Sea, to avoid a Norwegian tax of about 50 US$ per ton of CO_2 emitted offshore (Baklid et al., 1996; Torp and Gale, 2004). An other example is constituted by the Weyburn Oil Field in south central Saskatchewan, Canada which is currently the site of a large CO_2 injection project for enhanced oil recovery (Emberley et al., 2004, 2005). Although this is not a geological CO_2 sequestration project, approximately 30% of the injected CO_2 will be trapped in the reservoir at the end of project activities (Gunter et al., 2000). Besides, the geological CO_2 sequestration in sedimentary basins has been the subject of feasibility studies in The Netherlands (Lohuis, 1993) and in the Alberta Basin, Canada (Gunter et al., 1993; Bachu et al., 1994; Law and Bachu, 1996; Gunter et al., 1996, 1997, 2000).

Among the numerous modelling investigations on the geological CO_2 sequestration in sedimentary basins, some selected examples are discussed in the following sections. Related experimental and field studies are treated in Sections 7.7 and 7.8.

7.6.1. The glauconitic sandstone aquifer of the Alberta Sedimentary Basin

Reaction path modelling of geological CO_2 sequestration in a glauconitic sandstone aquifer of the Alberta Sedimentary Basin, Western Canada was studied by Gunter et al. (1997, 2000) by using the PATHARC.94 computer code (Perkins and Gunter, 1995).

This glauconitic sandstone aquifer has an average porosity of 12% and its average mineral composition comprises 87% quartz, 2% K-feldspar, 1% plagioclase (which was taken into account in the geochemical simulation introducing the two endmember components albite and anorthite), 5% glauconite (which was hypothesized to be represented by annite in reaction path modelling), 2% kaolinite, 1% calcite, 1% dolomite and 1% siderite. The reactive surface area was assumed to be equal to the geometrical surface area and individual mineral grains were considered constant-volume sphere with an average radius of 0.05 mm. The average temperature of the aquifer is 54°C. The average chemical composition of formation water is (concentration in mg L^{-1}) Na 28,800, K 690, Ca 2,970, Mg 578, HCO_3 198, Cl 51,600 and SO_4 366, with a pH of 7.2.

Gunter et al. (1997) simulated the injection of CO_2 into the glauconitic sandstone aquifer at a constant P_{CO_2} of 260 bar. Substantial trapping of CO_2 through precipitation of siderite (and subordinate calcite) and solubility trapping was predicted to take place in a time considerably smaller than the residence time of water in the sedimentary basin aquifer. The main occurring reaction was considered to be

$$3\text{annite} + 11CO_2 \rightarrow \text{muscovite} + 9\text{siderite} + 6\text{quartz} + H_2O + 2K^+ + 2\,HCO_3^-. \quad (7\text{-}50)$$

Gunter et al. (2000) chosen instead to dissolve a finite amount of CO_2, 1 mol, in 1,000 g of water at the beginning of the simulation. In this way, P_{CO_2} decreases with ξ, simulating the conditions prevailing upon termination of CO_2 injection. They found that equilibrium is expected to be attained in hundreds of years with a drop of P_{CO_2} from 87 to 0.02 bar and the generation of 1 mol of siderite, 0.05 mol of calcite and 0.02 mol of dolomite. These figures confirm the relatively high CO_2-trapping capacity of siliciclastic rocks containing Fe(II) minerals, such as annite.

However, these findings were questioned by Xu et al. (2000, 2004) who modelled the geological sequestration of CO_2 in the same glauconitic sandstone aquifer of the Alberta Sedimentary Basin by means of TOUGHREACT. They recognized that use of annite [$KFe(II)_3AlSi_3O_{10}(OH)_2$] as a proxy for glauconite [$K_{0.75}Mg_{0.25}Fe(II)_{0.25}Fe(III)_{1.25}Al_{0.5}Si_{3.75}O_{10}(OH)_2$] leads to overestimate the amount of precipitating siderite and, consequently, the CO_2-sequestration capacity. Therefore they considered glauconite in reaction path modelling instead of annite. They also introduced oligoclase instead of albite plus anorthite and illite instead of muscovite. Note that the latter choice determines a moderate increase in the CO_2-sequestration capacity, owing to the presence in illite of some Mg

$[K_{0.6}Mg_{0.25}Al_{1.8}(Al_{0.5}Si_{3.5}O_{10})(OH)_2]$ which is absent in muscovite. Besides, they assumed the presence of organic matter, namely type II kerogen.

To keep the description of precipitation kinetics to a simple level, backward rate constants were assumed to be one order of magnitude larger than forward (dissolution) rate constants. The total surface area was set at 10 $m^2\ dm^{-3}$ of medium, corresponding to a surface of $8.33 \times 10^5\ cm^2$ in contact with 1,000 g of water based on the average porosity of 12%. The surface area of secondary minerals was set to 0.25 $m^2\ dm^{-3}$ throughout the simulation, apart from clay minerals, whose surface area was assumed to be one order of magnitude higher, for glauconite, and two orders of magnitude higher, for all the other clay minerals. Note that some of these phases are not initially present in the considered system, but they are considered to have an initial surface area different from zero. This arbitrary assumption (or an alternative one, e.g., instantaneous equilibrium as we assumed above) is needed to describe the precipitation of secondary minerals in a kinetic framework.

The initial aqueous solution was assumed to be a 1 M NaCl solution. The geological CO_2 sequestration was simulated by setting the CO_2 injection pressure at 260 bar and temperature at 54°C at all times. The main precipitating solid phases resulted to be illite, K-feldspar, siderite, and ankerite accompanied by minor amounts of dawsonite and dolomite. Carbon dioxide is mainly incorporated in ankerite and siderite. After ~15,000 years, the total CO_2 sequestration capacity (solubility trapping plus mineral trapping) increases close to linearly with time, attaining the value of 17 kg m^{-3} of medium in 100,000 years. This corresponds to approximately 140 g kg^{-1} water, a value which is attained much earlier in the cases of dunite, serpentinite, tholeiitic basalt and basaltic glass that were discussed above. This indicates that the sequestration capacity of glauconitic sandstones is much lower than those of previously considered lithotypes, in line with the smaller total contents of Mg-, Ca-, Fe(II)- and Na-bearing silicates. Porosity is expected to decrease very slightly with time from the initial value of 12% to 11.6% in 100,000 years.

A possible uncertainty relates to the exclusion of reduced sulphur species from geochemical modelling. Indeed, Xu and coworkers recognize that even low P_{H_2S} would determine the production of pyrite, which could fix a significant proportion of Fe^{2+}, thus decreasing the amount of precipitating siderite (and ankerite, but calcite and dolomite would form instead of it) and, consequently, the CO_2 sequestration capacity of the process.

7.6.2. The sediments of the Gulf Coast

After a preliminary study of Apps (1996), reaction path modelling of geological CO_2 sequestration in the Gulf Coast sediments was investigated by Xu et al. (2000, 2004), again by means of TOUGHREACT.

These sediments have been considered to have an average porosity of 10% and to be made up of 49.7% (by volume) quartz, 1.8% kaolinite, 1.7% calcite, 0.85% illite, 2.6% organic matter (type II kerogen), 17.8% oligoclase, 7.4% K-feldspar, 3.5% Na-smectite, 2.7% clinochlore-14Å, 1.4% daphnite-14Å, 0.44% haematite. The total surface area was assumed to be 10 $m^2\ dm^{-3}$ of medium, corresponding to a surface of $1 \times 10^6\ cm^2$ in contact with 1,000 g of water.

Again, a 1 M NaCl solution was taken into account as initial aqueous solution. The geological CO_2 sequestration was simulated by setting the CO_2 injection pressure at 260 bar and temperature at 80°C. The most important product solid phases resulted to be illite, dawsonite and ankerite, whereas calcite and siderite initially precipitated and then re-dissolved. Carbon dioxide is mainly incorporated in dawsonite and ankerite. Precipitation of carbonate minerals determines an increase in the total CO_2 sequestration capacity up to a maximum of 90 kg m^{-3} of medium (900 g kg^{-1} water) after 100,000 years. Porosity decreases with time from the initial value of 10% to a value somewhat higher than 6% after 100,000 years. The sequestration capacity of these sediments is significantly higher than that of the glauconitic sandstones, mainly owing to the abundance of oligoclase and chlorite minerals.

The geochemical processes occurring in a sandstone–shale system representative of the Gulf Coast sediments upon injection of high-pressure CO_2 at depths of ~2 km were investigated further by Xu et al. (2005). They performed numerical simulations of reactive fluid flow and geochemical transport by fully exploiting the capabilities of the computer code TOUGHREACT. Although the theoretical treatment of this subject goes beyond the aims of this book (the interested reader is referred to Lichtner et al., 1997), it is very interesting to give a look at the results of this type of advanced geochemical–hydrogeological modelling.

Xu et al. (2005) adopted a simplified one-dimensional model made up of a 10-m-thick layer of shale overlying a layer of sandstone of the same thickness. The model grid (comprising all the blocks in this 1-D space) is finer in the shale near the sandstone/shale boundary, to appreciate the concentration gradients that are expected to develop in the shale, owing to its low permeability. In contrast, perfect mixing was assumed to occur in the sandstone layer, which was modelled by using a single block.

The same primary minerals listed above were considered to be present in the sandstone and in the shale layers, but in different amounts. Initial porosity was assumed to be 30% in the sandstone and 10% in the shale. Initial water chemistry was computed by reacting a 1 M NaCl solution with the primary minerals at 75°C under a P_{CO_2} of 0.01 bar. The obtained solution composition was taken as an initial condition in reactive transport modelling.

At the beginning of the reactive transport simulation, CO_2 was added to the sandstone block to achieve a gas phase saturation of 0.5 and a P_{CO_2} of 201 bar, whereas no CO_2 was added to the blocks of the shale layer, which were considered to be water saturated.

Changes with time in both water chemistry and the amounts of primary and secondary minerals along the sandstone–shale transect are computed by TOUGHREACT. For instance, the H^+ ion and aqueous CO_2 are found to diffuse from the sandstone to the shale, in line with the high P_{CO_2} imposed to the sandstone, which increases the H^+ activity of the aqueous phase. Initially carbonate mineral precipitation takes place chiefly in the sandstone, but afterwards it extends to the shale (though in lower amounts) owing to diffusive transport of CO_2 from the sandstone to the shale. Total dissolved Fe and Ca are expected to diffuse from the shale to the sandstone, since precipitation of siderite and ankerite causes a decrease in Fe^{2+} and Ca^{2+} concentrations in the sandstone. Precipitation of carbonate minerals (mainly dawsonite and ankerite) brings about a decrease in porosity from

30% to 24% after 100,000 years in the sandstone and rather complex changes in porosity in the shale.

The impacts of NO_x, SO_x and H_2S (which are typically present together with CO_2 in coal-fired waste streams) on the geological disposal of CO_2 in the sands of the Frio Formation was investigated by Knauss et al. (2005, see also Knauss et al., 2001). They carried out reactive transport modelling by using the reactive transport simulator CRUNCH and showed that the presence of H_2S should not interfere with CO_2 injection, whereas oxidation of small amounts of SO_2 could determine a significant decrease in pH and substantial differences in the process. This study has a very important applied component as the separation and compression of CO_2 is very expensive and represent the 75% of the total cost of the whole geological disposal process.

7.6.3. The White Rim Sandstone

This case history was investigated by White et al. (2005) along a cross section beneath the Hunter Power Plant in Central Utah, emitting yearly 15×10^6 tons of CO_2. The Permian White Rims Sandstone with a thickness of 61–244 m, high porosity and permeability and lying at over 1 km depth beneath the power plant was selected as the best potential reservoir. It is overlied by some potential seals, represented by shale and evaporite levels. In the investigated area, the sedimentary sequence is a sort of gently dipping cuesta, constituting the western flank of the San Rafael asymmetrical anticline. Owing to the regional dip, the White Rims Sandstone crops out at the surface some 40 km to the SE, posing a possible problem of lateral migration and approach to the surface of the CO_2-charged aqueous solution after prolonged CO_2 disposal.

Particular attention was devoted to the sealing properties of potential cap rocks, which were measured through mercury injection porosimetry by White et al. (2005). As an example, it was found that the second most effective cap rock (Mancos Shale) could withhold a CO_2 column of the order of 500–900 m in thickness.

Reactive transport modelling of CO_2 injection in the White Rims Sandstone was carried out by using the ChemTOUGH code (White, 1995), which is based on the porous media multiphase mass and energy flow code TOUGH2 (Preuss, 1991).

The reservoir rock was assumed to be made up of quartz (77% by volume), kaolinite (2.25%), Na-smectite (2.25%), calcite (1.8%), anorthite (0.66%), albite (0.60%), K-feldspar (0.60%). Note that the complement to 100% is represented by porosity, approximately 15%. Dolomite, siderite and dawsonite are not initially present in the reservoir rock but were considered as possibly precipitating solid phases. Aquifer temperature was fixed at 54°C. Initial water chemistry was computed by reacting a 3 M NaCl brine with the reservoir minerals for 1,000 years.

In the simulation, injection of CO_2 was terminated after 30 years. It was found that 1,000 years after the end of the injection period ~21% of the injected CO_2 was trapped in carbonate minerals (calcite and dawsonite), ~52% was present underground as a separate gas phase or dissolved in groundwaters and ~17% had leaked to the ground surface, a process that was still going on.

7.6.4. The shales of the North Sea

Gaus et al. (2005) carried out reaction path modelling (both in stoichiometric mode and in time frame) and reactive transport modelling to investigate the impact of high-pressure CO_2 injection on the porosity of the cap rock at the Sleipner site, Norwegian part of the North Sea, where CO_2 has been injected since 1996. The code PHREEQC v2.6 (Parkhurst and Appelo, 1996) was used to this purpose.

Carbon dioxide is currently injected in the sands of the Utsira Formation but it migrates upward, due to buoyancy effects, and accumulates below the overlying cap rock, represented by the shales of the 250-m-thick Nordland Group. Then CO_2 dissolves into the formation water of the cap rock and diffuse upward into it. At Sleipner, the migration of injected CO_2 is predicted and monitored by means of seismic surveys and reservoir modelling (Arts et al., 2004; Torp and Gale, 2004).

The mineral constituents of the cap rock considered in geochemical modelling are illite (24.7 wt.%), chalcedony (21.5%), kaolinite (18.0%), plagioclase (12.4%), a high-Fe–Mg smectite (8.8%), 14 Å-clinochlore (4.1%), pyrite (2.8%), K-feldspar (2.1%), siderite (1.6%) and calcite (1.0%). In three separate runs plagioclase was considered to be represented by either albite, or 50% albite plus 50% anorthite, or oligoclase (An_{20}).

Initial porosity was set to 5%. Initial chemistry of formation water was computed equilibrating an aqueous solution containing 0.479 M chloride with cap rock minerals at 37°C.

Isothermal, isobaric injection of CO_2 was assumed at P_{CO_2} of 101 bar (f_{CO_2}=52 bar) and temperature of 37°C. In addition to primary minerals, dawsonite, disordered-dolomite and magnesite were selected as possible secondary minerals, defining suitable rate constants and reactive surface areas different from zero (these are arbitrary but necessary assumptions, as already noted above).

Reaction path modelling in time frame shows that the main secondary carbonate mineral is calcite accompanied by a limited amount of dawsonite in all the three runs with different plagioclase composition. Small changes in porosity (<0.05 %) are calculated for the albite case, whereas a certain decrease in porosity (up to 2.8% in the first meter) is predicted for the albite plus anorthite case. This is attributed to the dissolution of anorthite and the subsequent precipitation of calcite and kaolinite by Gaus et al. (2005). It must be noted that the volume of the reaction:

$$CaAl_2Si_2O_{8(s)} + CO_{2(g)} + 2H_2O \rightarrow CaCO_{3(s)} + Al_2Si_2O_5(OH)_{4(s)}, \quad (7\text{-}51)$$

ΔV_r, is 35.7 cm^3 (per mole of anorthite, that is referring to the reaction as it is written) and the corresponding $\Delta V/V$ parameter is 35.4% (*see* equation (5-125)), if consideration is restricted to the solid phases only, as was done in the previous sections of this chapter and in Section 5.5. The ΔV_r reduces to -0.37 cm^3 and the $\Delta V/V$ parameter becomes -0.27% if water is also included in calculations as evidently assumed by Gaus et al. (2005). In my opinion, this assumption is reasonable as long as water is forced to remain in the considered system, which is rather unrealistic in natural systems. Therefore, porosity changes have to be taken with some caution. In spite of this divergent opinion, the study by

Gaus et al. (2005) is very important as it focuses the attention on the integrity of the cap rock, which is a very important issue for the safe geological disposal of CO_2.

The processes relevant for the geological sequestration of CO_2 at the Sleipner site were also investigated in a very interesting study by Johnson et al. (2001), who used an integrated toolbox comprising NUFT, a software package for the numerical simulation of multiphase/multicomponent flow and reactive transport (Nitao, 1998), GEMBOCHS, a thermodynamic/kinetic database and dedicated software library (Johnson and Lundeen, 1994a, b, 1995), and the graphic utility Xtool (Daveler, 1998). The Sleipner environment is represented by a spatial domain comprising an Utsira-like 200-m-thick saline aquifer confined by a 25-m-thick shale cap-rock, which in turn is overlain by a 25-m-thick aquifer. Three distinct aquifer scenarios were considered, differing for the absence or presence of intra-aquifer shales of different lateral continuity and permeability (the latter parameter was varied to simulate the absence or presence of microfractures). A 20-year-long time frame was adopted, comprising a prograde phase (in which the injection rate is kept constant at 10,000 ton CO_2/year), a transition phase (in which the injection rate is decreased to 0 in 3 months) and a retrograde phase (in which the injection rate is maintained at 0 for 9.75 years).

The study by Johnson et al. (2001) showed that the migration of immiscible CO_2 and the relative importance of solubility and mineral trapping processes are chiefly controlled by the distribution of intra-aquifer permeability, whereas the long-term storage security is mainly affected by cap-rock integrity. Intra-aquifer shales retard the vertical migration of the immiscible CO_2 plume and promote its lateral expansion (in agreement with the results of seismic surveys; Torp and Gale, 2004), thus increasing the spatial extent of plume–aquifer interaction and the efficacy of intraplume solubility and mineral trapping processes. The 25-m-thick shale cap-rock resulted to be an efficacious barrier prohibiting the vertical migration of the immiscible CO_2 plume (again, in agreement with seismic evidence; Torp and Gale, 2004). In the relatively short interval of time considered in the simulation, ~85% of the injected CO_2 constitutes an immiscible supercritical fluid phase, ~15% dissolves in the aqueous solution and <1% precipitates as carbonate minerals, in the form of dawsonite, magnesite/siderite and calcite. However, mineral trapping is enhanced during the retrograde phase and has very important consequences, in that it limits the lateral extent of plume migration and increases the cap-rock integrity, through carbonate sealing of microfracture porosity.

7.6.5. The carbonate rocks of the Alberta sedimentary basin

Similar to the glauconitic sandstone case (*see* Section 7.6.1), Gunter et al. (2000) modelled the effects of the dissolution of a finite amount of CO_2 (1 mol in 1,000 g of water) also on the Nisku carbonate aquifer of the Alberta Sedimentary Basin. These carbonate rocks consist of prevailing dolomite and/or calcite (both containing traces of Fe), possibly accompanied by small amounts (<1% in general) of anhydrite, illite, pyrite and quartz. Silicate and sulphide minerals were excluded from the simulation, which was carried out for a system comprising 14,000 g of dolomite, 1,100 g of calcite, 262 g of siderite and 1,000 g of water.

Equilibrium is predicted to be achieved in less than 1 day in the carbonate aquifer, owing to the occurrence of fast reactions, with a small reduction in P_{CO2} from 87 to 86 bar, the dissolution of small amounts of calcite (0.06 mol) and siderite (10^{-5} mol) and the precipitation of 0.02 mol of dolomite.

7.7. Water–rock reactions during geological CO_2 sequestration: the experimental evidence

7.7.1. Laboratory experiments on rocks from sedimentary basins

Precipitation of clay minerals and carbonate minerals as a consequence of dissolution of primary silicates and Al-silicates was generally not observed in laboratory experiments aimed at evaluating the interactions between CO_2, formation water and rocks from sedimentary basins. This is due to the slow kinetics of these processes. Exceptions are represented by the recent experimental works by Kaszuba et al. (2003, 2005).

Pearce et al. (1996) and Rochelle et al. (1996) injected supercritical CO_2, at pressures of 90 and 200 bar, into a sandstone at temperatures of 105° and 80°C, for periods of 1 and 8 months. Significant alteration of pre-existing calcite and dolomite was observed, as well as dissolution of anhydrite accompanied by precipitation of calcite and possible corrosion of detrital feldspars with precipitation of Na-smectite, although the evidence supporting this last process was not conclusive.

Gunter et al. (1997) performed 1-month-long experimental tests at 105°C and 90 bar of P_{CO_2} on glauconitic sandstones from the Alberta sedimentary basin. Under these conditions, only a small amount of CO_2 was trapped through reactions with Al-silicate minerals, whereas dissolution of mineral carbonates took place.

Sass et al. (2001) reacted anorthite and glauconite with CO_2 and a Na–Ca–Cl synthetic brine (containing other major components and several traces) at 50°C, 41 bar P_{CO_2} and 150°C, 138 bar P_{CO_2}, under a total pressure of 138 bar. The water–rock–gas interaction experiments were carried out in vessels placed on a shaker table or rocker for a duration of approximately 1 month. Anorthite incorporated some Na in the lattice and released a corresponding amount of Ca to the aqueous solution but precipitation of carbonate minerals did not occurred. In the experimental runs with glauconite, the Na content of the solid phase increased, K, Fe and Si decreased, whereas Al, Mg and Ca did not change. Also in this case, no carbonate mineral formed as a reaction product.

Kaszuba et al. (2003, 2005) simulated a brine–aquifer–aquitard system in a flexible cell hydrothermal reactor. A synthetic arkose, assumed to represent the aquifer, was prepared by mixing equal amounts of quartz, oligoclase ($An_{17–21}$) and microcline ($Or_{91–97}$) showing perthitic lamellae ($Or_{3–9}$) and adding some biotite. An argillaceous shale, made up of clay minerals (65%, by volume) quartz (27%), feldspar (5%), chlorite (2%, possibly including some kaolinite) and trace amounts of framboidal pyrite, was used for the aquitard. Kaszuba et al. (2003) used a synthetic brine containing Na^+, Mg^{2+}, K^+ and Cl^- as prevailing constituents whereas a pure Na–Cl brine was employed by Kaszuba et al. (2005), both with ionic strength of 5.5 mol kg^{-1}. In a first step, the system was maintained at 200°C, 200 bar for several days (59 in the 2003 run, 32 in the 2005 run) to attain steady state. Then CO_2 was added to the reactor. In a second step, the system was kept again at 200°C, 200 bar, for 80 days in the 2003 run and 77 days in the 2005 run. Kaszuba et al. (2005) performed also a reference experiment on the same system, at the same P, T conditions and for the same time, without CO_2 injection.

Kaszuba et al. (2003) documented (i) etching of potassic lamellae in perthitic microcline with respect to sodic lamellae; (ii) growth of clay minerals on oligoclase; (iii) coatings of magnesite associated with clay minerals on biotite grains; (iv) growth of analcime with skeletal habit.

Kaszuba et al. (2005) observed the (i) presence of corroded magnesite and euhedral siderite, suggesting the early occurrence of magnesite precipitation followed by magnesite dissolution and siderite precipitation; (ii) growth of clay minerals and analcime. Analcime was found to be more abundant in the CO_2-free experiment where it shows an euhedral texture. In contrast, analcime crystals with skeletal texture grew in the CO_2–brine–rock experiment, as also found by Kaszuba et al. (2003).

Significant variations were also detected in brine chemistry. In particular, the significant increase in the concentrations of several components, initially absent in the Na–Cl brine used by Kaszuba et al. (2005), undoubtedly indicate the occurrence of rock dissolution through reaction with CO_2 and brine. All in all, these observations are consistent with the occurrence of dissolution of primary silicate minerals and precipitation of both carbonates and silicates.

Kaszuba et al. (2005) underscored that "precipitation of the mixed hydroxyl carbonate mineral dawsonite ($NaAlCO_3(OH)_2$), predicted as a stable carbonate phase in reactive transport models of carbon sequestration (Johnson et al., 2002; Knauss et al., 2002) was not observed in this study." They estimated that the aqueous solution was probably undersaturated with dawsonite, although the thermodynamic affinity with respect to this mineral and other Al-bearing solid phases could not be computed because the Al concentration in the brine was below detection limit throughout the CO_2–brine–rock experiment. To try to elucidate this matter, it is worth taking into account the equilibrium reaction between dawsonite and analcime, an ubiquitous phase in the experiments of Kaszuba and coworkers:

$$\begin{aligned} Na_{0.96}Al_{0.96}Si_{2.04}O_6 \cdot H_2O + 0.96\ CO_{2(g)} &\Leftrightarrow 0.96\ NaAlCO_3(OH)_2 \\ &+ 2.04\ SiO_2 + 0.04\ H_2O. \end{aligned} \tag{7-52}$$

Assuming that the activity of aqueous SiO_2 is fixed by equilibrium with chalcedony and the activity of water and solid phases are equal to 1, it results $\log f_{CO_2} = -\log K_{7-52} / -0.96$. Using the thermodynamic equilibrium constants of EQ3/6, it turns out that at 200°C, dawsonite is stable for $f_{CO_2} > 7.8$ kbar and analcime is stable below this f_{CO_2} value. Since CO_2 fugacities were much lower than 7.8 kbar in the experiments of Kaszuba and coworkers, the absence of dawsonite as a solid product phase is not surprising. We may also note that the f_{CO_2} of analcime–dawsonite–chalcedony equilibrium coexistence is 0.23 bar at 60°C, 7.2 bar at 100°C and 296 bar at 150°C. In spite of possible uncertainties on the thermodynamic data of dawsonite, these figures confirm that dawsonite likes cool environments, in line with what was inferred in Section 5.1.6 on the albite–dawsonite–quartz coexistence. Therefore, high temperature (200°C) has certainly enhanced the kinetics of dissolution/precipitation reactions, but has also stabilized analcime at the expense of dawsonite, in the experiments by Kaszuba and coworkers.

7.7.2. Laboratory experiments on forsterite and serpentine

The carbonation reactions of forsterite and serpentine have been the subject of several studies, after Lackner et al. (1995, 1997) underscored that these Mg silicates obtained from ultramafic rocks are abundant and thermodynamically convenient primary minerals for mineral carbonation.

McKelvy et al. (2001) showed that dehydroxylated samples of serpentine (meta-serpentine), obtained through heat treatment, undergo carbonation more readily than untreated samples. The optimum temperature of heat treatment was found to be 600–650°C by O'Connor et al. (2001), with little increase in reactivity upon heat treatment at higher temperatures. O'Connor et al. (2001) investigated also the mineral carbonation of olivine and meta-serpentine in aqueous slurries and found optimum results at 155°C under a P_{CO_2} of 185 bar for a 0.64 M $NaHCO_3$ and 1.0 M NaCl aqueous solution containing 15% of solids. Under these conditions a 78% conversion of the Mg-silicates to carbonate was attained in 30 min.

Guthrie et al. (2001) carried out experiments of serpentine, meta-serpentine and olivine carbonation in a stirred reaction vessel typically containing ~0.8 L of water plus solids and ~1 L of supercritical CO_2 at temperatures up to 185°C and pressures up to ~160 bar. They proposed a geochemical model assuming (i) occurrence of dissolution and precipitation processes in the aqueous phase and not in the supercritical CO_2 phase and (ii) occurrence of carbonation via dissolution–precipitation and not through direct carbonation of the Mg-silicate. The geochemical model comprises four steps: (i) dissolution of CO_2 into the aqueous phase and of H_2O into the CO_2-rich phase; (ii) release of Mg^{2+} ion to the aqueous phase, which is considered to be the rate limiting step; (iii) speciation reactions in the aqueous phase; (iv) precipitation of magnesium carbonate. Magnesite resulted to be the most typical product, but hydromagnesite was observed in several runs upon either alteration of fluid chemistry or CO_2 venting before quenching of the reaction vessel.

Based on the findings of O'Connor et al. (2001), Wolf et al. (2004) performed the carbonation of meta-serpentine in a solution 0.64 M $NaHCO_3$ and 1.0 M NaCl under a P_{CO_2} of 150 bar, by using a suitably implemented externally-controlled microreaction system. This allows *in situ* monitoring of the reaction process by synchrotron X-ray diffraction and Raman spectroscopy. By progressively heating the system, they were able to detect the first X-ray reflections of magnesite at 150°C, which is the temperature at which carbonation reaction begins. Further carbonation of meta-serpentine to magnesite occurred upon continuous heating to 180°C. However, no reflection caused by Si-bearing minerals was detected, indicating that SiO_2-like products are highly disordered.

Further studies of this type were carried out by McKelvy et al. (2004), who used *in situ* synchrotron X-ray diffraction to investigate the reactivity of meta-serpentine (obtained through heat activation of lizardite at different temperatures) under the optimal conditions of carbonation established by O'Connor et al. (2001). Trace carbonation was observed at 120°C for the meta-serpentine obtained at 580°C, but further heating did not cause any further carbonation, indicating that the meta-serpentine was only partially activated. Substantial carbonation was found at 125°C for the 610°C meta-serpentine. The best results were obtained for the 650 and 720°C meta-serpentine in terms of progress and onset temperature of the carbonation reaction, which resulted to be 100 and 105°C, respectively. Besides, McKelvy et al. (2004) investigated the structure and composition of meta-serpentine by using TGA/DTA, XRD and FTIR analyses. Longer-range ordering, possibly due to doubling of the basal spacing from the characteristic value of 7.3 Å (*see* Section 5.3.2.1) to 14.7 Å, was observed in the material treated at 580°C, whereas the crystalline features disappear at 650°C indicating formation of amorphous meta-serpentine.

Schulze et al. (2004) used XPS and TEM to investigate the surface of ground lizardite, untreated, heat-treated at 630°C in a CO_2 flux for 3 h and heat-treated and reacted with a Na–Cl–HCO_3 solution and CO_2 at 155°C for 2 h, thus achieving a 22% conversion to carbonate. The latter treatment determined the formation of a 40–100 nm surface layer constituted by amorphous silica containing Mg. According to Schulze et al. (2004), the rate-limiting step of lizardite carbonation is the diffusion of Mg^{2+} ion from the amorphous phase.

Since heat treatment is a rather expensive operation, a different approach based on the physical activation of serpentine at low temperature coupled with a pH swing process was investigated by Park and Fan (2004). Fluidation of a serpentine slurry with 2 mm glass beads resulted to be the most effective method for removing the silica-rich surface layer and promoting further serpentine dissolution.

Giammar et al. (2005) investigated forsterite dissolution and magnesite precipitation through a series of batch experiments. Forsterite dissolution tests were carried out in dilute solutions (I 0.003 to 0.018 mol kg^{-1}) at 30 and 95°C under initial P_{CO_2} of 1 or 100 bar, whereas magnesite precipitation experiments were run in $MgCl_2$–$NaHCO_3$ solutions (I 0.07 to 1.25 mol kg^{-1}) under initial P_{CO_2} of 100 bar. Several analytical methods (elemental analysis of the aqueous solution, SEM, FTIR and XRD for the solids) and geochemical modelling were used to elucidate reaction products and progress.

The extent of forsterite dissolution was found to increase with increasing P_{CO_2} and temperature. The release of Mg and Si during forsterite dissolution was stoichiometric, but the concentration of dissolved Si was limited by saturation with respect to amorphous silica. Even though the aqueous phase attained conditions of oversaturation with respect to magnesite (saturation index up to 0.25) during forsterite dissolution experiments, $MgCO_3$ did not precipitated.

Magnesite precipitation occurred in ad hoc experiments designed to estimate the influence of suspended solid particles (of either forsterite or magnesite), time and saturation index with respect to magnesite. The composition of the aqueous solution, $MgCl_2$–$NaHCO_3$, was selected to simulate extensive forsterite dissolution. Magnesite precipitation was found to be controlled by nucleation, which requires a critical saturation index between 0.25 and 1.14 at 95°C and P_{CO_2} of 100 bar. Magnesite seeds accelerate magnesite precipitation, whereas forsterite particles do not, although magnesite forms on their surfaces.

7.8. Water–rock reactions during geological CO_2 sequestration: the field evidence

Emberley et al. (2004, 2005) evaluated the gas–water–rock interactions that occurred upon miscible CO_2 injection into the oil reservoir of Weyburn, Saskatchewan (Canada) for enhanced oil recovery (EOR). Oil is hosted into the carbonate rocks known as Midale beds, which are sealed by an anhydrite cap. The reservoir has an average temperature of 60°C and initial pressure was 146 bar on average. The CO_2 is generated by the Great Plains Synfuel coal gasification plant in Beulah, North Dakota (USA) and is transported at Weyburn by a 320-km-long pipeline. Note that CO_2 is injected together with water to retard CO_2 mobility. Since CO_2 is more soluble in oil than in water, it is released from water to oil. Owing to absorption of CO_2, oil experiences a decrease in viscosity and an increase in mobility. Besides, the low pH of the CO_2-rich aqueous solution can cause dissolution of carbonate minerals and increase in reservoir permeability. Although the aim of the miscibile CO_2 injection project is EOR, there are also important environmental aspects since up to 30% of the injected CO_2 is expected to be trapped in the reservoir.

To evaluate gas–water–rock interactions, a baseline fluid sampling was carried out before the onset of the CO_2 injection project and a second sampling survey (known as Monitoring-1) was performed 7 months after this date, when a total of approximately 150.6×10^6 standard m^3 of CO_2 (corresponding to ~3×10^5 tons) were disposed in the oil reservoir. Data on water and gas chemistry and isotope compositions were obtained in both surveys.

The samples collected during Monitoring-1 differ from the baseline data for lower pH values with differences of 0.5–0.6 pH units, higher alkalinity which has doubled compared to the baseline values (although a significant fraction of titration alkalinity is due to dissolved sulphide), higher Ca concentrations and higher $\delta^{13}C$ values of dissolved CO_2 by one to some ‰ units. All these differences were attributed to: (i) dissolution of the injected CO_2 in the aqueous phase as both carbonic acid and bicarbonate ion and (ii) concurrent dissolution of carbonate minerals, chiefly calcite and dolomite.

Nevertheless other chemical changes (e.g., in chloride and sulphate concentrations) also took place, indicating possible flushing of waters from other geological formations to the Midale beds and entrance in the monitored wells. Therefore, the overall picture is probably somewhat more complicated than that of simple CO_2 absorption in the aqueous phase and consequent dissolution of carbonate minerals. Data obtained during subsequent monitoring phases add further complexities to the overall picture rather than clarifying it.

7.9. The need for a synergistic approach

Based on what was discussed in previous sections, it is quite evident that the reactions of dissolution of primary silicates and precipitation of secondary minerals, including carbonates, proceed to relatively limited extents in some laboratory experiments and field tests and do not occur at all in others. Geochemical modelling generally predicts that comparatively long lapses of time, in the order of hundreds, thousands, or even tens of thousands years, are needed for the significant progress of carbonation reactions. Therefore, there is no contrast between the results of geochemical modelling and those of laboratory experiments and field tests. Experiments and tests are entirely compatible with the computer simulation of geochemical processes and, in our opinion, a synergistic use of the two approaches is recommendable, as already done in some high-quality studies (e.g., Gunter et al., 1997 and Giammar et al., 2005 among the others).

Geochemical models are very attractive as it takes a comparatively short lapse of time to obtain results that would be achieved after much longer intervals of time, sometimes not accessible to humans, by means of real experiments either in the laboratory or in the field. However, users of geochemical models must be aware of possible dangers.

One of these is to rely too much on computer experiments or to expect too much from them. This was clearly discussed by Giggenbach (1984), by means of the following words.

> *As the reality of these models increases there is a danger, however, that their complexity and intractability will soon rival those of the natural systems studied. The emerging gap between human and machine understanding of these systems is illustrated by a remark by Reed (1982) in which he stresses the utility of the computer approach "when one considers how difficult it would be to apply intuitive reasoning and partial calculations to determine the course of a water–rock reaction. The large number of unknowns and the non-linearity of the equations make it difficult to understand the complexly intertwined processes." Taking into account the fact that intuitive thinking, a prerogative of human mind, is the basis of all scientific progress, relegating the process of understanding our environment to machines may eventually stifle rather than further scientific development.*

Besides, it must be noticed that the validity of the results of computer experiments depends chiefly on how we have set up the model at the beginning of the simulation. No good answer can be expected from an ill-posed question whereas a good answer can be obtained if we succeed to formulate a correct question. Again, *setting-up a valid model requires an at least general, but conceptually accurate understanding of the processes active*, as underscored by Giggenbach (1984).

The outputs of computer experiments are also influenced by the quality of the thermodynamic and kinetic data, the validity of the used algorithms and many other "details". For instance, the application of geochemical models at high pressures and salinities is not straightforward, as discussed by Allen et al. (2005). See also Section 7.3.3 for the salinity effects.

Last but not least, presently available models are far from perfect as already discussed above and summarized in the next section.

7.9.1. Current limitations and future developments of reaction path modelling

The chemical reactions between minerals, the aqueous solution and CO_2 occurring upon high-pressure CO_2 injection are very complex and the theoretical framework needed for their description is not fully incorporated in existing computer codes. Some of these deficiencies are summarized below.

(1) Available codes incorporate at best the extended (B-dot) Debye–Hückel equation to compute *individual ion activity coefficients*, γ_i, although its application to brines is not advisable. Ideally, one should use the Pitzer's approach to compute the γ_i of solute species, but needed parameters are generally not available at high temperature and are also lacking at all temperatures for pivotal aqueous species such as dissolved SiO_2, Al^{3+} and related complex species.

(2) The *thermodynamic properties of some mineral phases* of interest are unknown or poorly known. This is especially true for clay minerals, which are usually described in terms of idealized, limiting compositions.

(3) The *kinetic database* is even less complete than the thermodynamic database and it is difficult to fill this gap as *no general laws of overall reaction kinetics exist in parallel with the general laws of thermodynamics, and no necessary genetic relationship with which to connect kinetic species to thermodynamic species is known* (Sposito, 1994).

(4) In particular, the *influence of CO_2* on the dissolution rate has been investigated for a limited number of mineral phases up to 1 bar P_{CO2} and only for labradorite, calcite, dolomite and magnesite up to 50–68 bar P_{CO2}. Although available data indicate that the effect of CO_2 is that of controlling the pH of the aqueous solution rather than participating directly to the dissolution mechanism of solid phases, this matter deserves further consideration.

(5) The processes occurring during the *precipitation of solid phases* (e.g., nucleation and crystal growth) are not described adequately. The backward (precipitation) rate laws incorporated in available codes require the knowledge of the initial surface area, but this is zero for a phase that is not present in the system at the beginning of the simulation. Adoption of a value different from zero or assumption of instantaneous precipitation are two possible choices for circumventing this difficulty but this way to proceed is far from satisfactory.

(6) Possible variations of the *reactive surface areas* with reaction progress, due to armouring of mineral phases by other phases and/or other reasons, cannot be described.

(7) In natural systems, the formation of a secondary phase rather than another often depends on their different kinetics of growth, which lead to *different reaction paths*. This kind of problem cannot be tackled by means of available codes.

For all these deficiencies, the geochemical evolution of the systems of interest upon high-pressure CO_2 injection cannot be predicted with sufficient accuracy by means of computer experiments at present. However, geochemical modelling is the only available tool for the evaluation of long-term geochemical processes, including those of interest for geological CO_2 sequestration. In spite of possible improvements in the experimental approach, there is little hope to extend their duration to multiples of the life time of humans even from a very optimistic point of view. In contrast, computer experiments can be improved in future to understand the long-term evolution of the systems of interest.

7.10. A final note

I tried to be consistent throughout this book in the use of symbols and unit of measurements, and, even more important, in the adoption of chemical species. For what concerns the symbols, I am aware that I was not able to fulfil my initial purposes and, more important, the expectations of possible readers.

For what concerns the units, I preferred to use the bar and sometimes the atmosphere (1 atm = 1.01325 bar) instead of the recommended unit, i.e. the Pascal (1 MPa = 1×10^6 Pa = 10 bar), as I think that bar and atm are more immediate and related to the real world than Pa. I expressed the thermodynamic affinity *A* in J mol^{-1} (or kJ mol^{-1}) in Chapter 6 but I used cal mol^{-1} (or kcal mol^{-1}) in Chapter 7, as this is the unit adopted by EQ3/6. After all, one has to be able to do conversions. This is especially important for energies, as they are also involved in dietetic suggestions. Energy conversions are done by multiplying the *A* value expressed in cal mol^{-1} (or kcal mol^{-1}) times 4.184 to obtain the corresponding value in J mol^{-1} (or kJ mol^{-1}).

Unfortunately, the adoption of chemical species is not always the same in the geochemical literature and, in particular, there is some confusion on the species adopted to describe neutral dissolved CO_2 species. As already recalled (see Chapter 2), both true dissolved molecular carbon dioxide, $CO_2°$, and true carbonic acid, $H_2CO_3°$, coexist in the aqueous solution and, at equilibrium, the $m_{H_2CO_3°}/m_{CO_2°}$ ratio is 0.0026 at 25°C, 1 bar (Sposito, 1994). However, in the geochemical literature the equilibrium between $CO_2°$ and $H_2CO_3°$ is usually neglected and these two species are considered together in a lumped parameter. This is called apparent carbonic acid, $H_2CO_3^*$, by some authors and aqueous carbon dioxide, $CO_{2(aq)}$, by others.

In this book, I have decided to follow the approach of each author, although this brings about some inconsistencies, of which the careful reader has already been aware at this point.

References

Aagaard, P. and Helgeson, H.C. (1982) Thermodynamic and kinetic constraints on reaction rates among minerals and aqueous solutions. I. Theoretical considerations. *Am. J. Sci.*, **282**, 237–285.

Aagaard, P. and Helgeson, H.C. (1983) Activity/composition relations among silicates and aqueous solutions. II. Chemical and thermodynamic consequences of ideal mixing of atoms on homological sistes in montmorillonites, illites, and mixed-layer clays. *Clays Clay Miner.*, **31**, 207–217.

Acker, J.G. and Bricker, O.P. (1992) The influence of pH on biotite dissolution and alteration kinetics at low temperature. *Geochim. Cosmochim. Acta*, **56**, 3073–3092.

Agricola, G. (1556) *De Re Metallica*. Hieronimo Frobenio et Nicolao Episcopio, Basel.

Alekseyev, V.A., Medvedeva, L.S., Prisyagina, N.I., Meshalkin, S.S. and Balabin, A.I. (1997) Change in the dissolution rates of alkali feldspars as a result of secondary mineral precipitation and approach to equilibrium. *Geochim. Cosmochim. Acta*, **61**, 1125–1142.

Alkattan, M., Oelkers, E.H., Dandurand, J.-L. and Schott, J. (1998) An experimental study of calcite and limestone dissolution rates as a function of pH from −1 to 3 and temperature from 25 to 80°C. *Chem. Geol.*, **151**, 199–214.

Alkattan, M., Oelkers, E.H., Dandurand, J.-L. and Schott J. (2002) An experimental study of calcite dissolution rates at acidic conditions and 25°C in the presence of $NaPO_3$ and $MgCl_2$. *Chem. Geol.*, **190**, 291–302.

Allen, D.E., Strazisar, B.R., Soong, Y. and Hedges, S.W. (2005) Modeling carbon dioxide sequestration in saline aquifers: Significance of elevated pressures and salinities. *Fuel Process. Technol.*, **86**, 1569–1580.

Allison, J.D., Brown, D.S. and Novo-Gradac, K.J. (1991) MINTEQA2/PRODEFA2, A Geochemical Assessment Model for Environmental Systems: Version 3.0 User's Manual, Report EPA/600/3-91/021, U. S. Environmental Protection Agency, Office of Research and Development, Environmental Research Laboratory, Athens, GA, 106pp.

Amrhein, C. and Suarez, D.L. (1992) Some factors affecting the dissolution kinetics of anorthite at 25°C. *Geochim. Cosmochim. Acta*, **56**, 1815–1826.

Anbeek, C. (1992) Surface roughness of minerals and implications for dissolution studies. *Geochim. Cosmochim. Acta*, **56**, 1461–1469.

Anderson, G.K. (2002) Solubility of carbon dioxide in water under incipient clathrate formation condition. *J. Chem. Eng. Data*, **47**, 219–222.

Anderson, G.M. and Crerar, D.A. (1993) *Thermodynamics in Geochemistry. The Equilibrium Model*. New York: Oxford University Press, 588pp.

Anderson, M.S. (1991) Reactivity of San Andreas dolomite, *SPE 20115, 1990 SPE Permian Basin Oil and Gas Recovery Conference*, March 8–9, Midland, TX.

Angell, C.A. (1983) Supercooled water. *Annu. Rev. Phys. Chem.*, **34**, 593–630.

Angus, S., Armstrong, B. and de Reuck, K.M. (1976) *International Thermodynamic Tables of the Fluid State – 3 Carbon Dioxide*. New York: Pergamon.

Appelo, C.A.J. and Postma, D. (1996) *Geochemistry, Groundwaters and Pollution*. Rotterdam: A.A. Balkema, 536pp.

Apps, J.A. (1996) An approach to modeling of the chemistry of waste fluid disposal in deep saline aquifers. In: Apps J.A. and Tsang C.F. (eds.), *Deep Injection Disposal of Hazardous and Industrial Waste: Scientific and Engineering Aspects*. San Diego, CA: Academic Press, pp. 465–488.

Arakaki, T. and Mucci, A. (1995) A continuous and mechanistic representation of calcite reaction-controlled kinetics in dilute solutions at 25°C and 1 atm total pressure. *Aquat. Geochem.*, **1**, 105–130.

Arakaki, T. and Mucci, A. (1999) Errata. A continuous and mechanistic representation of calcite reaction-controlled kinetics in dilute solutions at 25°C and 1 atm total pressure. *Aquat. Geochem.*, **5**, 225.

Arisi Rota, F., Brondi, A., Dessau, G., Branzini, M., Stea, B. and Vighi, L. (1971) I giacimenti minerari. In: "*La Toscana Meridionale. Fondamenti geologico-minerari per una prospettiva di valorizzazione delle risorse naturali*". *Rendiconti S.I.M.P.*, **27**, 357–544.

Arrhenius, S. (1889) Ober die reacktionsgeschwindigkeit bei der inversion von rohrzucker durch sauren. *Z. Physik. Chem.*, **4**, 226–248.

Arrhenius, S. (1896) On the influence of carbonic acid in the air upon the temperature of the ground. *Philos. Mag., Ser.* **5**, 237–275.

Arts, R., Eiken, O., Chadwick, A., Zweigel, P., van der Meer, L. and Zinszner, B. (2004) Monitoring of CO_2 injected at Sleipner using time-lapse seismic data. *Energy*, **29**, 1383–1392.

Arvidson, R.S., Ertan, I.E., Amonette, J.E. and Luttge, A. (2003) Variation in calcite dissolution rates: A fundamental problem? *Geochim. Cosmochim. Acta*, **67**, 1623–1634.

Arvidson, R.S. and Mackenzie, F.T. (2000) Temperature dependence of mineral precipitation rates along the $CaCO_3$–$MgCO_3$ join. *Aquat. Geochem.*, **6**, 249–256.

Awad, A., Koster van Groos, A. F. and Guggenheim, S. (2000) Forsteritic olivine: Effect of crystallographic direction on dissolution kinetics. *Geochim. Cosmochim. Acta*, **64**, 1765–1772.

Bacastow, R.B. (1976) Modulation of atmospheric carbon dioxide by the southern oscillation. *Nature*, **261**, 116–118.

Bachu, S., Gunter, W.D. and Perkins, E.H. (1994) Aquifer disposal of CO_2: Hydrodynamic and mineral trapping. *Energy Convers. Manage.*, **35**, 269–279.

Bailey, A. (1974) Effects of temperature on the reaction of silicates with aqueous solutions in the low temperature range. *First International Symposium on Water–Rock Interaction*. 1976, Praha, pp. 375–380.

Bailey, S.W. (ed.) (1988) *Hydrous Phyllosilicates (exclusive of micas). Reviews in Mineralogy*. Washington, DC: Mineralogical Society of America, **19**, 725pp.

Baker, J.C., Bai, G.P., Hamilton, P.J., Golding, S.D. and Keene, J.B. (1995) Continental-scale magmatic carbon dioxide seepage recorded by dawsonite in the Bowen–Gunnedah–Sydney basin system, Eastern Australia. *J. Sedimentary Res.*, **A65**, 522–530.

Baklid, A., Korbøl, R. and Owren, G. (1996) Sleipner vest CO_2 disposal. CO_2 injection into a shallow underground aquifer. SPE paper no. 36600.

Baldi, P., Bellini, S., Ceccarelli, A., Fiordalisi, A., Squarci, P. and Taffi, L. (1994) Correlazioni fra le anomalie termiche ed altri elementi geofisici e strutturali della Toscana Meridionale. *Studi Geol. Camerti*, **1**, 139–149.

Bales, R.C. and Morgan, J.J. (1985) Dissolution kinetics of chrysotile at pH 7 to 10. *Geochim. Cosmochim. Acta*, **49**, 2281–2288.

Bamberger, A., Sieder, G. and Maurer, G. (2000) High-pressure (vapor + liquid) equilibrium in binary mixtures of (carbon dioxide + water or acetic acid) at temperatures from 313 to 353 K. *J. Supercrit. Fluids*, **17**, 97–100.

Banwart, S., Davies, S. and Stumm, W. (1989) The role of oxalate in accelerating the reductive dissolution of hematite (α-Fe_2O_3) by ascorbate. *Colloid Surf.*, **39**, 303–309.

Barnola, J.-M., Anklin, M., Porcheron, J., Raynaud, D., Schwander, J. and Stauffer, B. (1995) CO_2 evolution during the last millenium as recorded by Antarctic and Greenland ice. *Tellus Ser. B-Chem. Phys. Meteorol.*, **47**, 264–272.

Barnola, J.M., Raynaud, D., Lorius, C. and Barkov, N.I. (2003) Historical CO_2 record from the Vostok ice core. In *Trends: A Compendium of Data on Global Change. Carbon Dioxide Information Analysis Center*. Oak Ridge, Tenn., USA: Oak Ridge National Laboratory, U.S. Department of Energy.

Barta, L. and Bradley, D.J. (1985) Extension of the specific interaction model to include gas solubilities in high temperature brines. *Geochim. Cosmochim. Acta*, **49**, 195–203.

Bartholomé, E. and Friz, H. (1956) *Chem. Ing. Technol.*, **28**, 706–708 (cited in Diamond and Akinfiev, 2003).

Battle, M., Bender, M., Tans, P.P., White, J.W.C., Ellis, J.T., Conway, T. and Francey, R.J. (2000) Global carbon sinks and their variability, inferred from atmospheric O_2 and $\delta^{13}C$. *Science*, **287**, 2467–2470.

Bauer, A. and Berger, G. (1998) Kaolinite and smectite dissolution rate in high molar KOH solutions at 35° and 80°C. *Appl. Geochem.*, **13**, 905–916.

Bayer, G. (1971) Thermal expansion anisotropy of dolomite-type borates $Me^{2+}Me^{4+}B_2O_6$. *Zeit. Kristallogr.*, **133**, 85–90.

Bender, M., Ellis, T., Tans, P., Francey, R. and Lowe, D. (1996) Variability in the O_2/N_2 ratio of southern hemisphere air, 1991–1994 – Implications for the carbon cycle. *Global Biogeochem. Cycles*, **10**, 9–21.

Bender, M., Sowers, T. and Labeyrie, L. (1994) The Dole effect and its variations during the last 130,000 years as measured in the VOSTOK ice core. *Global Biogeochem. Cycles*, **8**, 363–376.

Bennett, P.C. (1991) Quartz dissolution in organic-rich aqueous systems. *Geochim. Cosmochim. Acta*, **55**, 1781–1797.

Benson, S.W. (1968) *Thermochemical Kinetics*. New York: Wiley, 223pp.

Berg, A. and Banwart, S.A. (2000) Carbon dioxide mediated dissolution of Ca-feldspar: Implications for silicate weathering. *Chem. Geol.*, **163**, 25–42.

Berger, G., Cadore, E., Schott, J. and Dove, P.M. (1994a) Dissolution rate of quartz in lead and sodium electrolyte solutions between 25 and 300°C: Effect of the nature of surface complexes and reaction affinity. *Geochim. Cosmochim. Acta*, **58**, 541–551.

Berger, G., Claparols, C., Guy, C. and Daux, V. (1994b) Dissolution rate of a basalt glass in silica-rich solutions: Implications for long-term alteration. *Geochim. Cosmochim. Acta*, **58**, 4875–4886.

Berman, R.G. (1988) Internally-consistent thermodynamic data for minerals in the system $Na_2O–K_2O–CaO–MgO–FeO–Fe_2O_3–Al_2O_3–SiO_2–TiO_2–H_2O–CO_2$. *J. Petrol.*, **29**, 445–522.

Berner, R.A. (1978) Rate control of mineral dissolution under earth surface conditions. *Am. J. Sci.*, **278**, 1235–1252.

Berner, R.A. (1983) Kinetics of weathering and diagenesis. In: Lasaga, A.C. and Kirkpatrick, R.J. (eds.), *Kinetics of Geochemical Processes*: *Reviews in Mineralogy*, Mineralogical Society of America, 2nd printing, **8**, 111–134.

Berner, R.A. (1997) The rise of plants and their effect on weathering and atmospheric CO_2. *Science*, **276**, 544–546.

Berner, R.A. and Holdren, G.R., Jr. (1977) Mechanism of feldspar weathering: Some observational evidence. *Geology*, **5**, 369–372.

Berner, R.A. and Holdren, G.R., Jr. (1979) Mechanism of feldspar weathering – II. Observations of feldspars from soils. *Geochim. Cosmochim. Acta*, **43**, 1173–1186.

Berner, R.A., Sjöberg, E.L., Velbel, M.A. and Krom, M.D. (1980) Dissolution of pyroxenes and amphiboles during weathering. *Science*, **207**, 1205–1206.

Bevan, J. and Savage, D. (1989) The effect of organic acids on the dissolution of K-feldspar under conditions relevant to burial diagenesis. *Mineral. Mag.*, **53**, 415–425.

Beyer, A. (1871) Über die Zersetzung des Feldspathes unter dem Einfluss von Salzlösungen und einigen anderen Agentien. *Landwirtschaftliche Versuchs-Sta.*, **14**, 314–322.

Biber, M.V., Dos Santos Afonso, M. and Stumm, W. (1994) The coordination chemistry of weathering: IV. Inhibition of the dissolution of oxide minerals. *Geochim. Cosmochim. Acta*, **58**, 1999–2010.

Binning, G., Quate, C.F. and Gerber, Ch. (1986) Atomic force microscopy. *Phys. Rev. Lett.*, **56**, 930–933.

Bjerrum, N. (1929) Neuere auschauungen über elektrolyte. *Deutsche Chem. Gesell Ber.*, **62**, 1091–1103.

Blake, R.E. and Walter, L.M. (1999) Kinetics of feldspar and quartz dissolution at 70–80°C and near-neutral pH: Effect of organic acids and NaCl. *Geochim. Cosmochim. Acta*, **63**, 2043–2059.

Blanchard, I. G. (1994) *Epidote dissolution kinetics: An experimental study at 250°C and 500 bars between pH 3.4 and pH 12*. M. Sc. Thesis, University of Leeds, UK.

Blatt, H., Middleton, G. and Murray, R. (1980) *Origin of Sedimentary Rocks*, 2nd ed. NJ: Prentice-Hall, 782pp.

Bloom (1983) The kinetics of gibbsite dissolution in nitric acid. *Soil Sci. Soc. Am. J.*, **47**, 164–168.

Bloom, P.R. and Erich, M.C. (1987) Effect of solution composition on the rate and mechanism of gibbsite dissolution in acid solutions. *Soil Sci. Soc. Am. J.*, **51**, 1131–1136.

Blum, A.E. and Lasaga, A.C. (1986) Forsterite dissolution kinetics. *EOS*, **67**, 1278–1279.

Blum, A.E. and Lasaga, A.C. (1988) Role of surface speciation in the low-temperature dissolution of minerals. *Nature,* **331**, 431–433.

Blum, A.E. and Lasaga, A.C. (1991) The role of surface speciation in the dissolution of albite. *Geochim. Cosmochim. Acta*, **55**, 2193–2201.

Blum, A.E. and Stillings, L.L. (1995) Feldspar dissolution kinetics. In: White, A.F. and Brantley, S.L. (eds.), *Chemical Weathering Rates of Silicate Minerals, Reviews in Mineralogy*, Mineralogical Society of America, **31**, 291–351.

Blum, A.E., Yund, R.A. and Lasaga, A.C. (1990) The effect of dislocation density on the dissolution rate of quartz. *Geochim. Cosmochim. Acta*, **54**, 283–297.

Blum, E.H. and Luus, R. (1964) Thermodynamic consistency of reaction rate expressions. *Chem. Eng. Sci.*, **19**, 322–323.

Boettcher, A.L. and Wyllie, P.J. (1968) The calcite–aragonite transition measured in the system CaO–CO_2–H_2O. *J. Geol.*, **76**, 314–330.

Bolin, B., Sukumar, R., Ciais, P., Cramer, W., Jarvis, P., Kheshgi, H., Nobre, C., Semenov, S. and Steffen, W. (2000) Global perspective. In: Watson, R.T., Noble, I.R., Bolin, B., Ravindranath, N.H., Verardo, D.J. and Dokken, D.J. (eds.), *IPCC, Land Use, Land-Use Change, and Forestry. A Special Report of the IPCC*. Cambridge, UK: Cambridge University Press, pp. 23–51.

Bondam, J. (1967) Structural changes in adularia in hydrolytic environments. *Medd. Dansk Geol. For.*, **17**, 357–370.

Bondietti, G., Sinniger, J. and Stumm, W. (1993) The reactivity of Fe(III) (hydr)oxides: Effects of ligands in inhibiting the dissolution. *Colloids Surf. A*, **79**, 157–167.

Born, M. (1920) Volumen und hydratationswärme der ionen. *Zeitschr. Phys.*, **21**, 45–48.

Borodin, V.L., Lyntin, V.I., Ilyukhin, V.V. and Belov, N.V. (1979) Isomorphous calcite–otavite series. *Soviet Phys. Dokl.*, **24**, 226–227.

Bottinga, Y., Weill, D. and Richet, P. (1982) Density calculations for silicate liquids. I. Revised method for aluminosilicate compositions. *Geochim. Cosmochim. Acta*, **46**, 909–919.

Bowers, T.S., Jackson, K.J. and Helgeson, H.C. (1984). *Equilibrium Activity Diagrams for Coexisting Minerals and Aqueous Solutions at Pressures and Temperatures to 5 kb and 600°C*. New York: Springer, 397pp.

Boyd, F.R. and England, J.L. (1960) The quartz-coesite transition. *J. Geophys. Res.*, **65**, 749–756.

Brady, P.V. (1992) Silica surface chemistry at elevated temperatures. *Geochim. Cosmochim. Acta*, **56**, 2941–2946.

Brady, P.V. and Carrol, S.A. (1994) Direct effect of CO_2 and temperature on silicate weathering: Possible implications for climate control. *Geochim. Cosmochim. Acta*, **58**, 1853–1856.

Brady, P.V. and Gislason, S.R. (1997) Seafloor weathering controls on atmospheric CO_2 and global climate. *Geochim. Cosmochim. Acta*, **61**, 965–973.

Brady, P.V. and Walther, J.V. (1989) Controls on silicate dissolution rates in neutral and basic pH solutions at 25°C. *Geochim. Cosmochim. Acta*, **53**, 2823–2830.

Brady, P.V. and Walther, J.V. (1990) Kinetics of quartz dissolution at low temperatures. *Chem. Geol.*, **82**, 253–264.

Bragg, W.L. (1914) The analysis of crystals by the x-ray spectrometer. *Proc. R. Soc. London*, **A89**, 468–489.

Brandon, A.D., Creaser, R.A. and Chacko, T. (1996) Constraints on rates of granitic magma transport from epidote dissolution kinetics. *Science*, **271**, 1845–1848.

Brandt, F., Bosbach, D., Krawczyk-Bärsch, E., Arnold, T. and Bernhard, G. (2003) Chlorite dissolution in the acid pH-range: A combined microscopic and macroscopic approach. *Geochim. Cosmochim. Acta*, **67**, 1451–1461.

Brantley, S.L. and Chen, Y. (1995) Chemical weathering rates of pyroxenes and amphiboles. In: White, A. F. and Brantley, S.L. (eds.), *Chemical Weathering Rates of Silicate Minerals*, *Reviews in Mineralogy*, Mineralogical Society of America, **31**, 119–172.

Brantley, S.L. and Mellott, N.P. (1999) Surface area and porosity of primary silicate minerals. *Am. Mineral.*, **85**, 1767–1783.

Brantley, S.L. and Stillings, L.L. (1996) Feldspar dissolution at 25°C and low pH. *Am. J. Sci.*, **296**, 101–127.

Brindley, G.W. and Brown, G. (eds.) (1980) *Crystal Structures of Clay Minerals and their X-ray Identification*. London: Mineralogical Society Monograph No. **5**, Mineralogical Society, 495pp.

Briones, J.A., Mullins, J.C. and Thies, M.C. (1987) Ternary phase equilibria for acetic acid-water mixtures with supercritical carbon dioxide. *Fluid Phase Equilib.*, **36**, 235–246.

Broecker, W.S. and Peng, T.H. (1982) *Tracers in the Sea*. Palisades: Eldigio, 691pp.

Brönsted, J.N. (1928) Acid and basic catalysis. *Chem. Rev.*, **5**, 231–338.

Brown, C.A., Compton, R.G. and Narramore, C.A. (1993) The kinetics of calcite dissolution/precipitation. *J. Colloid Interface Sci.*, **160**, 372–379.

Brown, I.D. and Shannon, R.D. (1973) Empirical bond length–bond strength curves for oxides. *Acta Crystallogr.*, **A29**, 266–282.

Brunauer, S., Emmett, P. H. and Teller, E. (1938) Adsorption of gases in multimolecular layers. *J. Am. Chem. Soc.*, **60**, 309–319.

Bruni, J., Canepa, M., Cipolli, F., Marini, L., Ottonello, G., Vetuschi Zuccolini, M., Chiodini, G., Cioni, R. and Longinelli, A. (2001) Reactions governing the chemistry of waters interacting with serpentinites. In: Cidu, R. (ed.), *Water-Rock Interaction 2001.* Lisse: Balkema, pp. 145–148.

Bruni, J., Canepa, M., Cipolli, F., Marini, L., Ottonello, G., Vetuschi Zuccolini, M., Chiodini, G., Cioni, R. and Longinelli, A. (2002) Irreversible water–rock mass transfer accompanying the generation of the neutral, Mg-HCO_3 and high-pH, Ca-OH spring waters of the Genova province, Italy. *Appl. Geochem.*, **17**, 455–474.

Bruno, J., Stumm, W., Wersin, P. and Brandberg, F. (1992) On the influence of carbonate in mineral dissolution: I. The thermodynamics and kinetics of hematite dissolution in bicarbonate solutions at T = 25°C. *Geochim. Cosmochim. Acta*, **56**, 1139–1147.

Budyko, M.I. and Ronov, A.B. (1979) Chemical evolution of the atmosphere in the Phanerozoic. *Geochem. Int.*, **16**, 1–9.

Burton, W.K., Cabrera, N. and Frank, F.C. (1951) The growth of crystals and the equilibrium structure of their surfaces. *Philos. Trans. R. Soc. London,* **A243**, 299–358.

Busenberg, E. and Clemency, C.V. (1976) The dissolution kinetics of feldspars at 25°C and 1 atm CO_2 partial pressure. *Geochim. Cosmochim. Acta*, **40**, 41–49.

Busenberg, E. and Plummer, L.N. (1982) The kinetics of dissolution of dolomite in CO_2–H_2O systems at 1.5 to 65°C and 0 to 1 atm pCO_2. *Am J. Sci.*, **282**, 45–78.

Busenberg, E. and Plummer, L.N. (1986) A comparative study of the dissolution and crystal growth kinetics of calcite and aragonite. In: Mumpton, F.A. (ed.) *Studies in Diagenesis.* U.S.G.S. Bull., **1578**, 139–168.

Callendar, G.S. (1938) The artificial production of carbon dioxide and its influence on temperature. *R. Meteorol. Soc. Quart. J.*, **64**, 224–237.

Cama, J., Ganor, J., Ayora, C. and Lasaga, A.C. (2000) Smectite dissolution kinetics at 80°C and pH 8.8. *Geochim. Cosmochim. Acta*, **64**, 2701–2717.

Cama, J., Metz, V. and Ganor, J. (2002) The effect of pH and temperature on kaolinite dissolution rate under acidic conditions. *Geochim. Cosmochim. Acta*, **66**, 3913–3926.

Carlson, W.D. (1990) The polymorphs of $CaCO_3$ and the aragonite-calcite transformation. In: Reeder, R.J. (ed.), *Carbonates: Mineralogy and Chemistry, Reviews in Mineralogy*, Mineralogical Society of America, 2nd printing, **11**, 191–225.

Carpenter, M.A., Salje, E.K.H., Graeme-Barber, A., Wruck, B., Dove, M.T. and Knight, K.S. (1998) Calibration of excess thermodynamic properties and elastic constant variations associated with the $\alpha \leftrightarrow \beta$ phase transition in quartz. *Am. Mineral.*, **83**, 2–22.

Carroll, J.J. and Mather, A.E. (1992) The system carbon dioxide-water and the Krichevsky-Kasarnovsky equation. *J. Solut. Chem.*, **21**, 607–621.

Carroll, S.A. and Knauss, K.G. (2001) Experimental determination of Ca-silicate dissolution rates: A source of calcium for geologic CO_2 sequestration. *First National Conference on Carbon Sequestration*, National Energy Technology Laboratory, Washington DC, http://www.netl.doe.gov.

Carroll, S.A. and Knauss, K.G. (2005) Dependence of labradorite dissolution kinetics on $CO_{2(aq)}$, $Al_{(aq)}$, and temperature. *Chem. Geol.*, **217**, 213–225.

Carroll, S.A. and Walther, J.V. (1990) Kaolinite dissolution at 25°, 60°, and 80°C. *Am. J. Sci.*, **290**, 797–810.

Carroll-Webb, S.A. and Walther, J.V. (1988) A surface complex reaction model for the pH-dependence of corundum and kaolinite dissolution rates. *Geochim. Cosmochim. Acta*, **52**, 2609–2623.

Casey, W. H. (1991) On the relative dissolution rates of some oxide and orthosilicate minerals. *J. Colloid Interface Sci.*, **146**, 586–589.

Casey, W. H. and Bunker, B. (1990) Leaching of mineral and glass surfaces during dissolution. In: Hochella, M.F., Jr and White, A.F. (eds.), *Mineral–Water Interface Geochemistry, Reviews in Mineralogy*, Mineralogical Society of America, **23**, 397–424.

Casey, W.H. and Ludwig, C. (1995) Silicate mineral dissolution as a ligand-exchange reaction. In: White, A.F. and Brantley, S.L. (eds.), *Chemical Weathering Rates of Silicate Minerals, Reviews in Mineralogy*, Mineralogical Society of America, **31**, 87–117.

Casey, W.H. and Ludwig, C. (1996) The mechanism of dissolution of oxide minerals. *Nature*, **381**, 506–509.

Casey, W.H. and Westrich, H.R., (1992) Control of dissolution rates of orthosilicate minerals by divalent metal–oxygen bonds. *Nature*, **355**, 157–159.

Casey, W.H., Carr, M.J. and Graham, R.A. (1988) Crystal defects and the dissolution kinetics of rutile. *Geochim. Cosmochim. Acta*, **52**, 1545–1556.

Casey, W.H., Westrich, H.R., Arnold, G.W. and Banfield, J.F. (1989) The surface chemistry of dissolving labradorite feldspar. *Geochim. Cosmochim. Acta*, **53**, 821–832.

Casey, W.H., Westrich, H.R., Banfield, J.F., Ferruzzi, G. and Arnold, G.W. (1993) Leaching and reconstruction at the surface of dissolving chain-silicate minerals. *Nature*, **366**, 253–256.

Casey, W.H., Westrich, H.R. and Holdren, G.R. (1991) Dissolution rates of plagioclase at pH = 2 and 3. *Am. Mineral.*, **76**, 211–217.

Casey, W.H., Ludwig, C. and Holmén, B. (1998) Toward understanding the rates of reactions at mineral surfaces. In: Marini, L. and Ottonello, G. (eds.), *Proceedings of the Rome Seminar of Environmental Geochemistry*, Pacini, Pisa, 197pp.

Castellan, G.W. (1971) *Physical Chemistry*, 2nd ed.. Reading, MA: Addison-Wesley, 866pp.

Cellai, D. Carpenter, M.A., Wruck, B. and Salje, E.K.H. (1994) Characterization of high-temperature phase transitions in single crystals of Steinbach tridymite. *Am. Mineral.*, **79**, 606–614.

Chamberlin, T.C. (1898) The influence of great epochs of limestone formation upon the constitution of the atmosphere. *J. Geol.*, **6**, 609–621.

Chang, B.-T., Pak, L.-H. and Li, Y.-S. (1979) Solubilities and rates of dissolution of diaspore in NaOH aqueous solutions. *Bull. Chem. Soc. Japan*, **52**, 1321–1326.

Chang, L.L.Y. (1965) Subsolidus phase relations in the systems $BaCO_3$–$SrCO_3$, $SrCO_3$–$CaCO_3$, and $BaCO_3$–$CaCO_3$. *J. Geol.*, **73**, 346–368.

Chen, Y. and Brantley, S. L. (1997) Temperature- and pH-dependence of albite dissolution rate at acid pH. *Chem. Geol.*, **135**, 275–290.

Chen, Y. and Brantley, S. L. (1998) Diopside and anthophyllite dissolution at 25°C and 90°C and acid pH. *Chem. Geol.*, **147**, 233–248.

Chen, Y. and Brantley, S. L. (2000) Dissolution of forsteritic olivine at 65°C and $2 < pH < 5$. *Chem. Geol.*, **165**, 267–281.

Chessin, N., Hamilton, W.C. and Post, B. (1965) Position and thermal parameters of oxygen atoms in calcite. *Acta Crystallogr.*, **18**, 689–693.

Chiodini, G., Cioni, R., Guidi, M. and Marini, L. (1991) Chemical geothermometry and geobarometry in hydrothermal aqueous solutions: A theoretical investigation based on a mineral–solution equilibrium model. *Geochim. Cosmochim. Acta*, **55**, 2709–2727.

Chou, L. and Wollast, R. (1984) Study of the weathering of albite at room temperature and pressure with a fluidized bed reactor. *Geochim. Cosmochim. Acta*, **48**, 2205–2217.

Chou, L. and Wollast, R. (1985) Steady-state kinetics and dissolution mechanisms of albite. *Am. J. Sci.*, **285**, 963–993.

Chou, L., Garrels, R.M. and Wollast, R. (1989) Comparative study of the kinetics and mechanisms of dissolution of carbonate minerals. *Chem. Geol.*, **78**, 269–282.

Ciais, P., Denning, A.S., Tans, P.P., Berry, J.A., Randall, D.A., Collatz, G.J., Sellers, P.J., White, J.W.C., Trolier, M., Meijer, H.A.J., Francey, R.J., Monfray, P. and Heimann, M. (1997) A three-dimensional synthesis study of $\delta^{18}O$ in atmospheric CO_2. 1. Surface fluxes. *J. Geophys. Res.–Atmos.*, **102**, 5857–5872.

Cipolli, F., Gambardella, B., Marini, L., Ottonello, G. and Vetuschi Zuccolini, M. (2004) Geochemistry of high-pH waters from serpentinites of the Gruppo di Voltri (Genova, Italy) and reaction path modeling of CO_2 sequestration in serpentinite aquifers. *Appl. Geochem.*, **19**, 787–802.

Clay, J.P. and Thomas, A.W. (1938) The catalytic effect of anions upon the rate of dissolution of hydrous alumina by acids. *J. Am. Chem. Soc.*, **60**, 2384–2390.

Clegg, S.L. and Brimblecombe, P. (1990) Solubility of volatile electrolytes in multicomponent solutions with atmospheric applications. In: Melchior, D.C. and Bassett, R.L. (eds.), *Chemical Modeling of Aqueous Systems II*, American Chemical Society Symposium Series, **416**, 58–73.

Clemency, C.V. and Lin, F.-C. (1981) Dissolution kinetics of phlogopite. II. Open-system using an ion-exchange resin. *Clays Clay Miner.*, **29**, 107–112.

Coan, C.R. and King, A.D., Jr. (1971) Solubility of water in compressed carbon dioxide, nitrous oxide, and ethane. Evidence for hydration of carbon dioxide and nitrous oxide in the gas phase. *J. Am. Chem. Soc.*, **93**, 1857–1862.

Compton, R.G., Pritchard, K.L. and Unwin, P.R. (1989) The dissolution of calcite in acid waters: Mass transport versus surface control. *Freshwater Biol.*, **22**, 286–288.

Connaughton, L.M., Millero, F.J. and Pitzer, K.S. (1989) Volume changes for mixing the major sea salts: Equations valid to ionic strength 3.0 and temperature 95°C. *J. Solution Chem.*, **18**, 1007–1017.

Conway, T.J., Tans, P.P., Waterman, L.S., Thoning, K.W., Kitzis, D.R., Masarie, K.A. and Zhang, N. (1994) Evidence for interannual variability of the carbon cycle from the National Oceanic and Atmospheric Administration/Climate Monitoring and Diagnostics Laboratory global air sampling network. *J. Geophys. Res.*, **99**, 22831–22855.

Corazza, E., Sabelli, C. and Vannucci, S. (1977) Dawsonite: New mineralogical data and structure refinement. *Neues Jahrb. Mineral. Monatshefte*, **9**, 381–397.

Correns, C.W. (1940) Die Chemische Verwitterung der Silikate. *Naturw.*, **28**, 369–376.

Correns, C. W. and von Engelhardt W. (1938) Neue Untersuchungen über die Verwitterung des Kalifeldspates. *Chemie der Erde*, **12**, 1–22.

Corti, H.R., De Pablo, J.J. and Prausnitz, J.M. (1990) Phase equilibria for aqueous systems containing salts and carbon dioxide. Application of Pitzer's theory for electrolyte solutions. *J. Phys. Chem.*, **94**, 7876–7880.

Coveney, R.M. and Kelly, W.C. (1971) Dawsonite as a daughter mineral in hydrothermal fluid inclusions. *Contrib. Mineral. Petrol.*, **32**, 334–342.

Crovetto, R. (1991) Evaluation of solubility data for the system CO_2–H_2O from 273 K to the critical point of water. *J. Phys. Chem. Ref. Data*, **20**, 575–589.

Cygan, R.T., Casey, W.H., Roslough, M.B., Westrich, H.R., Carr, M.J. and Holdren, G.R. (1989) Dissolution kinetics of experimentally shocked silicate minerals. *Chem. Geol.*, **78**, 229–244.

D'Souza, R., Patrick, J.R. and Teja, A.S. (1988) High pressure phase equilibria in the carbon dioxide-*n*-hexadecane and carbon dioxide–water systems. *Can. J. Chem. Eng.*, **66**, 319–323.

Daubrée, A. (1867) Expériences sur les décompositions chimiques provoquées par les actions méchaniques dans divers minéraux tels que le feldspath. *Acad. Sci. Paris, Comptes Rendus*, **64**, 339–346.

Daux, V., Guy, C., Advocat, T., Crovisier, J.-L. and Stille P. (1997) Kinetic aspects of basaltic glass dissolution at 90°C: Role of aqueous silicon and aluminum. *Chem. Geol.*, **142**, 109–126.

Daveler, S.A. (1998) XTOOL 7.15 user's manual. Lawrence Livermore National Laboratory. Report UCRL-MA-130654, 32pp.

Debenedetti, P.G. and Stanley, H.E. (2003) Supercooled and glassy water. *Phys. Today*, **56**(6), 40–46.

Debye, P. and Hückel, E. (1923a) The theory of electrolytes. I. Lowering of freezing point and related phenomena. *Zeitschr. Physik*, **24**, 185–206.

Debye, P. and Hückel, E. (1923b) The theory of electrolytes. II. The limiting law of electrical conductivity. *Zeitschr. Physik*, **24**, 305–325.

Debye, P.J. (1954) *The Collected Papers of P.J. Debye*. New York: Wiley-Interscience, 700pp.

De Donder, T. (1920) *Leçons de Thermodynamique et de Chimie-Physique*. Paris: Gauthier-Villars.

Deer, W.A., Howie, R.A. and Zussman, J. (1992) *An Introduction to the Rock-forming Minerals*. London: Longman, 712pp.

Denbigh, K. (1971) *The Principles of Chemical Equilibrium*, 3rd ed. London: Cambridge University Press, 494pp.

Deng, Y. (1997) Effect of pH on the reductive dissolution rates of iron (III) hydroxide by ascorbate. *Langmuir*, **13**, 1835–1839.

Devidal, J.-L., Schott, J. and Dandurand, J.-L. (1997) An experimental study of kaolinite dissolution and precipitation kinetics as a function of chemical affinity and solution composition at 150°C, 40 bars, and pH 2, 6.8, and 7.8. *Geochim. Cosmochim. Acta*, **61**, 5165–5186.

Diamond, L.W. and Akinfiev, N.N. (2003) Solubility of CO_2 in water from -1.5 to 100°C and from 0.1 to 100 MPa: Evaluation of literature data and thermodynamic modelling. *Fluid Phase Equilib.*, **208**, 265–290.

Dibble, W.E., Jr. and Tiller, W.A. (1981) Non-equilibrium water/rock interactions – I. Model for interface-controlled reactions. *Geochim. Cosmochim. Acta*, **45**, 79–92.

Dietzel, M. and Böhme, G. (2005) The dissolution rates of gibbsite in the presence of chloride, nitrate, silica, sulfate, and citrate in open and closed systems at 20°C. *Geochim. Cosmochim. Acta*, **69**, 1199–1211.

Dinelli, E., Lombini, A., Simoni, A. and Ferrari, C. (1997) Heavy metals in serpentinite soils of selected outcrops of Piacenza and Parma provinces (Northern Apennines, Italy). *Miner. Petrogr. Acta*, **40**, 241–255.

Dohrn, R., Bünz, A.P., Devlieghere, F. and Thelen, D. (1993) Experimental measurements of phase equilibria for ternary and quaternary systems of glucose, water, CO_2 and ethanol with a novel apparatus. *Fluid Phase Equilib.*, **83**, 149–158.

Dollase, W.A. (1967) The crystal structure at 220 C of orthorhombic high tridymite from the Steinbach meteorite. *Acta Crystallogr.*, **23**, 617–623.

Dollase, W.A. and Baur, W.H. (1976) The superstructure of meteoritic low tridymite solved by computer simulation. *Am. Mineral.*, **61**, 971–978.

Dos Santos, M. and Stumm, W. (1992) Reductive dissolution of iron (III) (hydr)oxides by hydrogen sulfide. *Langmuir*, **8**, 1671–1675.

Dove, P.M. (1994) The dissolution kinetics of quartz in sodium chloride solutions at 25° to 300°C. *Am. J. Sci.*, **294**, 665–712.

Dove, P.M. (1995) Kinetic and thermodynamic controls on silica reactivity in weathering environments. In: White, A. F. and Brantley, S.L. (eds.), *Chemical Weathering Rates of Silicate Minerals*, *Reviews in Mineralogy*, Mineralogical Society of America, **31**, 235–290.

Dove, P.M. (1999) The dissolution kinetics of quartz in aqueous mixed cation solutions. *Geochim. Cosmochim. Acta*, **63**, 3715–3727.

Dove, P.M. and Crerar, D.A. (1990) Kinetics of quartz dissolution in electrolyte solutions using a hydrothermal mixed flow reactor. *Geochim. Cosmochim. Acta*, **54**, 955–969.

Dove, P.M. and Elston, S.F. (1992) Dissolution kinetics of quartz in sodium chloride solutions: Analysis of existing data and a rate model for 25°C. *Geochim. Cosmochim. Acta*, **56**, 4147–4156.

Dove, P.M. and Nix, C.J. (1997) The influence of the alkaline earth cations, magnesium, calcium, and barium on the dissolution kinetics of quartz. *Geochim. Cosmochim. Acta*, **61**, 3329–3340.

Drever, J.I. (1982) *The Geochemistry of Natural Waters*. NJ: Prentice-Hall, 437pp.

Duan, Z., Møller, N. and Weare, J.H. (1992) An equation of state for the CH_4–CO_2–H_2O system: II. Mixtures from 50 to 1000°C and 0 to 1000 bars. *Geochim. Cosmochim. Acta*, **56**, 2619–2631.

Duval, Y., Mielczarski, J.A., Pokrovsky, O.S., Mielczarski, E. and Ehrhardt, J.J. (2002) Evidence of the existence of three types of species at the quartz–aqueous solution interface at pH 0–10: XPS surface group quantification and surface complexation modeling. *J. Phys. Chem. B*, **106**, 2937–2945.

Dyne, J.R. (1996) *Sodium Carbonate Resources of the Green River Formation*, U.S. Geological Survey, Open-file report, pp. 96–729.

Effenberger, H., Mereiter, K. and Zemann, J. (1981) Crystal structure refinements of magnesite, calcite, rhodochrosite, siderite, smithsonite and dolomite, with discussion of some aspects of the stereochemistry of calcite-type carbonates. *Zeit. Kristallogr.*, **156**, 233–243.

Efremova, G.D. and Shvarts, A.V. (1972) Higher-order critical phenomena in ternary systems. The methanol–carbon dioxide–ethane system. *Russ. J. Phys. Chem.* (Engl. Transl.), **46**, 237–239.

Eggleston, C.M., Hochella, M.F., Jr. and Parks, G.A. (1989) Sample preparation and aging effects on the dissolution rate and surface composition of diopside. *Geochim. Cosmochim. Acta*, **53**, 797–804.

Ellis, A.J. and Golding, R.M. (1963) The solubility of carbon dioxide above 100°C in water and in sodium chloride solutions. *Am. J. Sci.* **261**, 47–60.

Elzinga, E.J., Reeder, R.J., Withers, S.H., Peale, R.E., Mason, R.A., Beck, K.M. and Hess, W.P. (2002) EXAFS study of rare-earth element coordination in calcite. *Geochim. Cosmochim. Acta*, **66**, 2875–2885.

Emberley, S., Hutcheon, I., Shevalier, M., Durocher, K., Gunter, W.D. and Perkins, E.H. (2004) Geochemical monitoring of fluid-rock interaction and CO_2 storage at the Weyburn CO_2-injection enhanced oil recovery site, Saskatchewan, Canada. *Energy*, **29**, 1393–1401.

Emberley, S., Hutcheon, I., Shevalier, M., Durocher, K., Mayer, B., Gunter, W.D. and Perkins, E.H. (2005) Monitoring of fluid–rock interaction and CO_2 storage through produced fluid sampling at the Weyburn CO_2-injection enhanced oil recovery site, Saskatchewan, Canada. *Appl. Geochem.*, **20**, 1131–1157.

Ernst, W.G. (1976) *Petrologic Phase Equilibria*. San Francisco: W.H. Freeman and Co., 333pp.

Etheridge, D.M., Steele, L.P., Langenfelds, R.L., Francey, R.J., Barnola, J.M. and Morgan, V.I. (1996) Natural and anthropogenic changes in atmospheric CO_2 over the last 1000 years from air in Antarctic ice and firn. *J. Geophys. Res.–Atmos.*, **101**, 4115–4128.

Etheridge, D.M., Steele, L.P., Langenfelds, R.L., Francey, R.J., Barnola, J.-M. and Morgan, V.I. (1998) Historical CO_2 records from the Law Dome DE08, DE08-2, and DSS ice cores. In *Trends: A Compendium of Data on Global Change*. Oak Ridge, TN, USA: Carbon Dioxide Information Analysis Center, Oak Ridge National Laboratory, U.S. Department of Energy.

Evans, B. (1965) Application of a reaction-rate method to the breakdown equilibria of muscovite and muscovite plus quartz. *Am. J. Sci.*, **263**, 647–667.

Eyring, H. (1935a) The activated complex in chemical reactions. *J. Chem. Phys.*, **3**, 107–115.

Eyring, H. (1935b) The activated complex and the absolute rate of chemical reactions. *Chem. Rev.*, **17**, 65–77.

Falkowski, P.G., Barber, R.T. and Smetacek, V. (1998) Biogeochemical controls and feedbacks on ocean primary production. *Science*, **281**, 200–206.

Fan, S.-S. and Guo, T.-M. (1999) Hydrate formation of CO_2-rich binary and quaternary gas mixtures in aqueous sodium chloride solutions. *J. Chem. Eng. Data*, **44**, 829–832.

Fei, Y. (1995) Thermal expansion. In: Ahrens, T.J. (ed.), *Mineral Physics and Crystallography: A Handbook of Physical Constants*, AGU Reference Shelf, **2**, 29–44.

Feigenbaum, S. and Shainberg, I. (1975) Dissolution of illite – A possible mechanism of potassium release. *Soil Sci. Soc. Am. Proc.*, **39**, 985–990.

Ferrante, M.J., Stuve, J.M. and Richardson, D.W. (1976) *Thermodynamic data for synthetic dawsonite*. 8129, Washington DC: U.S. Bureau of Mines Report Investigation, 13pp.

Ferruzzi, G.G. (1994) *The character and rates of dissolution of pyroxenes and pyroxenoids*. M. Sc. Thesis, University of California, Davis.

Finger, L.W. (1975) Least-squares refinement of the rigid-body motion parameters of CO_3 in calcite and magnesite and correlation with lattice vibrations. *Carnegie Inst. Washington, Year Book*, **74**, 572–575.

Fischer, A.G. (1983) The two Phanerozoic subcycles. In: Berggren, W. and Van Couvering, J. (eds.) *Catastrophies in Earth History; The New Uniformitarianism*. Princetown: Princetown University Press.

Fischer, H., Whalen, M., Smith, J., Mastroianni, D. and Deck, B. (1999) Ice core records of atmospheric CO_2 around the last three glacial terminations. *Science*, **283**, 1712–1714.

Fisher, D. (ed.) (1991) *Rules of Thumb for Engineers and Scientists*. Houston: Gulf Publishing Co., 242pp.

Flörke, O.W., Jones, J.B. and Schmincke, H.U. (1976) A new microcrystalline silica from Gran Canaria. *Zeit. Kristallogr.*, **143**, 156–165.

Fournier, R.O. (1973) Silica in thermal waters: Laboratory and field investigations. *Proceedings of the International Symposium on Hydrogeochemistry and Biogeochemistry, Tokyo*, pp. 122–139.

Fournier, R.O. (1991) Water geothermometers applied to geothermal energy. In: D'Amore, F. (co-ordinator), *Application of Geochemistry in Geothermal Reservoir Development*. Roma, Italy: UNITAR, pp. 37–69.

Francey, R.J., Tans, P.P. and Allison, C.E. (1995) Changes in oceanic and terrestrial carbon uptake since 1982. *Nature*, **373**, 326–330.

Frogner, P. and Schweda, P. (1998) Hornblende dissolution kinetics at 25°C. *Chem. Geol.*, **151**, 169–179.

Furmakova, L.N. (1981) The question of formation conditions for dawsonite. *Lithol. Miner. Res.*, **15**, 565–572.

Furrer, G. and Stumm, W. (1986) The coordination chemistry of weathering: I. Dissolution kinetics of δ-Al_2O_3 and BeO. *Geochim. Cosmochim. Acta*, **50**, 1847–1860.

Furrer, G., Zysset, M. and Schindler, P.W. (1993) Weathering kinetics of montmorillonite: Investigations in batch and mixed flow reactors. In: Manning, D.A.C., Hall, P.L. and Hughes, C.R. (eds.), *Geochemistry of Clay-Pore Fluid Interactions*. The Mineralogical Society Series, Berlin: Springer, **4**, 243–262.

Gambardella, B., Marini, L. and Baneschi, I. (2005) Dissolved potassium in the shallow groundwaters circulating in the volcanic rocks of central-southern Italy. *Appl. Geochem.*, **20**, 875–897.

Ganor, J., Mogollón, J.L. and Lasaga, A.C. (1995) The effect of pH on kaolinite dissolution rates and on activation energy. *Geochim. Cosmochim. Acta*, **59**, 1037–1052.

Ganor, J., Mogollón, J.L. and Lasaga, A.C. (1999) Kinetics of gibbsite dissolution under low ionic strength conditions. *Geochim. Cosmochim. Acta*, **63**, 1635–1651.

Gattuso, J.-P., Frankignoulle, M. and Wollast, R. (1998) Carbon and carbonate metabolism in coastal aquatic ecosystems. *Annu. Rev. Eco. Syst.*, **29**, 405–434.

Gaus, I., Azaroual, M. and Czernichowski-Lauriol, I. (2005) Reactive transport modelling of the impact of CO_2 injection on the clayey cap rock at Sleipner (North Sea). *Chem. Geol.*, **217**, 319–337.

Gautelier, M., Oelkers, E.H. and Schott, J. (1999) An experimental study of dolomite dissolution rates as a function of pH from −0.5 to 5 and temperature from 25 to 80°C. *Chem. Geol.*, **157**, 13–26.

Gautier, J.-M., Oelkers, E.H. and Schott, J. (1994) Experimental study of K-feldspar dissolution rates as a function of chemical affinity at 150°C and pH 9. *Geochim. Cosmochim. Acta*, **58**, 4549–4560.

Giammar, D.E., Bruant, R.G., Jr. and Peters, C.A. (2005) Forsterite dissolution and magnesite precipitation at conditions relevant for deep saline aquifer storage and sequestration of carbon dioxide. *Chem. Geol.*, **217**, 257–276.

Giggenbach, W.F. (1984) Mass transfer in hydrothermal alteration systems – A conceptual approach. *Geochim. Cosmochim. Acta*, **48**, 2693–2711.

Giggenbach, W.F. (1985) Construction of thermodynamic stability diagrams involving dioctahedral potassium clay minerals. *Chem. Geol.*, **49**, 231–242.

Giggenbach, W.F. (1987) Redox processes governing the chemistry of fumarolic gas discharges from White Island, New Zeland. *Appl. Geochem.*, **2**, 143–161.

Giggenbach, W.F. (1997) The origin and evolution of fluids in magmatic-hydrothermal systems. In: Barnes, H.L. (ed.), *Geochemistry of Hydrothermal Ore Deposits,* 3rd ed. New York: Wiley, 737–796.

Gislason, S.R. and Oelkers, E.H. (2003) Mechanism, rates, and consequences of basaltic glass dissolution: II. An experimental study of the dissolution rates of basaltic glass as a function of pH and temperature. *Geochim. Cosmochim. Acta*, **67**, 3817–3832.

Gislason, S.R., Heaney, P.J., Oelkers, E.H. and Schott, J. (1997) Kinetic and thermodynamic properties of moganite, a novel silica polymorph. *Geochim. Cosmochim. Acta*, **61**, 1193–1204.

Glass, B.P. (1984) Tektites. *J. Non-Cryst. Solids*, **67**, 333–344.

Golab, A. (2003) *The impact of igneous intrusion on coal, cleat carbonate, and groundwater composition*. Ph.D. Thesis, University of Wollongong, 327pp.

Goldberry, R and Loughnan, F.C. (1977) Dawsonite, alumohydrocalcite, nordstrandite and gorceixite in Permian marine strata of the Sydney Basin, Australia. *Sedimentology*, **24**, 565–579.

Goldich, S.S. (1938) A study in rock weathering. *J. Geol.*, **46**, 17–58.

Goldsmith, J.R. (1990) Phase relations of rhombohedral carbonates. In: Reeder, R.J. (ed.), *Carbonates: Mineralogy and Chemistry*, *Reviews in Mineralogy*, Mineralogical Society of America, 2nd printing, **11**, 49–76.

Goldsmith, J.R. and Heard, H.C. (1961) Subsolidus phase relations in the system $CaCO_3$–$MgCO_3$. *J. Geol.*, **69**, 45–74.

Golubev, S.V., Pokrosky, O.S. and Schott, J. (2005) Experimental determination of the effect of dissolved CO_2 on the dissolution kinetics of Mg and Ca silicates at 25°C. *Chem. Geol.*, **217**, 227–238.

Graetsch, H. and Topalovic-Dierdorf, I. (1996) ^{29}Si MAS NMR spectrum and superstructure of modulated tridymite L3-T_o(MX-1). *Eur. J. Mineral.*, **8**, 103–113.

Graf, D.L. (1961) Crystallographic tables for the rhombohedral carbonates. *Am. Mineral.*, **46**, 1283–1316.

Grandstaff, D.E. (1977) Some kinetics of bronzite orthopyroxene dissolution. *Geochim. Cosmochim. Acta*, **41**, 1097–1103.

Grandstaff, D.E. (1980) The dissolution rate of forsteritic olivine from Hawaiian beach sand. *Third International Symposium on Water–Rock Interaction*, Edmonton, Canada: Alberta Research Council, pp. 72–74.

Grandstaff, D.E. (1986) The dissolution rate of forsteritic olivine from Hawaiian beach sand. In: Colman, S.M. and Dethier, D.P. (eds), *Rates of Chemical Weathering of Rocks and Minerals*. Orlando: Academic Press.

Greenberg, J. and Tomson, M. (1992) Precipitation and dissolution kinetics and equilibria of aqueous ferrous carbonate vs temperature. *Appl. Geochem.*, **7**, 185–190.

Gunter, W.D., Bachu, S., Law D.H.S., Marwaha, V., Drysdale, D.L., McDonald, D.E. and McCann, T.J. (1996) Technical and economic feasibility of CO_2 disposal in aquifers within the Alberta Sedimentary Basin, Canada. *Energy Convers. Manage.*, **37**, 1135–1142.

Gunter, W.D., Perkins, E.H. and Hutcheon, I. (2000) Aquifer disposal of acid gases: Modeling of water–rock reactions for trapping acid wastes. *Appl. Geochem.*, **15**, 1085–1095.

Gunter, W.D., Perkins, E.H. and McCann, T.J. (1993) Aquifer disposal of CO_2-rich gases. Reaction design for added capacity. *Energy Convers. Manage.*, **34**, 941–948.

Gunter, W.D., Wiwchar, B. and Perkins, E.H. (1997) Aquifer disposal of CO_2-rich greenhouse gases: Extension of the time scale of experiment for CO_2-sequestering reactions by geochemical modeling. *Mineral. Petrol.*, **59**, 121–140.

Gustafsson, Å.B. and Puigdomenech, I. (2003) The effect of pH on chlorite dissolution rates at 25°C. *Mat. Res. Symp. Proc.*, **757**, 649–655.

Guthrie, G.D., Jr., Carey, J.W., Bergfeld, D., Byler, D., Chipera, S., Ziock, H.-J and Lackner, K. (2001) Geochemical aspects of the carbonation of magnesium silicates in an aqueous medium. First National Conference on Carbon Sequestration. National Energy Technology Laboratory, Washington DC, http://www.netl.doe.gov.

Gutjahr, A., Dabringhaus, H. and Lacmann, R. (1996) Studies of the growth and dissolution kinetics of the $CaCO_3$ polymorphs calcite and aragonite. I. Growth and dissolution rates in water. *J. Crystal Growth*, **158**, 296–309.

Guy, C. (1989) *Méchanismes de dissolution des solides dans les solutions hydrothermales déduits du comportement des verres basaltiques et de calcites déformées*. Ph.D. Thesis. Université Paul Sabatier.

Hajash, A., Jr., Carpenter, T. D. and Dewers, T. A. (1998) Dissolution and time-dependent compaction of albite sand: Experiments at 100°C and 160°C in pH-buffered organic acids and distilled water. *Tectonophysics*, **295**, 93–115.

Hall, A. (1987) *Igneous Petrology*. Harlow, UK: Longman Scientific & Technical, 573pp.

Hamer, M., Graham, R.C., Amrhein, C. and Bozhilov, K.N. (2003) Dissolution of ripidolite (Mg, Fe-Chlorite) in organic and inorganic acid solutions. *Soil Sci. Soc. Am. J.*, **67**, 654–661.

Hamilton, J.P., Brantley, S.L., Pantano, C.G., Criscenti, L.J. and Kubicki, J.D. (2001) Dissolution of nepheline, jadeite and albite glasses: Toward better models for aluminosilicate dissolution. *Geochim. Cosmochim. Acta*, **65**, 3683–3702.

Hamilton, J.P., Pantano, C.G. and Brantley, S.L. (2000) Dissolution of albite glass and crystal. *Geochim. Cosmochim. Acta*, **64**, 2603–2615.

Harned, H.S. (1920) The thermodynamic properties of the ions of some strong electrolytes and of the hydrogen ion in solutions of tenth molal hydrochloric acid containing uni-univalent salts. *J. Am. Chem. Soc.*, **42**, 1808–1832.

Harned, H.S. and Owen, B.B. (1958) *The Physical Chemistry of Electrolyte Solutions*, 3rd ed. New York: Reinhold.

Harvie, C.E. and Weare, J.H. (1980) The prediction of mineral solubilities in natural waters: The Na–K–Mg–Ca–Cl–SO_4–H_2O system from zero to high concentrations at 25°C. *Geochim. Cosmochim. Acta*, **44**, 981–997.

Harvie, C.E., Møller, N. and Weare, J.H. (1984) The prediction of mineral solubilities in natural waters: The Na–K–Mg–Ca–H–Cl–SO_4–OH–HCO_3–CO_3–CO_2–H_2O system to high ionic strengths at 25°C. *Geochim. Cosmochim. Acta*, **48**, 723–751.

Hayashi, H. and Yamada, M. (1990) Kinetics of dissolution of noncrystalline oxides and crystalline clay minerals in a basic Tiron solution. *Clays Clay Miner.*, **38**, 308–314.

Hayes, K.F. and Leckie, J.O. (1987) Modeling ionic strength effects on cation adsorption at hydrous oxide/solution interface. *J. Colloid Interface Sci.*, **115**, 564–572.

Heaney, P.J. (1994) Structure and chemistry of the low-pressure silica polymorphs. In: Heaney, P.J., Prewitt, C.T. and Gibbs, G.V. (eds.), *Silica: Physical Behavior, Geochemistry, and Materials Applications, Reviews in Mineralogy*, Mineralogical Society of America, **29**, 1–40.

Helgeson, H.C. (1967) Solution chemistry and metamorphism. In: Abelson, P.H. (ed.), *Researches in Geochemistry*, Vol. 2. New York: Wiley, pp. 362–404.

Helgeson, H.C. (1968) Evaluation of irreversible reactions in geochemical processes involving minerals and aqueous solutions: I. Thermodynamic relations. *Geochim. Cosmochim. Acta*, **32**, 853–877.

Helgeson, H.C. (1969) Thermodynamics of hydrothermal systems at elevated temperatures and pressures. *Am. J. Sci.*, **267**, 729–804.

Helgeson, H.C. (1971) Kinetics of mass transfer among silicates and aqueous solutions. *Geochim. Cosmochim. Acta*, **35**, 421–469.

Helgeson, H.C. (1979) Mass transfer among minerals and hydrothermal solutions. In: Barnes, H.L. (ed), *Geochemistry of Hydrothermal Ore Deposits*, 2nd ed. New York: Wiley, pp. 568–610.

Helgeson, H.C. and Kirkham, D.H. (1974a) Theoretical prediction of the thermodynamic behavior of aqueous electrolytes at high pressures and temperatures: I. Summary of the thermodynamic/electrostatic properties of the solvent. *Am. J. Sci.*, **274**, 1089–1198.

Helgeson, H.C. and Kirkham, D.H. (1974b) Theoretical prediction of the thermodynamic behavior of aqueous electrolytes at high pressures and temperatures: II. Debye–Hückel parameters for activity coefficients and relative partial molal properties. *Am. J. Sci.*, **274**, 1199–1261.

Helgeson, H.C. and Kirkham, D.H. (1976) Theoretical prediction of the thermodynamic behavior of aqueous electrolytes at high pressures and temperatures: III. Equation of state for aqueous species at infinite dilution. *Am. J. Sci.*, **276**, 97–240.

Helgeson, H.C. and MacKenzie, F.T. (1970) Silicate-sea water equilibria in the ocean system. *Deep-Sea Res.*, **17**, 877–892.

Helgeson, H.C., Brown, T.H., Nigrini, A. and Jones, T.A. (1970) Calculation of mass transfer in geochemical processes involving aqueous solutions. *Geochim. Cosmochim. Acta*, **34**, 569–592.

Helgeson, H.C., Delany, J.M., Nesbitt, H.W. and Bird, D.K. (1978) Summary and critique of the thermodynamic properties of rock-forming minerals. *Am. J. Sci.*, **278A**, 229.

Helgeson, H.C., Garrels, R.M. and Mackenzie, F.T. (1969) Evaluation of irreversible reactions in geochemical processes involving minerals and aqueous solutions: II. Applications. *Geochim. Cosmochim. Acta*, **33**, 455–481.

Helgeson, H.C., Kirkham, D.H. and Flowers, G.C. (1981) Theoretical prediction of the thermodynamic behavior of aqueous electrolytes at high pressures and temperatures: IV. Calculation of activity coefficients, osmotic coefficients and relative partial molal properties to 600°C and 5 kb. *Am. J. Sci.*, **281**, 1249–1516.

Helgeson, H.C., Murphy, W.M. and Aagaard, P. (1984) Thermodynamic and kinetic constraints on reaction rates among minerals and aqueous solutions. II. Rate constants, effective surface area, and the hydrolysis of feldspar. *Geochim. Cosmochim. Acta*, **48**, 2405–2432.

Hellmann, R. (1994) The albite–water system: Part I. The kinetics of dissolution as a function of pH at 100, 200 and 300°C. *Geochim. Cosmochim. Acta*, **58**, 595–611.

Hellmann, R., Schott, J., Dran, J.-C., Petit, J.-C. and Della Mea, G. (1990) A comparison of the dissolution behavsior of albite and albite glass under hydrothermal conditions. *Geol. Soc. Am. Ann. Meeting*, A292.

Henley, R.W. and Ellis, A.J. (1983) Geothermal systems, ancient and modern: A geochemical review. *Earth Sci. Rev.*, **19**, 1–50.

Heydemann, A. (1966) Über die chemische Verwitterung von Tonmineralen (Experimentelle Untersuchungen). *Geochim. Cosmochim. Acta*, **30**, 995–1035.

Higgins, S.R., Jordan, G. and Eggleston, C.M. (2002) Dissolution kinetics of magnesite in acidic aqueous solution: A hydrothermal atomic force microscopy study assessing step kinetics and dissolution flux. *Geochim. Cosmochim. Acta*, **66**, 3201–3210.

Hitchon, B. (ed.) (1996) *Aquifer Disposal of Carbon Dioxide*. Sherwood Park, Alberta, Canada: Geoscience Publishing Ltd.

Hochella, M.F., Jr. (1990) Atomic structure, microtopography, composition, and reactivity of mineral surfaces. In: Hochella, M.F., Jr. and White, A. F. (eds.), *Mineral–Water Interface Geochemistry, Reviews in Mineralogy*, Mineralogical Society of America, **23**, 87–132.

Hochella, M.F., Jr. and Banfield, J.F. (1995) Chemical weathering of silicates in nature: A microscopic perspective with theoretical considerations. In: White, A. F. and Brantley, S.L. (eds.), *Chemical Weathering Rates of Silicate Minerals, Reviews in Mineralogy*, Mineralogical Society of America, **31**, 353–406.

Hodson, M.E., Lee, M.R. and Parsons, I. (1997) Origins of the surface roughness of unweathered alkali feldspar grains. *Geochim. Cosmochim. Acta*, **61**, 3885–3896.

Högbom, A.G. (1894) On the probability of secular variations of atmospheric carbon dioxide (in Swedish). *Svensk Chemisk Tidsskrift*, **6**, 169–176.

Holdren, G.R., Jr. and Berner, R.A. (1979) Mechanism of feldspar weathering – I. Experimental studies. *Geochim. Cosmochim. Acta*, **43**, 1161–1171.

Holdren, G.R., Jr. and Speyer, P.M. (1987) Reaction rate-surface area relationships during the early stages of weathering. II. Data on eight additional feldspars. *Geochim. Cosmochim. Acta*, **51**, 2311–2318.

Hollingsworth, C.A. (1952) Equilibrium and the rate laws for forward and reverse reactions. *J. Chem. Phys.*, **20**, 921–922.

Hu, Y., Liu, X. and Xu Zhenghe (2003) Role of crystal structure in flotation separation of diaspore from kaolinite, pyrophyllite and illite. *Miner. Eng.*, **16**, 219–227.

Huang, Y.C., Fowkes, F.M., Lloyd, T.B. and Sanders, N.D. (1991) Adsorption of calcium ions from calcium chloride solutions onto calcium carbonate particles. *Langmuir*, **7**, 1742–1748.

Hückel, E. (1925) Zur theorie konzentrierterer wässeriger Lösungen starker elektrolyte. *Zeitschr. Physik*, **26**, 93–147.

Huertas, F.J., Caballero, E., Jiménez de Cisneros, C., Huertas, F. and Linares, J. (2001) Kinetics of montmorillonite dissolution in granitic solutions. *Appl. Geochem.*, **16**, 397–407.

Huertas, F.J., Chou, L. and Wollast, R. (1999) Mechanism of kaolinite dissolution at room temperature and pressure. Part II: Kinetic study. *Geochim. Cosmochim. Acta*, **63**, 3261–3275.

Huggins, C.W. and Green, T.E. (1973) Thermal decomposition of dawsonite. *Am. Mineral.*, **58**, 548–550.

Hume, L.A. and Rimstidt, J.D. (1992) The biodurability of chrysotile asbestos. *Am. Mineral.*, **77**, 1125–1128.

Indermühle, A., Stocker, T.F., Joss, F., Fischer, H., Smith, H.J., Wahlen, M., Deck, B., Mastroianni, D., Tschumi, J., Blunier, T., Meyer, R. and Stauffer, B. (1999) Holocene carbon-cycle dynamics based on CO_2 trapped in ice at Taylor Dome, Antarctica. *Nature*, **398**, 121–126.

Inkinen, O. and Lahti, L. (1964) X-ray crystal analysis of calcite. The "correct crystal model". *Ann. Acad. Sci. Fennicae*, Ser. A., VI, No. 142.

Inskeep and Bloom (1985) An evaluation of rate equations for calcite precipitation kinetics at pCO_2 less than 0.01 atm and pH greater than 8. *Geochim. Cosmochim. Acta*, **49**, 2165–2180.

Jackson, J., Jr., Huggins, C.W. and Ampian, S.G. (1972) Synthesis and characterization of dawsonite. 7664, Washington DC: US Bureau of Mines Report Investigation, 19pp.

Jackson, K., Bowman, L.E. and Fulton, J.L. (1995) Water solubility measurements in supercritical fluids and high-pressure liquids using near-infrared spectroscopy. *Anal. Chem.*, **67**, 2368–2372.

Johnson, J.W. and Lundeen, S.R. (1994 a) Jewel: A graphical-user interface for generating custom GEMBOCHS thermodynamic datafiles for use with geochemical modeling software. LLNL-YMP Milestone report MOL63, 23pp.

Johnson, J.W. and Lundeen, S.R. (1994 b) GEMBOCHS thermodynamic datafiles for use with the EQ3/6 software package. LLNL-YMP Milestone report MOL72, 99pp.

Johnson, J.W. and Lundeen, S.R. (1995) Facet: A graphical-user interface for viewing and updating the GEMBOCHS thermodynamic database. LLNL-YMP Milestone report MOL208, 28pp.

Johnson, J.W., Nitao, J.J., Steefel, C.I. and Knauss, K.G. (2001) Reactive transport modeling of geologic CO_2 sequestration in saline aquifers: The influence of intra-aquifer shales and the relative effectiveness of structural, solubility, and mineral trapping during prograde and retrograde sequestration. *First National Conference on Carbon Sequestration*. National Energy Technology Laboratory, Washington DC, http://www.netl.doe.gov.

Johnson, J.W., Oelkers, E.H. and Helgeson, H.C. (1992) SUPCRT 92: A software package for calculating the standard molal thermodynamic properties of minerals, gases, aqueous species, and reactions from 1 to 5000 bars and 0 to 1000°C. *Comput. Geosci.*, **18**, 899–947.

Johnson, J.W., Steefel, C.I. and Knauss, K.G. (2002) Reactive transport modeling of geologic CO_2 sequestration. Abstr. *Programs – Geol. Soc. Am.*, **34**, 390.

Jonckbloedt, R.C.L. (1998) Olivine dissolution in sulphuric acid at elevated temperatures – Implications for the olivine process, an alternative waste acid neutralizing process. *J. Geochem. Explor.*, **62**, 337–346.

Jones, C.F., Segall, R.L., Smart, S.C. and Turner, P.S. (1981) Initial dissolution kinetics of ionic oxides. *Proc. R. Soc. London A*, **374**, 141–153.

Jones, P., Haggett, M.L., Longridge, J.L. and Jethro, L. (1964) The hydration of carbon dioxide. *J. Chem. Educ.*, **41**, 610–612.

Jordan, G. and Rammensee, W. (1996) Dissolution rates and activation energy from dissolution of brucite (001): A new method based on the microtopography of crystal surfaces. Geochim. Cosmochim. Acta, **60**, 5055–5062.

Jordan, G. and Rammensee, W. (1998) Dissolution rates of calcite $(10\bar{1}4)$ obtained by scanning force microscopy. Microtopography-based dissolution kinetics on surface with anisotropic step velocities. *Geochim. Cosmochim. Acta*, **62**, 941–947.

Jordan, G., Higgins, S.R. and Eggleston, C.M. (1999) Dissolution of the periclase (001) surface: A scanning force microscopy study. *Am. Mineral.*, **84**, 144–151.

Jordan, G., Higgins, S.R., Eggleston, C.M., Knauss, K.G. and Schmahl, W.W. (2001) Dissolution kinetics of magnesite in acidic aqueous solution, a hydrothermal atomic force microscopy (HAFM) study: Step orientation and kink dynamics. *Geochim. Cosmochim. Acta*, **65**, 4257–4266.

Kalinowski, B.E. and Schweda, P. (1996) Kinetics of muscovite, phlogopite, and biotite dissolution and alteration at pH 1–4, room temperature. *Geochim. Cosmochim. Acta*, **60**, 367–385.

Kalinowski, B. E., Faith-Ell, C. and Schweda, P. (1998) Dissolution kinetics and alteration of epidote in acidic solutions at 25°C. *Chem. Geol.*, **151**, 181–197.

Kaszuba, J.P., Janecky, D.R. and Snow, M.G. (2003) Carbon dioxide reaction processes in a model brine aquifer at 200°C and 200 bars: Implications for geological sequestration of carbon. *Appl. Geochem.*, **18**, 1065–1080.

Kaszuba, J.P., Janecky, D.R. and Snow, M.G. (2005) Experimental evaluation of mixed fluid reactions between supercritical carbon dioxide and NaCl brine: Relevance to the integrity of a geologic carbon repository. *Chem. Geol.*, **217**, 277–293.

Keeling, C.D. and Whorf, T.P. (2003) Atmospheric CO_2 records from sites in the SIO air sampling network. In: *Trends: A compendium of data on global change*. Oak Ridge, TN, USA: Carbon Dioxide Information Analysis Center, Oak Ridge National Laboratory.

Keeling, C.D., Bacastow, R.B. and Tans, P.P. (1980) Predicted shift in the 13C–12C ratio of atmospheric carbon-dioxide. *Geophys. Res. Lett.*, **7**, 505–508.

Keeling, C.D., Whorf, T.P., Wahlen, M. and Vanderplicht, J. (1995) Interannual extremes in the rate of rise of atmospheric carbon dioxide since 1980. *Nature*, **375**, 666–670.

Keeling, R.F. (1988) Measuring correlations between atmospheric oxygen and carbon-dioxide mole fractions – A preliminary study in urban air. *J. Atmos. Chem.*, **7**, 153–176.

Keeling, R.F. and Shertz, S.R. (1992) Seasonal and interannual variations in atmospheric oxygen and implications for the global carbon cycle. *Nature*, **358**, 723–727.

Keeling, R.F., Najjar, R.P. and Bender, M.L. (1993) What atmospheric oxygen measurements can tell us about the global carbon-cycle. *Global Biogeochem. Cycles*, **7**, 37–67.

Keeling, R.F., Piper, S.C. and Heimann, M. (1996) Global and hemispheric CO_2 sinks deduced from changes in atmospheric O_2 concentration. *Nature*, **381**, 218–221.

Kerrick, D.M. and Jacobs, G.K. (1981) A modified Redlich–Kwong equation for H_2O, CO_2 and H_2O–CO_2 mixtures at elevated pressures and temperatures. *Am. J. Sci.*, **281**, 735–767.

Kihara, K. (1977) An orthorhombic superstructure of tridymite existing between about 105 and 180 C. *Zeit. Kristallogr.*, **146**, 185–203.

Kihara, K. (1978) Thermal change in unit-cell dimensions, and hexagonal structure of tridymite. *Zeit. Kristallogr.*, **148**, 237–253.

King, M.B., Mubarak, A., Kim, J.D. and Bott, T.R. (1992) The mutual solubilities of water with supercritical and liquid carbon dioxide. *J. Supercrit. Fluids*, **5**, 296–302.

Kitahara, S. (1960) The solubility equilibrium and the rate of solution of quartz in water at high temperatures and pressures. *Rev. Phys. Chem., Japan*, **30**, 122–130.

Knauss, K.G. and Copenhaver, S.A. (1995) The effect of malonate on the dissolution kinetics of albite, quartz, and microcline as a function of pH at 70°C. *Appl. Geochem.*, **10**, 17–33.

Knauss, K.G. and Wolery, T.J. (1986) Dependence of albite dissolution kinetics on pH and time at 25°C and 70°C. *Geochim. Cosmochim. Acta*, **50**, 2481–2497.

Knauss, K.G. and Wolery, T.J. (1988) The dissolution kinetics of quartz as a function of pH and time at 70°C. *Geochim. Cosmochim. Acta*, **52**, 43–53.

Knauss, K.G. and Wolery, T.J. (1989) Muscovite dissolution kinetics as a function of pH and time at 70°C *Geochim. Cosmochim. Acta*, **53**, 1493–1501.

Knauss, K.G., Nguyen, S.N. and Weed, H.C. (1993) Diopside dissolution kinetics as a function of pH, CO_2, temperature, and time. *Geochim. Cosmochim. Acta*, **57**, 285–294.

Knauss, K.G., Johnson, J.W. and Steefel, C.I. (2005) Evaluation of the impact of CO_2, co-contaminant gas, aqueous fluid and reservoir rock interactions on the geologic sequestration of CO_2. *Chem. Geol.*, **217**, 339–350.

Knauss, K.G., Johnson, J.W., Steefel C.I. and Nitao, J.J. (2001) Evaluation of the impact of CO_2, aqueous fluid, and reservoir rock interactions on the geologic sequestration of CO_2, with special emphasis on economic implications. *First National Conference on Carbon Sequestration*. National Energy Technology Laboratory, Washington, DC, http://www.netl.doe.gov.

Knauss, K.G., Steefel, C.I., Johnson, J.W. and Boram, L.H. (2002) Impact of CO_2, contaminant gas, aqueous fluid, and reservoir rock interactions on the geologic sequestration of CO_2. Abstr. *Programs – Geol. Soc. Am.*, **34**, 306.

Knoche, R., Dingwell, D.B. and Webb, S.L. (1995) Melt densities for leucogranites and granitic pegmatites: Partial molar volumes for SiO_2, Al_2O_3, Na_2O, K_2O, Li_2O, Rb_2O, Cs_2O, MgO, CaO, SrO, BaO, B_2O_3, P_2O_5, F_2O_{-1}, TiO_2, Nb_2O_5, Ta_2O_5, WO_3. *Geochim. Cosmochim. Acta*, **59**, 4645–4652.

Kodama, H. and Schnitzer, M. (1973) Dissolution of chlorite minerals by fulvic acid. *Can. J. Soil Sci.*, **53**, 240–243.

Köhler, S.J., Dufaud, F. and Oelkers, E.H. (2003) An experimental study of illite dissolution kinetics as a function of pH from 1.4 to 12.4 and temperature from 5 to 50°C. *Geochim. Cosmochim. Acta*, **67**, 3583–3594.

Koziol, A.M. and Newton, R.C. (1995) Experimental determination of the reactions magnesite + quartz = enstatite + CO_2 and magnesite = periclase + CO_2 and the enthalpies of formation of enstatite and magnesite. *Am. Mineral.*, **80**, 1252–1260.

Kuwahara, Y. and Aoki, Y. (1995) Dissolution process of phlogopite in acid solution. *Clays Clay Miner.*, **43**, 39–50.

Laçin, O., Dönmez, B. and Demir, F. (2005) Dissolution kinetics of natural magnesite in acetic acid solutions. *Int. J. Miner. Process.*, **75**, 91–99.

Lackner, K.S., Butt, D.P. and Wendt, C.H. (1997) Progress on binding CO_2 in mineral substrates. *Energy Convers. Manage.*, **38**, S259–S264.

Lackner, K.S., Wendt, C.H., Butt, D.P., Joyce, E.L., Jr. and Sharp, D.H. (1995) Carbon dioxide disposal in carbonate minerals. *Energy*, **20**, 1153–1170.

Lagache, M. (1965) Contribution à l'étude de l'altération des feldspaths, dans l'eau, entre 100 et 200°C, sous diverses pressions de CO_2, et application à la synthèse des minéraux argileux. *Bull. Soc. Française Minéral. Cristallogr.*, **88**, 223–253.

Lagache, M. (1976) New data on the kinetics of the dissolution of alkali feldspars at 200°C in CO_2 charged water. *Geochim. Cosmochim. Acta*, **40**, 157–161.

Lagache, M., Wyart, J. and Sabatier, G. (1961) Dissolution des feldspaths alcalins dans l'eau pure ou chargée de CO_2 à 200°C. *Acad. Sci. Paris, Comptes Rendus*, **253**, 2019–2022.

La Mer, V.K. (1932) Reaction velocity in ionic systems. *Chem. Rev.*, **10**, 179–212.

Lange, R.L. (1997) A revised model for the density and thermal expansivity of K_2O–Na_2O–CaO–MgO–Al_2O_3–SiO_2 liquids from 700 to 1900 K: extension to crustal magmatic temperatures. *Contrib. Mineral. Petrol.*, **130**, 1–11.

Lange, R.L. and Carmichael, I.S.E. (1987) Densities of Na_2O–K_2O–CaO–MgO–FeO–Fe_2O_3–Al_2O_3–TiO_2–SiO_2 liquids: New measurements and derived partial molar properties. *Geochim. Cosmochim. Acta*, **51**, 2931–2946.

Lange, R.L. and Carmichael, I.S.E. (1990) Thermodynamic properties of silicate liquids with emphasis on density, thermal expansion and compressibility. In: Nicholls J. and Russel J.K. (eds.), *Modern Methods of Igneous Petrology, Reviews in Mineralogy*, **24**, 25–64.

Langenfelds, R.L., Francey, R.J. and Steele, L.P. (1999) Partitioning of the global fossil CO_2 sink using a 19-year trend in atmospheric O_2. *Geophys. Res. Lett.*, **26**, 1897–1900.

Langmuir, D. (1965) Stability of carbonates in the system MgO–CO_2–H_2O. *J. Geol.*, **73**, 730–754.

Langmuir, D. (1984) Physical and chemical characteristics of carbonate water. In: Lamoreaux, P.E., Wilson, B.M. and Memeon, B.A. (eds.), *Guide to the Hydrology of Carbonate Rocks*. Paris: UNESCO, pp. 60–105, 116–130.

Langmuir, D. (1997) *Aqueous Environmental Geochemistry*. Upper Saddle River, NJ: Prentice-Hall, 600pp.

Larsen, O. and Postma, D. (2001) Kinetics of reductive bulk dissolution of lepidocrocite, ferrihydrite, and goethite. *Geochim. Cosmochim. Acta*, **65**, 1367–1379.

Lasaga, A.C. (1983a) Rate laws of chemical reactions. In: Lasaga, A. C. and Kirkpatrick, R.J. (eds.), *Kinetics of Geochemical Processes, Reviews in Mineralogy*, Mineralogical Society of America, 2nd printing, **8**, 1–68.

Lasaga, A.C. (1983b) Transition state theory. In: Lasaga, A. C. and Kirkpatrick R.J. (eds.), *Kinetics of Geochemical Processes, Reviews in Mineralogy*, Mineralogical Society of America, 2nd printing, **8**, 135–169.

Lasaga, A.C. (1984) Chemical kinetics of water–rock interactions. *J. Geophys. Res.*, **89**, 4009–4025.

Lasaga, A.C. (1995) Fundamental approaches in describing mineral dissolution and precipitation rates. In: White, A. F. and Brantley, S.L. (eds.), *Chemical Weathering Rates of Silicate Minerals, Reviews in Mineralogy*, Mineralogical Society of America, **31**, 23–86.

Lasaga, A.C. (1998) *Kinetic Theory in the Earth Sciences.* Princeton, NJ: Princeton University Press, 811pp.

Laubengayer, A.W. and Weisz, R.S. (1943) A hydrothermal study of equilibria in the system alumina-water. *J. Am. Chem. Soc.*, **65**, 247–250.

Law, D.H.S. and Bachu, S. (1996) Hydrogeological and numerical analysis of CO_2 disposal in deep aquifers in the Alberta Sedimentary Basin. *Energy Convers. Manage.*, **37**, 1167–1174.

Lawlor, D.W. (1993) *Photosynthesis: Molecular, Physiological and Environmental Processes.* Harlow, UK: Longman.

Le Chatelier, H. (1889) Sur la dilatation du quartz. *Comptes Rendus*, **108**, 1046.

Le Maitre, R.W., (ed.) (2002) *Igneous Rocks. A Classification and Glossary of Terms.* 2nd ed. Cambridge: Cambridge University Press, 236pp.

Lea, A.S., Amonette, J.E., Baer, D.R., Liang, Y. and Colton, N.G. (2001) Microscopic effects of carbonate, manganese, and strontium ions on calcite dissolution. *Geochim. Cosmochim. Acta*, **65**, 369–379.

Lebrón, I. and Suárez, D.L. (1998) Kinetics and mechanisms of precipitation of calcite as affected by P_{CO_2} and organic ligands at 25°C. *Geochim. Cosmochim. Acta*, **62**, 405–416.

Leicester, H.M. (1956) *The Historical Background of Chemistry.* New York: Wiley.

Lemmon, E.W., McLinden, M.O. and Friend, D.G. (2003) Thermophysical Properties of Fluid Systems. In: Linstrom, P.J. and Mallard, W.G. (eds.), *NIST Chemistry WebBook, NIST Standard Reference Database Number 69*, National Institute of Standards and Technology, Gaithersburg, MD, http://webbook.nist.gov/.

Levich, V.G. (1962) *Physicochemical Hydrodynamics*, Englewood Cliffs, NJ: Prentice-Hall, 700pp.

Levitus, S., Antonov, J.I., Boyer, T.P. and Stephens, C. (2000) Warming of the world ocean. *Science*, **287**, 2225–2229.

Lewis, G.N. and Randall, M. (1921) The activity coefficient of strong electrolytes. *J. Am. Chem. Soc.*, **43**, 1112–1154

Lewis, G.N. and Randall, M. (1923) *Thermodynamics.* New York: McGraw-Hill.

Li, L. and Kiran, E. (1988) Gas–liquid critical properties of methylamine + nitrous oxide and methylamine + ethylene binary mixtures. *J. Chem. Eng. Data*, **33**, 342.

Lichtner, P.C. (1991) The quasi-stationary state approximation for fluid/rock reaction: Local equilibrium revisited. In: Ganguly, J. (ed.), *Diffusion, Atomic Ordering, and Mass Transport, Advances in Physical Geochemistry,* New York, Springer, **8**, 452–560.

Lichtner, P.C. (1996) Continuum formulation of multicomponent-multiphase reactive transport. In: Lichtner, P.C., Steefel, C.I. and Oelkers, E.H. (eds.). *Reactive Transport Porous Media, Rev. Miner.*, **34**, 1–81.

Lichtner, P.C. (1998) Modeling reactive flow and transport in natural systems. In: Marini, L. and Ottonello, G. (eds.), *Proceedings of the Rome Seminar on Environmental Geochemistry*, Pisa: Pacini, pp. 5–72.

Lichtner, P.C., Steefel, C.I. and Oelkers, E.H. (eds.) (1997) Reactive transport in porous media. *Rev. Miner.*, **34**, 438.

Liebau, F. (1985) *Structural Chemistry of Silicates.* Berlin: Springer.

Lietzke, M.H. and Stoughton, R.W. (1962) The calculation of activity coefficients from osmotic coefficient data. *J. Phys. Chem.*, **66**, 508–509.

Lin, F.-C. and Clemency, C.V. (1981a) The dissolution kinetics of brucite, antigorite, talc, and phlogopite at room temperature and pressure. *Am. Mineral.*, **66**, 801–806.

Lin, F.-C. and Clemency, C.V. (1981b) The kinetics of dissolution of muscovites at 25°C and 1 atm CO_2 partial pressure. *Geochim. Cosmochim. Acta*, **45**, 571–576.

Lin, F.-C., and Clemency, C.V. (1981c) Dissolution kinetics of phlogopite. I. Closed-system. *Clays Clay Minerals*, **29**, 101–106.

Lindberg, R.D. and Runnells, D.D. (1984) Ground water redox reactions: An analysis of equilibrium state applied to Eh measurements and geochemical modeling. *Science*, **225**, 925–927.

Lippmann, F. (1973) *Sedimentary Carbonate Minerals. Minerals, Rocks and Inorganic Materials, Monograph Series of Theoretical and Experimental Studies*, Vol. 4. Berlin: Springer, 228pp.

Lohuis, J.A.O. (1993) Carbon dioxide disposal and sustainable development in The Netherlands. *Energy Convers. Manage.*, **34**, 815–821.

Longhurst, A., Sathyendranath, S., Platt, T. and Caverhill, C. (1995) An estimate of global primary production in the ocean from satellite radiometer data. *J. Plankton Res.*, **17**, 1245–1271.

Löns, J. and Hoffmann, W. (1987) Zur Kristallstruktur der inkommensurablen Raumtemperaturphase des Tridymits. *Zeit. Kristallogr.*, **178**, 141–143.

Lowson, R.T., Comarmond, M.-C.J., Rajaratnam, G. and Brown, P.L. (2005) The kinetics of the dissolution of chlorite as a function of pH and at 25°C. *Geochim. Cosmochim. Acta*, **69**, 1687–1699.

Luce, R.W., Bartlett, R.W. and Parks, G.A. (1972) Dissolution kinetics of magnesium silicates. *Geochim. Cosmochim. Acta*, **36**, 35–50.

Lund, K., Fogler, H.S. and McCune, C.C. (1973) Acidization – I. The dissolution of dolomite in hydrochloric acid. *Chem. Eng. Sci.*, **28**, 691–700.

Lundström, U. and Ohman, L.-O. (1990) Dissolution of feldspars in the presence of natural organic solutes. *J. Soil Sci.*, **41**, 359–369.

Lüttge, A., Winkler, U. and Lasaga, A.C. (2002) Interferometric study of the dolomite dissolution: A new conceptual model for mineral dissolution. *Geochim. Cosmochim. Acta*, **67**, 1099–1116.

MacInnis, I.N. and Brantley, S.L. (1992) The role of dislocations and surface morphology in calcite dissolution. *Geochim. Cosmochim. Acta*, **56**, 1113–1126.

Maier, C.G. and Kelley, K.K. (1932) An equation for the representation of high temperature heat content data. *Am. Chem. Soc. J.*, **54**, 3243–3246.

Malatesta, F. (2005) On the Rodil–Vera method for determining ion activity coefficients. *Fluid Phase Equilib.*, **233**, 103–109.

Malmström, M. and Banwart, S. (1997) Biotite dissolution at 25°C: The pH dependence of dissolution rate and stoichiometry. *Geochim. Cosmochim. Acta*, **61**, 2779–2799.

Malmström, M., Banwart, S., Lewenhagen, J., Duro, L. and Bruno, J. (1996) The dissolution of biotite and chlorite at 25°C in the near-neutral pH region. *J. Contam. Hydrol.*, **21**, 201–213.

Manes, M., Hofer, L.J.E. and Weller, S. (1950) Classical thermodynamics and reaction rates close to equilibrium. *J. Chem. Phys.*, **18**, 1355–1361.

Manley, E.P. and Evans, L.J. (1986) Dissolution of feldspars by low molecular-weight aliphatic and aromatic acids. *Soil Sci.*, **141**, 106–112.

Manning, D.C., Gestsdóttir, K. and Rae, E.I.C. (1991) Feldspar dissolution in the presence of organic acid anions under diagenetic conditions: An experimental study. *Org. Geochem.*, **19**, 483–492.

Marini, L. and Chiodini, G. (1994) The role of carbon dioxide in the carbonate-evaporite geothermal systems of Tuscany and Latium (Italy). *Acta Vulcanol.*, **5**, 95–104.

Marini, L. and Manzella, A. (2005) Possible seismic signature of the α–β quartz transition in the lithosphere of Southern Tuscany (Italy). *J. Volcanol. Geotherm. Res.* **148**, 81–97.

Marini, L. and Ottonello, G. (eds.) (2002). *Atlante degli acquiferi della Liguria. Volume III: Le acque dei complessi ofiolitici (bacini: Arrestra, Branega, Cassinelle, Cerusa, Erro, Gorzente, Leira, Lemme, Lerone, Orba, Piota, Polcevera, Rumaro, Sansobbia, Stura, Teiro, Varenna, Visone*. Pisa: Pacini Editore, 102pp.

Marini, L., Canepa, M., Cipolli, F., Ottonello, G. and Vetuschi Zuccolini, M. (2001). Use of stream sediment chemistry to predict trace element chemistry of groundwater. A case study from the Bisagno valley (Genoa, Italy). *J. Hydrol.*, **241**, 194–220.

Marini, L., Ottonello, G., Canepa, M. and Cipolli, F. (2000). Water–rock interaction in the Bisagno Valley (Genoa, Italy): Application of an inverse approach to model spring water chemistry. *Geochim. Cosmochim. Acta*, **64**, 2617–2635.

Markgraf, S.A. and Reeder, R.J. (1985) High-temperature structure refinements of calcite and magnesite. *Am. Mineral.*, **70**, 590–600.

Marland, B., Boden, T.A. and Andres, R.J. (2000) Global, regional, and national CO_2 emissions. In: *Trends: A Compendium of Data on Global Change*. Oak Ridge, TN, USA: Carbon Dioxide Information Analysis Center, Oak Ridge National Laboratory.

Marshall, C.E. (1962) Reactions of feldspars and micas with aqueous solutions. *Econ. Geol.*, **57**, 1219–1227.

Mast, M.A. and Drever, J.I. (1987) The effect of oxalate on the dissolution rates of oligoclase and tremolite. *Geochim. Cosmochim. Acta*, **51**, 2559–2568.

Matous, J., Sobr, J., Novak, J.P. and Pick, J. (1969). Solubilities of carbon dioxide in water at pressure up to 40 atm. *Collect. Czech. Chem. Commun.*, **34**, 3982–3985.

Mattigod, S.V. and Sposito, G. (1978) Improved method for estimating the standard free energies of formation of smectites. *Geochim. Cosmochim. Acta*, **42**, 1753–1762.

May, H., Acker, J.G., Smyth, J.R., Bricker, O.P. and Dyar, M.D. (1995) Aqueous dissolution of low-iron chlorite in dilute acid solutions at 25°C. *Clay Miner. Soc. Prog. Abstr.*, **32**, 88.

May, H.M., Kinniburgh, D.G., Helmke, P.A. and Jackson, M.L. (1986) Aqueous dissolution, solubilities and thermodynamic stabilities of common aluminosilicate minerals: Kaolinite and smectites. *Geochim. Cosmochim. Acta*, **50**, 1667–1677.

Mayer, J.E. and Mayer, M.G. (1940) *Statistical Mechanics*. New York: Wiley, 495pp.

McKelvy, M.J., Chizmeshya, A.V.G., Diefenbacher, J., Béarat, H. and Wolf, G.H. (2004) Exploration of the role of heat activation in enhancing serpentine carbon sequestration reactions. *Environ. Sci. Technol.*, **38**, 6897–6903.

McKelvy, M.J., Sharma, R., Chizmeshya, A.V.G., Carpenter, R. and Streib, K. (2001) Magnesium hydroxide dehydroxylation: *In situ* nanoscale observations of lamellar nucleation and growth. *Chem. Mater.*, **13**, 921–926.

Metz, V. and Ganor, J. (2000) Stirring effect on kaolinite dissolution rate. *Geochim. Cosmochim. Acta*, **65**, 3475–3490.

Meybeck, M. (1993) Riverine transport of atmospheric carbon – Sources, global typology and budget. *Water, Air Soil Pollut.*, **70**, 443–463.

Michard, G., Fouillac, C., Grimaud, D. and Denis, J. (1981) Une méthode globale d'estimation des températures des resérvoirs alimentant les sources thermales. Exemple du Massif Central Français. *Geochim. Cosmochim. Acta*, **45**, 1199–1207.

Mogollón, J.L., Ganor, J., Soler, J.M. and Lasaga, A.C. (1996) Column experiments and the full dissolution rate law of gibbsite. *Am. J. Sci.*, **296**, 729–765.

Mogollón, J.L., Pérez-Diaz, A. and Lo Monaco, S. (2000) The effect of ion identity and ionic strength on the dissolution rate of a gibbsitic bauxite. *Geochim. Cosmochim. Acta*, **64**, 781–795.

Monnin, C. (1989) An ion interaction model for the volumetric properties of natural waters: Density of the solution and partial molal volumes of electrolytes to high concentrations at 25°C. *Geochim. Cosmochim. Acta*, **53**, 1177–1188.

Moore, J., Adams, M., Allis, R., Luts, S. and Rauzi, S. (2005) Mineralogical and geochemical consequences of the long-term presence of CO_2 in natural reservoirs: An example from the Springerville-St. Johns Field, Arizona and New Mexico, U.S.A. *Chem. Geol.*, **217**, 365–385.

Morrison, G. (1981) Effect of water on the critical points of carbon dioxide and ethane. *J. Phys. Chem.*, **85**, 759–61.

Morse, J.W. (1974) Dissolution of calcium carbonate in seawater: III. A new method for the study of carbonate reaction kinetics. *Am. J. Sci.*, **274**, 97–107.

Morse, J.W. (1978) Dissolution kinetics of calcium carbonate in seawater: IV. The near-equilibrium dissolution kinetics of calcium carbonate-rich deep sea sediments. *Am. J. Sci.*, **278**, 344–353.

Morse, J.W. and Arvidson, R.S. (2002) The dissolution kinetics of major sedimentary carbonate minerals. *Earth-Sci. Rev.*, **58**, 51–84.

Morse, J.W. and Berner, R.A. (1972) Dissolution kinetics of calcium carbonate in seawater: II. A kinetic origin for lysocline. *Am. J. Sci.*, **272**, 840–851.

Morse, J.W. and Berner, R.A. (1979) The chemistry of calcium carbonate in the deep oceans. In: Jenne, E. (ed.), *Chemical Modeling – Speciation, Sorption, Solubility, and Kinetics in Aqueous Systems*. Washington, DC: American Chemical Society, pp. 499–535.

Muir, I.J., Bancroft, G.M. and Nesbitt, H.W. (1989) Characteristics of altered labradorite surfaces by SIMS and XPS. *Geochim. Cosmochim. Acta*, **53**, 1235–1241.

Muir, I.J., Bancroft, G.M., Shotyk, W. and Nesbitt, H.W. (1990) A SIMS and XPS study of dissolving plagioclase. *Geochim. Cosmochim. Acta*, **54**, 2247–2256.

Müller, G., Bender, E. and Maurer, G. (1988) Das Dampf-Flüssigkeits-gleichgewicht des Ternären Systems Ammoniak-Kohlendioxid-Wasser bei Hohen Wassergehalten im Bereich Zwischen 373 und 473 Kelvin. *Berichte Bunsen-Gesellshaft Phys. Chem.*, **92**, 148–160.

Murakami, T., Ito, J.-I., Utsunomiya, S., Kasama, T., Kozai, N. and Ohnuki, T. (2004) Anoxic dissolution processes of biotite: Implications for Fe behavior during Archean weathering. *Earth Planet. Sci. Lett.*, **224**, 117–129.

Murphy, W.M. (1989) Dislocations and feldspar dissolution. *Eur. J. Mineral.*, **1**, 315–326.

Murphy, W.M. and Helgeson, H.C. (1987) Thermodynamic and kinetic constraints on reaction rates among minerals and aqueous solutions. III. Activated complexes and the pH-dependence of the rates of feldspar, pyroxene, wollastonite, and olivine hydrolysis. *Geochim. Cosmochim. Acta*, **51**, 3137–3153.

Murphy, W.M., Pabalan, R.T., Prikryl, J.D. and Goulet, C.J. (1996) Reaction kinetics and thermodynamics of aqueous dissolution and growth of analcime and Na-clinoptilolite at 25°C. *Am. J. Sci.*, **296**, 128–186.

Nagy, K.L. (1995) Dissolution and precipitation kinetics of sheet silicates. In: White, A. F. and Brantley, S.L. (eds.), *Chemical Weathering Rates of Silicate Minerals*, *Reviews in Mineralogy*, Mineralogical Society of America, **31**, 173–233.

Nagy, K.L. and Lasaga, A.C. (1992) Dissolution and precipitation kinetics of gibbsite at 80°C and pH 3: The dependence on solution saturation state. *Geochim. Cosmochim. Acta*, **56**, 3093–3111.

Nancollas, G.H. and Reddy, M.M. (1971) The crystallization of calcium carbonate: II. Calcite growth mechanism. *J. Colloid Interface Sci.*, **37**, 824–830.

Navrotsky, A. and Loucks, D. (1977) Calculation of subsolidus phase relations in carbonates and pyroxenes. *Phys. Chem. Miner.*, **1**, 109–127.

Neftel, A., Friedli, H., Moor, E., Lötscher, H., Oeschger, H., Siegenthaler, U. and Stauffer, B. (1994) Historical CO_2 recoed from the Simple station ice core. In: Boden, T.A., Kaiser, D.P., Sepanski, R.J. and Stoss, F.W. (eds.) *Trends '93: A Compendium of Data on Global Change*. TN, USA: Carbon Dioxide Information Analysis Center, Oak Ridge National Laboratory, Oak Ridge, pp. 11–14.

Nernst, W. (1904) Theorie der reaktionsgeschwindigkeit in heterogenen systemen. *Zeitschr. Phys. Chem.*, **47**, 52–55.

Nesbitt, H.W. and Muir, I.J. (1988) SIMS depth profiles of weathered plagioclase and processes affecting dissolved Al and Si in some acidic soils. *Nature*, **334**, 336–338.

Newman, A.C.D. (ed.) (1987) *Chemistry of Clays and Clay Minerals*. Mineralogical Society Monograph No. **6**. New York: Wiley, 480pp.

Ng, H.-J. and Robinson, D.B. (1985) Hydrate formation in systems containing methane, ethane, propane, carbon dioxide or hydrogen sulfide in the presence of methanol. *Fluid Phase Equilib.*, **21**, 145–155.

Nickalls, R.W.D. (1993) A new approach to solving the cubic: Cardan's solution revealed. *Math. Gazette*, **77**, 354–359.

Nickel, E. (1973) Experimental dissolution of light and heavy minerals in comparison with weathering and intrastratal solutions. *Contrib. Sedimentol.*, **1**, 3–68.

Nielsen, A.E. (1964) *Kinetics of Precipitation*. New York: MacMillan, 151pp.

Nilsson, O. and Sternbeck, J. (1999) A mechanistic model for calcite crystal growth using surface speciation. *Geochim. Cosmochim. Acta*, **63**, 217–225.

Nitao, J.J. (1998) Reference Manual for the NUFT Flow and Transport Code, version 2.0. Lawrence Livermore National Laboratory. Report UCRL-MA-130651, 55pp.

Nordstrom, D.K. and Munoz, J.L. (1985) *Geochemical Thermodynamics*. Menlo Park, CA: The Benjamin/Cummings Publishing Co., 477pp.

Nowack, B. and Sigg, L. (1997) Dissolution of Fe(III) (hydr)oxides by metal–EDTA complexes. *Geochim. Cosmochim. Acta*, **61**, 951–963.

Nriagu, J.O. (1975) Thermochemical approximations for clay minerals. *Am. Mineral.*, **60**, 834–839.

Nukui, A., Nakazawa, H. and Akao, M. (1978) Thermal changes in monoclinic tridymite. *Am. Mineral.*, **63**, 1252–1259.

O'Connor, W.K., Dahlin, D.C., Nilsen, D.N., Rush, G.E., Walters, R.P. and Turner, P.C. (2001) Carbon dioxide sequestration by direct mineral carbonation: Results from recent studies and current status. *First National Conference on Carbon Sequestration. National Energy Technology Laboratory*, Washington, DC, http://www.netl.doe.gov.

Odin, G.S., Desprairies, A., Fullagar, P.D., Bellon, H., Decarreau, A., Fröhlich, F. and Zelvelder, M. (1988) Nature and geological significance of celadonite. In: Odin, G.S. (ed.), *Green Marine Clays*, *Developments in Sedimentology*. Amsterdam: Elsevier, **45**, pp. 337–398.

Oelkers, E.H. (1996) Physical and chemical properties of rocks and fluids for chemical mass transport calculations. In: Lichtner, P.C., Steefel, C.I. and Oelkers, E.H. (eds.), *Reactive Transport in Porous Media*, *Reviews in Mineralogy*, Mineralogical Society of America, **34**, 131–191.

Oelkers, E.H. (2001a) General kinetic description of multioxide silicate mineral and glass dissolution. *Geochim. Cosmochim. Acta*, **65**, 3703–3719.

Oelkers, E.H. (2001b) An experimental study of forsterite dissolution rates as a function of temperature and aqueous Mg and Si concentrations. *Chem. Geol.*, **175**, 485–494.

Oelkers, E.H. (2002) The surface areas of rocks and minerals. In: Buccianti, A., Marini, L., Ottonello G. and Vaselli, O. (eds.), *Proceedings of the Arezzo Seminar on Fluid Geochemistry*, Arezzo, August 29th–September 1st, 2000, Pisa: Pacini editore, pp. 18–30.

Oelkers, E.H. and Gislason, S.R. (2001). The mechanism, rates, and consequences of basaltic glass dissolution: I. An experimental study of the dissolution rates of basaltic glass as a function of aqueous Al, Si, and oxalic acid concentration at 25°C and pH = 3 and 11. *Geochim. Cosmochim. Acta*, **65**, 3671–3681.

Oelkers, E.H. and Helgeson, H.C. (1991) Calculation of activity coefficients and degrees of formation of neutral ion pairs in supercritical electrolyte solutions. *Geochim. Cosmochim. Acta*, **55**, 1235–1251.

Oelkers, E.H. and Schott, J. (1995) Experimental study of anorthite dissolution and the relative mechanism of feldspar hydrolysis. *Geochim. Cosmochim. Acta*, **59**, 5039–5053.

Oelkers, E.H. and Schott, J. (1998) Does organic acid adsorption affect alkali-feldspar dissolution rates? *Chem. Geol.*, **151**, 235–245.

Oelkers, E.H. and Schott, J. (1999) Experimental study of kyanite dissolution rates as a function of chemical affinity and solution composition. *Geochim. Cosmochim. Acta*, **63**, 785–797.

Oelkers, E.H. and Schott, J. (2001) An experimental study of enstatite dissolution rates as a function of pH, temperature, and aqueous Mg and Si concentration, and the mechanism of pyroxene/pyroxenoid dissolution. *Geochim. Cosmochim. Acta*, **65**, 1219–1231.

Oelkers, E.H., Schott, J. and Devidal, J.-L. (1994) The effect of aluminum, pH, and chemical affinity on the rates of aluminosilicate dissolution reactions. *Geochim. Cosmochim. Acta*, **58**, 2011–2024.

Oh, K.D., Morikawa, H., Iwai, S., and Aoki, H. (1973) The crystal structure of magnesite. *Am. Mineral.*, **58**, 1029–1033.

Ortoleva, P., Chadman, J., Merino, E. and Moore, C. (1987) Geochemical self-organization. I: Feedback mechanisms and modeling approach. *Am. J. Sci.*, **287**, 979–1007.

Orton, R. and Unwin, P.R. (1993) Dolomite dissolution kinetics at low pH: A channel-flow study. *J. Chem. Soc., Faraday Trans.*, **89**, 3947–3954.

Ottonello, G. (1997) *Principles of Geochemistry.* New York: Columbia University Press, 897pp.

Oxburgh, R., Drever, J.I. and Sun, Y.-T. (1994) Mechanism of plagioclase dissolution in acid solution at 25°C. *Geochim. Cosmochim. Acta*, **58**, 661–669.

Pabalan, R.T. and Pitzer, K.S. (1987) Thermodynamics of concentrated electrolyte mixtures and the prediction of mineral solubilities to high temperatures for mixtures in the system Na–K–Ca–Mg–Cl–SO_4–OH–H_2O *Geochim. Cosmochim. Acta*, **51**, 2429–2443.

Pabalan, R.T. and Pitzer, K.S. (1988) Heat capacity and other thermodynamic properties of $Na_2SO_{4(aq)}$ in hydrothermal solutions and the solubilities of sodium sulfate minerals in the system Na–Cl–SO_4–OH–H_2O. *Geochim. Cosmochim. Acta*, **52**, 2393–2404.

Pačes, T. (1973) Steady-state kinetics and equilibrium between ground water and granitic rock. *Geochim. Cosmochim. Acta*, **37**, 2641–2663.

Packter, A. and Dhillon, H.S. (1969) The heterogeneous reactions of gibbsite powder with aqueous inorganic acid solutions; kinetics and mechanism. *J. Chem Soc.* A, **17**, 2588–2592.

Packter, A. and Dhillon, H.S. (1973) The kinetics and mechanism of the heterogeneous reactions of crystallized gibbsite powders with aqueous sodium hydroxide solutions. *J. Phys. Chem.*, **77**, 2942–2947.

Packter, A. and Dhillon, H.S. (1974) Studies on recrystallized aluminium hydroxide precipitates: Kinetics and mechanism of dissolution by sodium hydroxide solutions. *Colloid Polym. Sci.*, 252, 249–256.

Pagani, M., Arthur, M.A. and Freeman, K.H. (1999a) Miocene evolution of atmospheric carbon dioxide. *Paleoceanography*, **14**, 273–292.

Pagani, M., Freeman, K. and Arthur, M.A. (1999b) Late Miocene atmospheric CO_2 concentrations and the expansion of C4 grasses. *Science*, **285**, 876–879.

Palandri, J.L. and Kharaka, Y.K. (2004) A compilation of rate parameters of water-mineral interaction kinetics for application to geochemical modeling. U. S. Geological Survey, Open File Report 2004–1068.

Park, A.-H.A. and Fan, L.-S. (2004) CO_2 mineral sequestration: Physically activated dissolution of serpentine and pH swing process. *Chem. Eng. Sci.*, **59**, 5241–5247.

Parkhurst, D.L. and Appelo, C.A.J. (1996) User's guide to PHREEQC (version 2) – A computer program for speciation, batch-reaction, one-dimensional transport, and inverse geochemical calculations. U.S.G.S. Water-Resources Investigations Report 99-4259, 312pp.

Parks, G.A. (1990) Surface energy and adsorption at mineral/water interfaces: An introduction. In: Hochella, M.F., Jr. and White, A.F. (eds.), *Mineral–Water Interface Geochemistry, Reviews in Mineralogy*, Mineralogical Society of America, **23**, 133–175.

Pauling, L. (1940) *The Nature of the Chemical Bond and the Structure of Molecules and Crystals*, 2nd ed. Ithaca, New York: Cornell University Press.

Pauling, L. (1970) *General Chemistry*. New York: Dover, 959pp.

Pearce, J.M., Holloway, S., Wacker, H., Nelis, M.K., Rochelle, C. and Bateman, K. (1996) Natural occurrences as analogues for the geological disposal of carbon dioxide. *Energy Convers. Manage.*, **37**, 1123–1128.

Pearson, P.N. and Palmer, M.R. (1999) Middle Eocene seawater pH and atmospheric carbon dioxide concentrations. *Science*, **284**, 1824–1826.

Peric', J., Krstulovic, R. and Feric, T. (1985) Kinetics aspects of the gibbsite digestion process in aqueous solution of sodium hydroxide. *Croat. Chem. Acta*, **58**, 255–264.

Perkins, E.H. and Gunter, W.D. (1995) A user's manual for β PATHARCH.94: A reaction path-mass transfer program. Alberta Research Council Report ENVTR 95-11, 179pp.

Petit, J.R., Jouzel, J., Raynaud, D., Barkov, N.I., Barnola, J.M., Basile, I., Bender, M., Chappellaz, J., Davis, M., Delaygue, G., Delmotte, M., Kotlyakov, V.M., Legrand, M., Lipenkov, V.Y., Lorius, C., Pepin, L., Ritz, C., Saltzman, E. and Stievenard, M. (1999) Climate and atmospheric history of the past 420,000 years from the Vostok ice core, Antarctica. *Nature*, **399**, 429–436.

Petrović, R. (1976) Rate control in feldspar dissolution – II. The protective effect of precipitates. *Geochim. Cosmochim. Acta*, **40**, 1509–1521.

Petrović, R., Berner, R.A. and Goldhaber, M.B. (1976) Rate control in dissolution of alkali feldspars – I. Study of residual grains by X-ray photoelectron spectroscopy. *Geochim. Cosmochim. Acta*, **40**, 537–548.

Peybernès, N., Michel, E., Barnola, J.-M., Delmotte, M., Chappellaz, J. and Raynaud D. (2000) Information on carbon cycle during the last 8000 years deduced from CO_2, $\delta^{13}CO_2$, and CH_4 profiles obtained on a Vostok ice core. *EGS, Nice, France*, 25–26 April 2000.

Pfaff, F. cited in Mellor, J.W. (1935) *A Comprehensive Treatise on Inorganic and Theoretical Chemistry*, Vol. 14. London: Longmans, Green & Co., 359pp.

Pfeifer, H.-R. (1977) A model for fluids in metamorphosed ultramafic rocks. Observations at surface and subsurface conditions (high pH spring waters). *Schweiz. Mineral. Petrogr. Mitt.*, **57**, 361–396.

Pitzer, K.S. (1973) Thermodynamics of electrolytes. I. Theoretical basis and general equations. *J. Phys. Chem.*, **77**, 268–277.

Pitzer, K.S. (1975) Thermodynamics of electrolytes. V. Effects of higher-order electrostatic terms. *J. Solution Chem.*, **4**, 249–265.

Pitzer, K.S. (1979) Theory: Ion interaction approach. In: Pytkowicz, R.M. (ed.), *Activity Coefficients in Electrolyte Solutions*. Boca Raton, FL: CRC Press, pp. 157–208.

Pitzer, K.S. (1987) Thermodynamic model for aqueous solutions of liquid-like density. In: Carmichael, I.S.E. and Eugster, H.P. (eds.), *Thermodynamic Modeling of Geological Materials: Minerals, Fluids and Melts*. Mineralogical Society of America, *Rev. Mineral.*, **17**, 97–142.

Pitzer, K.S. (1992) Ion interaction approach: Theory and data correlation. In: Pitzer, K.S. (ed.), *Activity Coefficients in Electrolyte Solutions*, 2nd ed. Boca Raton, FL: CRC Press, pp. 75–153.

Pitzer, K.S. and Kim, J.J. (1974) Thermodynamics of electrolytes. IV. Activity and osmotic coefficients for mixed electrolytes. *J. Am. Chem. Soc.*, **96**, 5701–5707.

Plass, G.N. (1956) The carbon dioxide theory of climatic change. *Tellus*, **8**, 140–154.

Plummer, L.N. and Wigley, T.M.L. (1976) The dissolution of calcite in CO_2-saturated solutions at 25°C and 1 atmosphere total pressure. *Geochim. Cosmochim. Acta*, **40**, 191–202.

Plummer, L.N., Wigley, T.M.L. and Parkhurst D.L. (1978) The kinetics of calcite dissolution in CO_2-water systems at 5° and 60°C and 0.0 to 1.0 atm CO_2. *Am. J. Sci.*, **278**, 179–216.

Pokrovsky, O.S. and Schott, J. (1999a). Processes at the magnesium-bearing carbonates/solution interface. II. Kinetics and mechanism of magnesite dissolution. *Geochim. Cosmochim. Acta*, **63**, 881–897.

Pokrovsky, O.S. and Schott, J. (1999b). Processes at the magnesium-bearing carbonates/solution interface. I. A surface speciation model for magnesite. *Geochim. Cosmochim. Acta*, **63**, 863–880.

Pokrovsky, O.S. and Schott, J. (2000a) Forsterite surface composition in aqueous solutions: A combined potentiometric, electrokinetic, and spectroscopic approach. *Geochim. Cosmochim. Acta*, **64**, 3299–3312.

Pokrovsky, O.S. and Schott, J. (2000b) Kinetics and mechanism of forsterite dissolution at 25°C and pH from 1 to 12. *Geochim. Cosmochim. Acta*, **64**, 3313–3325.

Pokrovsky, O.S. and Schott J. (2001) Kinetics and mechanism of dolomite dissolution in neutral to alkaline solutions revisited. *Am. J. Sci.*, **301**, 597–626.

Pokrovsky, O.S. and Schott, J. (2002) Surface chemistry and dissolution kinetics of divalent metal carbonates. *Environ. Sci. Technol.*, **36**, 426–432.

Pokrovsky, O.S. and Schott, J. (2004) Experimental study of brucite dissolution and precipitation in aqueous solutions: Surface speciation and chemical affinity control. *Geochim. Cosmochim. Acta*, **68**, 31–45.

Pokrovsky, O.S., Golubev, S.V. and Schott, J. (2005b) Dissolution kinetics of calcite, dolomite and magnesite at 25°C and 0 to 50 atm pCO_2. *Chem. Geol.*, **217**, 239–255.

Pokrovsky, O.S., Mielczarski, J.A., Barres, O. and Schott, J. (2000) Surface speciation models of calcite and dolomite/aqueous solution interfaces and their spectroscopic evaluation. *Langmuir*, **16**, 2677–2688.

Pokrovsky, O.S., Schott, J. and Castillo, A. (2005a) Kinetics of brucite dissolution at 25°C in the presence of organic and inorganic ligands and divalent metals. *Geochim. Cosmochim. Acta*, **69**, 905–918.

Pokrovsky, O.S., Schott, J. and Thomas, F. (1999) Dolomite surface speciation and reactivity in aquatic systems. *Geochim. Cosmochim. Acta*, **63**, 3133–3143.

Prausnitz, J.M., Lichtenthaler, R.N. and Gomes de Azevedo, E. (1999) *Molecular Thermodynamics of Fluid-Phase Equilibria*, 3rd ed. Upper Saddle River, NJ: Prentice-Hall, 860pp.

Prentice, I.C., Farquhar, G.D., Fasham, M.J.R., Goulden, M.L., Heimann, M., Jaramillo, V.J., Kheshgi, H.S., Le Quéré, C., Scholes, R.J., Wallace, D.W.R., Archer, D., Ashmore, M.R., Aumont, O., Baker, D., Bender, M., Bopp, L.P., Bousquet, P., Caldeira, K., Ciais, P., Cox, P.M., Cramer, W., Dentener, F., Enting, I.G., Field, C.B., Friedlingstein, P., Holland, E.A., Houghton, R.A., House, J.I., Ishida, A., Jain, A.K., Janssens, I.A., Joos, F., Kaminski, T., Keeling, C.D., Keeling, R.F., Kicklighter, D.W., Kohfeld, K.E., Knorr, W., Law, R., Lenton, T., Lindsay, K., Maier-Reimer, E., Manning, A.C., Matear, R.J., McGuire, A.D., Melillo, J.M., Meyer, R., Mund, M., Orr, J.C., Piper, S., Plattner, K., Rayner, P.J., Sitch, S., Slater, R., Taguchi, S., Tans, P.P., Tian, H.Q., Weirig, M.F., Whorf, T. and Yool, A. (2001) The carbon cycle and atmospheric carbon dioxide. In: Houghton, J.T., Ding, Y., Griggs, D.J., Noguer, M., van der Linden, P.J., Dai, X., Maskell, K. and Johnson, C.A. (eds.) *Climate Change 2001: The Scientific Basis.* Cambridge, UK: Cambridge University Press, pp. 183–238 (available at http://www.ipcc.ch).

Presnall, D.C. (1995) Phase diagrams of earth-forming minerals. In: Ahrens, T.J. (ed.), *Mineral Physics and Crystallography: A Handbook of Physical Constants.* AGU Reference Shelf 2, pp. 248–268.

Preuss, K. (1991) TOUGH2 – A General Purpose Numerical Simulator for Multiphase Fluid and Heat Flow. Report LBL-29400. Lawrence Berkeley Laboratory, Berkeley, CA.

Prigogine, I. (1955) *Introduction to Thermodynamics of Irreversible Processes*. New York: Wiley.

Pryde, A.K.A. and Dove, M.T. (1998) On the sequence of phase transitions in tridymite. *Phys. Chem. Miner.*, **26**, 171–179.

Pulfer, K., Schindler, P.W., Westall, J.C. and Grauer, R. (1984) Kinetics and mechanism of dissolution of bayerite (γ-$Al(OH)_3$) in HNO_3–HF solutions at 298.2°K. *J. Colloid Interface Sci.*, **101**, 554–564.

Rao, K.V.K. and Murthy, K.S. (1970) Thermal expansion of manganese carbonate. *J. Mater. Sci.*, **5**, 82–83.

Reddy, M.M. and Nancollas, G.H. (1971) The crystallization of calcium carbonate: I. Isotopic exchange and kinetics. *J. Colloid Interface Sci.*, **36**, 166–172.

Reddy, M.M, Plummer, L.N. and Busenberg, E. (1981) Crystal growth of calcite from calcium bicarbonate solutions at constant P_{CO2} and 25°C: A test of a calcite dissolution model. *Geochim. Cosmochim. Acta*, **45**, 1281–1289.

Redfield, A.C., Ketchum, B.H. and Richards, F.A. (1963) The influence of organisms on the composition of seawater. In: Hill, M.N. (ed.), *The Sea*. New York: Wiley, pp. 26–77.

Redlich, O. (1976) *Thermodynamics: Fundamentals, Applications*. New York: Elsevier, 277pp.

Redlich, O. and Kwong, J. (1949) The thermodynamics of solutions. V. An equation of state. Fugacities of gaseous solutions. *Chem. Rev.*, **44**, 233–244.

Reed, M.H. (1982) Calculation of multicomponent chemical equilibria and reaction processes in systems involving minerals, gases, and aqueous phase. *Geochim. Cosmochim. Acta*, **46**, 513–528.

Reed, M.H. (1997) Hydrothermal alteration and its relationship to ore fluid composition. In: Barnes, H. L. (ed.), *Geochemistry of Hydrothermal Ore Deposits*. 3rd ed. New York: Wiley, pp. 517–611.

Reeder, R.J. (1990) Crystal chemistry of the rhombohedral carbonates. In: Reeder, R.J. (ed.), *Carbonates: Mineralogy and Chemistry, Reviews in Mineralogy*, Mineralogical Society of America, 2nd printing, **11**, 1–47.

Reeder, R.J. and Markgraf, S.A. (1986) High-temperature crystal chemistry of dolomite. *Am. Mineral.*, **71**, 795–804.

Richet, P. and Polian, A. (1998) Water as a dense icelike component in silicate glasses. *Science*, **281**, 396–398.

Rickard, D.T. and Sjöberg, E.L. (1983) Mixed kinetic control of calcite dissolution rates. *Am. J. Sci.*, **283**, 815–830.

Ridley, M.K., Wesolowski, D.J., Palmer, D.A., Bénézeth, P. and Kettler, R. (1997) Effect of sulfate on the release rate of Al^{3+} from gibbsite in low-temperature acidic waters. *Environ. Sci. Technol.*, **31**, 1922–1925.

Riebesell, U., Wolf-Gladrow, D.A. and Smetacek, V. (1993) Carbon-dioxide limitation of marine phytoplankton growth-rates. *Nature*, **361**, 249–251.

Rimstidt, J.D. and Barnes, H.L. (1980) The kinetics of silica-water reactions. *Geochim. Cosmochim. Acta*, **44**, 1683–1699.

Rimstidt, J.D. and Dove, P.M. (1986) Mineral/solution reaction rates in a mixed flow reactor: Wollastonite hydrolysis. *Geochim. Cosmochim. Acta*, **50**, 2509–2516.

Rimstidt, J.D., Balog, A. and Webb, J. (1998) Distribution of trace elements between carbonate minerals and aqueous solutions. *Geochim. Cosmochim. Acta*, **62**, 1851–1863.

Robie, R.A. and Hemingway, B.S. (1995) *Thermodynamic Properties of Minerals and Related Substances at 298.15 K and 1 bar (10^5 pascals) Pressure and at Higher Temperatures*. U.S. Geological Survey Bulletin 2131, 461pp.

Robinson, G.P. and Haas, J.L. (1983) Heat capacity, relative enthalpy, and calorimetric entropy of silicate minerals: An empirical method of prediction. *Am. Mineral.*, **68**, 541–553.

Robinson, R.A. and Stokes, R.H. (1968) *Electrolyte Solutions*. London: Butterworths, 571pp.

Rochelle, C., Bateman, K. and Pearce, J.M. (1996) Fluid-rock interactions resulting from the underground disposal of carbon dioxide. In: Bottrell, S.H. (ed.), *Proc. 4th Int. Symp. Gochem. Earth's Surf.*, University of Leeds, Dept. of Earth Sciences, Leeds, pp. 448–452.

Rochelle, C.A., Bateman, K., MacGregor, R., Pearce, J.M., Savage, D. and Wetton, P.D. (1996) Experimental determination of chlorite dissolution rates. In: Murakami, T and Ewing, R.C. (ed.), *Materials Research Society Symposium, No 353. Scientific Basis for Nuclear Waste Management XVIII*, Kyoto 1994, pp. 149–156.

Rose, N.M. (1991) Dissolution rates of phrenite, epidote and albite. *Geochim. Cosmochim. Acta* **55**, 3273–3286.

Rosenbauer, R.J., Bischoff, J.L. and Koksalan, T. (2001) An experimental approach to CO_2 sequestration in saline aquifers: Application to Paradox Valley, CO. *EOS: Trans. Am. Geophys. Union*, **82**, 47, Fall Meeting Suppl. Abstract V32B-0974.

Ross, G.J. (1967) Kinetics of acid dissolution of an orthochlorite mineral. *Can. J. Chem.*, **45**, 3031–3034.

Rosso, J.J. and Rimstidt, J.D. (2000) A high resolution study of forsterite dissolution rates. *Geochim. Cosmochim. Acta*, **64**, 797–811.

Rowlinson, J.S. (1988) *Studies in Statistical Mechanics*, Vol. 14. Amsterdam: North-Holland.

Sako, T., Sugeta, T., Nakazawa, N., Obuko, T., Sato, M., Taguchi, T. and Hiaki, T. (1991) Phase equilibrium study of extraction and concentration of furfural produced in reactor using supercritical carbon dioxide. *J. Chem. Eng. Jpn.*, **24**, 449–454.

Salmon, S.U. and Malmström, M. (2001) Mineral weathering rates in mill tailings from Kristinenberg, Northern Sweden. *11th Annual V.M. Goldsmith Conference*, Abstract No. 3375, LPI Contribution No. 1088, Lunar and Planetary Institute, Houston.

Samson, S.D., Nagy, K.L. and Cotton, W.B., III (2005) Transient and quasi-steady-state dissolution of biotite at 22–25°C in high pH, sodium, nitrate, and aluminate solutions. *Geochim. Cosmochim. Acta*, **69**, 399–413.

Sanemasa, I., Yoshida, M. and Ozawa, T. (1972) The dissolution of olivine in in aqueous solutions of inorganic acids. *Bull. Chem. Soc. Jap.*, **45**, 1741–1746.

Sarmiento, J.L. and Gruber, N. (2002) Sinks for anthropogenic carbon. *Phys. Today*, **55**(8), 30–36.

Sarmiento, J.L. and Toggweiler, J.R. (1984) A new model for the role of the oceans in determining atmospheric pCO_2. *Nature*, **308**, 621–624.

Sass, B.M., Gupta, N., Ickes, J.A., Engelhard, M.H., Baer, D.R., Bergman, P. and Byrer, C. (2001) Interaction of rock minerals with carbon dioxide and brine: A hydrothermal investigation. *1st National Conference on Carbon Sequestration. National Energy Technology Laboratory*, Washington DC, http://www.netl.doe.gov.

Sass, R.L., Vidale, R. and Donohue, J. (1957) Interatomic distances and thermal anisotropy in sodium nitrate and calcite. *Acta Crystallogr.*, **10**, 567–570.

Saugier, B. and Roy, J. (2001) Estimations of global terrestrial productivity: Converging towards a single number? In: Roy, J., Saugier, B. and Mooney, H.A. (eds.), *Global Terrestrial Productivity: Past, Present and Future.* San Diego: Academic Press, pp. 541–555.

Saxena, S.R. (1973) *Thermodynamics of Rock-Forming Crystalline Solutions.* New York: Springer.

Scatchard, G. (1936) Concentrated solutions of strong electrolytes. *Chem. Rev.*, **19**, 309–327.

Schlitzer, R. (2000) Applying the adjoint method for biogeochemical modeling: Export of particuate organic matter in the world ocean. In: Kasibhatla, P., Heimann, M., Rayner, P., Mahowald, N., Prinn, R.G. and Hartley, D.E. (eds.), *Inverse Methods in Global Biogeochemical Cycles. Geophysical Monograph Series*, **114**, 107–124.

Schnoor, J.L. (1990) Kinetics of chemical weathering: A comparison of laboratory and field weathering rates. In: Stumm, W. (ed.), *Aquatic Chemical Kinetics*, New York: Wiley, 475–504.

Schott, J. and Berner, R.A. (1983) X-ray photoelectron studies of the mechanism of iron silicate dissolution during weathering. *Geochim. Cosmochim. Acta*, **47**, 2233–2240.

Schott, J. and Berner, R.A. (1985) Dissolution mechanisms of pyroxenes and olivines during weathering. In: Drever, J.I. (ed.), *The Chemistry of Weathering, NATO ASI Series C, Mathematical and Physical Sciences*, **149**, 35–53.

Schott, J., Berner, R.A. and Sjöberg, E.L. (1981) Mechanism of pyroxene and amphibole weathering. I. Experimental studies of iron-free minerals. *Geochim. Cosmochim. Acta*, **45**, 2123–2135.

Schott, J., Brantley, S.L., Crerar, D., Guy, C., Borcsik, M. and Willaime C. (1989) Dissolution kinetics of strained calcite. *Geochim. Cosmochim. Acta,* **53**, 373–382.

Schulze, E.D. and Heimann, M., (1998) Carbon and water exchange of terrestrial ecosystems. In: Galloway, J.N. and Melillo, J. (eds.), *Asian Change in the Context of Global Change.* Cambridge: Cambridge University Press.

Schulze, R.K., Hill, M.A., Field, R.D., Papin, P.A., Hanrahan, R.J. and Byler, D.D. (2004) Characterization of carbonated serpentine using XPS and TEM. *Energy Convers. Manage.*, **45**, 3169–3179.

Schweda, P. (1989) Kinetics of alkali feldspar dissolution at low temperature. In: Miles, D.L. (ed.), *Proceeding of the 6th International Symposium on Water–Rock Interaction.* Rotterdam: AA Balkema, pp. 609–612.

Schweda, P. (1990) *Kinetics and mechanisms of alkali feldspar dissolution at low temperature.* PhD Thesis, Stockholm University, 99pp.

Scotford, R.F. and Glastonbury, J.R. (1971) Effect of temperature on the rates of dissolution of gibbsite and boehmite. *Can. J. Chem. Eng.*, **49**, 611–616.

Scotford, R.F. and Glastonbury, J.R. (1972) The effects of concentration on the rates of dissolution of gibbsite and boehmite. *Can. J. Chem. Eng.*, **50**, 754–758.

Segall, R.L., Smart, S.C. and Turner, P.S. (1978) Ionic oxides: Distinction between mechanisms and surface roughening effects in the dissolution of magnesium oxide. *J. Chem. Soc. Faraday Trans. I*, **74**, 2907–2912.

Seifritz, W. (1990) CO_2 disposal by means of silicates. *Nature*, **345**, 486.

Setchénow, (1892) Action de l'acide carbonique sur les solutions des sels à acides forts. *Ann. Chim. Phys.*, **25**, 225–270.

Shackleton, N.J. (1977) Carbon-13 in Uvigerina: Tropical rainforest history and the equatorial Pacific carbonate dissolution cycles. In: Anderson, N.R. and Malahoff, A. (eds.), *The Fate of Fossil Fuel CO_2 in the Oceans*, New York: Plenum, 401–427.

Shade, J.W. (1981) Comparison of glass and ceramic leaching behavior by natural analogs. *Nucl. Chem. Waste Manage.*, **2**, 219–228.

Shannon, R.D. (1976) Revised effective ionic radii and systematic studies of interatomic distances in halides and chalcogenides. *Acta Crystallogr.*, **A32**, 751–767.

Shannon, R.D., Prewitt, C.T. (1969) Effective ionic radii in oxides and fluorides. *Acta Crystallogr.*, **B25**, 925–946.

Shiraki, R. and Brantley, S.L. (1995) Kinetics of near-equilibrium calcite precipitation at 100°C: An evaluation of elementary reaction-based and affinity-based rate laws. *Geochim. Cosmochim. Acta*, **59**, 1457–1471.

Shiraki, R., Rock, P.A. and Casey, W.H. (2000) Dissolution kinetics of calcite in 0.1 M NaCl solution at room temperature: An atomic force microscopy (AFM) study. *Aquat. Geochem.*, **6**, 87–108.

Shock, E.L. (1995) Organic acids in hydrothermal solutions: Standard molal thermodynamic properties of carboxylic acids and estimates of dissociation constants at high temperatures and pressures. *Am. J. Sci.*, **295**, 496–580.

Shock, E.L. and Helgeson, H.C. (1988) Calculation of the thermodynamic and transport properties of aqueous species at high pressures and temperatures: Correlation algorithms for ionic species and equation of state predictions to 5 kb and 1000°C. *Geochim. Cosmochim. Acta*, **52**, 2009-2036.

Shock, E.L. and Helgeson, H.C. (1990) Calculation of the thermodynamic and transport properties of aqueous species at high pressures and temperatures: Standard partial molal properties of organic species. *Geochim. Cosmochim. Acta*, **54**, 915–945.

Shock, E.L. and Koretsky, C.M. (1993) Metal-organic complexes in geochemical processes: Calculation of standard partial molal thermodynamic properties of aqueous acetate complexes at high pressures and temperatures. *Geochim. Cosmochim. Acta*, **57**, 4899–4922.

Shock, E.L. and Koretsky, C.M. (1995) Metal–organic complexes in geochemical processes: Estimation of standard partial molal thermodynamic properties of aqueous complexes between metal cations and monovalent organic acid ligands at high pressures and temperatures. *Geochim. Cosmochim. Acta*, **59**, 1497–1532.

Shock, E.L., Helgeson, H.C. and Sverjensky, D.A. (1989) Calculation of the thermodynamic and transport properties of aqueous species at high pressures and temperatures: Standard partial molal properties of inorganic neutral species. *Geochim. Cosmochim. Acta*, **53**, 2157–2183.

Shock, E.L., Oelkers, E.H., Johnson, J.W., Sverjensky, D.A. and Helgeson, H.C. (1992) Calculation of the thermodynamic properties of aqueous species at high pressures and temperatures: Effective electrostatic radii, dissociation constants, and standard partial molal properties to 1000°C and 5 kb. *J. Chem. Soc. (Lond.) Faraday Trans.*, **88**, 803–826.

Shotyk, W. and Nesbitt, H.W. (1990) A ligand-promoted dissolution of plagioclase feldspar: A comparison of the surface chemistry of dissolving labradorite and bytownite using SIMS. *Chem. Geol.* **84**, 320–321.

Shyu, G.-S., Hanif, N.S.M., Hall, K.R. and Eubank, P.T. (1997) Carbon dioxide–water phase equilibria results from the Wong-Sandler combining rules. *Fluid Phase Equilib.*, **130**, 73–85.

Siebert H., Youdelis W.V., Leja J. and Lilge E. (1963) The kinetics of the dissolution of crystalline quartz in water at high temperatures and pressures. *Unit Processes in Hydrometallurgy: Metall. Soc. AIME Conf.*, **24**, 284–299.

Siegel, D.I. and Pfannkuch, H.O. (1984) Silicate mineral dissolution at pH 4 and near standard temperature and pressure. *Geochim. Cosmochim. Acta*, **48**, 197–201.

Siegenthaler, U., Friedli, H., Lötscher, H., Moor, E., Neftel, A., Oeschger, H. and Stauffer, B. (1988) Stable-isotope ratios and concentration of CO_2 in air from polar ice cores. *Annal. Glaciol.*, **10**, 1–6.

Sjöberg, E.L. (1976) A fundamental equation for calcite dissolution kinetics. *Geochim. Cosmochim. Acta*, **40**, 441–447.

Sjöberg, E.L. (1978) Kinetics and mechanism of calcite dissolution in aqueous solutions at low temperatures. *Stockholm Contrib. Geol.*, **32**, 1–96.

Sjöberg, E.L. and Rickard, D. (1983) The influence of experimental design on the rate of calcite dissolution. *Geochim. Cosmochim. Acta*, **47**, 2281–2285.

Sjöberg, E.L. and Rickard, D.T. (1984a) Calcite dissolution kinetics: Surface speciation and the origin of the variable pH dependence. *Chem. Geol.*, **42**, 119–136.

Sjöberg, E.L. and Rickard, D.T. (1984b) Temperature dependence of calcite dissolution kinetics between 1 and 62°C at pH 2.7 to 8.4 in aqueous solutions. *Geochim. Cosmochim. Acta*, **48**, 485–493.

Sjöberg, E.L. and Rickard, D.T. (1985) The effect of added dissolved calcium on calcite dissolution kinetics in aqueous solutions at 25°C. *Chem. Geol.*, **49**, 405–413.

Skinner, B.J. (1966) Thermal expansion. In: Clark S. (ed.), *Handbook of Physical Constants, Geol. Soc. Am. Mem.*, pp. 75–96.

Smith, H.J., Fischer, H., Wahlen, M., Mastroianni, D. and Deck, B. (1999) Dual modes of the carbon cycle since the Last Glacial Maximum. *Nature*, **400**, 248–250.

Smith, W.R. and Missen, R.W. (1982) *Chemical Reaction Equilibrium Analysis: Theory and Algorithms.* New York: Wiley, 364pp.

Song, K.Y. and Kobayashi, R. (1987) Water content of CO_2 in equilibrium with liquid water and/or hydrates. *SPE Form. Eval.*, **2**, 500–508.

Sonnenfeld, P. (1984) *Brines and Evaporites*. Orlando, FL: Academic Press, 613pp.

Span, P. and Wagner, W. (1996) A new equation of state for carbon dioxide covering the fluid region from the triple-point temperature to 1100K at pressures up to 800Mpa. *J. Phys. Chem. Ref. Data*, **25–6**, 1509–1596.

Speedy, R.J. (1982) Stability-limit conjecture. An interpretation of the properties of water. *J. Phys. Chem.*, **86**, 982–991.

Speedy, R.J. and Angell, C.A. (1976) Isothermal compressibility of supercooled water and evidence for a thermodynamic singularity at −45°C. *J. Chem. Phys.*, **65**, 851–858.

Sposito, G. (1994) *Chemical Equilibria and Kinetics in Soils*. New York: Oxford University Press, 268pp.

Spycher, N. and Pruess, K. (2005) CO_2–H_2O mixtures in the geological sequestration of CO_2. II. Partitioning in chloride brines at 12–100°C and up to 600 bar. *Geochim. Cosmochim. Acta*, **69**, 3309–3320.

Spycher, N.F. and Reed, M.H. (1988) Fugacity coefficients of H_2, CO_2, CH_4, H_2O, and H_2O–CO_2–CH_4 mixtures: A virial equation treatment for moderate pressures and temperatures applicable to hydrothermal boiling. *Geochim. Cosmochim. Acta*, **52**, 739–749.

Spycher, N., Pruess, K. and Ennis-King, J. (2003) CO_2–H_2O mixtures in the geological sequestration of CO_2. I. Assessment and calculation of mutual solubilities from 12 to 100°C and up to 600 bar. *Geochim. Cosmochim. Acta*, **67**, 3015–3031.

Steinmann, G. (1905) Geologische Beobachtungen in den Alpen. *Ber naturf Ges Freiburg*, **16**, 18–67.

Stillings, L.L. and Brantley, S.L. (1995) Feldspar dissolution at 25°C and pH 3: Reaction stoichiometry and the effect of cations. *Geochim. Cosmochim. Acta*, **59**, 1483–1496.

Stillings, L.L., Drever, J.I., Brantley, S.L., Sun, Y. and Oxburgh, R. (1996) Rates of feldspar dissolution at pH 3−7 with 0−8 mM oxalic acid. *Chem. Geol.*, **132**, 79–89.

Stipp, S.L. and Hochella, M.F., Jr. (1991) Structure and bonding environments at the calcite surface observed with X-ray photoelectron spectroscopy (XPS) and low energy electron diffraction (LEED). *Geochim. Cosmochim. Acta*, **55**, 1723–1736.

Stoessell, R.K. (1981) Refinements in a site-mixing model for illites: Local electrostatic balance and the quasi-chemical approximation. *Geochim. Cosmochim. Acta*, **45**, 1733–1741.

Stumm, W. (1992) *Chemistry of the Solid-Water Interface*. New York: Wiley.

Stumm, W. (1997) Reactivity at the mineral–water interface: Dissolution and inhibition. *Colloids Surf. A*, **120**, 143–166.

Stumm, W. and Morgan, J.J. (1996) *Aquatic Chemistry. Chemical Equilibrium and Rates in Natural Waters.* New York: Wiley.

Stumm, W., Wehrli, B. and Wieland, E. (1987) Surface complexation and its impact on geochemical kinetics. *Croatica Chem. Acta*, **60**, 429–456.

Suehiro, Y., Nakajima, M., Yamada, K. and Uematsu, M. (1996) Critical parameters of $\{x\mathrm{CO2} + (1 - x)\mathrm{CHF3}\}$ for $x = (1.0000, 0.7496, 0.5013$, and $0.2522)$. *J. Chem. Thermodyn.*, **28**, 1153–1164.

Sulzberger, B., Suter, D., Siffert, C., Banwart, S. and Stumm, W. (1989) Dissolution of Fe(III) (hydr)oxides in natural waters; laboratory assessment on the kinetics controlled by surface coordination. *Marine Chem.*, **28**, 127–144.

Sun, Y. (1994) *The effect of pH and oxalate ion on the dissolution rates of plagioclase feldspar*. M. Sc. Thesis, University of Wyoming, Laramie, 136pp.

Suter, D., Banwart, S. and Stumm, W. (1991) Dissolution of hydrous iron(III) oxides by reductive mechanisms. *Langmuir*, **7**, 809–813.

Sverdrup, H. U. (1990) *The Kinetics of Base Cation Release due to Chemical Weathering*. Lund: Lund University Press, 246pp.

Sverjensky, D.A. (1987) Calculations of the thermodynamic properties of aqueous species and the solubilities of minerals in supercritical electrolyte solutions. In: Carmichael, I.S.E. and Eugster, H.P. (eds.), *Thermodynamic Modeling of Geological Materials: Minerals, Fluids and Melts. Reviews in Mineralogy*, **17**, 177–209.

Sveriensjky, D.A., Shock, E.L. and Helgeson, H.C. (1997) Prediction of the thermodynamic properties of aqueous metal complexes to 1000°C and 5 kb. *Geochim. Cosmochim. Acta*, **61**, 1359–1412.

Swoboda-Colberg, N.G. and Drever, J.I. (1993) Mineral dissolution rates in plot-scale field and laboratory experiments. *Chem. Geol.* **105**, 51–69.

Takashima, Y. and Ohashi, S. (1968) The Mossbauer spectra of various natural minerals. *Bull. Chem. Soc. Japan*, **41**, 88–93.

Talman, S.J., Wiwchar, B., Gunter, W.D. and Scarge, C.M. (1990) Dissolution kinetics of calcite in the H_2O–CO_2 system along the steam saturation curve to 210°C. In: Spencer, R.J. and Chou, I.M. (eds.), *Fluid–Mineral Interactions: A Tribute to H.P. Eugster.* Geochem. Soc. Spec. Publ. Series, **2**, 41–55.

Tan, K.H. (1980) The release of silicon, aluminum, and potassium during decomposition of soil minerals by humic acid. *Soil Sci.*, **129**, 5–11.

Tanger, J.C. and Helgeson, H.C. (1988) Calculation of the thermodynamic and transport properties of aqueous species at high pressures and temperatures: Revised equations of state for the standard partial molal properties of ions and electrolytes. *Am. J. Sci.*, **288**, 19–98.

Tardy, Y. and Fritz, B. (1981) An ideal solid solution model for calculating solubility of clay minerals. *Clay Miner.*, **16**, 361–373.

Tardy, Y. and Garrels, R.M. (1974) A method of estimating the Gibbs energy of formation of layer silicates. *Geochim. Cosmochim. Acta*, **38**, 1101–1116.

Taylor, A.S., Blum, J.D. and Lasaga, A.C. (2000b) The dependence of labradorite dissolution and Sr isotope release rates on solution saturation state. *Geochim. Cosmochim. Acta*, **64**, 2389–2400.

Taylor, A.S., Blum, J.D., Lasaga, A.C. and MacInnis, I.N. (2000a) Kinetics of dissolution and Sr release during biotite and phlogopite weathering. *Geochim. Cosmochim. Acta*, **64**, 1191–1208.

Tchoubar, C. (1965) Formation de la kaolinite à partir d'albite altérée par l'eau à 200°C. Étude en microscopie et diffraction électroniques. *Soc. Française Minéralogie Cristallographie Bull.*, **88**, 483–518.

Tchoubar, C. and Oberlin, A. (1963) Altération de l'albite par action de l'eau. Étude en microscopie et microdiffraction électroniques de la précipitation et de l'évolution des fibres de boehmite formée. *J. Microsc.*, **2**, 415–432.

Teng, H., Yamasaki, A., Chun, M.-K. and Lee, H. (1997) Solubility of liquid CO_2 in water at temperatures from 278 K to 293 K and pressures from 6.44 MPa to 29.49 MPa and densities of the corresponding aqueous solutions. *J. Chem. Thermodyn.*, **29**, 1301–1310.

Teng, H.H., Dove, P.M. and De Yoreo, J.J. (2000) Kinetics of calcite growth: Surface processes and relationships to macroscopic rate laws. *Geochim. Cosmochim. Acta*, **64**, 2255–2266.

Terjesen, S.G., Erga, O., Thorsen, G. and Ve, A. (1961) I1. Phase boundary processes as rate determining steps in reactions between solids and liquids. The inhibitory action of metal ions on the formation of calcium bicarbonate by the reaction of calcite with aqueous carbon dioxide. *Chem. Eng. Sci.*, **74**, 277–288.

Tester, J.W., Worley, W.G., Robinson, B.A., Grigsby, C.O. and Feerer, J.L. (1994) Correlating quartz dissolution kinetics in pure water from 25 to 625°C. *Geochim. Cosmochim. Acta*, **58**, 2407–2420.

Thomassin, J.H., Goni, J., Baillif, P., Touray, J.C. and Jaurand, M.C. (1977) An XPS study of the dissolution kinetics of chrysotile in 0.1 N oxalic acid at different temperatures. *Phys. Chem. Miner.*, **1**, 385–398.

Thompson, J.B., Jr. (1955) The thermodynamic basis for the mineral facies concept. *Am. J. Sci.*, **53**, 65–103.

Tödheide, K. and Frank, E.U. (1963) Das Zweiphasengebiet und die Kritische Kurve im system Kohlendioxid-Wasser bis zu Drucken von 3500 bars. *Z. Phys. Chem., Neue Folge*, **37**, 387–401.

Tole, M.P., Lasaga, A.C., Pantano, C.G. and White, W.B. (1986) The kinetics of dissolution of nepheline ($NaAlSiO_4$). *Geochim. Cosmochim. Acta*, **50**, 379–392.

Tomazawa, M., Erwin, C.Y., Takata, M. and Watson, E.B. (1982) Effect of water content on the chemical durability of $Na_2O \cdot 3\ SiO_2$ glass. *J. Am. Ceram. Soc.*, **65**, 182–183.

Toplis, M.J. and Richet, P. (2000) Equilibrium density and expansivity of silicate melts in the glass transition range. *Contrib. Mineral. Petrol.*, **139**, 672–683.

Torp, T.A. and Gale, J. (2004) Demonstrating storage of CO_2 in geological reserovirs: The Sleipner and SACS projects. *Energy*, **29**, 1361–1369.

Trolier, M., White, J.W.C., Tans, P.P., Masarie, K.A. and Gemery, P.A. (1996) Monitoring the isotopic composition of atmospheric CO_2 – measurements from the NOAA Global Air Sampling Network. *J. Geophys. Res. – Atmos.*, **101**, 25897–25916.

Tunell, G. (1931) The definition and evaluation of the fugacity of an element or compound in the gaseous state. *J. Phys. Chem.*, **35**, 2885–2913.

Turpault, M.-P. and Trotignon, L. (1994) The dissolution of biotite single crystals in dilute HNO_3 at 24°C: Evidence of an anisotropic corrosion process of micas in acidic solutions. *Geochim. Cosmochim. Acta*, **58**, 2761–2775.

Tuttle, O.F. and Bowen, N.L. (1958) Origin of granite in the light of experimental studies in the system $NaAlSi_3O_8$– $KAlSi_3O_8$–SiO_2–H_2O. *Geol. Soc. Am. Mem.*, **74**, 153pp.

Urch, D.S. and Wood, P.R. (1978) The determination of the valency of manganese in minerals by x-ray fluorescence spectroscopy. *X-ray Spectrom.*, **7**, 9–11.

Van Cappellen, P., Charlet, L., Stumm, W. and Wersin, P. (1993) A surface complexation model of the carbonate mineral-aqueous solution interface. *Geochim. Cosmochim. Acta*, **57**, 3505–3518.

Van Herk, J., Pietersen, H.S. and Schuiling, R.D. (1989) Neutralization of industrial waste acids with olivine. The dissolution of forsteritic olivine at 40–70°C. *Chem. Geol.* **76**, 341–352.

Van Lier, J.A., De Bruyn, P.L., Overbeek, J. and Th, G. (1960) The solubility of quartz. *J. Phys. Chem.*, **64**, 1675–1682.

Van Zeggeren, F. and Storey, S.H. (1970) *The Computation of Chemical Equilibrium*. London: Cambridge University Press, 176pp.

Velbel, M.A. (1999) Bond strength and the relative weathering rates of simple orthosilicates. *Am. J. Sci.*, **299**, 679–696.

Vermilyea, D.A. (1969) The dissolution of MgO and $Mg(OH)_2$ in aqueous solutions. *J. Electrochem. Soc.*, **116**, 1179–1183.

Vernadsky, V.I. (1924) *La Géochimie*. Paris: Alcan, 404pp.

Wagman, D.D., Evans, W.H., Parker, V.B., Schumm, R.H., Halow, I., Bailey, S.M., Churney, K.L. and Nuttall, R.L. (1982) The NBS tables of chemical thermodynamic properties: Selected values for inorganic and C_1 and C_2 organic substances in SI units. *J. Phys. Chem. Ref. Data*, **11**(2), 392pp.

Walther, J. V. (1996) Relation between rates of aluminosilicate mineral dissolution, pH, temperature, and surface charge. *Am. J. Sci.*, **296**, 693–728.

Walther, J.V. and Helgeson, H.C. (1977) Calculation of the thermodynamic properties of aqueous silica and the solubility of quartz and its polymorphs at high pressures and temperatures. *Am. J. Sci.*, **277**, 1315–1351.

Watson, A.J., Nightingale, P.D. and Cooper, D.J. (1995) Modeling atmosphere ocean CO_2 transfer. *Philos. Trans. Royal Soc. Lond. Ser. B-Biol. Sci.*, **348**, 125–132.

Weare, J.H. (1987) Models of mineral solubility in concentrated brines with application to field observations. In: Carmichael, I.S.E. and Eugster, H.P. (eds.), *Thermodynamic Modeling of Geological Materials: Minerals, Fluids and Melts* . Mineralogical Society of America, *Reviews in Mineralogy*, **17**, 143–176.

Weber, L.A. (1989) Simple apparatus for vapor–liquid quilibrium measurements with data for the binary systems of carbon cioxide with *n*-butane and isobutane. *J. Chem. Eng. Data*, **34**, 171.

Weill, D.F. and Fyfe, W.S. (1964) The solubility of quartz in the range 1000–4000 bars and 400–500°C. *Geochim. Cosmochim. Acta*, **28**, 1243–1255.

Weissbart, E.J. and Rimstidt, J.D. (2000) Wollastonite: Incongruent dissolution and leached layer formation. *Geochim. Cosmochim. Acta*, **64**, 4007–4016.

Welch, S.A. and Ullman, W.J. (1993) The effect of organic acids on plagioclase dissolution rates and stoichiometry. *Geochim. Cosmochim. Acta*, **57**, 2725–2736.

Welch, S.A. and Ullman, W.J. (1996) Feldspar dissolution in acidic and organic solutions: Compositional and pH dependence of dissolution rate. *Geochim. Cosmochim. Acta*, **60**, 2939–2948.

Welch, S.A. and Ullman, W.J. (2000) The temperature dependence of bytownite feldspar dissolution in neutral aqueous solutions of inorganic and organic ligands at low temperature (5–35°C). *Chem. Geol.*, **167**, 337–354.

Wendland, M., Hasse, H. and Maurer, G. (1999) Experimental pressure-temperature data on three- and four-phase equilibria of fluid, hydrate, and ice phases in the system carbon dioxide-water. *J. Chem. Eng. Data*, **44**, 901–906.

Wenk, H.-R., Barber, D.J. and Reeder, R.J. (1990) Microstructures in carbonates. In: Reeder, R. J. (ed.), *Carbonates: Mineralogy and Chemistry, Reviews in Mineralogy*, Mineralogical Society of America, 2nd printing, **11**, 301–367.

Westrich, H.R., Cygan, R.T., Casey, W.H., Zemitis, C. and Arnold, G.W. (1993) The dissolution kinetics of mixed-cation orthosilicate minerals. *Am. J. Sci.*, **293**, 869–893.

White, A.F. (1984) Weathering characteristics of natural glass and influences on associated water chemistry. *J. Non-Cryst. Solids*, **67**, 225–244.

White, A. F. and Peterson, M. L. (1990) Role of reactive surface area characterization in geochemical models. In: Melchior, D. C. and Bassett, R. L. (eds.), *Chemical Modelling of Aqueous Systems II. Am. Chem. Soc. Symp. Ser.*, **416**, 416–475.

White, A.F., Peterson, M.L. and Hochella, M.F., Jr. (1994) Electrochemistry and dissolution kinetics of magnetite and ilmenite. *Geochim. Cosmochim. Acta*, **58**, 1859–1875.

White, S.P. (1995) Multiphase non-isothermal transport of systems of reacting chemicals. *Water Resour. Res.*, **31**, 7.

White, S.P., Allis, R.G., Moore, J., Chidsey, T., Morgan, C., Gwynn, W. and Adams, M. (2005) Simulation of reactive transport of injected CO_2 on the Colorado Plateau, Utah, USA. *Chem. Geol.*, **217**, 387–405.

Wiebe, R. and Gaddy, V.L. (1939) The solubility in water of carbon dioxide at 50, 75, and 100°, at pressures to 700 atmospheres. *J. Am. Chem. Soc.*, **61**, 315–318.

Wiebe, R. and Gaddy, V.L. (1940) The solubility of carbon dioxide in water at various temperatures from 12 to 40° and at pressures to 500 atmospheres: Critical phenomena. *J. Am. Chem. Soc.*, **62**, 815–817.

Wiebe, R. and Gaddy, V.L. (1941) Vapor phase composition of carbon dioxide–water mixtures at various temperatures and at pressures to 700 atmospheres. *J. Am. Chem. Soc.*, **63**, 475–477.

Wieland, E. and Stumm, W. (1992) Dissolution kinetics of kaolinite in acidic aqueous solutions at 25°C. *Geochim. Cosmochim. Acta*, **56**, 3339–3355.

Wieland, E., Wehrli, B. and Stumm, W. (1988) The coordination chemistry of weathering: III. A generalization on the dissolution rates of minerals. *Geochim. Cosmochim. Acta*, **52**, 1969–1981.

Wildeman, T.R. (1970) The distribution of Mn^{2+} in some carbonates by electron paramagnetic resonance. *Chem. Geol.*, **5**, 167–177.

Wilson, M. (1989) *Igneous Petrogenesis. A Global Tectonic Approach*. London: Unwin Hyman, 466pp.

Wogelius, R.A. and Walther, J.V. (1991) Olivine dissolution at 25°C: Effects of pH, CO_2, and organic acids. *Geochim. Cosmochim. Acta*, **55**, 943–954.

Wogelius, R.A. and Walther, J.V. (1992) Olivine dissolution kinetics at near-surface conditions. *Chem. Geol.* **97**, 101–112.

Wogelius, R.A., Refson, K., Fraser, D.G., Grime, G.W. and Goff, J.P. (1995) Periclase surface hydroxylation during dissolution. *Geochim. Cosmochim. Acta*, **59**, 1875–1881.

Wolery, T.J. (1978) *Some chemical aspects of hydrothermal processes at mid-oceanic ridges – A theoretical study. I. Basalt-sea water reaction and chemical cycling between the oceanic crust and the oceans. II. Calculation of chemical equilibrium between aqueous solutions and minerals.* Unpub. Ph. D. Diss., Northwestern Univ., Evaston, IL, 263pp.

Wolery, T.J. (1979) Calculation of chemical equilibrium between aqueous solutions and minerals: The EQ3/6 software package. Report UCRL-52658, Lawrence Livermore National Laboratory, Livermore.

Wolery, T.J. (1992) EQ3NR, A computer program for geochemical aqueous speciation-solubility calculations: Theoretical manual, user's guide and related documentation (version 7.0). Report UCRL-MA-110662 PT III. Lawrence Livermore National Laboratory, Livermore.

Wolery, T.J. and Daveler, S.A. (1992) EQ6, A computer program for reaction path modeling of aqueous geochemical systems: Theoretical manual, user's guide, and related documentation (version 7.0). Report UCRL-MA-110662 PT IV. Lawrence Livermore National Laboratory, Livermore.

Wolf, G.H., Chizmeshya, A.V.G., Diefenbacher, J. and McKelvy, M.J. (2004) In situ observation of CO_2 sequestration reactions using a novel microreaction system. *Environ. Sci. Technol.*, **38**, 932–936.

Wolff-Boenisch, D., Gislason, S.R. and Oelkers, E.H. (2004a) The effect of fluoride on the dissolution rates of natural glasses at pH 4 and 25°C. *Geochim. Cosmochim. Acta*, **68**, 4571–4582.

Wolff-Boenisch, D., Gislason, S.R., Oelkers, E.H. and Putnis, C.V. (2004b) The dissolution rates of natural glasses as a function of their composition at pH 4 and 10.6, and temperatures from 25 to 74°C. *Geochim. Cosmochim. Acta*, **68**, 4843–4858.

Wollast, R. (1967) Kinetics of the alteration of K-feldspar in buffered solutions at low temperature. *Geochim. Cosmochim. Acta*, **31**, 635–648.

Wollast, R. and Chou, L. (1985) Kinetic study of the dissolution of albite with a continuous flow-through fluidized bed reactor. In: Drever, J.I. (ed.), *The Chemistry of Weathering, NATO ASI Series C, Mathematical and Physical Sciences*, **149**, 75–96.

Worden, R.H. (2006) Dawsonite cement in the Triassic Lam Formation, Shabwa Basin, Yemen: A natural analogue for a potential mineral product of subsurface CO_2 storage for greenhouse gas reduction. *Marine Petrol. Geol.*, **23**, 67–77.

Wyart, J., Oberlin, A. and Tchoubar, C. (1963) Étude en microscopie et microdiffraction électroniques de la boehmite forméelors de l'altération de l'albite. *Acad. Sci. Paris, Comptes Rendus*, **256**, 554–555.

Xiao, Z., Jonckbloedt, R.C.L. and Schuiling, R.D. (1999) Neutralization of waste acids with olivine dissolution in organically contaminated acids. *Chem. Erde*, **59**, 33–45.

Xie, Z. (1994) *Surface properties of silicates, their solubility and dissolution kinetics*. PhD Thesis, Northwestern University, Evanston, IL.

Xie, Z. and Walther, J.V. (1994) Dissolution stoichiometry and adsorption of alkali and alkaline earth elements to the acid-reacted wollastonite surface at 25°C. *Geochim. Cosmochim. Acta*, **58**, 2587–2598.

Xu, T. and Pruess, K. (1998) Coupled modeling of non-isothermal multiphase flow, solute transport and reactive chemistry in porous and fractured media: 1. Model development and validation. Lawrence Berkeley National Laboratory Report LBNL-42050, Berkeley, CA.

Xu, T., Apps, J.A. and Pruess, K. (2000) Analysis of mineral trapping for CO_2 disposal in deep aquifers. Lawrence Berkeley National Laboratory Report LBNL-47315, Berkeley, CA.

Xu, T., Apps, J.A. and Pruess, K. (2004) Numerical simulation of CO_2 disposal by mineral trapping in deep aquifers. *Appl. Geochem.*, **19**, 917–936.

Xu, T., Apps, J.A. and Pruess, K. (2005) Mineral sequestration of carbon dioxide in a sandstone-shale system. *Chem. Geol.*, **217**, 295–318.

Yasunishi, A. and Yoshida, F. (1979) Solubility of carbon dioxide in aqueous electrolyte solutions. *J. Chem. Eng. Data*, **24**, 11–14.

Zawisza, A. and Malesinska, B. (1981). Solubility of carbon dioxide in liquid water and of water in gaseous carbon dioxide in the range 0.2–5 MPa and at temperatures up to 473K. *J. Chem. Eng. Data*, **26**, 388–391.

Zellmer, L.A. (1986) Dissolution kinetics of crystalline and amorphous albite. M.Sc. Thesis, The Pennsylvania State University, University Park, PA.

Zellmer, L.A. and White, W.B. (1986) The comparative dissolution kinetics of albite crystals and albite glass in aqueous solutions at 70°C. *5th International Symposium on Water–Rock Interaction*, **5**, 652–655.

Zemann, J. (1981) Zur Stereochimie der Karbonate. *Fortschr. Mineral.*, **59**, 95–116.

Zhang, H., Bloom, P.R. and Nater, E.A. (1993) Change in surface area and dissolution rates during hornblende dissolution at pH 4.0. *Geochim. Cosmochim. Acta*, **57**, 1681–1689.

Zhang, X., Wen, Z., Gu, Z., Xu, X. and Lin Z. (2004) Hydrothermal synthesis and thermodynamic analysis of dawsonite-type compounds. *J. Solid. State Chem.*, **177**, 849–855.

Zhang, Y. and Dawe, R. (1998) The kinetics of calcite precipitation from a high salinity water. *Appl. Geochem.*, **13**, 177–184.

Zinder, B., Furrer, G. and Stumm, W. (1986) The coordination chemistry of weathering: II. Dissolution of Fe(III) oxides. *Geochim. Cosmochim. Acta*, **50**, 1861–1869.

Zuddas, P. and Mucci, A. (1998) Kinetics of calcite precipitation from seawater: II. The influence of the ionic strength. *Geochim. Cosmochim. Acta*, **62**, 757–766.

Zysset, M. and Schindler, P.W. (1996) The proton promoted dissolution kinetics of K-montmorillonite. *Geochim. Cosmochim. Acta*, **60**, 921–931.

Subject Index